T0179732

OPTICAL AND MICROWAVE TECHNOLOGIES FOR TELECOMMUNICATION NETWORKS

OPTICAL AND MICROWAVE TECHNOLOGIES FOR TELECOMMUNICATION NETWORKS

Prof. Dr.-Ing. Prof. h.c. Dr. h.c. Otto Strobel
Esslingen University of Applied Sciences

WILEY

This edition first published 2016
© 2016 John Wiley & Sons, Ltd

Registered office
John Wiley & Sons Ltd, The Atrium, Southern Gate, Chichester, West Sussex, PO19 8SQ, United Kingdom

For details of our global editorial offices, for customer services and for information about how to apply for permission to reuse the copyright material in this book please see our website at www.wiley.com.

Library of Congress Cataloging-in-Publication Data

Names: Strobel, Otto, 1950- author.
Title: Optical and microwave technologies for telecommunication networks / Otto Strobel.
Description: Chichester, UK ; Hoboken, NJ : John Wiley & Sons, 2016. | Includes bibliographical references and index.
Identifiers: LCCN 2015047969| ISBN 9781119971900 (cloth) | ISBN 9781119154587 (Adobe PDF) | ISBN 9781119154594 (epub)
Subjects: LCSH: Optical fiber communication. | Microwave communication systems. | Telecommunication systems–Design and construction.
Classification: LCC TK5103.592.F52 S77 2016 | DDC 621.382/1–dc23
LC record available at http://lccn.loc.gov/2015047969

A catalogue record for this book is available from the British Library.

Cover image: Courtesy of the Author
Cover picture: Free Space-Optical Communication, Fiber-Optic Communication, WLAN-Communication

ISBN: 9781119971900

Set in 10/12pt Times by Aptara Inc., New Delhi, India
Printed and bound in Malaysia by Vivar Printing Sdn Bhd

1   2016

This book is dedicated to my whole family
in particular
to my very beloved grandchildren
Sevi, Jamie and Clara

Contents

Preface

After human beings solved their most elementary problems of nutrition and availability of warming and protective clothes, they felt the need to communicate between each other. Even then, this communication improved the results of their labor. People first started by talking to each other at distances our ears are able to understand acoustically. The next step was visible communication limited by the resolution and focusing abilities of our eyes. Smoke signals, for example, were used during the day and fire beacons at night. The oldest written proof of optical communication is presented in AESCHYLOS's (Αισχυλος) play *Agamemnon*, written in the 5th century BC [1.25]. The news of the fall of Troy in 1200 BC, after years of siege by the Greeks, was reported to Agamemnon's wife Clytemnestra by fires which were lit on hills all the way from Asia Minor to Argos in Greece.

The first development of a useful optical telegraph happened to be during the time of the French Revolution. CLAUDE CHAPPE, a former Abbé, invented the *semaphore*. On top of a building a moveable beam was arranged, which carried a moveable arm at both ends; 192 different positions could be realized. In 1880, Alexander Graham Bell invented the *photophone*. The idea was that a light beam was modulated by acoustic vibrations of a thin mirror. The demodulation of the optical signal could be realized, for example, by utilizing the photoelectric effect in selenium.

All free space transmissions depend on good weather and undisturbed atmosphere. Some methods work only during the daytime, some only at night. An exception is free space transmission in outer space because, outside of the Earth's atmosphere, typical problems like natural disturbances by fog, rain or snow or artificially caused impurities do not inherently exist. However, even on Earth, it was desirable that communication is independent of environmental conditions. Therefore, some form of guidance of the light beam in a protective environment was necessary. There were ideas of guiding the light within a tube, whose inner walls reflect the light.

The development of the laser by Theodor Maiman, at the beginning of the 1960s, provided a light source which yields an entirely different behavior compared to the sources we had before. A short time after this very important achievement, diode lasers for usage as optical transmitters were developed. Parallel to that accomplishment in the early 1970s, researchers and engineers accomplished the first optical glass fiber with sufficiently low attenuation to transmit electromagnetic waves in the near infrared region. The photodiode as detector already worked, and thus, systems could be developed using optoelectric (O/E) and electrooptic (E/O) components for transmitters and receivers, as well as a fiber in the center of the arrangement. In 1966, Charles K. Kao and G.A. Hockam of Standard Telecommunications Laboratories in Harlow, England, published a paper in which they proposed the guidance of light within dielectric glass fibers. The immediate problem was the optical attenuation in fibers. Whereas, on a clear day, atmospheric attenuation is about 1 dB/km, the best glass then available showed an attenuation of about 1000 dB/km. To illustrate this, the optical power is reduced to 1‰ after a path of only 30 m. Kao and Hockham's main thesis was that if the attenuation could be reduced to 20 dB/km at a convenient wavelength, then practical fiber-optic communication should be possible.

In 1970, Corning Glass Works, USA, achieved this goal. By further refinement of fiber production, the attenuation coefficient could be reduced to below 0.2 dB/km in 1982. Fibers of commercial mass production today show an attenuation of approximately 0.2 dB/km. The optical power in such a fiber still amounts to about 1% after traveling a distance of 100 km.

In the 1970s and 1980s, reliable semiconductor light sources and detectors were developed. First field trials of fiber-optic links were very successful during the 1980s.

People often discuss the quality of systems in simple terms, such as good or bad. From the physical point of view, nothing is good or bad; it is as it is – the only question is what you need it for. For example, are we discussing a high-speed long distance system in the order of one-tenth of Gbit up to 100 Gbit/s (or more) with nearly no cost restriction, or are we talking about application in cars with 150 Mbit/s and about 10 m link lengths at low cost demands? These are completely different worlds and thus, for each demand, we have to find the proper solution.

In the last five decades, landline network communication has mainly been considered for application in telecom areas. The most well-known use is for high-speed, middle and long distance systems, as well as MAN and LAN networking; any last-mile application, including in-house communication to a single user's desk, needs to be connected to the rest of the world. Most recently, mobile communication, in particular cell phones (more recently smart phones), tablets, tablet PCs, laptops, PCs, etc., have been developed to replace cable-based phone calls, emails and Internet communication.

For about 20 years, Fiber-to-the-home (FTTH) has become the phrase on everybody's lips – the efforts to also bring optical communication into a single-family house. This did not happen until now for reasons of economy. However, because of the soaring use of the Internet, higher data rate needs increasingly occur in single-family houses, too. In order to permit a corresponding quantum leap, it remains absolutely essential to reduce costs for the participants. The keyword is "opening up the last mile". Latest developments can help to achieve this aim.

In the last ten years, communication in transportation systems has become more and more in demand – for communication within a vehicle, from one vehicle to another and to land-line networks too. Development started in high-end cars with application in the infotainment area and has already reached airplanes and ships where sensor-relevant needs were also addressed. These techniques began with low data rates. Car communication technologies for the coming decade will also include high bit-rate systems up to the level of Gbit/s. Moreover, a new industry-standard, named communication in automation engineering, has been developed. By applying this technology, new perspectives could be opened up for data linking between tooling machines and central control units.

The idea of this book is to address a broad scope of readers, in order to give them an introduction to optical and microwave communication systems. For this reason, we not only present articles on state-of-the-art methods but also promising techniques for the future are discussed as well. On the one hand, it is important that the key differences between optical and non-optical systems are appreciated, yet on the other hand, similarities can be also seen. Moreover, a combination of these different physical techniques might lead to excellent results, which cannot be reached using them separately. Taking all these optical and microwave techniques, as well as GPS, together with high-speed high-data processing devices and appropriate software, may mean that the old human dream of easy worldwide communication (involving nearly unlimited data consumption), be it listening, seeing or reading, could be realized in the not too distant future.

For readers not familiar with all these topics, there is coverage of many subjects of optical and microwave fundamentals. The book is intended to help undergraduate, graduate and PhD students with a basic knowledge of the subjects studying communication technologies. In addition, R&D engineers in companies should also find this book interesting and useful. This is true for novices as well as for experts checking certain facts or dealing with areas of expertise peripheral to their normal work.

I would like to express my appreciation to my former colleagues at Alcatel-Lucent Research Center (now Bell Labs Germany) for numerous helpful discussions. I also gratefully acknowledge my current colleagues at Esslingen University for much help, in particular Prof. Dr. Dr. h.c. R. Martin. Moreover, I have to mention my staff member Dipl.-Phys. H. Bletzer for active support in lab and manuscript preparation. For the latter, also many thanks to M.Sc. Marko Cehovski, who as my student also co-authored several publications – also Dipl.-Ing. Daniel Seibl, M.Sc. Jan Lubkoll, now with ASML Veldhoven/Niederlande. For this book, I was able to find a variety of R&D contributors from companies and universities all over the world: MSc. Werner Auer, FOP Faseroptische Produkte GmbH, Crailsheim, Germany (Chapter 4.1), Dr. Krzysztof Borzycki, National Institute of Telecommunications, Warsaw, Poland (Chapter 4.2 and 10.1), Dr. Ronald Freund, Dr. Markus Nölle, Fraunhofer Heinrich Hertz Institute, Berlin, Germany (Chapter 9.2), Dr. Ronald Freund, Dr. Nicolas Perlot, Fraunhofer Heinrich Hertz Institute, Berlin, Germany (Chapter 9.3), MSc. Marko Čehovski, Institut für Hochfrequenztechnik, Technische Universität Braunschweig, Germany (Chapter 9.5), Thorsten Ebach, eks Engel GmbH & Co. KG, Wenden, Germany (Chapter 9.6), Dr. Alicia López, Dr. M. Ángeles Losada, Dr. Javier Mateo, GTF, Aragón Institute of Engineering Research (i3A), University of Zaragoza, Spain (Chapter 10.2), Dr. Joaquín Beas, Dr. Gerardo Castañón, Dr. Ivan Aldaya, Dr. Alejandro Aragón-Zavala, Tecnológico de Monterrey, Mexico (Chapter 10.3), Prof. Dr. Zabih Ghassemlooy, Dr. Hoa Le Minh, Dr. Muhammad Ijaz, Northumbria University, Newcastle, UK, (Chapter 10.4), Dr. Riccardo Scopigno, MSc. Daniele Brevi, Multi-Layer Wireless Research Area, Istituto Superiore Mario Boella, Torino, Italy, (Chapter 10.5), Dr. Paolo Monti, Dr. Lena Wosinska, Dr. Richard Schatz, KTH Royal Institute of Technology, Stockholm, Sweden, Dr. Luca Valcarenghi, Dr. Piero Castoldi, ScuolaSuperiore Sant'Anna, Pisa, Italy,Aleksejs Udalcovs, Institute of Telecommunications, Riga Technical University, Riga, Latvia, (Chapter 10.6), Prof. Dr. Kira Kastell, Frankfurt University of Applied Sciences, Germany (Chapter 11.1), Prof. Dr. Vladimir Rastorguev, Dr. Andrey Ananenkov, Engineer Anton Konovaltsev, Prof. Dr. Vladimir Nuzhdin; Engineer Pavel Sokolov, Moscow Aviation Institute, National Research University, Russia (Chapter 11.2). Also Dr. rer. nat. Sebastian Döring from TU Braunschweig, Germany has to be acknowledged for actual contributions to recent research in organic Lasers (OLASERs). All R&D projects were carried out in cooperation with companies such as Alcatel-Lucent (Bell Labs Germany), HP, Agilent, Mercedes-Benz Technology, Siemens, Diehl Aerospace and Balluff Germany. The same holds for national and international universities, in particular concerning Bachelor, Master and PhD theses. Grateful acknowledgement to all university and company members involved.

Last but not least, a very big thank you to my family, especially to my wife Dorothee (who has to date suffered more than 40 years under an often absent professor), as well as my children Sven, Jana and Jasmin.

Otto A. Strobel
Esslingen/Germany, Summer 2015

1

Introduction

In this book, we present state-of-the-art and next-decade technologies for optical and related microwave transmission in telecom applications, high-speed long distance as well as last-mile and in-house communication. Furthermore, we have learned a lot about the needs of companies producing state-of-the-art systems. They use practical systems in order to compete in the market for such products. They want a complete solution to their demands, and they do not care if fiber optics are part of the solution or not. Consequently, nowadays, further physical techniques have to be developed: Wireless applications open up a new field of data transmission. High-speed wireless LED transmission offers short-range data transmission without EMI/EMC problems. Visible light communication using high power LEDs is an interesting technique. The first aim of using these light sources is to illuminate a room. However, at the same time, they can be modulated to transmit a data signal (Figure 1.6). Thus, optical and non-optical solutions, microwave-, Radio over Fiber- or even RADAR-systems have to be developed.

All these systems working in common will offer high-speed up- and downloads for offices, labs and private homes, and also for transportation systems such as cars, airplanes and ships.

Since the beginning of the 1960s, there has been a light source which yields a completely different behavior compared to the sources we had before. This light source is the LASER, **L**ight **A**mplification by **S**timulated **E**mission of **R**adiation. Basic work had been published already, in 1917 by Albert Einstein [1.1]. The first laser realized was the bulk-optic ruby laser, a solid state laser [1.2] developed by Theodor Maiman in 1960. A short time after this very important achievement, diode lasers for use as optical transmitters were developed [1.3]. At the beginning they were difficult to operate. They had to be cooled using nitrogen at −169°C. It took until 1970 to drive semiconductor diode lasers in a continuous wave (CW) mode at room temperature [1.4]. Parallel to that accomplishment, the use of dielectric optical waveguides as media for transmission systems was suggested. Charles Kao and George Hockam [1.5] can be regarded as inventors of fiber-optic transmission systems, as well as Manfred Börner [1.6]. Nowadays, their invention would not be regarded as anything remarkable. Take a light source as a transmitter, an optical fiber as a transmission medium, and a photodiode as a detector (Figure 1.1)! Yet, in the 1960s it was a revolution, because the attenuation of optical glass was in the order of 1000 dB/km corresponding to a factor of 10^{-100} over one kilometer. This was totally unrealistic for use in practical systems.

Optical and Microwave Technologies for Telecommunication Networks, First Edition. Otto Strobel.
© 2016 John Wiley & Sons, Ltd. Published 2016 by John Wiley & Sons, Ltd.

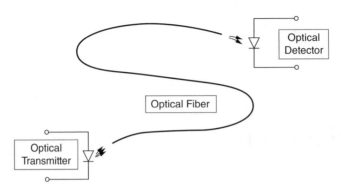

Figure 1.1   Basic arrangement of a fiber-optic system

In the early 1970s, physicists and engineers in research laboratories developed the first optical glass fiber with sufficient low attenuation to transmit electromagnetic waves in the near infrared region (Corning Glass Works [1.7]). They achieved a value of about 20 dB/km, that is, after one kilometer there is still 1% of light at the detector from the light which was coupled into the fiber by the transmitter. Today's fibers have an attenuation of below 0.2 dB/km, which means the 1% value is still achieved after 100 km of the fiber.

The development of optical detectors started much earlier. Already in 1876, W. Adams and R. Day proved the separation of charge carriers in Selenium [1.8] by illumination of light. They discovered the inner photo effect, also named the photovoltaic effect. This effect is still the fundamental process exploited in modern photo detectors. First experiments with silica solar cells took place in the early 1950s, and new developments for practical use to transmit a data signal started in 1970 [1.9]. Thus, systems could be developed using optoelectric (O/E) and electrooptic (E/O) components for transmitters and receivers, as well as a fiber in the center of the arrangement. The main fields of application of such systems are found in the area of fiber-optic transmission and fiber-optic sensors (Figure 1.2).

In the beginning, in both areas exclusively, the intensity of light was of interest. In analog as well as in digital systems, signal transmission is realized by modulation of the laser power. At the fiber end, an intensely sensitive receiver is used exclusively to detect the data signal. Regarding sensor systems, the measurement of the physical quantity of interest is also exclusively concerned with the power of the light. In this case the optical power is varied by exploiting a change of the fiber attenuation or other system components. Modern more sophisticated systems make use of the fact that an electromagnetic wave also carries information on frequency, phase and polarization.

The media to transport the propagation of light is named the waveguide. The most common waveguide is a glass fiber. The geometrical shape does not need to be strictly circular like a fiber, but can also be rectangular forming a planar waveguide. This was suggested by Stewart Miller in 1969 and thus the new field of integrated optics was born [1.10]. The aim of this technology is to integrate a maximum of components onto a single chip such as an integrated circuit – elements with a variety of different functions within the smallest space to avoid macro-optic devices like mirrors and lenses. Some integrated optic components are able to influence all parameters of an optical wave: amplitude (intensity), frequency, phase and polarization. However, the first optical transmission is much older. Native Americans for instance, were

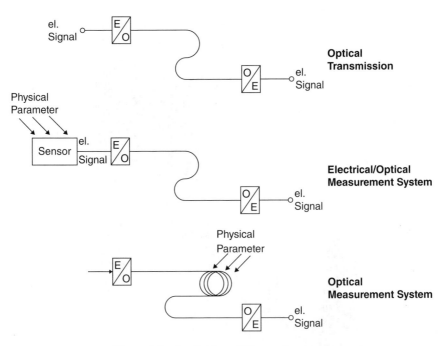

Figure 1.2   Application of fiber-optic systems

already communicating with smoke signals a long time ago [1.12,1.13]. Furthermore, it was a very sophisticated and modern system, because it was already a digital system, consisting of "binary 1" and "binary 0" (smoke/no smoke) (Figure 1.3).

Thus digital systems have already existed for a long time, providing the basics for information technologies, information processing and transmission. However, analog techniques are still of interest – physical quantities at the origin and the reception (e.g. human reception). But as soon as they are processed or transmitted nowadays, almost exclusively digital techniques are used. A further free space transmission has also been developed, the optical telegraph realized by Claude Chappe. He invented a semaphore set-up by means of movable bars able to produce several signs. But free space transmission on Earth suffers from atmospheric disturbance [1.11]. This also holds for free space laser transmission on Earth, which came to light in the 1960s after the invention of the laser. There are exceptions in outer space applications and for short distance air transmissions (Chapter 9.3).

The real breakthrough of optical data transmission systems came with the glass fiber, which had sufficient low attenuation for propagation of electromagnetic waves in the near infrared region. This low value of attenuation is one of the most attractive advantages of fiber-optic systems compared to conventional electrical ones.

Figure 1.4 depicts the attenuation behavior. In particular, we observe independence of the modulation frequency of fiber-optic systems in contrast to electrical ones which suffer from the skin effect. Yet it has to be confessed that there are different problems leading to a frequency limit, the dispersions: modal, chromatic and polarization mode dispersions must be mentioned (Chapter 3). Solutions of how to deal with these problems with sufficient success will be

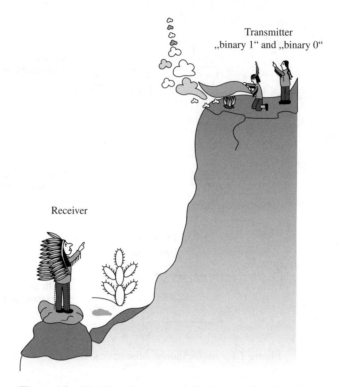

Figure 1.3   Digital optical transmission by use of smoke signals

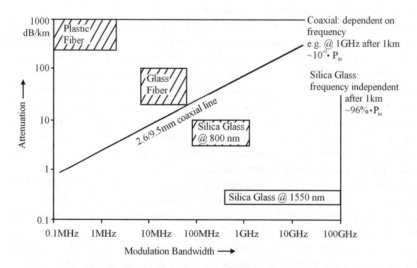

Figure 1.4   Attenuation of coaxial cables and optical fibers

presented. In the end, the enormous achievable bandwidth must be highlighted. That leads to a high transmission capacity in terms of the product of achievable fiber bandwidth B and length L, also named *transmission capacity,* C_t. It is a figure of merit, as one of the most important goals is to maximize this product for every kind of data transmission with respect to the demands concerning its application.

$$C_t = B \cdot L = Max! \tag{1.1}$$

where:

B maximum achievable bandwidth and
L maximum achievable link length

In addition, low weight, small size, insensitivity against electromagnetic interference (EMI, EMC), electrical insulation and low crosstalk must be mentioned.

Glass fiber systems are used in the near infrared range, right above the wavelengths we can see with our eyes. As optoelectronic components for light sources, we apply GaAlAs (Gallium-Aluminium-Arsenide) LEDs and laser diodes for wavelengths in the 850 nm region and InGaAsP (Indium-Gallium-Arsenide-Phosphide) devices for the long wavelengths of about 1200 nm to over 1600 nm. Photodiode materials of interest are well known, Si for 850 nm, and Ge and InGaAsP for the long wavelength range (Figure 6.40). However, an optical communication system is more than a light source, a fiber, and a photodiode. There is a laser driver circuit necessary to provide a proper high-bit-rate electric signal; this driver, combined with a laser or an LED, builds the optical transmitter. Also the photodiode (pin or APD – Avalanche Photodiode), together with the front-end amplifier, forms the optical detector, also called the optical receiver. This front-end amplifier consists of a very highly sophisticated electric circuit. It has to detect a high bandwidth operating with very few photons due to a large fiber length and it is struggling with a variety of noise generators. In addition, there are further electric circuits to be taken into account, such as circuits for coding, scrambling, error correction, clock extraction, temperature power-level, and gain controls. If the desired link length cannot be realized, a repeater consisting of a front-end amplifier and a pulse regenerator will be inserted. This pulse regenerator is necessary to restore the data signal before it is fed to a further laser driver followed by another laser (Figure 9.2). Alternatively, an optical amplifier can be used, in particular the Erbium-doped fiber amplifier is a great success (Figure 8.2).

Moreover, instead of unidirectional systems, we need bi-directional ones (Figure 9.9); that is, it is not sufficient that for a telephone link a person at one side of the link is able to speak, but at the other side another person can only listen; the system does not operate the other way round. To overcome this inconvenient situation, optical couplers on both sides of the link are inserted (Figures 4.53 and 4.54). The two counter propagating optical waves superimpose undisturbed, and they separate at the optical couplers on the other side of the link and reach the according receivers. To improve the transmission capacity drastically, wavelength selective couplers are applied, called multiplexers and demultiplexers. Several laser diodes operating at different wavelengths are used as transmitters; their light waves are combined by the multiplexer and on the other link end they are separated by the demultiplexer. This set-up is named the wavelength division multiplex system (WDM). If we apply this arrangement again in the two counter propagating directions, we achieve a bi-directional WDM system. The transmission capacity then rises by the number N of the channels transmitted over one single fiber.

For about 20 years now, last mile communication has been discussed. The idea of fiber to the home (FTTH) has also been discussed, but until now this did not happen because it was too expensive. Latest improvements could help to achieve this special communication. All-plastic PMMA-fibers (poly methyl methacrylate) have been developed, named *Polymer Optical Fibers* (POF), which feature in contrast to the previous PMMA-fibers at a considerably lower attenuation [1.17] (Chapter 10.2). As an economic alternative to the application of semiconductor lasers, another important step is the development of cheap high-speed LEDs [1.18] or low-cost VCSELs (Vertical Cavity Surface Emitting Lasers) [3.1], which can be modulated fast enough.

Combinations of fiber-optic with mm-waves systems or coaxial cable systems have been developed for the local area network as a possible alternative. As another alternative to cable systems, the declared dead free space transmission could be also revived with distances in the 100 m range. For this purpose, the light emerging from a fiber is fed to a lens and formed into a parallel beam. At the reception site it is again coupled into a fiber or directly to a detector.

This conjunction of fiber-optic transmission and free space may successfully be used in sky scraper areas, where air distances lie in the 100 m range and cable systems would need to be in kilometres. Wireless data transmission is also an interesting option for distances in the 10 m range. Connections between, for example, a PC, printer, scanner or adjacent participants in intranets (LANs, *Local Area Networks*, see below) should not be bridged by interfering electrical cables. Further developments in the microwave range have been developed, such as the recent well-known *Bluetooth systems*.

Moreover, besides typical point-to-point connections, network systems are necessary. In nearby zones, for example inside a business house, the commonly used term is LANs (*Local Area Networks*); in the local net or metro region it is MANs (*Metropolitan Area Networks*), [9.85–9.87]. The network topologies are bus, star or ring structures (Chapter 9.5.3).

Furthermore, free space transmission is gaining a particular renaissance in outer space, because outside of the Earth's atmosphere typical problems such as natural disturbances by fog, rain, snow or artificially caused impurities do not inherently exist. Therefore, laser free space connections between satellites have been already tried and tested successfully (Chapter 9.3).

In the last 10 years, communication in automotive systems became of great interest (Chapter 11.1). Currently, optical data buses in vehicles are almost exclusively used for infotainment (information and entertainment) applications. The Media Oriented Systems Transport (MOST) is the optical data bus technology used nowadays in cars with a data rate up to 150 MBit/s (Figure 11.5 [1.14]). The development of infotainment applications in cars began with a radio and simple loudspeakers. Today's infotainment systems in cars include but are not limited to ingenious sound systems, DVD-changers, amplifiers, navigation and video functions. Voice input and Bluetooth interfaces complement these packages. Important and basic logical links of these single components are already well known from a simple car radio. Everybody probably knows the rise of volume in the case of road traffic announcements. However, the integration of more and more multimedia and telematic devices in vehicles led to a large increase in data traffic demands. In particular, for luxurious classes, a huge need for network capacity and higher complexity by integration of various applications have to be taken into account. MOST

Figure 1.5 Lightning strike in an airplane [1.22]. Source: Reproduced with permission of Denny Both, Piranha.dl 3d Animation, Berlin

150 operates with LEDs, a POF and silicon photodiodes. However, to enable the next step towards autonomous driving, new bus systems with higher data rates will be required.

Another serious challenge arises in protecting new generation aircrafts, particularly against lightning strikes [1.15]. This is because new airplanes will be built using carbon-fibers to reduce the weight of the fuselage. Therefore, these airplanes will lose their inherent Faraday cage protection against lightning, cosmic radiation and further electrostatic effects (Figure 1.5).

In order to avoid failures in signal transmission in the physical layer, the electrical copper wires should be well protected. But this solution is too expensive and increases the weight of the cables [1.15]. A reasonable solution is to use glass or plastic fibers as transmission media in these new airplanes. Since the FlexRay bus protocol [1.16] is more adequate for avionic applications, it should be adapted for this kind of transmission. Thus, this solution is cost-efficient and offers more safety in the aviation domain. A promising solution for higher sophisticated systems could be the use of optical data transmission based on new laser types, such as VCSELs and the application of Polymer Cladded Silica (PCS) fibers. This enables EMC compatibility and paves the way for the future.

In this book we mainly give an introduction to optical transmission. Emphasis is on fiber transmission systems, working with basic components. The reader should be familiar with

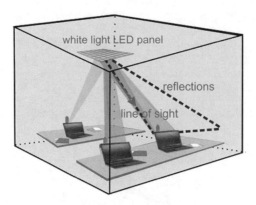

Figure 1.6 Room lighting with inherent data modulation and transmission

the fundamental optical techniques for communication systems. Moreover, for more comprehensive considerations there are further components to be dealt with, for example the optical amplifier to enhance the link length. In order to achieve this, Erbium and Raman amplifiers (Chapter 8) [8.1,8.2] have been developed to overcome the problem of attenuation in fibers. Due to the above-mentioned dispersions, there are signal distortions in optical fibers. The systems suffer from pulse broadening (Figure 3.21) leading to bandwidth reduction with impact on the transmission capacity, the product of bandwidth and fiber length. Solutions to overcome or at least to reduce these problems are discussed in Chapter 3.2.

Furthermore, wireless applications open new fields for data transmission [1.19]. High speed wireless LED transmissions offer short and middle range data transmission without EMI/EMC problems (Figure 1.6). Higher bandwidths than non-optical wireless applications will offer high-speed up- and downloads in offices, labs and private homes.

Also, car-to-car communication could be an interesting scenario (Figure 11.1 [1.20]). In particular, safety relevant applications are of great interest (Iizuka 2008, [1.20]). The catchword is "Pre-crash safety" by VL-ISC: Visible Light Image Sensor Communication. Human reaction time is a problem in security, as in difficult circumstances it could last much longer than a sensor does. The vehicle in front might suddenly start braking. Thus, depending on the brake pressure, immediately its stoplights will give information to the following vehicle. In critical cases the following vehicle will automatically start emergency braking to avoid a collision or at least to avoid serious damage.

A further new development is shown in Figure 11.2, a red-light-to-car communication. In this case, the red light tells a waiting driver that he will get a green signal in 50 seconds. Another example would be that a car is approaching a red light and the driver gets the information 10 seconds before he can even see it. While the driver is waiting at the red light, the system receives the red interval from the traffic light in order to take the decision to stop the engine. The outcome would be a gas-mileage and CO_2 reduction of more than 5% (November 2008). In addition, the driver does not have to fix his eyes on the traffic light permanently. These systems can predict idling intervals with accuracy to solve unnecessary engine stops and starts. Including information about the green, amber and red time zones, the traffic volume can be better regulated.

Moreover, a new industry standard, named communication in automation engineering, has been developed. Applying this technology has opened up new perspectives for data linking between tooling machines and a central control unit: Industry 4.0 is a collective term for technologies and concepts of value chain organization (Chapter 9.6).

Finally, non-optical techniques have to be taken into account (Chapters 10.5 and 11.2). For example, WLAN and even RADAR systems can be used in automotive application, to guarantee more safety in limited optical visibility situations like heavy rain, fog or snow.

As mentioned above, it has to be underlined that nowadays the user wants a complete solution for his demands and he does not stop to ask if that is fiber optics or whatever. Moreover, he also wants combinations of other physical techniques with or without fibers, so Optical wireless communications, Optical and Non-Optical Solutions, and Microwaves in Radio over Fiber (RoF) have to work together.

2

Optical and Microwave Fundamentals

The propagation of electromagnetic waves in transmission media is very important for optical transmission techniques as well as for fiber-optic sensor applications. The spectrum of electromagnetic waves varies from long-wave radio waves to short-wave cosmic radiation. The area which is interesting in fiber optics and sensor techniques spans from visible light to the near infrared region (Figure 2.1).

The related physical area is called *Optics*. In a closer sense, almost exclusively, electromagnetic waves in the visible area are named light but often the IR- and UV-regions are included in the term too. The propagation can take place in free space, air and outer space or in guided media. The electromagnetic wave propagation device is called the *optical waveguide*. The most well-known type of optical waveguide is a *glass fiber*. Instead of glass, it is also possible to use a transparent plastic material, a polymer optical fiber (POF). Moreover, it is not imperative that the waveguide shows a round cross-sectional shape like the fiber does. For example, it can be inserted in a plane substrate. Thus, a two-dimensional waveguide structure will be achieved, which finds its application in the field of *integrated optics*.

2.1 Free Space Propagation of Electromagnetic Waves

Electromagnetic waves appear as a periodic spatiotemporal excitation of field quantities of a physical field transporting physical energy. The electric field vector ($\vec{E}$) oscillates perpendicularly to the magnetic field vector ($\vec{H}$) and moreover, both fields are perpendicular to the wave propagation direction (Figure 2.2); such a wave is called a *transversal wave*.

Optical and Microwave Technologies for Telecommunication Networks, First Edition. Otto Strobel.
© 2016 John Wiley & Sons, Ltd. Published 2016 by John Wiley & Sons, Ltd.

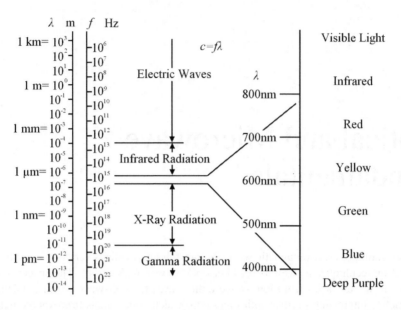

Figure 2.1   Infrared (IR) and visible region (VIS) including ultraviolet (UV) is called the field of optical radiation

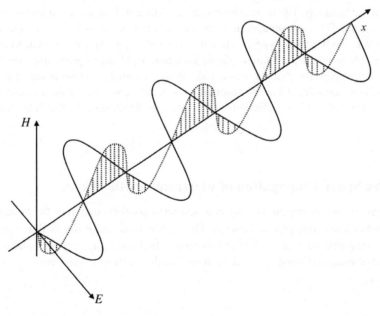

Figure 2.2   Electric ($\vec{E}$) and magnetic field vector ($\vec{H}$) of an optical wave at a certain time with propagation in the x-direction

The mathematical description of electromagnetic wave propagation is based on Maxwell's theory of the electromagnetic field [2.1]. For *Maxwell's equations* include:

$$rot\,\vec{E} = -\frac{\partial \vec{B}}{\partial t} \tag{2.1}$$

$$rot\,\vec{H} = \frac{\partial \vec{D}}{\partial t} + \vec{j} \tag{2.2}$$

$$div\,\vec{D} = \rho \tag{2.3}$$

$$div\,\vec{B} = 0 \tag{2.4}$$

where:

$\vec{B}$ Magnetic induction
$\vec{D}$ Dielectric displacement
$\vec{j}$ Current density and
     Electric charge density

There are material equations which take into account the media properties that the waves are propagating:

$$\vec{D} = \varepsilon\vec{E}, \varepsilon = \varepsilon_0\varepsilon_\tau \tag{2.5}$$

$$\vec{B} = \mu\vec{H}, \mu = \mu_0\mu_\tau \tag{2.6}$$

$$\vec{j} = \kappa\vec{E} \tag{2.7}$$

where:

κ Specific conductivity
ε Dielectric constant (Permittivity)
ε_0 Free space dielectric constant (Permittivity)
ε_r Relative dielectric number
μ Permeability
μ_0 Induction constant
μ_r Relative permeability

Several helpful simplifications are obtained in optics and consequently for wave propagation in optical waveguides. Concerning the used wavelength areas, the attenuation is very small, particularly in glass fibers (Chapter 3.1). Thus it can be neglected for actual considerations and thus the waveguide will be treated as free of absorption. Furthermore, glass is a non-conductive material, and there are no charge carriers either. Moreover, the magnetic induction in the waveguide is approximately the same in a vacuum, so it follows that:

$$\kappa = 0, j = 0, \rho = 0 \text{ and } \mu_r = 1$$

Therefore a simplification of Maxwell's Eqs (2.1) to (2.4) is gained by the application of the laws of vector analysis following from Eqs (2.1) to (2.4) [2.2]:

$$\Delta \vec{E} - \mu\varepsilon \frac{\partial^2 \vec{E}}{\partial t^2} = 0 \tag{2.8}$$

$$\Delta \vec{H} - \mu\varepsilon \frac{\partial^2 \vec{H}}{\partial t^2} = 0 \tag{2.9}$$

Applying the Laplace operator Δ to each Cartesian component of the $\vec{E}$ and $\vec{H}$ vector, respectively:

$$\Delta = \frac{\partial^2}{\partial x^2} + \frac{\partial^2}{\partial y^2} + \frac{\partial^2}{\partial z^2}$$

the propagation velocity v (phase velocity) of an electromagnetic wave in the media is given by:

$$v = \frac{1}{\sqrt{\mu\varepsilon}} = \frac{1}{\sqrt{\mu_0\varepsilon_0\varepsilon_r}} = c/n = \omega/(k_0 n) \tag{2.10}$$

where:

c Light velocity in vacuum
n Refraction index
ω Angular frequency
f Frequency
k_0 Value of the wave vector $\vec{k}_0$ in vacuum ($k_0 = 2\pi/\lambda$)
λ Wavelength of the electromagnetic wave

Now Eqs (2.8) and (2.9) can be recalculated and regarding the electric and the magnetic field strength the following differential equations, *the wave equations,* are given by:

$$\Delta \vec{E} - \frac{n^2 k_0^2}{\omega^2} \frac{\partial^2 \vec{E}}{\partial t^2} = 0 \tag{2.11}$$

$$\Delta \vec{H} - \frac{n^2 k_0^2}{\omega^2} \frac{\partial^2 \vec{H}}{\partial t^2} = 0 \tag{2.12}$$

Next, solutions to these differential equations have to be found for the vectors $\vec{E}$ and $\vec{H}$. The most general solution is (e.g. the electric field strength):

$$\vec{E} = \vec{E}(\vec{r}, t) = \vec{E}(\vec{k}\vec{r} - \omega t)$$

where:

$\vec{k}\vec{r} - \omega t$ is the phase of the searched wave, t is the time and vector $\vec{r} = (x/y/z)$ is the position vector.

For the time dependence applies:

$$\vec{E}(t) \sim \exp(j\,\omega t) \text{ or } \sim \sin(\omega t) \quad \text{or} \sim \cos(\omega t) \tag{2.13}$$

Such a solution is called a harmonic wave. The spatial design of the wave can have different forms, for example a cylindrical wave or a spherical wave. In practice, plane waves are important because adequate approximations are often allowed. A *scalar wave* will be obtained if the vector of the electric field exclusively oscillates in the direction of one local coordinate (linearly polarized wave, see below). The following equation applies to a planar harmonic scalar wave propagating in the x-direction (Figure 2.3, the $\vec{H}$-field is not illustrated):

$$E(x, t) = A\, e^{j(kx - \omega t)} \tag{2.14}$$

where:

A amplitude of the wave

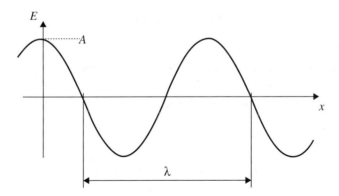

Figure 2.3   Intensity of the scalar electric field E of an optical wave versus local coordinate x

where:

λ Wavelength
A Amplitude

2.2 Interference

Waves having fixed phase relations between themselves are called coherent waves. A spatial and time coincidence of two or several waves with adequate polarization leads to a characteristic superposition which is called interference [2.3,2.4]. In the following, the superposition of two waves (E_1 and E_2) at a fixed position are considered. For each of both waves it holds that:

$$E_i = A_i e^{j(\omega_i t + \phi_i)} \tag{2.15}$$

where:

A_i Amplitude of the wave i
ω_i Angular frequency of the wave i
ϕ_i Phase of the wave i

At a certain point of the superposition area, the total electric field strength E results from the sum of the single fields. A detector is used for observation, for example, with the eye, a photographic film or a photodiode. The visual sense is attained by the recognition of a specific intensity. Concerning a photographic film, an optical density appears and regarding the photodiode, we achieve a corresponding photocurrent. All detectors mentioned measure the *absolute value of the square of the electric field strengths,* often called intensity. It is not the field strength itself that will be observed. The reason is that the frequency of the wave is in the order of 200 to 300 THz (terahertz), which is about 3 to 4 orders of magnitude too high for any detector to be resolved. What we see is something similar, like the effective value (root mean square) of a 230 VAC current coming out of a socket. This intensity I is precisely described in physics as the energy flux density, which is the absolute value S of the pointing vector:

$$\vec{S} = \vec{E} \times \vec{H}, S = |\vec{S}|, \tag{2.16a}$$

The unit is W/m^2.

For I follows by time-related averaging of the pointing vector:

$$I = <\vec{S}> = \tfrac{1}{2}(\vec{E} \times \vec{H}^*) \cdot (\vec{E}^* \times \vec{H}) \tag{2.16b}$$

and thus:

$$I \sim |E|^2 = (E_1 + E_2) \cdot (E_1 + E_2)^* = (E_1 + E_2) \cdot (E_1^* + E_2^*) \tag{2.16c}$$

where:

E^* conjugate complex of E

Thus follows:

$$I \sim A_1^2 + A_2^2 + 2 A_1 A_2 \cos[(\omega_1 - \omega_2)t + \phi_1 - \phi_2] \tag{2.17}$$

The first two terms in Eq. (2.17) represent information regarding the amplitude. In contrast, the third term, *the interference term,* gives us information about the *amplitude* as well as *frequency* and *phase.* Thus, the last term can be used for extraordinary applications in data transmission (Chapters 3.2.5 and 9.4). In particular, Eq. (2.17) tells us that in the absence of a second wave ($A_2 = 0$), there is no information about frequency and phase. Exclusively information about amplitude could be gained by applying such absolute values of the square detectors mentioned above. A lot of applications superimpose two (or more) partial waves coming from the same light source, that is, their frequencies are equal. Therefore Eq. (2.17) can be simplified to:

$$I \sim A_1^2 + A_2^2 + 2 A_1 A_2 \cos(\phi_1 - \phi_2) \qquad (2.18)$$

Figure 2.4 illustrates the intensity as a function of the phase difference $\Delta\phi = \phi_1 - \phi_2$ of two waves with adequate amplitude ($A_1 = A_2 = A$). For $\Delta\phi = 0, 2\pi, 4\pi, \dots$ a maximum of intensity can be seen. This is called *constructive interference.* For $\Delta\phi = \pi, 3\pi, 5\pi, \dots$ both waves are annihilated. This is a case of *destructive interference.*

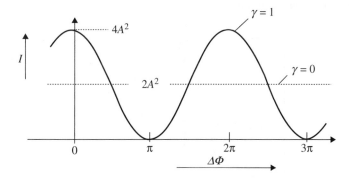

Figure 2.4   Intensity of two superimposing waves versus phase difference $\Delta\phi$

where:

A Amplitude of each wave

2.3   Coherence

Concerning the derivation of Eq. (2.17), there is the assumption that the superposing waves are completely coherent. However, this does not generally hold. The interference capability [2.5] of waves which come from light sources of finite elongation and finite spectral width can be described by the idea of *partial coherence.* The quantitative degree for partial coherence, the coherence function γ, indicates the conditions between complete random phase correlation and fixed phase coupling of the two waves. There is $0 \leq \gamma \leq 1$; thus, incoherent and coherent superposition is described as borderline cases of the coherence function, where $\gamma = 0$ is the incoherent superposition, $0 < \gamma < 1$ is the partial coherent superposition and $\gamma = 1$ is the coherent superposition

Hence, considering the coherence function, the following is derived from Eq. (2.17):

$$I \sim A_1^2 + A_2^2 + 2A_1 A_2 \cos(\Phi_1 - \Phi_2) \cdot \gamma(\Delta\phi, \Delta f) \qquad (2.19)$$

where:

Δf Spectral width of the light source

For $\gamma = 1$ and $A_1 = A_2 = A$ follows for the maximum intensity $I_{max} \sim 4A^2$, and for the minimal intensity, $I_{min} = 0$. For $\gamma = 0$ and $A_1 = A_2 = A$, the result is $I_{max} = I_{min} \sim 2A^2$ (Figure 2.4). Thus, γ can be interpreted as the contrast K from the interference phenomenon:

$$\gamma = \frac{I_{max} - I_{min}}{I_{max} + I_{min}} = K \qquad (2.20)$$

If a high interference contrast should be achieved, $\gamma \rightarrow 1$ must be obtained. The narrower the spectral width of the used light source, the larger the value of the coherence function γ. This means the spectral width of the light source should be as narrow as possible; it would be ideal if $\gamma = 1$. Thus, it follows that the light source is allowed to oscillate in a single frequency. That is called *rigorous monochromasia*.

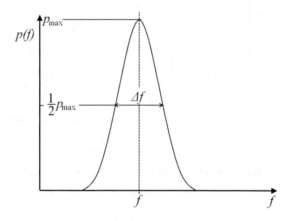

Figure 2.5   Spectral power density $p(f)$ versus frequency f

where:

f_0 Center frequency
Δf Spectral width of the light source (FWHM: full width at half maximum)

In this case, the distribution of the spectral power density $p(f)$ can be described by a delta function (Figure 2.5, broken line):

$$\int_{-\infty}^{+\infty} p(f)\delta(f - f_0)df = p(f_0) \qquad (2.21)$$

with $p(f) = dP/df$

where:

P Optical power of the light source

Real light sources, also high coherent lasers, at best are *quasi monochromatic*. The light source has to be considered as time coherent. The corresponding monochromasia requirement then results from the *center frequency* f_0 and the *spectral width* Δf (Figure 2.5, continuous line). The radiation of such a light source over this major time interval can be treated as coherent, as long as the induced phase difference due to the cut-off-frequency $f_0 + \Delta f/2$ (Figure 2.5) is not larger than π compared to the center frequency:

$$(\Delta\omega/2)\, t_{coh} \leq \pi \tag{2.22}$$

where:

t_{coh} Coherence time

The coherence length L_{coh} is given by:

$$L_{coh} = c/\Delta f \tag{2.23}$$

The contrast of an interference phenomenon disappears if the optical path difference of two partial waves is equal to its coherence length (minor changes of this definition are known in literature). The optical path difference Δg relates to the phase difference $\Delta\phi$ by the following:

$$\Delta\phi = \Delta g 2\pi/\lambda \tag{2.24}$$

The *optical path length* g results from the product of the geometrical length L and the refraction index n of the media in which the waves propagate:

$$g = nL. \tag{2.25}$$

Equation (2.23) shows that the larger the coherence of a light source the narrower is its spectral width. In light sources which exhibit a continuous spectrum, the coherence function $\gamma(\Delta\phi)$ decays according to a 1/e-function. If there is no continuous spectrum, such as in a bulb or an LED, but a mode structure such as in a multimode laser or a super luminescent diode, again a mode structure is obtained for the coherence function $\gamma(\Delta\phi)$ [2.6] (coherence function and spectrum are linked by the Fourier transformation [1.13,2.7].), that is, with increasing phase difference $\Delta\phi$, the amount of the coherence function varies from high to low values and completely disappears at large phase differences. Figure 2.6 shows the measured mode spectrum of a super luminescent diode, and Figure 2.7 the corresponding coherence function.

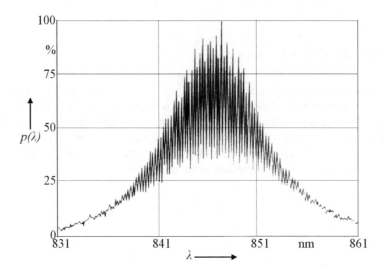

Figure 2.6   Spectral (longitudinal) modes of a super luminescence diode

where:

$p(f)$ spectral power density.

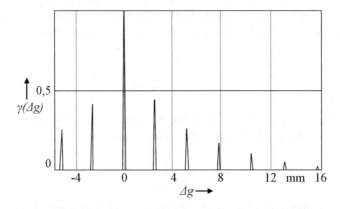

Figure 2.7   Coherence function γ of the luminescence diode according to Figure 2.6

Δg optical path difference:

$$\Delta\lambda = (\lambda^2/c)\Delta f \qquad (2.26)$$

Examples for spectral width and coherence length are shown in Table 2.1. Regarding large spectral widths, the units of wavelengths are chosen, because otherwise large numerical values would be generated:

Table 2.1   Spectral width and coherence length

	$\Delta\lambda$ resp. Δf	L_{coh}
Light bulb	some 100 nm	some µm
LED	some 10 nm	some 10 µm
Laser	some MHz to GHz	some m to km

2.4   Polarization

Hitherto, there was the assumption that the vector of the electromagnetic field perpetually oscillates in the direction of a single local coordinate (Figure 2.8). However, this limitation is not generally accepted.

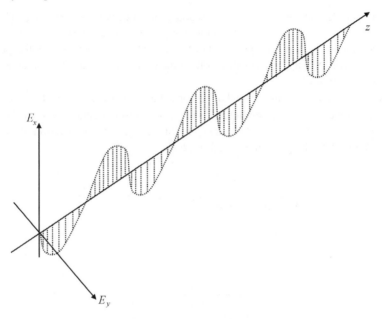

Figure 2.8   Electromagnetic wave propagating in the z-direction, while the $\vec{E}$-vector oscillates constantly in the x-direction

The peak of the $\vec{E}$-vector can be oriented in a different direction at a later point in time. Hence, light as an electromagnetic wave has to be treated like a vector [2.8]:

$$E_i \rightarrow \vec{E}_i = \vec{A}_i \exp(j\Delta\psi_i)\exp(j\omega_i t) \tag{2.27}$$

with $\vec{A}_i = (A_{xi}, A_{yi})$
where:

A_{xi}, A_{yi} Amplitudes of the wave i in x- and y-direction, respectively
ω_i Angular frequency of the wave i
$\Delta\psi_i$ Phase difference between E_{xi} and E_{yi}

The waves propagate in the z-direction. Thus, the total scalar treatment in Eq. (2.16) has to be described as a vector:

$$I = |\vec{E}|^2 = (\vec{E}_1 + \vec{E}_2) \cdot (\vec{E}_1 + \vec{E}_2)^+ = (\vec{E}_1 + \vec{E}_2^+) \cdot (\vec{E}_1^+ + \vec{E}_2) \tag{2.28}$$

where:

$\vec{E}^+$ Hermetic conjugate vector of $\vec{E}$

It follows that:

$$I = |\vec{E}_1|^2 + |\vec{E}_2|^2 + 2\text{Re}(\vec{E}_1\vec{E}_2^+) \tag{2.29}$$

Within the interference term appears the scalar product of two vectors $\vec{E}_1\vec{E}_2^+$. This product becomes zero if both vectors are perpendicular. To realize a constructive interference, a polarization adaption of the superposing waves must take place. Therefore, the polarization plays an important role in all fiber-optic communication and sensor challenges which work with coherent techniques.

The polarization describes the special behavior of the $\vec{E}$-vector. By projecting the arrowhead of the $\vec{E}$-vector onto a plane perpendicular to the propagation direction over a sufficient long-time period, an ellipse will usually be gained (Figure 2.9). Form and position of the ellipse in this plane are given by the general ellipse Eq. [2.9]:

$$\frac{E_x^2}{A_x^2} + \frac{E_y^2}{A_y^2} - \frac{2E_xE_y}{A_xA_y} \cdot \cos\Delta\psi - \sin^2\Delta\psi = 0 \tag{2.30}$$

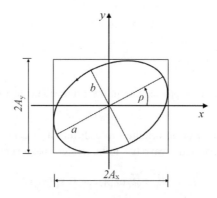

Figure 2.9   Ellipse of polarization

where:

A_x, A_y Amplitudes in x- and y-direction, respectively
a, b Major and minor ellipse half-axis, respectively

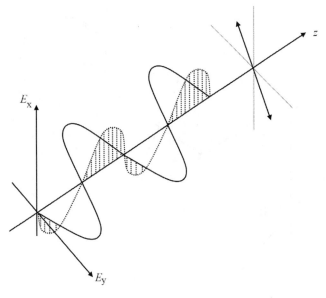

Figure 2.10   Electromagnetic wave propagating in the z-direction., The $\vec{E}$-vector oscillates constantly in a plane given by the amplitudes in the x-direction and y-direction respectively

The inclination (inclination angle ρ) of the ellipse half-axes and the axial ratio b/a are determined by the amplitudes of both waves A_x and A_y, as well as by their optical path difference $\Delta\psi$:

$$-\pi/2 \leq \rho \leq \pi/2 \tag{2.31}$$

with a, b as major and minor ellipse half-axis, respectively:

$$\text{For } 0 < \Delta\psi_i < \pi \tag{2.32}$$

Clockwise polarization states are attained, that is the arrowhead of the electromagnetic field vector shows a clockwise rotation on the ellipse:

$$\text{For } \pi < \Delta\psi < 2\pi \tag{2.33}$$

Counter clockwise polarization states are gained. *Linearly polarized light* will be obtained as a special case of the polarization ellipse (Figure 2.10):

$$\Delta\psi = n\pi, \quad \text{with } n = 0, 1, 2, 3, \ldots$$

With Eq. (2.29), it follows that:

$$E_y = (-1)^n (A_y/A_x) \cdot E_x \tag{2.34}$$

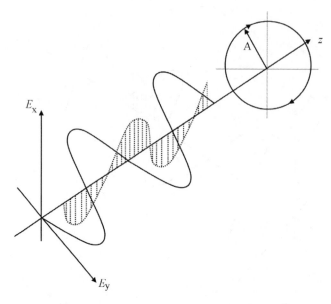

Figure 2.11   Electromagnetic wave propagating in the z-direction. The $\vec{E}$-vector rotates while the propagation takes place on a circle with radius A around the z-axes

Thus, a straight line with positive and negative slopes, respectively according to $b/a = 0$ or $b/a \to \infty$.

As another special case, *circularly polarized light* will be gained (Figure 2.11):

$$\Delta\psi = (2n + 1)\,\pi/2 \quad \text{and} \quad A_x = A_y = A.$$

With Eq. (2.29), it follows that:

$$E_x^2 + E_y^2 = A^2 \tag{2.35}$$

Thus, it is a circle according to $b/a = 1$.

In addition, non-polarized light has to be mentioned, such as during wave propagation in the z-direction the $\vec{E}$-vector statistically oscillates in the x- and y-direction, respectively. All further polarization states can be reduced to combinations of the states described above. It is also possible that light is *partially polarized*. This will be characterized by the degree of polarization, which specifies the relation between polarized light intensity and the sum of polarized and non-polarized intensity.

Positioning a linear polarizer in an optical path exclusively with an oscillation direction of the transverse wave fitting to the polarizer inclination direction will be transmitted [2.10] (Figure 2.12a).

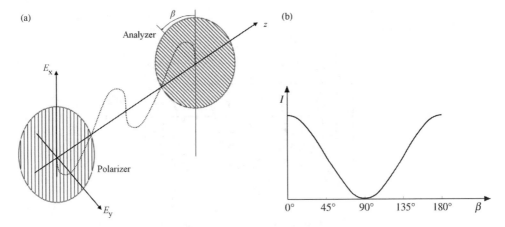

Figure 2.12 (a) Linearly polarized light wave with polarizer and analyzer (b) Transmitted intensity $I_t = I$ after the analyzer according to *Malus' law*

β angle between transmission direction of polarizer and analyzer.

To determine the direction of the oscillation plane, a second polarizer will be used, which in this case is called the *analyzer*. If the transmission direction will be rotated by an angle β compared to the angle of the polarizer exclusively, the projection E_t of the $\vec{E}$-vector will be transmitted by the analyzer:

$$E_t = E_0 \cos \beta \tag{2.36}$$

where:

E_0 Value of the $\vec{E}$-vector in front of the analyzer

The transmitted intensity I_t after the analyzer is following from Eq. (2.16). The result is given by the *Malus' Law* shown in Figure 2.12b:

$$I_t = I_0 \cos^2 \beta \tag{2.37}$$

where:

I_0 Intensity in front of the analyzer

β angle between transmission direction of polarizer and analyzer.

By introducing a transparent-optical inactive substance [2.1] between two crossed polars, there is no light transmission through the system in Eq. (2.37). An important effect, which is strongly connected to polarization, is the *Faraday effect* [2.10].

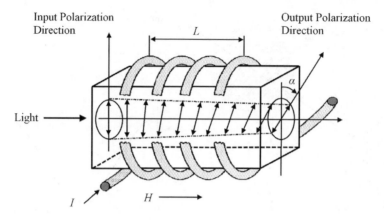

Figure 2.13   Faraday rotator

where:

L   Length
I   Current
H   Magnetic field
α   Angle between input and output polarization direction

Concerning certain substances, the polarization direction of the linearly polarized wave is rotated by an angle α by the aid of the Faraday effect applying a magnetic field parallel to the propagation direction of the light wave; a *Faraday rotator* is gained (Figure 2.13), which applies to:

$$\alpha = VLH \tag{2.38}$$

where:

H   Value of the magnetic field
L   Length along the magnetic field takes effect
V   *Verdet constant*

The Verdet constant is the material constant of the Faraday effect; it shows how much the oscillation direction will be rotated by the applied magnetic field. The magnetic field can be generated by an electric coil as well as by a permanent magnet. For the magnetic field strength, which is excited by means of an electric coil:

$$H = NI/L \tag{2.39}$$

where:

N   Number of coil windings
I   Value of the current through the coil

In addition, introducing a Faraday rotator into the optical path according to the arrangement referred to Figure 2.12, the polarization direction of a linearly polarized light wave rotates by angle α (Figure 2.13). The transmitted intensity by the analyzer varies according to Malus' Law (Eq. (2.37)).

The Faraday effect as the *non-reciprocal effect* can occasionally be used for a few applications in optical communication and sensor techniques. Both polarizers will be adjusted to an angle of 45 degree to each other, according to the input and output polarization directions of the Faraday rotator, respectively (Figure 2.13). For this arrangement, the total intensity will be transmitted if the Faraday rotator rotates about the polarization plane by 45 degree in the correct transmission direction of the analyzer.

Assuming a reflection of the light wave at a following optical component only, light having the correct polarization direction passes the analyzer in the reverse direction, that is, the reflected light wave and the analyzer must have the same polarization orientation. Due to its non-reciprocity, the Faraday rotator does not turn back the oscillation plane, but continues the rotation by an angle of 45 degrees. The oscillation plane of the reflected light is then oriented by an angle of 90 degree with respect to the transmission direction of the input polarizer. Thus, the light wave cannot pass the polarizer; an *optical isolator* is gained. Instead of an electric coil, a permanent magnet can also be applied. Optical isolators with a suppression of about 60 dB are available.

2.5 Refraction and Reflection

Refraction and reflection occur when a light wave transition from a media having a refraction index n_1 to a media with a different refraction index $n_2 \neq n_1$ takes place.

Figure 2.14 illustrates the incidence of a plane wave on an interface between such media.

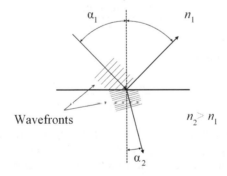

Figure 2.14 Refraction and reflection of a light wave at an interface

where:

n_1, n_2	Refraction indices in front and behind of an interface, respectively
α_1	Angle of incidence and reflection, respectively
α_2	Angle of refraction

With respect to the interface, the wave front in media 1 is inclined by an angle α_1. The wave vector simplified by the light ray is oriented perpendicularly to the wave front and forms the angle α_1 with the vertical of the interface (Figure 2.14). If the wave approaches the interface in this case, the left-hand side of the wave front penetrates earlier into media 2 in comparison to the right-hand side (view from wave front). From Eq. (2.10), the wave velocity in media 1 is given by $v_1 = c/n_1$. In Figure 2.14, it is assumed that the refraction index of media 2 is higher than that of media 1. Therefore, it can be written as $n_2 > n_1$ and $v_2 = c/n_2 < v_1$. The part of the wave front which has already penetrated into media 2 moves slower than the one left behind in media 1. The wave front inclines and changes its propagation direction. The distance of the following phase fronts has become smaller. The wave vector and the vertical to the interface then include the angle α_2.

Regarding the transition from an optical thinner to a denser media ($n_1 < n_2$), the light ray is refracted toward the vertical plane of the interface. Vice versa, in the case that $n_1 > n_2$, it is refracted off. The correlation between angles and refraction indices is described by the law of refraction, *Snell's law* [2.10]:

$$n_1 \sin \alpha_1 = n_2 \sin \alpha_2 \tag{2.40}$$

However, in general, a light wave does not completely penetrate into media 2, because a part of it is reflected at the interface. The reflected light ray together with the vertical planer of the interface encloses the angle α_1, too. This reflection is called the *Fresnel reflection* [2.10]. The part of the reflected light depends on the wave angle α_1 and the polarization of the incident light (this also applies for the transmitted part). Incident and reflected light rays, as well as the vertical plane to the interface, form an incidence plane. A light wave oscillating parallel to this plane is named the parallel polarized light. In this case, the following reflection coefficient applies for amplitudes:

$$r_{par} = \frac{n_2 \cos \alpha_1 - n_1 \cos \alpha_2}{n_1 \cos \alpha_2 + n_2 \cos \alpha_1} \tag{2.41}$$

Perpendicular to this it follows that:

$$r_{per} = \frac{n_1 \cos \alpha_1 - n_2 \cos \alpha_2}{n_1 \cos \alpha_1 + n_2 \cos \alpha_2} \tag{2.42}$$

Relating to an air–glass transition ($n_1 \approx 1, n_2 \approx 1,5$), Figure 2.15 illustrates the dependency of the reflection coefficients r_{par} and r_{per} on the angle of incidence α_1.

However, the perception is the square of the reflection coefficients for amplitudes (Chapter 2.2). For intensities I and optical powers $P(P \sim I)$, respectively follows $R \sim r^2$. Figure 2.16 shows the dependency of the reflection coefficients R_p and R_s to the angle of incidence, respectively.

Perpendicular incidence to the interface incident light is a special case which often occurs. It holds for: $\alpha_1 = 0 \rightarrow \alpha_2 = 0$ and

$$R_{par} = R_{per} = \left(\frac{n_2 - n_1}{n_2 + n_1} \right)^2 \tag{2.43}$$

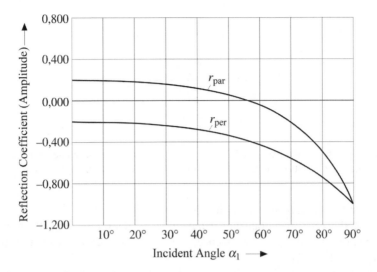

Figure 2.15 Reflection coefficients for amplitudes r_{par} and r_{per} versus angle α_1 for an air–glass transition $(n_1 \approx 1, n_2 \approx 1,5)$

Thus, an air–glass transition yields a reflected intensity part of about 4%. Concerning parallel polarized light with respect to the plane of incidence, the reflected part disappears at a certain angle, the *Brewster angle* (Figure 2.15 and 2.16, respectively): $\alpha_1 = \alpha_B$.

In that case, the reflected and refracted rays enclose an angle of 90°(Figure 2.17). Snell's law follows in this case:

$$\tan \alpha_B = n_2/n_1 \tag{2.44}$$

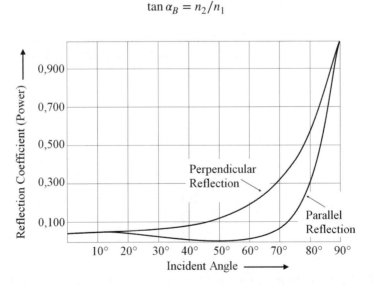

Figure 2.16 Reflection coefficients for powers R_p and R_s versus angle α_1 for an air–glass transition $(n_1 \approx 1, n_2 \approx 1,5)$

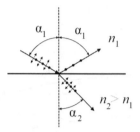

Figure 2.17   Refraction and reflection of incident polarized light waves at an interface, parallel (arrows) and perpendicular (dots) to the incidence plane in the case of the Brewster angle $\alpha_1 = \alpha_B$

According to the air–glass transition, it is $\alpha_B = 56^0$.

Considering the case of a transition from optically denser to optically thinner media $(n_2 < n_1)$, regarding small angles of incidence, a certain part of light is reflected and transmitted (Figure 2.18, rays 1 and 2).

By enhancing the angle of incidence α_1 to the cut-off angle, the refracted ray (α_2) runs increasingly flat and finally completely disappears into the boundary line which separates both media (Figure 2.18, ray 3). In case no light leaves media 1 (just the reflected wave exists), this case is named *total internal reflection*. From Snell's law of refraction, it follows that:

$$\sin \alpha_1 = \sin \alpha_C = n_2/n_1 < 1 \tag{2.45}$$

where:

α_C Cut-off angle of total internal reflection

From Eq. (2.45) it can be seen that total internal reflection only happens if there is a light transition from media 1 with a higher refraction index n_1 to media 2 with a lower refraction

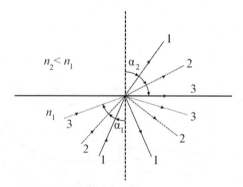

Figure 2.18   Refraction and reflection at a transition of light from an optically denser to an optically thinner media. The waves 1 and 2 experience partial refraction and reflection, but wave 3 exists exclusively as a reflection, the total internal reflection

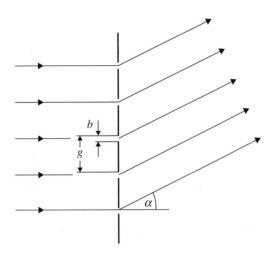

Figure 2.19   Diffraction at a grating

index n_2, such as a transition from an optically denser to an optically thinner media – not the other way round.

2.6   Diffraction

Diffractive elements are used for several applications, in particular when frequency shifts are of interest [2.10], for example, the Bragg effect in integrated optics or implemented in fibers (Chapter 5.2). There is use for these elements for optical multiplexers and demultiplexers or inside and outside semiconductor lasers. They are also needed for fiber and laser measurement techniques to investigate their spectral behavior. Finally, we mention the use in the sensor area. Moreover, also undesirable *diffraction* can produce serious problems in all areas where elements with a size in the order of the wavelength are implemented.

Effects have been studied for a slit, a double slit, and a pinhole, but the most common and also important device is an optical grating. Figure 2.19 shows the diffraction at a grating.

A plane wave is propagating towards a slit. After the slit, we see circular waves instead of plane waves, such that a change of the propagation direction occurs. This holds in general for all diffraction elements, as a diffraction means a propagation direction change:

$$\sin \alpha_\mathrm{m} = \pm m \frac{\lambda}{g} \tag{2.46}$$

where:

$m = 0,1,2$ Order of diffraction
b Width of the slit
g Grating constant (distance between two slits)

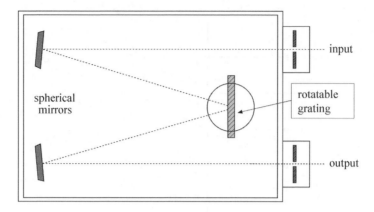

Figure 2.20   The Czerny-Turner Monochromator

The structure of all diffraction formulas is the same: in the dominator you find the wavelength, in the denominator there is always a geometrical size in place. In this case, it is the grating constant, because the distance of two slits appears in a double slit experiment as well as the width for a single split or the diameter of a pinhole.

Figure 2.20 shows a Czerny-Turner Monochromator, also named as a spectrometer. Such a grating is used to investigate the spectral behavior of the light to be investigated.

A spherical mirror is used in the formation of a parallel beam coming from the divergent light of the input slit, for example from a tungsten lamp. This parallel white light beam is then fed to a rotatable grating. Thus the different wavelengths (colors) appear arranged in front of

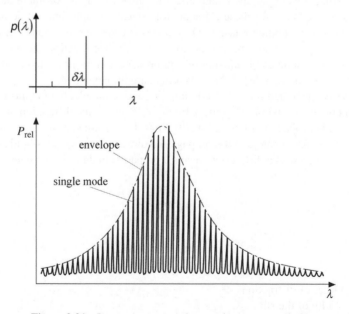

Figure 2.21   Laser spectrum and monochromator resolution

the output slit. Exclusively wavelengths fitting to the slit placement will leave the set-up. A grating rotation then allows the scanning of the whole spectrum of the incoming light. The width of the output slit determines the spectral width fed to a detector. Moreover, the resolution $\Delta\lambda$ of the investigated light is given by the wavelength λ [1.13,1.23], the grating order m, and the whole number p of the slits being illuminated (Figure 2.21, insert), to give:

$$\frac{\Delta\lambda}{\lambda} = m \cdot p \qquad (2.47)$$

Figure 2.21 shows the measured laser diode spectrum of a gain guided laser.

3

Optical Fibers

The propagation of waves in a light waveguide no longer happens only in a homogenous media infinitely expanded in all local coordinates. First we consider waveguiding in slab waveguides, also named film waveguides. Interfaces exist with refraction index changes. A guidance mechanism is generated resulting in guided waves. A general helpful and simple description is that a waveguide is a waveguide if the guiding media is surrounded by a media with a lower refraction index.

It is assumed that a slab waveguide consisting of three slabs is infinitely expanded in the propagation direction z, as well as in one perpendicular coordinate [2.2] (Figure 3.1a).

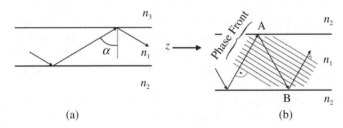

(a)   (b)

Figure 3.1 Slab waveguide (a) and model for transversal mode (b); the original wave interferes with a twice reflected one

where:

n_1 Refraction index of the waveguiding layer
n_2 Refraction index of the layer below
n_3 Refraction index of the layer above
α Angle of incidence into the waveguiding layer

The middle slab is surrounded by media with a smaller refraction index. The refraction indices of upper and lower slabs may differ from each other (n_2, $n_3 < n_1$ and $n_2 \neq n_3$). In the sidewise coupling light into the middle slab of such a waveguide, there is no light transition to media 2 and 3, thus the wave is guided. This is true, assuming the light ray in media 1 does not

Optical and Microwave Technologies for Telecommunication Networks, First Edition. Otto Strobel.
© 2016 John Wiley & Sons, Ltd. Published 2016 by John Wiley & Sons, Ltd.

run too steeply, that is it has been already sufficiently flat injected. Regarding the inclination angle, it follows from Snell's law of refraction that:

$$\alpha > \arcsin(n_2/n_1) \text{ and } \alpha > \arcsin(n_3/n_1) \tag{3.1}$$

Total internal reflection again occurs at every further reflection at the interface to media 2 or 3 (Chapter 2.4). The light ray can no longer leave media 1, thus a waveguide is generated. Based on hitherto existing considerations, wave propagation in the light waveguide could be assumed as follows: All light rays may accept any angle, starting from the light ray along the optical axis to the cut-off angle of the total internal reflection of $\alpha_G \leq \alpha \leq 90°$.

However, it appears that the angles are not continuously represented between these limit values, but exclusively specific discrete angles are allowed. Thus, the corresponding light waves are capable of propagating. This phenomenon can be demonstrated by the aid of superposition of multi-reflected waves.

Considering the wave in Figure 3.1b, first in point A and then, after the double reflection in point B, it is obvious that the wave has covered a definite distance Δg between these both points. Referring to Eq. (2.23), this yields to an optical path difference which takes effect on the interference of the original wave with the twice reflected one. Moreover, phase shifts have to be taken into account regarding the reflection at a denser media. According to Eq. (2.17), this leads to constructive or destructive interference being dependent on the path difference. In the latter case, a transportation of light energy in the desired propagation direction is inhibited. Only definite optical path differences and consequently exclusively specific distance differences are permitted. The distance differences depend on the inclination angle α of the light rays. This again means that only waves having light rays with specific angles are capable of propagation. The residual waves in the waveguide which propagate corresponding to the allowed light paths are named *modes*. In this connection, it deals with properties in space, thus they are also named *spatial* or *transversal modes*.

Concerning a fiber, the interface is self-contained, whereas a slab waveguide has two parallel interface layers. The geometrical form of a fiber is a cylinder in the center surrounded by a concentric hollow cylinder. Different materials are used to construct a fiber. Large-core fibers are typically all-plastic fibers. All-plastic PMMA fibers (poly methyl methacrylate) have been developed, nowadays named *Polymer Optical Fibers*, POF (Figure 3.2a). The outer diameter of the cladding is typically 1000 µm, and the core is nearly the same at 980 µm. The core refraction index in this example is 1.492 and the cladding refraction index is some percent lower. Due to the material, the attenuation is relative high. Best values amount to about 0.2 dB/m. A more detailed description is given in Chapter 10.2. Besides polymer optical fiber (POF), polymer-cladded silica (PCS) fibers are also of interest. In this case, the core is made of silica and the cladding of poly methyl methacrylate (PMMA). Therefore, the attenuation is much lower compared to POF fibers, at about 0.08 dB/m.

As shown in Figure 1.4, standard silica fibers reach a significantly lower attenuation of about 0.2 dB/km. But for cheap applications in cars or other vehicles, PCS fibers could be an attractive alternative. This is due to the advantages this fiber type has in common with the POF compared to a silica fiber, for example its robustness, easy connector fitting and low price. The diameter of the silica core of such fibers is typically 200 µm. This polymer-clad silica fiber combines the advantages of the glass fiber and the POF. Figure 3.2b shows a comparison of these three different fiber constructions [3.1]. In particular, a significant difference in core

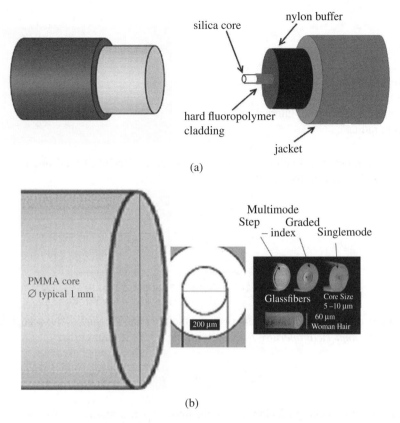

(a)

(b)

Figure 3.2 Comparison of polymer optical fibers, polymer-cladded silica fibers (a) and pure glass fibers (b)

sizes can be realized, for example the core diameter of a POF is about a factor of 100 larger compared to a single-mode glass fiber (Figure 3.2).

Figure 3.3 shows the general construction of a fiber. The cylinder exhibits the fiber *core* which features the refraction index n_1. The hollow cylinder forms the fiber cladding with the refraction index n_2. In particular, concerning the light coupling into a fiber, the acceptance angle δ is important; δ is the largest possible angle leading to waveguiding. Exceeding δ, total internal reflection in the core region will no longer be achieved; the light waves then penetrate into the fiber cladding.

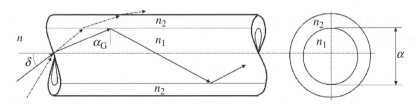

Figure 3.3 General construction of a step index fiber

where:

n_1 Core refraction index
n_2 Cladding refraction index
n Refraction index outside of the fiber
δ Angle of acceptance
α_C Cut-off angle of total internal reflection
a Core radius

From Snell's law of refraction and from the conditions to fulfill total internal reflection, according to Eq. (2.44) it follows that:

$$n \sin \delta = n_1 \cos \alpha_C \tag{3.2}$$

and thus:

$$n \sin \delta = n_1 \sqrt{1 - \sin^2 \alpha_C} = n_1 \sqrt{1 - (n_2/n_1)^2} \tag{3.3}$$

The product of the refraction index n outside the fiber and the sine of the acceptance angle δ is named the *numerical aperture* A_N. Applying Eq. (3.3), it follows that:

$$A_N = n \sin \delta = \sqrt{n_1^2 - n_2^2} \tag{3.4}$$

To gain a high coupling efficiency of light into a fiber, a numerical aperture as large as possible has to be attained. For further considerations, the knowledge of the *normalized refraction index difference* Δ is necessary:

$$\Delta = \frac{n_1^2 - n_2^2}{2 \, n_1^2} \tag{3.5}$$

Regarding a small relative refraction index difference $(n_1 - n_2)$, it is possible to expand Eq. (3.5) as a Taylor series. The approximated result is given by:

$$\Delta = \frac{n_1 - n_2}{n_1} \tag{3.6}$$

Due to the introduction of the relative refraction index difference for the numerical aperture, it follows that:

$$A_N = n_1 \sqrt{2\Delta} \tag{3.7}$$

Concerning the hitherto existing considerations, it has been assumed that the refraction indices n_1 and n_2 are constant through the fiber cross-sectional area. In general, this does not

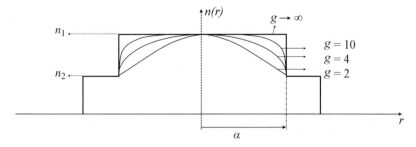

Figure 3.4   Refraction profile, refraction index n versus radial coordinate r and profile parameter g

apply. Thus, the numerical aperture varies as a function of the radial coordinate r (see also Figure 4.43):

$$A_N = \sqrt{n^2(r) - n_2^2} \qquad (3.8)$$

where $n(r)$ is called the refraction index profile, that is in the core area of the fiber the refraction index varies as a function of the radial coordinate, r. Furthermore, the cladding is $n_2 = \text{const.}$ For special fibers, there are modifications [3.2–3.4]).

For numerical calculations, refraction index profiles are assumed as follows:

$$n(r) = n_1 \sqrt{1 - 2\Delta(r/a)^g} \quad \text{for } r < a \qquad (3.9)$$

$$n(r) = n_2 \quad \text{for } r \geq a$$

where:

a Core radius

g Profile exponent.

Figure 3.4 shows a diagram of $n(r)$. Different fiber types can be described by variation of the profile parameter. For $g \to \infty$, it follows for $r < a$: $(r/a)^g \to 0$ and thus $n(r) = n_1 = \text{const.}$ The original form of the refraction indices is again achieved. Such a fiber is called the *step index fiber*. For $r = a$, a significant step in the refraction index profile takes place. Regarding all other profile parameters, a continuous variation of the refraction index with the radius occurs. A gradient is generated. Hence, such fibers are named *graded index fibers*. A profile parameter $g = 2$ is often chosen (Chapter 3.2.1). A parabolic refraction index profile is approximately gained.

Considering step index fibers, the following aspects have to be taken into account. The value of the wave vector is also called the *phase propagation constant*; in a media it holds that:

$$k = k_0 \, n_1 \qquad (3.10)$$

with regard to the phase propagation constant in the z-direction β, it follows Figure 3.5.

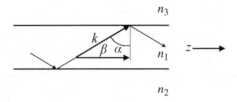

Figure 3.5 Phase propagation constant β in propagation direction and general phase propagation constant k (values of each vector)

$$\beta = k_0 n_1 \sin \alpha \qquad (3.11)$$

where is $90° > \alpha > \alpha_G$.

Thus, depending on the inclination angle, it follows from Eq. (3.11) and the condition for total internal reflection Eq. (2.44):

$$k_0 \, n_1 > \beta > k_0 \, n_2 \qquad (3.12)$$

The following wave equation, which is essential for further considerations, can be deduced from the Maxwell equations (2.1) to (2.4) by the use of vector calculus laws. Provided that the refraction index varies by a small value only, it follows that approximately (it also holds analog for $\vec{H}$-field):

$$\Delta \vec{E} + k_0^2 n_i^2 \vec{E} = 0 \qquad (3.13)$$

where $n_i = n_1$ for $r < a$ and $n_i = n_2$ for $r \geq a$, (Δ is the Laplacian (Eq. (2.8)). To solve this differential equation, it can be written as:

$$\vec{E}(x, y, z) = \vec{E}(x, y) \exp(-j\beta z) \qquad (3.14)$$

Thus, an equation is obtained which describes the spatial wave propagation in the z-direction. The field strength vector oscillates in the x- and y-directions, respectively. Regarding the fiber geometry, the introduction of cylindrical coordinates (r, Φ) is obvious:

$$\vec{E}(x, y, z) \rightarrow \vec{E}(r, \Phi) \exp(-j\beta z) \qquad (3.15)$$

Because the boundary conditions that the field strengths have to be continuously differential concerning the transition from the core area into the cladding, solutions for the wave equation can be achieved. Solutions are gained which can be described in the core area by Bessel functions [3.16], and in the cladding by Hankel functions [3.16]. Regarding graded index fibers, the relations are more complicated. In this case, analytically closed solutions can no longer be found. Approximation methods are the only tools to work with [3.5–3.7] (WKM method, WKB: Wentzel, Kramers, Brillouin).

LP_{01} LP_{21} LP_{83}

Figure 3.6   Mode figures on a fiber surface

where:

LP_{01} Fundamental modus
LP_{21}, LP_{83} Higher order modes

Possible solutions, the existing modes of oscillation, are denoted as modes of a fiber. They are named LP_{mn} modes, where m is the azimuthal mode number and n the radial one. Figure 3.6 shows mode figures on the surface of a fiber; m can be interpreted as half the number of spots on a circle, and n is consistent with the number of circles.

Moreover, there are modes leading their energy from the core to the cladding. Losses result from a relation between waveguiding and evanescent radiation; *evanescent waves* arise [3.9–3.11]. The tendency to create evanescent waves increases if the fiber is bent. Figure 3.7 shows an intersection of the power distribution ($P \sim |\vec{E}|^2$) regarding a fiber cross-section. The figure shows that a certain power part will be guided, not only in the core but also in the cladding.

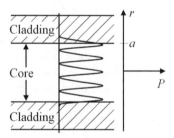

Figure 3.7   Modes as power distribution across a fiber cross section, optical power P versus radial coordinate r

where:

a Fiber core radius

Furthermore, at a given wavelength, λ modes propagating in the z-direction depend on fiber parameters diameter a and numerical aperture A_N. These quantities are combined as *V-number*, also the normalized frequency as:

$$V = 2\pi a A_N / \lambda \qquad\qquad (3.16)$$

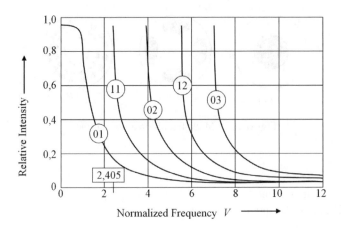

Figure 3.8   Power distribution $P(r)$ vs. V-parameter

Figure 3.8 presents the power distribution between the core and cladding area as a function of the V parameter of a step index fiber for a few low modes [3.13]. It can be observed that for $V < 2.405$, exclusively mode LP_{01} is guided in the core. A *single-mode fiber* is generated; LP_{01} is denoted as a fundamental modus. For higher V-parameter values ($V > 2.405$), higher modes are guided too. We receive a *multimode fiber*. For a high number of modes, the number of guided modes is given by [3.14]:

$$N = \frac{V^2}{2} \cdot \frac{g}{g+2} \tag{3.17}$$

Thus, regarding a step index fiber ($g \to \infty$), $N = V^2/2$ is obtained, and for a graded fiber ($g = 2$), $N = V^2/4$ is achieved. Thus, a step index fiber guides twice as many modes as a graded one if we assume the same conditions. A typical step index fiber with a core radius of 50 µm and a numerical aperture of 0.5 guides approximately 17 000 modes at a wavelength of 850 nm. For the same wavelength, about 500 modes are attained concerning a graded index fiber with typical values of core radius and numerical aperture (25 µm/0.25).

From Eq. (3.17) it follows that boundary conditions exist for the number of modes capable of propagation. This occurs, for example, if N is decreased by means of a corresponding modification in such a way that a certain mode is no longer guided. This is named the cut-off of the corresponding mode. Regarding a specified fiber (physically realized fiber), the parameters a, g and A_N, are given and unchangeable. A decrease of the V parameter thus results from an increase in the wavelength in the manner that a specific mode which has been guided up until now will no longer be guided (Figure 3.8). The wavelength at which this happens is named the *cut-off wavelength*, λ_C. By further increase of the wavelength or corresponding modification of the V parameter in the design phase, a single-mode fiber is again achieved ($V \leq 2.405$). If the parameters are chosen in such a way that the V parameter becomes too small, the electromagnetic field will be too weakly concentrated in the fiber core. A too large a part is then guided into the cladding (Figures 3.8 and 3.9), therefore, the fiber attenuation increases (Chapter 3.1). Concerning a single-mode fiber, $1.5 \leq V \leq 2.405$ is reasonably chosen. Wave forms of single-mode fibers can be described as *Gaussian rays* [3.15,2.1].

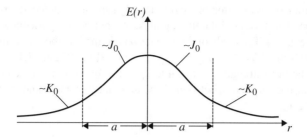

Figure 3.9 Field distribution of a single-mode fiber; electric filed strength $\vec{E}(r)$ versus radial coordinate r

where:

   a Fiber core radius
   J_0 Bessel function of the first type and zero order
   K_0 Hankel function of zero order

Figure 3.9 shows the field distribution $\vec{E}(r)$ of the fundamental modus. The correct mathematical description of the field distribution is given by Bessel and Hankel functions, respectively [3.15,3.16]. A *Gaussian function* is a good approximation (Figure 3.10):

$$\vec{E}(r) = E_0 e^{-\left(\frac{r}{w_0}\right)^2} \tag{3.18}$$

w_0 is the Gaussian beam width of the fundamental modus (also named beam waist or spot size (Figure 3.11). w_0 is a measure for the radial field dimension in the fiber (in general $w_0 \neq a$, but similar if reasonable parameters are chosen). A detector measures the following power distribution:

$$P(r) = P_0 e^{-2\left(\frac{r}{w_0}\right)^2} \tag{3.19}$$

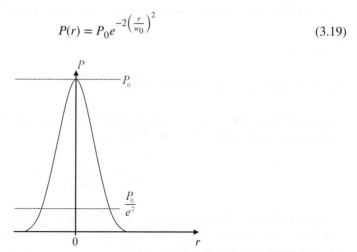

Figure 3.10 Power distribution $P(r)$ in a single-mode fiber versus radial coordinate r

Figure 3.10 shows a cross-section of the rotationally symmetric power distribution. Measurements are performed at the end of the fiber, that is, in a non-guiding media with homogeneous refraction index (e.g. air). The angle of divergence δ is defined by the line which corresponds to a power reduction from $P(\delta = 0) = P_0 = P_{max}$ to $P(\delta) = P_0/e^2$ (Figure 3.11). This corresponds to a decrease of the field strength to the value E_0/e. For measurement techniques concerning single-mode fibers, the correlation between the Gaussian beam width and the angle of radiation is given by [3.15,2.1]:

$$w_0 = \frac{\lambda}{\pi n \tan \delta} \qquad (3.20)$$

Attention: n is the refraction index in the media outside of the fiber!

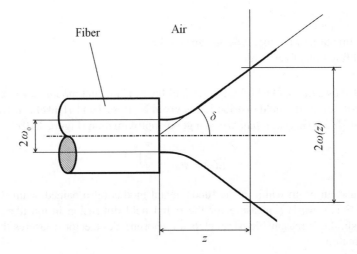

Figure 3.11 Gaussian beam width $w(z)$ at a distance z from fiber end and angle of divergence δ

where:

w_0 Gaussian beam width on fiber surface

It is sufficient to measure the power distribution at a certain distance behind the fiber. Thus, the Gaussian beam width $w(z)$ can be determined:

$$w(z) = \frac{\lambda z}{\pi n w_0} \qquad (3.21)$$

In particular, regarding coupling challenges, the knowledge of these correlations is important, for example for coupling of fibers among themselves as well as coupling between fibers and light sources or detectors (Chapter 4.3).

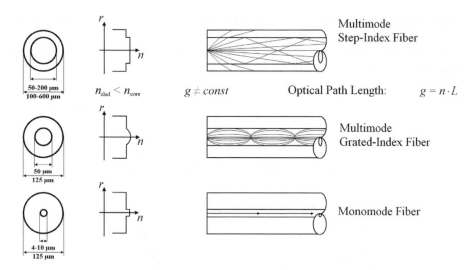

Figure 3.12   Multimode fiber with step-index and graded-index and single-mode fiber in comparison

Figure 3.12 depicts a comparison of the three most important fiber types. Figure 3.13 shows the corresponding scanning micrographic recordings in comparison with a human hair. The fiber surfaces have been etched to visualize the refraction index profile [3.17].

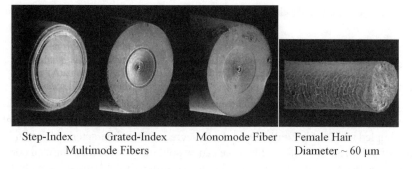

Step-Index   Grated-Index   Monomode Fiber   Female Hair
            Multimode Fibers   Diameter ~ 60 µm

Figure 3.13   Scanning electron microscope (SEM) recording of etched fiber ends and human hair corresponding to Figure 3.12

Concerning optical wave propagation in fibers, polarization is a further fundamental subject. Strictly speaking, the term "single-mode fiber" is not correct. There are two polarization modes of propagation, two fundamental modes with vertical polarization to each other. As already mentioned in Chapter 2.4, maintenance of the polarization state of a fundamental modus propagating in a fiber is necessary for specific applications. However, the standard single-mode fibers described above do not preserve polarization. They feature a low *birefringence*, which is not allowed to be negligible for certain applications. Considering the fiber in two directions perpendicular to each other (both perpendicular to the propagation direction), there are small refraction index differences. This effects the propagation of two linear polarized light waves which are perpendicular to each other. They propagate with different velocity in the fiber, which causes different delay times. Consequently, the polarization state of the light permanently varies during the propagation in the fiber.

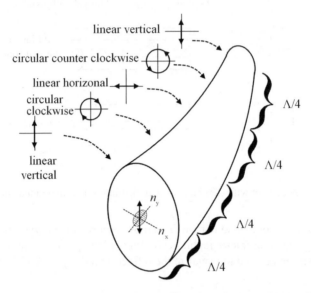

Figure 3.14   Polarization development of a light wave in a linear birefringent single-mode fiber

where:

Λ Beat wavelength (distance between two equal polarization states

Figure 3.14 depicts a polarization development, provided that a linear polarized light wave has been coupled into the fiber with respect to 45 degrees of both characterized directions (principal axes of polarization [3.18]). This wave can be split into two linear polarized components along the principal axes of polarization. Due to the different delay times, a path difference between both components is generated. Passing a certain distance, the phase distance $\Delta\Psi$ is assumed to be $\pi/2$ (Figure 2.11). According to Eqs. (2.29) and (2.34), right-hand circularly polarized light is consequently gained. After passing the double distance in the fiber, the path difference $\Delta\Psi$ is π. The superposition of both components yields a linear polarized light wave (Eq. (2.33)), but with 90-degree inclination to the initial polarization. A further increase to $\Delta\Psi = 3\pi/2$ results in left-hand circularly polarized light. Regarding $\Delta\Psi = 2\pi$, the original polarization state as coupled into the fiber is again achieved.

The distance between two equal polarization states of a single-mode fiber is named the *beat wavelength*. It is obvious that at the end of the fiber the polarization state depends on the fiber length. Moreover, external effects such as temperature and pressure also influence polarization development. Therefore, for several applications, measures have to be taken to stabilize the polarization (Chapter 3.2.2).

One opportunity to control this problem is the application of *polarization preserving fibers* [3.19]. In this case, fibers will be used that have a strong birefringence. Polarization is preserved if linear polarized light is coupled in the manner whereby the polarization direction is chosen parallel to the major or minor polarization axes of the fiber. The coupling of linear polarized light to the major or minor polarization axes of high-birefringent fibers is much stronger as compared to standard single-mode fibers. External effects like moderate mechanical forces no longer cause an over coupling from the major to minor polarization axes or vice versa. Thus no change of the polarization state will take place, except for very strong pressures skew to the axes. Fortunately, usually this does not happen. A further opportunity to solve this problem is the application of polarization control [3.19,3.21] (Chapter 3.2.2).

3.1 Attenuation in Glass Fibers

Propagating in an optical waveguide with a length increment of dz, an optical signal having the magnitude of the field strength E loses the value dE. The following ansatz is chosen:

$$- d\mathrm{E} = \alpha' \, E \, dz \tag{3.22a}$$

The proportionality factor α' is a measure of attenuation. This differential equation can be solved by a simple integration:

$$\int dE/E = \alpha' \int dz \tag{3.22b}$$

The field strength, which is coupled into a waveguide, is named $E(0) = E_0$, thus it follows that:

$$E(z) = E_0 \exp(-\alpha' z) \tag{3.23}$$

The description of wave propagation by attenuation of the light wave is often combined to:

$$E(z) = E_0 \exp(j\beta z - \alpha' z) \tag{3.24}$$

where:

$$\beta = k_0 n' \text{ and } \alpha' = k_0 n'' \tag{3.25}$$

n' and n'' present real and imaginary parts of the refraction index n of a lossy dielectric material in which the light wave propagates:

$$n = n' - jn'' \tag{3.26}$$

The power can be written as $P \sim |E|^2$. Thus, the law of Lambert-Beer-Bouguer follows from Eq. (3.22):

$$-dP = 2 \, \alpha' \, P \, dz \tag{3.27}$$

and with Eq. (3.23):

$$P(z) = P_0 \exp(2\alpha' z) \tag{3.28}$$

the power decreases exponentially with increasing length. For the attenuation coefficient $2\alpha' = \tilde{\alpha}$, it follows that:

$$\tilde{\alpha} = -\frac{Np}{L} \ln \frac{P}{P_0} \tag{3.29}$$

where:

Np $Neper = 1/m$ is the unit of this attenuation coefficient

Concerning light waveguide transmission, it is common practice to calculate in decibels (dB). For the attenuation coefficient α, it follows that:

$$\alpha = -\frac{10\,\text{dB}}{L} \lg \frac{P}{P_0} \tag{3.30}$$

By transformation of $\tilde{\alpha}$ into α, we achieve:

$$\alpha = \tilde{\alpha}\frac{10\,\text{dB}}{Np \ln 10} \tag{3.31}$$

The specification in a data sheet of a fiber is typically given in dB/km.

3.1.1   Attenuation Mechanisms in Glass Fibers

Scattering, absorption and *radiation losses* are the most important attenuation mechanisms that glass fibers suffer from. Scattering and absorption are intrinsic loss mechanisms due to material properties (Figure 3.16). Radiation losses are extrinsic ones due to effects during and after the fiber manufacturing process.

Statistical refraction index fluctuations determine linear scattering, the *Rayleigh scattering* (named after Lord Rayleigh). That is due to the stochastic molecular structure of the used waveguides (mainly SiO_2). The refraction index varies within local domains, which are small in comparison to the light wavelength. The incident light wave is scattered at these spatial inhomogeneities. The scattered power distribution corresponds to that of a *Hertzian dipole* (Figure 3.15):

$$P_S \sim P_0 \cos^2 \delta \tag{3.32}$$

where:

P_S Scattered power
P_0 Power of the incident light wave

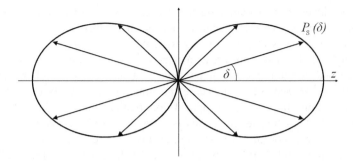

Figure 3.15 Spatial characteristic of Rayleigh scattering

where:

P_S Scattered power
δ Angle with respect to the wave propagation direction (rotational symmetric aligned to the z-axis)

Note that it is three-dimensional with rotational symmetry perpendicular to the propagation direction. The three-dimensional view looks like a tire in the center. Figure 3.15 shows a section through this tire. Scattering takes place in a forward direction as well as in the reverse direction. Only a minor angle area of the radiation lies within the numerical aperture of the fiber (Figure 3.15). Thus, a very small power part is scattered in the propagation direction in such a way that it can be guided by the fiber core. This is also valid for the reverse direction. The scattered power P_S is strongly dependent on the wavelength of the used light source (Figure 3.16):

$$P_S \sim \frac{1}{\lambda^4} \tag{3.32}$$

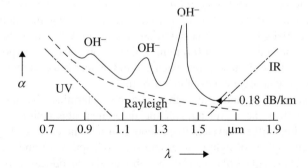

Figure 3.16 Spectral attenuation of a glass fiber: Rayleigh scattering, OH absorption, UV- and IR-self absorption, which is obviously superposed by Rayleigh scattering, the edge of the IR absorption (principal sketch for visualization)

P_S has to be considered as a loss and reduces the optical power for the transmission signal. If light waves having very high power densities propagate in the fiber, non-linear scattering effects arise in addition to linear scattering, mainly Raman and Brillouin scattering. In particular, concerning single-mode fibers with core diameters of a few micrometers, high power densities are quickly reached. Assuming power in the range of about 10 mW in a fiber, power densities of about 10 kW/cm^2 are achieved. In this case, the input power cannot be enhanced arbitrarily. One of the most important targets in optical communication is realizing high data rates and simultaneously high link lengths. This is possible if sufficient high optical power at the receiver is available. For this purpose, high-energy lasers can be used with a few 10 mW output power. Their possible field of application is to date limited by non-linear effects in fibers [3.22–3.25].

Absorption is conversion of radiation energy in heat caused by material impurities. In particular, undesired OH ions in SiO_2 have a dominating influence on absorption properties. The degree of effects depends on the wavelength of the radiation. In particular, certain absorption areas lie at 950 nm, 1250 nm and 1390 nm. Even very small concentrations in the range of 1 ppb (parts per billion) have a noticeable effect. One ppb corresponds to a value of 10^{-9}. At 1 ppm (part per million, i.e. 10^{-6}) the attenuation caused by OH ions at a wavelength of 1390 nm is about 50 dB/km (Figure 3.16). Further impurities result from metal ions (copper, iron, cobalt, manganese, nickel, chromium, vanadium). In contrast to the OH ions, these materials play a secondary role. Additional molecular absorptions in SiO_2 itself (even without impurities) occur approximately in the ultraviolet region, as well as in the infrared above 1.5 µm.

Radiation losses due to the conversion from guided modes to non-guided modes exist in addition to scattering and absorption. This is due to fluctuations of the V-parameter (Eq. (3.16)). Such fluctuations may be already generated during the fiber production process as a consequence of variations regarding the radius a, the profile parameter g and the refraction indices from core and cladding. But they can subsequently develop by coating or wiring of the fiber. *Microbending* occurs, which results in local fluctuations of the refraction index. This is notably clarified if the attenuation of a fiber, which is wound on a cable reel, is compared with the attenuation value after the freewheeling of the fiber. By means of the winding process, bending is generated if the superimposed fibers interlace. That leads to an increase of attenuation which disappears after the freewheeling. Moreover, radiation losses also occur due to *macrobending*, transformation from guided modes in evanescent waves.

Figure 3.16 shows the general *spectral attenuation behavior* of a glass fiber, including all intrinsic attenuation mechanisms. According to Rayleigh scattering, the attenuation decreases with increasing wavelengths, superimposed by attenuation maxima due to the OH absorption. In contrast to the molecular UV absorption, which is obviously superposed by Rayleigh scattering, the edge of the IR absorption extends into the near-infrared region and dominates the Rayleigh scattering at wavelengths above 1.6 µm.

Thus, IR-self-absorption and Rayleigh scattering determine the minimum value of 0.18 dB/km, as well as the wavelength of the theoretical attenuation minimum at about 1550 nm. Concerning further wavelengths, attenuation minima exist, as Figure 3.16 shows. Taking into account aspects of light sources and sinks (Chapters 6.1 and 6.2), the term *three windows* appears. Thus, a glass fiber can be favorably used as an optical waveguide, at approximately 850 nm, 1300 nm and 1550 nm (Figure 9.1). At 850 nm, the attenuation coefficients of single-mode fibers lie at about 2 dB/km, and approximately 0.35 dB/km is gained at about 1300 nm.

Nowadays, smaller values of OH absorption can be achieved (Figure 9.1). Moreover, regarding modern fibers, the increase of the attenuation due to the OH ions has been limited to values below 1 dB/km. Therefore, also wavelengths of 950 nm, 1250 nm and 1390 nm can be used for optical transmission in a glass fiber. Thus, modern fibers can be applied over the whole wavelength range from about 800 nm to more than 1600 nm, enabling an enormous transmission capacity by use of wavelength division multiplex systems (WDM, Chapter 9.1.1).

3.1.2   Attenuation Measurement Techniques

The attenuation mechanisms partially provide the fundamental knowledge for measurement techniques to determine the attenuation in fibers. The *cut-back method* directly results from the definition of the attenuation coefficient (Eq. (3.29)). First, the optical power P_E is measured at the end of a fiber with a length that lies in the hundred meter- or kilometer-range (Figure 3.17).

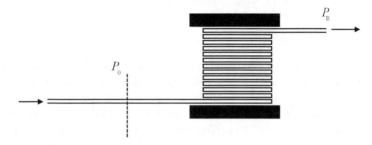

Figure 3.17   Attenuation measurement by the cut-back method

where:

P_0 Power into the fiber
P_E Power at the fiber end

Afterwards the fiber will be cut back to about 1 m in length behind the input coupling. The optical power is then measured again. Thus, the total attenuation D of the fiber can be expressed as:

$$D = -10 \text{ dB } \lg(P_E/P_0) \tag{3.33}$$

However, a precise measurement presupposes a constant input power between both measurements. Thus, the power of the light source has to remain constant and the conditions of the coupling into the fiber must not change. Measurement errors often arise due to problems caused by *cladding modes*. By coupling of light into the fiber, it cannot be avoided that a certain power part will be coupled into the fiber cladding. The transition from the optical fiber cladding to the plastic primary coating as a protective layer (Chapter 4.1) induces a refraction index jump. The refraction index of the used polymer material is smaller than that of the fiber cladding, such that a further waveguide is developed on the interface glass-optic cladding to plastic primary coating. Cladding modes are generated, as shown in Figure 3.18.

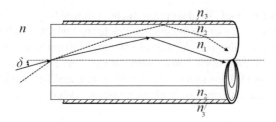

Figure 3.18   Cladding mode formation

where:

n_1, n_2 Glass optical core and cladding refraction index
n_3 Refraction index of the surrounding plastic cladding
n Refraction index outside of the fiber
δ Angle of acceptance

However, these modes do not contribute to the transmission of the signal over a link length in the kilometer range. After some hundreds of meters, these modes are absorbed due to the high attenuation polymer material. Therefore, this power part has to be suppressed for the measurement when the fiber is cut back after some meters to determine the coupled input power. This happens by means of a *cladding modes stripper*. Therefore the primary plastic coating is removed. Instead of the coating, a material is deposited with a refraction index with a real part (Eq. (3.26)) as equal as possible to the one of the optical fiber cladding. Hence, according to the Fresnel reflection (Chapter 2.5), the reflection disappears at this interface. The whole power part traveling in the optical cladding now penetrates into the domain of the mode stripper. The imaging part of the material should be as large as possible. Thus, the undesired power part will be absorbed within a fiber length of several centimeters. Epoxy adhesives and oil, as well as camera and nail polish, are used as the materials, which in practice prove non-complicated applications. Regarding multimode fibers, the attenuation also depends on groups of modes which are excited in the fiber. An *equilibrium mode distribution* is intended to get relevant information about the attenuation. This kind of distribution can be achieved with the aid of special input coupling optics or *mode mixers* [3.26].

A further measurement technique can be used by applying the *insertion-loss method*. In this case, a launching fiber with a length of several hundreds of meters is used to eliminate the problem of equilibrium mode distribution. After measuring the optical power at the end of the launching fiber, the latter is spliced with the fiber under test (Chapter 4.3.3). Afterwards the power is measured again at the end of this fiber. The attenuation then is given by Eq. (3.33). The unknown attenuation due to the fiber-to-fiber coupling after the splice is the disadvantage of this method. This also holds for the conservation of the equilibrium mode distribution investigating multimode fibers.

Both techniques, the cut-back method as well as insertion-loss method, merely give integral information of the fiber attenuation. In contrast to these techniques, the *backscattering method* also gives local information along the fiber [3.27,3.28.]. Basic knowledge of this measurement technique is Rayleigh scattering in a fiber (Chapter 3.1.1).

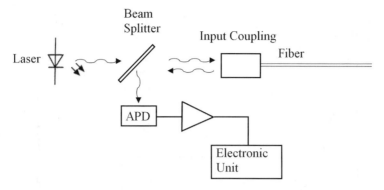

Figure 3.19   Attenuation measurement by use of the backscattering method

Figure 3.19 shows a schematic presentation of the corresponding test set-up. A laser emits short light pulses which are coupled into the fiber [3.28]. While traveling through the fiber, a minor power part of the light pulse is scattered back at each location along the fiber. This power part is fed to an optical detector near the fiber input by means of a beam splitter. The detector signal will be further processed in an electronic unit and an *optical time domain reflectometer* (OTDR) is achieved. The time interval τ between the emission and the reception of the light pulse depends on the location z in the fiber where the light pulse has been backscattered:

$$\tau = 2n_i z/c \tag{3.34}$$

The electronic unit takes a signal sample exactly after this time interval. Thus the light power results from a volume element at location z in the fiber. By varying the time interval, a change of the monitored location in the fiber follows. By means of further time variations, the whole fiber can be covered and the location exactly assigned according to Eq. (3.34). Figure 3.20 depicts the backscattered power as a function of the time interval τ. During this time, the light pulse propagating in the forward direction, as well as the power part which is scattered in the reverse direction, are attenuated by the fiber.
where:

τ Time interval between light emission and reception

P_{BSB}, P_{BSA} Backscattered power before and after the splice point

The attenuation coefficient α of the fiber can be written as:

$$\alpha = -\frac{10 \text{ dB}}{2z} \lg \frac{P_{BSE}}{P_{BS0}} = -\frac{10 \text{ dB} n_1}{c\tau} \lg \frac{P_{BSE}}{P_{BS0}} \tag{3.35}$$

where:

P_{BS0} Backscattered power at the beginning of the fiber

P_{BSE} Backscattered power at the end of the fiber

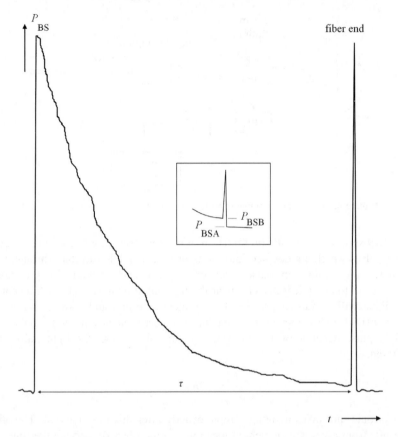

Figure 3.20   Backscattered power P_{BS} vs. time t (insertion: power behavior at a poor splice

Instead of P_{BS0} and P_{BSE}, values at different locations in the fiber can be used to determine α:

$$\alpha = -\frac{10 \text{ dB} n_1}{c \Delta \tau} \lg \frac{P(z_2)}{P(z_1)} \tag{3.36}$$

The time difference $\Delta \tau = \tau_2 - \tau_1$ results from both locations z_1 and z_2, according to Eq. (3.34). Thus, it is possible to achieve a step-by-step investigation of the attenuation in the fiber [3.28]. Different changes of the $P_{BS}(t)$ signal in the diagram shown in Figure 3.20 correspond to different attenuations. Also, discontinuous changes of the attenuation can be analyzed, for example a poor splice. First, there is a short increase due to the Fresnel reflection followed by a decrease significantly below the prior value due to the splice attenuation. Therefore, the attenuation of the splice point can be directly determined by the backscattered values before (P_{BSA}) and after (P_{BSB}) of the splice point (see insertion in Figure 3.20):

$$D_{Spl} = -5 \text{ } dB \lg \frac{P_{BSA}}{P_{BSB}} \tag{3.37}$$

A huge increase occurs at the end of the fiber due to the Fresnel reflection (Figure 3.20). The reflection at this glass–air transition almost happens perpendicularly to the propagation direction. Thus, according to Eq. (2.42), high reflection takes place with a reflection coefficient of approximately 4%. If the fiber is broken anywhere within the whole link, the same behavior will happen. Comparing the three methods, backscattering has a significant advantage in that it is not necessary to have both ends of the fiber accessible. Therefore, it is particularly suitable for practical application concerning already installed fibers, whereas cut-back and insertion-loss methods can only be used in the laboratory (exception: ring or loop structures).

3.2   Dispersions in Fibers

If a light pulse is propagating along a fiber, there is not only a reduction of its amplitude due to fiber attenuation, but also a broadening of the pulse. A narrow input pulse with a full width at half maximum t_1 finally appears as a broad output pulse with the full width at half maximum t_2 (Figure 3.21). This effect is denoted as the *pulse broadening* $\Delta\tau$. It holds that:

$$\Delta\tau = \sqrt{t_2^2 - t_1^2} \qquad\qquad (3.38)$$

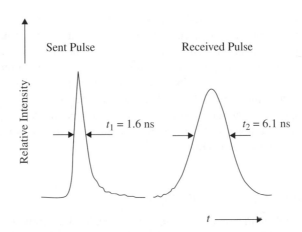

Figure 3.21   Optical pulse at the beginning and at the end of the fiber link

where:

t_1 Full with at half maximum (FWHM) at the beginning of the fiber link
t_2 Full with at half maximum (FWHM) at the end of the fiber link

Equation (3.38) is an approximation. Strictly speaking, instead of the full widths at half maximum (FWHM), the effective pulse widths (rms value) have to be inserted into Eq. (3.38) [3.30]. Due to pulse broadening, the edges of adjacent subsequent pulses move closer together. They are tangential to each other with an increasing fiber link and finally join to a single light pulse (Figure 3.22).

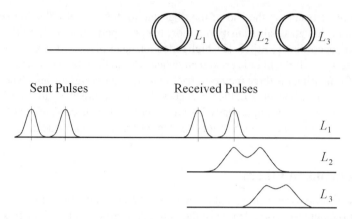

Figure 3.22   Merging of two adjacent pulses overlapping to a single pulse due to pulse broadening

where:

L_1, L_2, L_3 Different fiber link length

   Thus, it will be no longer possible to distinguish single pulses by means of a detector which is located at the end of the fiber. Accordingly, the pulse interval has to be enlarged at the access to the fiber, which is at the transmitter. In this case the detector is able to select the broadened pulses. Unfortunately, this procedure results in diminishing of the pulse repetition frequency. Thus, less information per time unit can be transmitted along the fiber link. If the pulse repetition frequency is still retained, the useful fiber link has to be shortened. The essential goal of optical communication engineering is to maximize the product of pulse repetition frequency and bridgeable fiber length or, in other words, the product of bandwidth times length should be a maximum (Eq. (1.1).

3.2.1   Dispersion Mechanisms in Fibers

The most significant phenomenon is *mode dispersion* regarding multimode fibers with a step index profile. Figure 3.23a illustrates the light paths of three modes in such a fiber, which are arbitrarily picked out. It is obvious that these modes strongly differ in their geometrical path lengths. Due to the constant refraction index within the core region, the three modes also differ in their optical path lengths (Eq. (2.24)). Each mode contributes its part to the power transportation of the optical light pulse. If it is possible to excite the modes discretely it can be measured so that the corresponding light pulses would arrive at the end of the fiber at different times. They show *delay times*. Due to the simultaneous presence of all modes, the pulses superimpose to a single broadened pulse. This leads to pulse broadening caused by mode dispersion (Figure 3.23b). Concerning the delay time between the fundamental modus and the highest order modulus, limited by total internal reflection, it follows from [3.31]:

$$\Delta t_{\text{mod}} = \frac{n_{1g}L}{2cn_1^2}A_N^2 = \frac{n_{1g}L}{c}\Delta \qquad (3.39a)$$

where:

n_1 Phase refraction index in the fiber core
n_{1g} Group refraction index in the fiber core
Δ Relative refraction index difference (Eq. (3.4)).

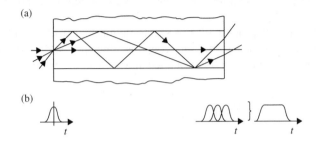

(a)

(b)

Figure 3.23   Visualization of mode dispersion in a step-index fiber

where:

a) Different optical paths of three selected modes
b) Input pulse and the according three output pulses merging to a single broadened pulse

   Considering typical step index fibers with a refraction index difference of a few percent, a delay time of about 50 ns is achieved after a fiber length of one kilometer. This delay time could be diminished by a reduction of the mode number N (Eq. (3.16) and (3.17)). Therefore the fiber core radius or the numerical aperture has to be minimized according to a step index fiber. But the disadvantage here is that the light coupling from a light source into a fiber becomes more difficult (Chapter 4.3.1). Relating the delay time to the fiber length, the mode dispersion coefficient is given by:

$$M_{\mathrm{mod}} = \Delta t_{\mathrm{mod}} / L \tag{3.39b}$$

   This physical quantity represents an important fiber parameter, particularly regarding the calculation of the performance of a transmission link (Chapter 9.1). Thus, it is registered in the fiber data sheet. The units are generally given in ns/km or ps/km.
   Figure 3.24 shows the light paths of several modes in a graded index fiber with a parabolic refraction index profile. Modes propagating near to the cladding have to run along a large geometrical path, as in step index fibers. But this will be balanced in such a way that its

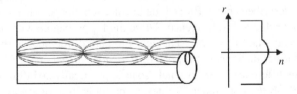

Figure 3.24   Equal optical path lengths of three selected modes of a gradient index fiber, refraction index n versus radial coordinate r, geometrical path differences and refraction index differences feature equal optical path lengths

propagation velocity becomes major in the outer core region due to the minor refraction index (Figure 3.4). In contrast, a shorter geometrical path is gained regarding the fundamental modus with a corresponding higher refraction index (Eq. (2.24)). Therefore, in an ideal case, we achieve isochronous optical light paths for all modes. Thus, the mode dispersion would disappear. However, this does not work entirely. Theoretical analyses have shown that the mode dispersion features a minimum when the profile exponent g is chosen to $g = 2 - 2\Delta$ [3.14,3.32].

In comparison with a step index fiber, a strong diminishment of the mode dispersion is achieved. Moreover, at the same time, the coupling efficiency suffers from an inessential decrease. Concerning graded index fibers, a good approximation can be written as [3.33]:

$$\Delta t_{\text{mod}} = \frac{n_{1g}L\Delta}{4cn_1^2}A_N^2 = \frac{n_{1g}L}{c}\frac{\Delta^2}{2} \tag{3.40}$$

Compared to step index fibers, we achieve an enhancement factor of $2/\Delta \approx 200$. Approximately 250 ps/km will be gained regarding a typical relative refraction index difference of $\Delta \approx 1\%$. Microbending (Chapter 3.1.1) and refraction index variations due to imperfect fiber production cause mode coupling with each other, for example higher-order modes convert to lower ones and vice versa. This holds for graded index fibers as well as for step index fibers.

Hitherto, it has been supposed that pulse broadening due to mode dispersion proportionally increases with the fiber length. However, the results demonstrate a minor increase in the fiber length in contrast to that expected from Eqs. (3.39a) and (3.40), respectively [3.34,3.35]. This is due to coupling of modes with each other. An *equilibrium mode distribution* results after a certain fiber length. The distribution of light power among the different modes after that length stays approximately constant along the fiber. A steady-state mode power distribution is then achieved. The according length is named as the coupling length. In this case the distribution is independent of the length. Considerably below the *coupling length*, the following can be written:

$\Delta\tau \sim L$, above this length it holds that $\Delta\tau \sim L^\gamma$ with $0.5 < \gamma < 1$, where γ usually has a value between 0.7 and 0.8. The length dependence of pulse broadening caused by mode dispersion has therefore to be taken into account regarding fiber information. The coupling length is between several 100 meters and a few kilometers of fiber length. For practical use, this effect is standardized to 1 km fiber length, for example 600 ps/km$^\gamma$, with $\gamma = 0.7$. This implies that such a fiber exhibits the following pulse broadening after 10 km: $600 \text{ ps} \cdot 10^{0.7} \approx 3\text{ns}$, instead of 6 ns without this effect.

A further reduction of pulse broadening is achieved if the fundamental modus is exclusively able to propagate in the fiber. Thus, a single-mode fiber is obtained which per se features no mode dispersion as described above. But, taking into account the polarization, two modes exist in a single-mode fiber due to birefringence. These modes propagate with different velocities (Figure 3.14): $v_x = c/n_x$ and $v_y = c/n_y$. This leads to path differences at the end of the fiber link and thus to a further dispersion, the *polarization mode dispersion, PMD* (Chapter 3.2.2). However, these path differences are very small, so have practically no influence on signals with data rates below 10 Gbit/s. For higher bit rates, as partially used nowadays and in particular in future high-bit rate transmission systems, these differences absolutely have to be taken into account.

A further kind of dispersion that all fiber types suffer from is *material dispersion*.

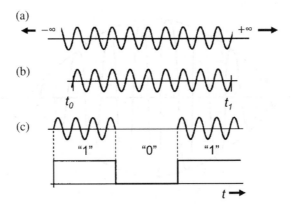

Figure 3.25   Time behavior of an optical wave

where:

a) Infinite time-considered wave lasting from past to future
b) Time-limited wave lasting from t_1 to t_2
c) Digitally modulated light wave

The effect would not occur if a strict monochromatic wave propagated in the light waveguide (Chapter 2.3). But time-considered, a strict monochromatic wave is a wave that must never stop. If a light source begins to irradiate at a certain time t_0 and ceases at a later moment t_1, the wave will no longer be monochromatic (Figures 3.25a and 3.25b). However, in optical communications, a power modulation is often carried out (Figure 3.25c). That means a modulated transmitted signal is no longer a monochromatic one, even though there would be a monochromatic light source. Real light sources, including high coherent lasers, are not yet strictly monochromatic, even without modulation. They emit no time-unlimited waves, but time-limited ones only, named *wave trains*, *wave packets* or *wave groups* (Figure 3.26). The oscillating part propagates with the phase velocity v_p, and the envelope with the group velocity v_g. For the group velocity, it holds that:

$$v_g = d\omega/dk \tag{3.41}$$

where:

ω Angular frequency of the light wave
k Absolute value of the wave vector

Applying Eq. (2.10), the correlation between group and phase velocity is given by:

$$v_g = v_p - \lambda dv_p/d\lambda \tag{3.42}$$

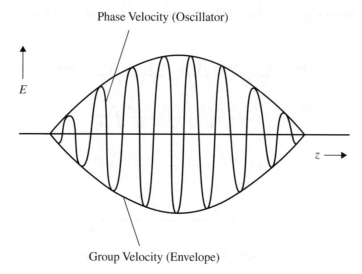

Figure 3.26 Time-limited wave train emitted by a real light source

where:

E Electric field strength
z Propagation direction

Regarding the relation between group refraction index n_g and phase refraction index n as a good approximation, it follows that:

$$n_g \approx n - \lambda dn/d\lambda \qquad (3.43)$$

Figure 3.27 shows phase and group refraction index of pure quartz glass (SiO_2) versus wavelength. In a vacuum, $n = 1 = $ const., thus it follows that with $v_p = c/n$ from Eqs. (3.42) and Eq. (3.43), $v_g = v_p$. In contrast, in a media where $dn/d\lambda \neq 0$ and $dv_p/d\lambda \neq 0$, respectively, it follows that $v_g \neq v_p$.

Figure 3.28 shows the superposition of two monochromatic waves with small frequency differences, so a beat is generated. By superimposing not only two but a multitude of monochromatic waves with adjacent frequencies corresponding to a real light source, a wave train with an oscillating part will be gained, as shown in Figure 3.26.

Each single wave with its according frequency and wavelength propagates with a different velocity in a media compared to the neighboring one. The wave train spreads out and thus suffers from a broadening. This process is called *dispersion*.

where:

E Electric field strength
z Propagation direction

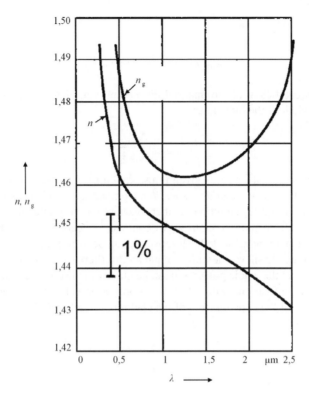

Figure 3.27   Phase refraction index n and group refraction index n_g of silica glass versus wavelength λ

In optical communications, the term *material dispersion* is used. The group delay in a fiber follows from Eq. (3.43):

$$t_g = \frac{(n - \lambda dn/d\lambda)}{c}L \qquad (3.44)$$

Concerning the delay time resulting from a spectral width $\Delta\lambda$ of a light source, it can be written as approximately:

$$\Delta t_{mat} = (dt_g/d\lambda)\Delta\lambda \qquad (3.45)$$

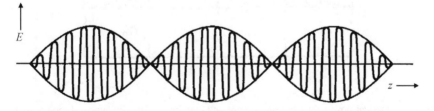

Figure 3.28   Superposition of two monochromatic waves with small frequency differences

Thus, it follows from Eq. (3.44):

$$\Delta t_{mat} = -\frac{(d^2 n/d\lambda^2)}{c} \lambda \Delta \lambda L \tag{3.46a}$$

The dispersion coefficient of the material dispersion is normalized along the fiber length and the spectral width of the light source:

$$M_{mat} = \frac{\Delta t_{mat}}{L \Delta \lambda} \tag{3.46b}$$

The information on a data sheet is given in ps/(km · nm). A further dispersion, which also occurs in all fiber types, is *waveguide dispersion*. Compared to mode dispersion, this kind of dispersion is often negligible in multimode fibers because it is very small (exception: regarding long fiber lengths due to mode coupling). In contrast, concerning single-mode fibers, this dispersion can no longer be neglected. Moreover, it can be even favorably used by parameter optimization. Figure 3.29 shows the fundamental modus (here single modus) of a single-mode fiber. It is obvious that the whole power will not be guided in the core, but a certain power part will penetrate into the fiber cladding. Thus, the propagation velocity v depends on the group refraction index n_{1g} of the core area and of the cladding refraction index n_{2g}.

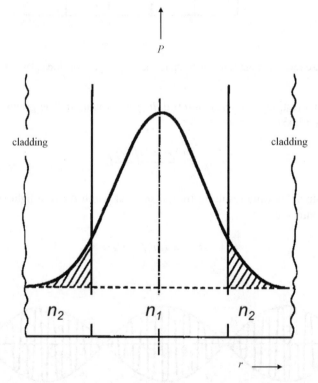

Figure 3.29   Optical power distribution on core and cladding of the fundamental modus (here single modus) of a single-mode fiber, Power P versus radial coordinate r

where:

n_1 Refraction index of the fiber core
n_2 Refraction index of the fiber cladding

The cladding part tries to attain a higher velocity than the core part, because the cladding refraction index is smaller than the core refraction index. An effective propagation velocity v is reached, depending on both parts. The penetration depth of the optical field in the cladding, and thus the power part guided in the cladding, depends on the V parameter (Figure 3.8). Therewith, the effective velocity depends on the wavelength of the used light source:

$$v = v(V(\lambda))$$ (3.47a)

This effect is called the waveguide dispersion because this dispersion is caused by the guiding behavior of the waveguide. The material dispersion of a dispersion coefficient of the waveguide dispersion is defined as:

$$M_w = \frac{\Delta t_w}{L \Delta \lambda}$$ (3.47b)

The information is given in ps/(km · nm).

3.2.2 Polarization Mode Dispersion in Single-Mode Fibers

Taking into account polarization, it is obvious that a standard single-mode fiber is a two-mode fiber because of the birefringence (Figure 3.14). An input pulse, which distributes on both polarization main axes, propagates in a fiber with different data velocities (Figure 3.30).

Again, this results in delay times $\Delta \tau$ at the end of the fiber link However, the delay times are not large enough to separate both pulses completely, but they overlap; this leads to a

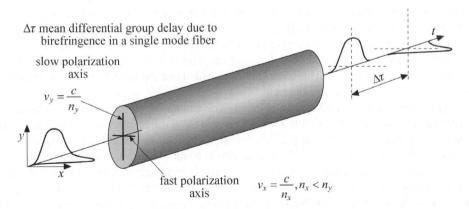

Figure 3.30 Polarization mode separation in a birefringent single-mode fiber

input puls

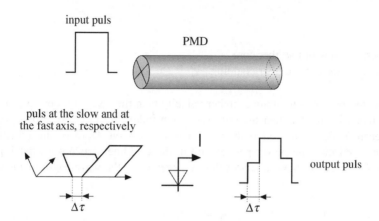

Figure 3.31   (a) Pulse separation by polarization mode dispersion, PMD; (b) Pulse development by PMD

pulse broadening and thus to further dispersion, the *polarization mode dispersion* (PMD). Figure 3.31 schematically shows a typical pulse at the end of a fiber, which arises by reason of the superposition of both polarization modes.

Due to the digital data, there is a typical shape to the receiver signals (Figures 3.31 and 3.33). This kind of shape is an enormous help for electronic signal restoration.

Thereby the delay times are very small and significantly influence the process at data rates only above 10 Gbit/s; in particular for high-bit rate long-distance traffic systems. Furthermore, similar to the mode conversion in multimode fibers, there is a polarization mode conversion, which slightly diminishes the PMD problem. Thus, the complexity of problems is slightly less. The dispersion effect also increases non-linearly, that is approximately with the square root of the fiber length; typical values lie below 0.5 ps/km$^{0.5}$.

Whereas classical dispersions, such as modal and chromatic dispersions are deterministic processes, PMD is a statistical process which is strongly time variable. It can happen that for several days the transmission system can feature the required bit error rate without any difficulty and suddenly a complete failure appears. Reducing the PMD does not mean the problem will be absolutely avoided, but the probability that a complete failure will occur is diminished. Thus, the effects of PMD are characterized by the loss of receiver sensitivity and by failure probabilities. A diminishing of the achievable transmission capacity of systems results from this fact (bandwidths-lengths product). Figure 3.32 depicts an estimation of the bit rate as a function of the transmission length, provided that the delay times due to the PMD are less than 10% of the bit periodic time.

Experimental studies corroborate this belief. The techniques to diminish the problem are classified into two categories [3.49]. One opportunity is to use an electronic signal equalizer to solve the problem. Regarding digital transmission, the typical cascaded output signal of the detector immediately suggests PMD. Taking into account the well-known bit periodic time and the regenerative sampling techniques, the signal is electrically regenerated (Figure 3.33).

Concerning optical signal equalization, a strong birefringent media is used which compensates for the delay times due to PMD. For this purpose however, it is necessary that the media gets the correct polarization component. The arrangement must work in a way that the pulse

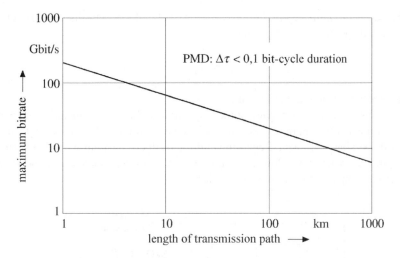

Figure 3.32   PMD-induced bit rate reduction versus link length

component traveling along the fast polarization axis coincidences with that of the slow one. This is attained by means of a polarization control [3.20,3.21].

3.2.3   Joint Action of Dispersion Mechanisms

PMD problems are exclusively relevant in single-mode fibers, if multimode fibers material and the mode dispersion are dominant. Material (M_{mat}) and waveguide (M_w) dispersion can

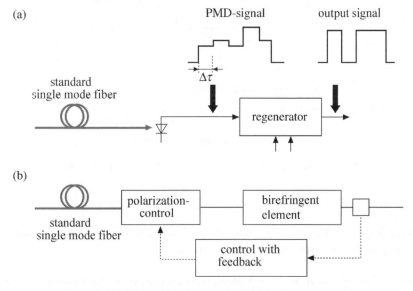

Figure 3.33   Electronic (a) and optical (b) PMD compensation

be combined as *chromatic dispersion* (Eqs. (3.56b) and (3.47b)). The dispersion coefficient of the chromatic dispersion M_{chr} is given by:

$$M_{chr} = M_{mat} + M_w \tag{3.48}$$

It is generally indicated in ps/(km · nm).

Figure 3.34 shows the dependency of the dispersion coefficients of material and waveguide dispersion versus wavelength, as well as their superposition to the wavelength dependent chromatic dispersion coefficient. Regarding multimode fibers, the joint action between material and waveguide dispersion is of low interest. In particular, in step index fibers, both dispersions are negligible compared to the mode dispersion.

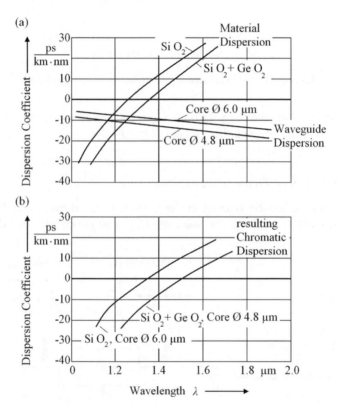

Figure 3.34   Wavelength dependence of dispersion coefficients

where:

a) Material dispersion for pure silica and waveguide dispersion for various core diameters
b) Resulting chromatic dispersion according to data in a)

In contrast, in single-mode fibers, the interaction between material and waveguide dispersion can be utilized for an interesting wavelength range [3.37]. By choosing an appropriate core

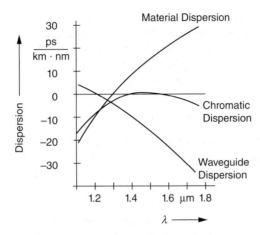

Figure 3.35 Dispersion flattened fiber by suitable combination of material and waveguide dispersion

diameter and refraction index profile of the fiber, respectively, it is possible to strongly influence the waveguide dispersion in a single-mode fiber. Moreover, by means of doping, the material dispersion can be varied. The resulting wavelength is where $M_{chr} = 0$ is named zero of dispersion. By these procedures, this wavelength can be modified to different values in order to optimize the applied fiber for the desired wavelength area, for example to a wavelength of 1.55 μm. At this wavelength the fiber attenuation features a minimum (Chapter 3.1); a dispersion shifted fiber is achieved. It is also possible to reverse the course of the chromatic dispersion curve and thus to create a wider optimized wavelength range.

Indeed, the chromatic dispersion does not disappear but remains within a low tolerable interval (Figure 3.35). Thus an optimum use of the fiber over a wide wavelength range is achieved. This fiber is called a *dispersion flattened fiber* [3.2,3.37].

Adding mode dispersion, an equation for practical use to determine the total dispersion of a fiber will be gained; it can be expressed as:

$$\Delta\tau = \sqrt{(M_{\text{mod}}\, L^{\gamma})^2 + (M_{chr}L\Delta\lambda)^2} \tag{3.49}$$

This equation explains the phenomenon of pulse broadening $\Delta\tau$, which can be determined by direct measurement of the pulse widths at the beginning and at the end of the fiber links (cf. Eq. (3.38)). Moreover, now the mechanisms to understand the formation of the dispersion are known. The expected pulse broadening can be well-predicted by the contents of a fiber data sheet.

Note that Eq. (3.49) only applies to wavelength ranges beyond the directional zero of dispersion [3.38]. Referring to Eq. (3.49), pulse broadening in single-mode fibers should completely disappear at the directional zero of dispersion. But that does not happen by reason of a non-disappearing spectral width of a light source. There is a detailed mathematical description in [3.46]. Concerning partial compensation of the chromatic dispersion, a dispersion compensating fiber (DCF, *Dispersion Compensating Fiber*) is used in the laboratory for research and development [9.81]. At the wavelength to be used, this fiber features a chromatic dispersion with an opposite sign as compared to the transmission fiber and a six times higher absolute value [3.47]. Intending to compensate for the effect completely, a DCF has to be used with a length of a sixth of the transmission fiber. However, this fiber exhibits a higher attenuation in contrast to off-the-shelf single-mode fibers. In addition, it is considerably more expensive and thus unsuitable for practical application in fiber links.

3.2.4   Dispersion Measurement Techniques

To achieve an integral information on pulse broadening without consideration of the formation along the fiber direct measurement of sent and received pulses is sufficient. The determination can be calculated according to Eq. (3.38) by comparison of the *input* and *output pulse* [3.39] (Figure 3.21). The FWHM t_1 of a preferably short optical pulse ($\sim$100 ps) of a light source is measured. Therefore, a semiconductor laser is applied, which is able to produce such short pulses (Chapter 6.1). The laser light is coupled into a fiber of a few meters length, a fiber pigtail fed to a photodetector. The measurement has to be carried out by a high speed photodiode able to detect such small pulses to avoid errors (Chapter 6.2). The generated electric pulse can be recorded and analyzed, for example by the use of a sampling oscilloscope with subsequent unit. Afterwards, the fiber link to be investigated will be inserted between transmitter and receiver and then the measurement repeated (measurement of t_2). Thus, the pulse broadening can be calculated for the total fiber link according to Eq. (3.38). This direct measurement is particularly applied if high values for t_2 are expected. This holds in particular for multimode fibers, especially with such a step index profile.

Concerning multimode fibers with a graded index profile, this kind of measurement could be difficult and lead to errors. The reason is that the values of the pulse widths are very small and thus high-speed electronics are required. This holds in particular for single-mode fibers typically used for very high-speed transmission links. Single-mode fibers inherently do not suffer from mode dispersion.

For single-mode fibers, the remaining dispersion coefficient of chromatic dispersion is therefore determined by means of time *delay measurements* [3.40,3.41]. Therefore, a short light pulse is emitted at a certain wavelength over the fiber link that is to be explored. Then the output pulse is observed. Afterwards the wavelength of the light source is varied and the process repeated. The output pulse will now be received at a different time. A delay is achieved due to the according wavelength. The wavelength will be varied over the whole measurement range that is to be explored. The according time delays are determined. Figure 3.36 shows the time delays as a function of the wavelength. The dispersion coefficient M_{chr} results from the derivative of the curve shown in Figure 3.36:

$$M_{chr} = (d\tau/d\lambda)/L \qquad (3.50)$$

Instead of measurements in the time domain, measurements in the frequency domain can be carried out [3.39]. Therefore the absolute value of the *optical transfer function* of a fiber can be calculated. Figure 3.37 depicts the corresponding test set-up. By means of a driver, a sweep generator (wobbler), the frequency continuously rises, controlling the electric current of a semiconductor laser in order to modulate the optical signal. This signal will then be coupled into the fiber. At the fiber end, it is has to be analyzed by a photodetector, again transforming the optical signals into electric ones and feeding them to a network analyzer.

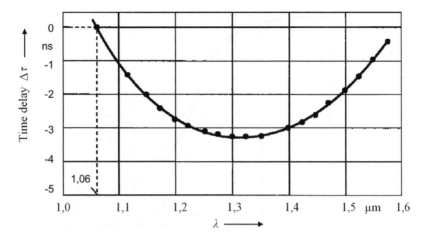

Figure 3.36   Time delay Δ*t versus* λ in a single-mode fiber (zero value of Δ*t* is arbitrarily chosen at a wavelength of 1.06 μm)

The ratio of the modulated optical output power to the modulated optical input power as a function of the laser frequency presents the absolute value of the modulation transfer function (MTF) [3.42].

Figure 3.38 shows the principal dispersion low-pass behavior. Regarding lower frequencies, the absolute value of the transfer function varies by a small extend. However, it decreases with increasing frequency by pulse broadening in the fiber. The modulation amplitude measured by the detector decreases. The frequency where a factor 1/2 results is defined as the *3 dB cut-off frequency*. The *bandwidth B* is determined by the frequency range from 0 Hz to the 3 dB cut-off frequency of the fiber link. A presupposition for correct measurement is that the transfer function of transmitter and receiver does not influence the test results or their transfer function is well-known, respectively. Thus it can be neutralized in the result. Concerning the

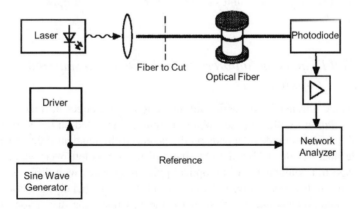

Figure 3.37   Measurement set-up for determination of the modulation transfer function

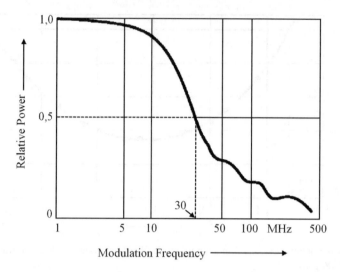

Figure 3.38 Modulation transfer function (MTF) of a step-index fiber

correlation between bandwidth B and the pulse broadening $\Delta\tau$, the following approximation is applied [3.44]:

$$B \approx \frac{1}{2\Delta\tau} \tag{3.51}$$

The exact value of the bandwidth depends on the type of the light pulse coupled into the fiber. Therefore, from Eq. (3.51), differing information is often found in the literature. Moreover, definition issues of the pulse widths often lead to different conversion factors (FWHM, effective width, 1/e width). A detailed description of the dispersion measurement technique is given in [3.45]. The bandwidth determined above also presents the limitation of an optical transmission system, provided that no further components (transmitter or receiver) affect the total bandwidth. Therewith, it can be considered as a quantity as to how much information per time unit can be transmitted over an optical transmission link (Chapter 9.1).

3.2.5 *Partial Dispersion Suppression by Soliton Transmission in Single-Mode Fibers*

In order to realize a transmission with solitons, a non-linear effect in fibers is utilized, the *self-phase modulation* [3.50]. This self-phase modulation is based on the Kerr effect, which implies that the refraction index depends on the power of the incident light wave. Regarding a single-mode fiber besides PMD, the chromatic dispersion is left over, which implies that the different spectral components of an optical pulse propagates with different velocities in the fiber, causing pulse broadening (Chapter 3.2.1). If we artificially increase the velocity for the slower components and decrease it for the fast ones, the dispersion effect can be compensated for. This principle of soliton propagation can be approximately described as the

exact view to be found in [3.50]. Due to the Kerr effect, the refraction index n in the fiber is given by:

$$n(P) = n_0 + n_2(P) \tag{3.52}$$

where:

n_0 Refraction index without the Kerr effect, independent of the optical power in the glass fiber

n_2 Refraction index change based on the Kerr effect, dependent on the optical power in the glass fiber

A periodic change of the power $P(t)$ implies a corresponding periodic change of the refraction index. Thus, the electromagnetic wave transmitted by the fiber features a time-dependent phase modification $\varphi(t)$ (Eqs. (2.23) and (2.24)), which is approximately proportional to the periodic change of the power $P(t)$ above a defined threshold. Hence, the phase modification of the electromagnetic wave was caused by itself; therefore, the commonly used term is *self-phase modulation*. A periodic change of the phase $\varphi = kr - \omega t$ (Chapter 2.1) leads to a frequency change:

$$\omega(t) = \frac{d\varphi}{dt} \sim -\frac{dP}{dt} \tag{3.53}$$

The correlations are depicted in Figure 3.39; however, the effect only occurs by high powers exceeding a certain threshold power, P_{th}. Hence, during pulse development, the frequency of the wave is shifted from small to high values and the wavelength changes from high to small values. Descriptively speaking, the head of the pulse is red and its end is blue (Figure 3.40). The red parts of the pulse reach the end of the fiber before the blue ones. Thus, the blue parts have to make up lost ground. This will be attained by decreasing the propagation velocity of the red parts and increasing that of the blue parts, respectively.

With Eqs. $v_g = c/n_g$ and $n_g = n_g(\lambda)$, it follows that a pulse compression is generated, provided that the operating wavelength is larger than the dispersion zero point (dot-dashed line in Figure 3.40, upper part).

With the aid of suitable dimensions, the effects caused by the chromatic dispersion and the self-phase modulation cancel each other out. As a result of this procedure, a pulse whose width remains unchanged travels through the fiber. In the case where the operating wavelength is larger, being on the wrong side of the dispersion zero point, an additional increase of pulse broadening will occur.

Thus, related to a single-mode fiber in service with high powers in a proper dispersion range, it is possible to generate an optical pulse which does not show any broadening during its propagation along the fiber. This effect is called the generation of *solitons* [3.50]. Based on spectacular results, experimental studies have always proved the feasibility of the transmission with solitons [3.51–3.59]. For instance, a transmission with solitons could be realized over a length of 1 million kilometers with a data rate of 10 Gbit/s. The link length of 1 million kilometers was simulated by means of multiple passages of a loop configuration. Erbium fiber amplifiers were suitable for the necessary optical amplification (Chapter 8.1) [3.47].

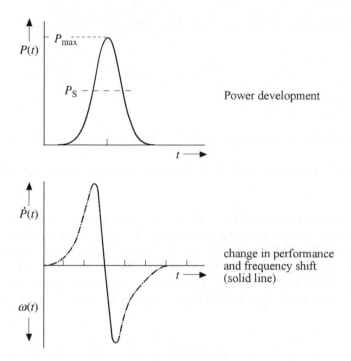

Power development

change in performance
and frequency shift
(solid line)

Figure 3.39   Power development and frequency shift of an optical pulse traveling through the fiber

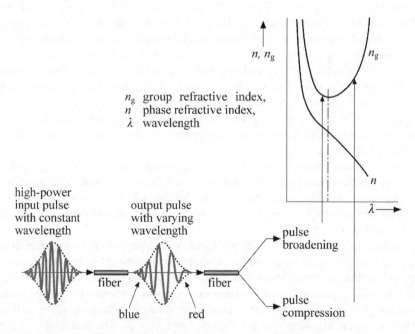

n_g group refractive index,
n phase refractive index,
λ wavelength

Figure 3.40   Soliton propagation in glass fibers

Figure 3.41   Comparison of runner competition and soliton propagation

Above all the adaption of the absolute power with respect to the implication for the size of the effect poses a problem regarding the realization of practical systems. As a result of the power change, in particular caused by fiber attenuation, specific methods of control engineering are therefore required. This fact has for a long time prohibited industrial application of new systems. However, there is work showing the possibility to solve these problems and demonstrate that the soliton technique can be successful [3.61].

A descriptive explanation could be a comparison with a group of runners with different power ratings [3.62]. They run on soft ground which will be pressed down by their weight. Thus the front runner, a strong runner, always has to run up the hill, whereas the last one, a weak runner, always runs downhill. Thus the power ratings of the different runners are equalized. Let us assume all runners are linked by a rubber tape. When they run on hard ground, the rubber tape will be extended because of the front runner, and the rear runner would gain a large difference in path length. But, in contrast, if they run on soft ground, the tape extension would vanish (Figure 3.41). The rubber length in our soliton transmission is nothing but the pulse width of our optical signal. Running on hard ground means pulse broadening, while running on soft ground means the pulse width remains constant by traveling along the fiber; this pulse then is named a soliton.

Instead of the self-phase modulation, the increase of the velocity for the slower parts and the decrease for the faster parts, respectively, can be gained by an adequate variation of the laser injection current [3.63]. This is caused by the fact that with a change of the injection current, except for the laser output power change, also a variation of the laser wavelength is generated (Chapter 6.1.2). Being arranged in a proper dispersion range, a pulse compression occurs. This technique is named as the *dispersion supporting transmission*, DST [8.16].

4

Fiber Manufacturing, Cabling and Coupling

4.1 Fiber Manufacturing

Werner Auer, FOP Faseroptische Produkte GmbH, Crailsheim, Germany

The optical waveguide as a general term includes both glass and plastic fibers for transmission from meters to kilometers, and planar waveguide substrate materials in geometrical lengths ranging from 100 microns to a few centimeters. The focus of this chapter is the glass fiber, with low loss and high bandwidth. The techniques for manufacturing optical fibers were invented in the 1970s and developed and optimized in the following decade for high volume production. Nowadays a wide spectrum of fibers is available, for example single mode (SM) in terrestrial networks – submarine, backbone, metro and access – optimized in attenuation and PMD (polarization mode dispersion) for specific wavelength applications [4.1] or multi-mode (MM) in enterprise and data center networks, optimized for high bandwidth (OM3/OM4), and bend insensitive fibers in MM and SM (Table 4.1). Fiber lengths on shipping spools vary (depending on the application requirements) from 4 to 62 km.

In addition a large number of special fibers are available for applications in different environments and markets, for example devices for laser medicine and endoscopy [4.2].

The manufacturing of a glass fiber is carried out in two or three steps, depending on the process used: lay down (deposition of glass material), consolidation and drying (realization of a vitreous preform) and fiber drawing. In one of the described processes, steps 1 and 2 are carried out simultaneously. Preform production is carried out mainly with the aid of a rotating lathe system and this preform is then drawn into a fiber by using a fiber draw tower. All manufacturing processes usually take place in a clean room area.

4.1.1 Preparation of a Preform

The installed production capacity for high-quality optical fibers is based on low-loss fused silica (quartz glass, SiO_2). The necessary change in the refraction index, to obtain a waveguide, is achieved by suitable doping of the basic raw material SiO_2, and examples of refraction index

Optical and Microwave Technologies for Telecommunication Networks, First Edition. Otto Strobel.
© 2016 John Wiley & Sons, Ltd. Published 2016 by John Wiley & Sons, Ltd.

Table 4.1 Fiber specification

Single Mode (SM)	Spec ITU/IEC	Fiber ϕ (μm)	Mode Field (μm)	
			1310 nm	1550 nm
Standard Low Loss 1310/1550 nm	ITU-T G.652	125	9,2±0,4	9,2±0,4
long-haul NZ-DSF (non zero dispersion shifted)	ITU-T G.655	125	–	9,6±0,4
Submarine (Customer made solutions)				
Bend insensitive fiber	ITU-T G.657.B3	125	8,6±0,4	

Multi Mode (MM)	Spec ITU/IEC	Fiber ϕ (μm)	Core (μm)
OM1	IEC 60793-2-10Alb	125	62,5±3,0
OM2 GB Ethernet 1000Base-SX/LX	IEC 60793-2-10A1a.1	125	50±2,5
OM3/OM4 10 GB Ethernet	IEC 60793-2-10Ala.2/3	125	50±2,5
Bend insensitive fiber OM3/OM4	IEC 60793-2-10Ala.2/3	125	50±2,5

MM Ethernet Link Length						
Type of Fiber	OM1	OM2	OM3	OM4	OM3BI	OM4BI
Length (m)						
GB Ethernet 1000 Base-SX (850 nm)	350	500	1000	1040	1000	1040
10 GB Ethernet 10 GBase-SX (850 nm)			350	550	350	550

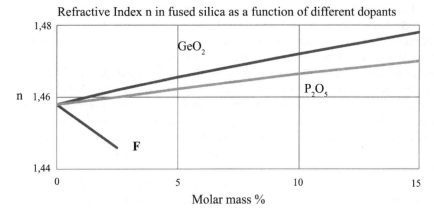

Figure 4.1   Suitable doping

variation are shown in Figure 4.1. As already demonstrated in Chapter 3, the refraction index of an optical waveguide in the core region has to be greater than that in the cladding region. This can be achieved either by keeping the refraction index in the cladding fixed and increasing it in the core or keeping it fixed in the core and reducing it in the cladding.

In order to decrease the refraction index, fluorine (F) is commonly used, to increase the refraction index germanium dioxide (GeO_2), or phosphorus pentoxide (P_2O_5) is applied. An important condition for the preparation of low-loss optical fiber is that the manufacturing process of the preform is carried out in a high purity environment. Quality preforms are based using the principle of chemical vapor deposition (CVD) (Eqs (4.1–4.3)). The raw materials belong to the group of metal halides. For pure quartz glass, the chemical process can be described as follows:

$$SiCl_4 + O_2 \rightarrow SiO_2 + 2Cl_2 \uparrow \tag{4.1}$$

For Germanium and Phosphorus as a dopant, apply:

$$GeCl_4 + O_2 \rightarrow GeO_2 + 2Cl_2 \uparrow \tag{4.2}$$
$$4POCl_3 + 3O_2 \rightarrow 2P2O_5 + 6Cl_2 \uparrow \tag{4.3}$$

For the chemical vapor deposition, there are three methods basically used:

- *Modified Chemical Vapor Deposition (MCVD)*
- *Outside Vapor Deposition (OVD)*
- *Vapor Axial Deposition (VAD)*

Preforms can be produced by all three methods. Such a preform already has the desired core-cladding ratio and the refraction index profile of the final fiber. When inside vapor deposition is used, the deposition of oxides (Eqs (4.1 and 4.2)) takes place in a rotating glass tube. For this, the quartz glass tube is clamped into a lathe system and a heat source is moved in the direction of the gas flow. Ultra-high purity components are introduced into a gas control cabinet and delivered via welded stainless steel pipes into the lathe system. At room temperature, the raw materials $SiCl_4$ and $GeCl_4$ are in liquid form. Oxygen is allowed to flow through a container tank housing these liquids. Bubbles form, and the carrier gas (oxygen) is then saturated with the chloride vapors. The heat source converts the gases into fine soot particles and this powder is vitrified at a temperature of about 1700°C on the inside of the glass tube. When the first layer is deposited, the heat source reverses back to the starting position and in between the gas mixture is changed and layer upon layer are deposited. Varying the composition of the gases, and thus the relative proportions of silica and germanium, the refraction index can be changed in a radial direction (final core structure in the optical fiber) [4.3]. An oxy/hydrogen burner (H_2, O_2) is generally used as a heat source. In this case, the method is known as the *modified CVD* process (Figure 4.2a,b).

This type of process was first reported by Bell Laboratories in 1973. Today, we can also find furnaces as the heat source in use, referred to as FCVD (*Furnace Chemical Vapor Deposition*). If a high frequency source is used for supplying the heat, the method is called *plasma-activated*

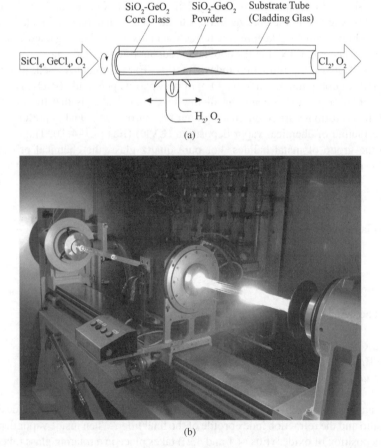

Figure 4.2   (a) Modified Chemical Vapor Deposition (MCVD) schematic diagram; (b) MCVD FCVD furnace

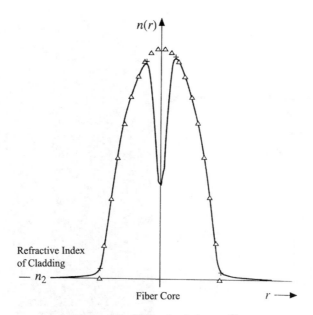

Figure 4.3   Refraction index profile

CVD (invented by Philipps Research Laboratories in 1976). In this process, the gas mixtures are fed into the substrate glass tube at a specific low pressure. The tube is completely enclosed in a furnace at a temperature of 1200°C and a non-isothermal plasma is created in a microwave cavity which traverses at a high speed (up to 24 m/min). In this case, very thin layers of pure or doped silica glass, without soot formation, are deposited. Several thousand layers can be made with precise control of the refraction index profile [4.4].

When the deposition process is completed, the coated glass tube is heated to about 2000°C. Then the tube softens and due to surface tension and an applied centerline vacuum it shrinks down into a rod; it collapses. The resulting bar is called a preform. Ge-dopants evaporate partially out of the center of the preform at this high temperature during the collapse. This can lead to a decrease in the refraction index in the core center (dip) and cause a significant bandwidth problem in a MM fiber due to modal delay. Figure 4.3 shows a typical refractive index profile of a fiber with a dip.

Nowadays this problem has been solved to a large extent and has overcome the evaporation of the Germanium and the decrease in the refraction index of the core. An example is described in a patent from the year 2001. The refraction index profile in the core center is modified by adding an index ridge [4.5].

A conventional preform made by the MCVD process has a diameter of up to 40 mm (a larger diameter is possible) and a length of about 100 cm. This gives a usable fiber length of up to 100 km. The finished product is cut into standard shipping spool lengths according to the fiber type. Longer fiber lengths are less important for cabled fibers, because only a few kilometers of cable can be laid in one piece, but fiber is cheap, and long fiber lengths can be produced more economically. This is due to the machine set-up time, when starting the fiber draw process.

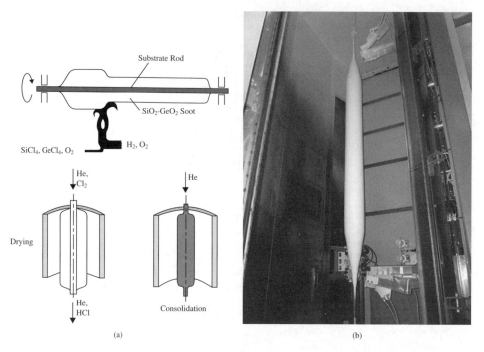

Figure 4.4   (a) Outside Vapor Deposition (OVD) schematic diagram; (b) OVD soot preform

To increase the achievable fiber length per preform and thus to improve the economics of the manufacturing process, sleeving technology is used for the MCVD method. In this process, a glass tube similar to a sleeve is placed over the preform and continuously melted down. The result is a preform with two to three times the fiber length compared to the original technique. As the core/clad ratio of the preform and the finished fiber type must be identical, it is necessary to deposit more core layers during the MCVD process. A further improvement in efficiency is achieved by a combination of the sleeving process with a stretching process that maintains the diameter of the preform. The sleeving process leads to a preform with a relatively large cross-sectional area. If this gets too large, it can cause problems during the draw process. To optimize the production process, it is therefore advantageous to stretch the preform before drawing.

When the *Outside Vapor Deposition (OVD)* process is used (Figures 4.4a and 4.4b), the chlorides are introduced directly into the flame of the oxy/hydrogen torch. The deposition of the powdery oxides is carried out onto a substrate rod. Again this is clamped in a lathe and rotates. The layer sequence now runs in the reverse direction as in the MCVD process. After completion of the deposition, the substrate rod is then removed, leaving a porous glass mold (glass soot). This can be cleaned, and in particular, the unwanted OH content can be reduced by dehydration (flushing with chlorine gas). The following chemical reaction occurs:

$$2H_2O + 2Cl_2 \rightarrow 4HCl + O_2 \uparrow \tag{4.4}$$

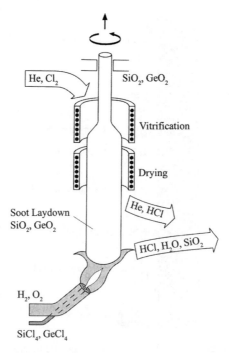

He, Cl$_2$

SiO$_2$, GeO$_2$

Vitrification

Drying

Soot Laydown
SiO$_2$, GeO$_2$

He, HCl

HCl, H$_2$O, SiO$_2$

H$_2$, O$_2$

SiCl$_4$, GeCl$_4$

Figure 4.5 Vapor Axial Deposition (VAD)

Temperatures of at least 1200°C are required for this process. Subsequently, at a higher temperature of about 1500°C the porous preform is sintered into a solid glass rod. Deposition and vitrification processes do not take place simultaneously, as in the MCVD technique. Therefore each process step can be optimized separately. Another advantage is the higher deposition rate per unit time. A disadvantage is the higher risk of contamination, because reactions no longer take place within a quartz tube.

In the *Vapor Axial Deposition* (Figure 4.5) method, in contrast to the two previously described methods, the deposition is carried out axially. The supply of chlorides is as in the OVD process, directly through the flame of oxy/hydrogen burners. Typically two burners are used, which are arranged asymmetrically. They produce different amounts of SiCl$_4$ and GeCl$_4$. By varying the proportions during deposition and by appropriate arrangement of the burner, the desired refraction index profile can be achieved. The deposition takes place axially on a rotating quartz rod, which is, however, only necessary for the start of the process. The rod is slowly pulled away from the torches and a porous soot preform grew in length (a typical soot preform size is 180 mm diameter × 1000 mm in length). The dehydration and vitrification in the next step is carried out in a similar way, as already described for the OVD technique. But in comparison to the OVD technique, no substrate has to be removed and preforms with a very low OH content can be produced. Values down to 1 ppb are possible, 1 OH ion to 1 billion SiO$_2$ molecules. As a result, the related OH attenuation increase at 1390 nm is negligible (Zero Water Peak Fiber – ZWP).

A further advantage is, in contrast to the MCVD/PCVD and OVD processes, that no refraction index dip in the fiber core occurs. Thus the desired refraction index profile can

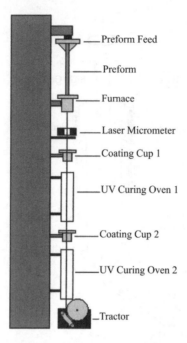

Figure 4.6   Drawing tower schematic diagram

be implemented and enables high fiber bandwidth. The VAD method leads to preforms from which fibers can be pulled up to 100 km in length.

Today modern production facilities for fiber preforms need an environmentally friendly concept for waste disposal. For example, in the MCVD process, the remaining solid components are treated in such a way that they can be disposed of in landfills. It is a non-toxic reaction product resulting in a fine white dust. This dust is concentrated in a mixer to form sludge before being removed to the landfill. The gaseous waste materials, in particular the chlorine compounds, are subjected to a washing process, so that the purified exhaust gas does not exceed the limit of the Cl_2 emission of 5 mg/m^3. Subsequently it can then be discharged into the atmosphere.

4.1.2 Fiber Drawing

In the final step of manufacturing glass fibers, the preform is drawn with the aid of a fiber drawing machine. This equipment is named, because of its height (over 12 m), as a drawing tower. A high drawing tower (Figure 4.6) is required, because most subsystems are arranged perpendicularly. The first unit (preform feed system) lowers the preform into the furnace. The preform end is heated up (to about 2200°C), and brought into a viscous state. At the softening point, the end tail of the preform drops down by its own weight under gravity. This then forms a thin filament, the fiber, which is pulled out by a capstan without the use of any additional molds. Before further processing, a certain length as a cooling section is required. This also contributes to the large height of the equipment. The fiber diameter is a function of both preform feed and drawing speed.

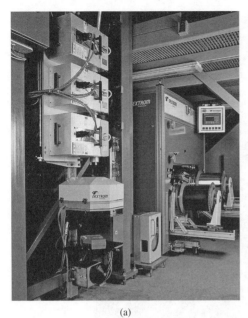
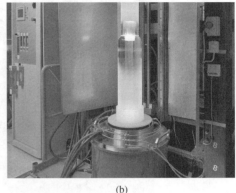

(a)   (b)

Figure 4.7   (a) Drawing tower; (b) Drawing tower furnace

The fiber diameter is measured directly under the furnace and controlled via a loop which regulates the speed of the capstan. The desired fiber diameter (e.g. 125 μm) can be controlled to a tight tolerance of less than 1 μm over the total fiber length. To prevent contamination on the fiber surface, and thus having adverse effects on the strength of glass fibers, protective coatings are applied during the drawing process. Currently a dual layer coating system is used. The first layer is a soft, cushioning one; the second is a hard, abrasion resistant one. Typical diameters are 190 μm and 245 μm respectively [4.6].These acrylates are cured by UV radiation. These first protective layers are called the primary coating. Draw speeds in the fiber production area of up to 1500 m/min are possible. Even a draw speed of 2500 m/min has been demonstrated in a laboratory [4.7].

4.1.3   Mechanical Properties of Optical Fibers

To guarantee a minimum mechanical strength during a 25 year in-service lifetime, the fiber is tested after drawing on a separate line to a specific proof test tension (Figure 4.8). Depending on the application, this test level can vary between 100 and 200 kpsi; a typical test level is 100 kpsi (100 kpsi => 0.689 GPa), and this corresponds for a glass fiber with 125 μm outer diameter approximately to 8.5 N. If the applied stress during operation is below 20% of the proof test level (1/5 rule), flaws will not grow to failure, so the fiber can be stressed to this level over its lifetime [4.8].

However, new fiber specifications for Fiber-to-the-Home and premises networks (ITU-T G.657) requires low bend loss fibers, because the fiber installation will be done in compact housing with smaller bending radii for the fiber.

Figure 4.8   Proof testing and rewinding

In general, the mechanical strength of a glass fiber is determined by the quality of the raw materials, such as the tubing, the properties of the preform, the conditions in the draw area and the contaminations on the surface during coating application. The purpose of this proof testing is to screen out flaws weaker than the test level. Fiber breaks at much higher tensile forces (>40 N) are due to micro cracks in the structure of the glass used and at the surface of the glass fiber. At medium tensile forces (~15 N to 40 N), breaks are typically caused by gas inclusions or foreign particles in the fiber cladding. Breaks at small tensile forces result primarily from defects directly on the fiber surface, as well as gas bubbles and foreign particles in the applied surface layer. This can lead to one-sided, or non-concentric coating, or poor coating adhesion. The strength of glass fibers is evaluated in a Weibull diagram, as this gives the percentage of fiber breaks at a given tension. It cannot be defined for any individual fiber at exactly what tension it will break, but instead a statistical assessment is possible. For a high strength fiber in a statistically determined tensile force, the mean value is about 60 N. This corresponds, at a fiber diameter of 125 microns (typical value), with a tensile strength of about 5 Giga Pascal = $5 \cdot 10^9$ N/m^2.

The fiber proof test level of 100 kpsi is normally sufficient for subsequent processing steps as cabling and installation, as well as exposure to operating conditions in service lifetimes.

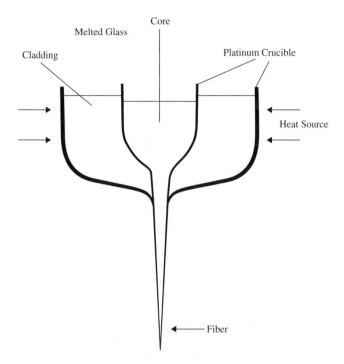

Figure 4.9   Double crucible process

4.1.4   Alternative Fiber Manufacturing Processes

For fibers with lower specification requirements, (low loss/high bandwidth, λ between 800 – 1650 nm), there are two other procedures in use which are not based on the principle of vapor deposition: the rod-in-tube and the double crucible method.

When the *rod-in-tube method* (Figure 4.9) is used, a core glass rod is inserted into a glass tube of lower refraction index, the assembly is then heated and a fiber is drawn out. If instead of the cladding glass, the fiber is directly drawn and just coated with a low loss plastic (with lower refraction index than the glass rod), a PCS fiber (Polymer-Cladded Silica) is produced. The quartz fiber represents the fiber core, the plastic material the fiber cladding. The attenuation of such a plastic layer is substantially higher than that of a quartz glass layer. Since the electromagnetic waves are not only carried in the core, but also partially in the cladding of the fiber, the attenuation in PCS fibers is greater than that in glass fibers. The attenuation of a PCS fiber at a wavelength of 780 nm is typically between 3 dB/km and 4 dB/km.

For cost improvement in single-mode fibers, it has been reported that the rod-in-tube method can be used [4.9].

A preform made in the VAD process is inserted into a large cylinder (OD 170 mm, ID 60 mm, length 3 m) and drawn directly into a fiber, yielding over 5000 km in length. The fiber loss is comparable to conventionally produced products and exceeds the requirements of G.652.D single-mode fibers.

The *double crucible process* (Figure 4.10) consists of at least two concentrically arranged melting vessels, each having a circular nozzle in the bottom and an inner nozzle that is slightly

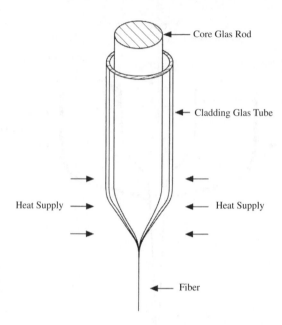

Core Glas Rod

Cladding Glas Tube

Heat Supply

Heat Supply

Fiber

Figure 4.10   Rod-in-tube method

above the outer one. The core glass in the inner vessel and the cladding glass in the surrounding vessel are heated to the drawing temperature, then the materials flow together and the fiber can be drawn. If the materials remain separated, the result is a step index fiber. The subsequent processing procedures are the same as already described in fiber drawing. The crucibles can be made of platinum if melting temperatures up to 900°C are required, or quartz if higher temperatures are necessary. In this process, low-melting multi-component glasses are used.

Based on the same principle, it is possible to produce pure synthetic plastic fibers (*Polymer Optical Fibers – POF*); the core consists of pure poly methyl methacrylate (PMMA) and the cladding is made of fluoropolymer. This results in significantly higher attenuation values. The attenuation of such a plastic fiber in the range of 450–1100 nm is given in Figure 4.11. For transmission in the field, the wavelength region from 640–670 nm is used. Values for attenuation of 160 dB/km can be obtained. These plastic fibers will be used in industrial networks (process field bus) or in high-speed media networks (MOST), for example the automotive industry.

4.2   Fiber Cabling

Dr Krzysztof Borzycki
National Institute of Telecommunications, Warsaw, Poland

4.2.1   Fibers for Telecom and Data Networks

Optical fibers serving as transmission medium are subject to worldwide standardization and made in huge quantities, approximately 240 million km in 2012 [4.10]. Today, terrestrial telecom and data transmission networks use almost exclusively single-mode, and multi-mode

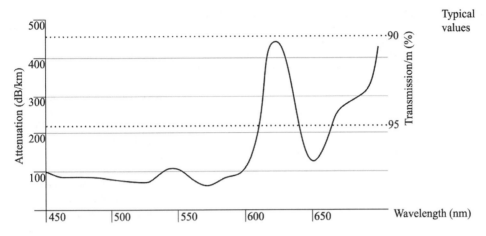

Figure 4.11   Attenuation in POF

fibers made of fused silica (SiO_2), listed in Tables 4.2 and 4.3, covered by synchronized standards issued by *IEC* [4.11,4.12] and ITU-T [4.13–4.16].

The telecom market has consistently preferred just one fiber type in both single-mode and multi-mode categories, despite wider selection offered by manufacturers and standards. Today, we see dominance of 9/125 µm dispersion-unshifted *single-mode fiber* (SMF). This fiber is used in all telecom and video networks, SANs, MANs and partly LANs. The 50/125 µm graded index multimode fiber (GI-MMF) is in particular applied in LANs (Chapter 9.5.1).

Both are upgraded designs introduced over 30 years ago. Other products, once very popular, in particular large-core *multi-mode fibers* (MMF) (62.5/125 µm, 85/125 µm, 100/140 µm, etc.) and several types of dispersion-shifted single-mode fibers, were heavily promoted for use in core and metropolitan networks between 1985 and 2005. Today, they have disappeared or are in decline. The main reasons are: (a) user preference for low-cost, easy-to-install fibers, (b) advances in active equipment. This was mostly introduced for 1 Gbit/s and 10 Gbit/s laser transponders in LANs, working best with 50/125 µm MMF (or SMF), and digital dispersion

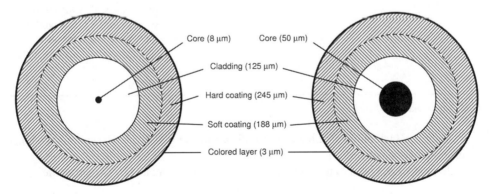

Figure 4.12   SMF (left) and 50/125 µm MMF (right) in primary coating. All parts to scale. Diameters of soft coating and SMF core are typical values.

Table 4.2 Single-mode optical fibers for telecom, CATV and data networks

Acronym	Diameter (MFD[1]/clad)	ITU-T designation	IEC designation	Main properties and applications
SMF	9/125 μm	G.652.A, B	B1.1	Older versions of dispersion unshifted single-mode fiber, with zero dispersion around 1310 nm. Works at 1310 nm and 1550 nm. Widely deployed, cheap.
SMF	9/125 μm	G.652.C, D	B1.3	Current versions of dispersion unshifted single-mode fiber. Wide band: 1270–1625 nm, and/or low PMD. Low cost, easy splicing. Dominant single-mode fiber, ~90% of sales in 2012.
DSF	≈ 7/125 μm	G.653	B2	Zero dispersion at 1550 nm, for long-distance networks and analog cable TV. Not suitable for DWDM networks. Expensive, discontinued.
NZDSF	≈ 8/125 μm	G.655.A-E, G.656	B4.a-e, B5	Optimized for DWDM networks, low dispersion in 1530–65 nm band. Several not fully compatible variants, expensive. Widely deployed in core/metro networks until 2009, now in decline (1% in 2012).
BIF	9/125 μm	G.657.A1, A2	B6.a1, a2	Bending-tolerant single-mode fiber, otherwise compatible with SMF. Originally developed for FTTH networks, now finds several other uses (4.5% of sales in 2012).
BIF	8–9/125 μm	G.657.B2, B3	B6.b2, b3	Bending-insensitive single-mode fiber, otherwise (mostly) compatible with SMF. Originally designed for FTTH networks. Old "exotic" versions were difficult to splice, new are more craft friendly. Usage low, but rising quickly.

[1] Mode Field Diameter

compensation, which eliminated the need for dispersion-shifted fibers. Exceptions are long-distance submarine networks, not covered here.

Single-mode fibers used as transmission medium are differentiated by two properties:

- Characteristics of chromatic dispersion (CD) (Figure 4.13)
- Bending tolerance (Table 4.3).

Table 4.3 Categories and requirements of dispersion-unshifted single-mode fibers [4.13]

Fiber category	Attenuation [dB/km]					PMD$_Q$ ps/$\sqrt{km}$
	1310 nm	1383 nm	1550 nm	1625 nm	1310–1625 nm	
G.652.A	≤0.50	N/A	≤0.40	N/A	N/A	≤0.50
G.652.B	≤0.40	N/A	≤0.35	≤0.40	N/A	≤0.20
G.652.C	≤0.40	≤0.40	≤0.30	≤0.40	≤0.40	≤0.50
G.652.D	≤0.40	≤0.40	≤0.30	≤0.40	≤0.40	≤0.20

Other important properties of single-mode fibers include mode field diameter (MFD) and effective mode area (A_{eff}), refraction index profile and dopants used, cable cutoff wavelength (λ_{cc}) and polarization mode dispersion (PMD). The PMD can be modified somewhat independently of the fiber type, but with varying degree of difficulty. However, several of them are inter-dependent and actual fiber design is always a compromise.

Chromatic dispersion D is a measure of dependence of group delay τ_g on wavelength λ, normalized with respect to fiber length L:

$$D(\lambda) = \frac{d\tau_g}{L \cdot d\lambda} \tag{4.5}$$

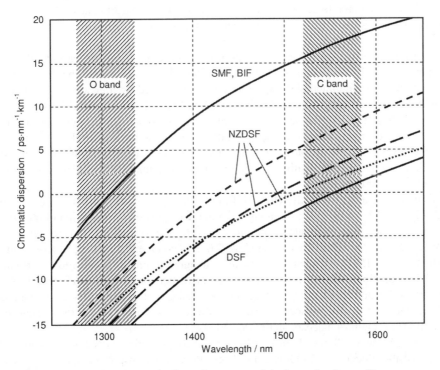

Figure 4.13 Chromatic dispersion curves of single-mode telecom fibers

In a single wavelength, high-speed, long link, the dispersion shall be close to zero to minimize inter-symbol interference and error rate. This produces a pulse spreading Δt proportional to signal spectral width $\Delta\lambda$ and length L:

$$\Delta t = \Delta\lambda \cdot D \cdot L \qquad (4.6)$$

Each system has a dispersion tolerance range, depending on bit rate, modulation and coding. Dispersion-unshifted single-mode fibers (SMF), introduced in 1982 and standardized in *ITU-T* Recommendation G.652 [4.13] have dispersion characteristics close to one of bulk doped silica glass. No efforts are made to change it, with zero value at 1310 nm (Figure 4.13), in the so called "O" (old) band. At this wavelength, the fiber *attenuation* is 0.3–0.35 dB/km, allowing transmission at distances up to 50–80 km. However, long-distance networks require the use of 1550 nm or "C" band wavelengths. At this wavelength the fiber attenuation is lowest with 0.18–0.25 dB/km.

Also erbium-doped optical fiber amplifiers (EDFA) are available for signal boosting and building purely optical repeaters; EDFA pass-band is approximately 1528–1570 nm. Unfortunately, chromatic dispersion of SMF at 1550 nm is high, 16–18 ps/nm·km; in most 10 Gbit/s and faster systems, costly dispersion compensation is required. To avoid this, *dispersion shifted fiber* (DSF), covered by ITU-T Recommendation G.653 [4.14], was introduced in 1985. By reducing the core diameter and modifying the refractive index profile, the value of the waveguide dispersion (negative) in this fiber was increased. But not only the wavelength of zero dispersion was shifted to 1550 nm, but also the fiber name changed. Unfortunately, DSF had its drawbacks compared to SMF: increased *attenuation* (≈ 0.25 dB/km at 1550 nm) and high cost due to the large concentration of the expensive GeO_2 dopant. Furthermore, manufacturing difficulties occurred at splicing, because of the smaller MFD.

DSF was deployed by several large operators before the advent of dense wavelength division multiplexing (DWDM) systems around 1998 to 2000. It was discovered that low chromatic dispersion lead to accumulation of spurious signals produced by four-wave mixing (FWM) of multiple signals propagating at closely spaced wavelengths due to optical nonlinearity of the fiber. This sort of interference can be mitigated by chromatic dispersion, because FWM products are a changing relative phase along the fiber and do not add up. From this point of view, SMF was good, but operators wanted to reduce cost of dispersion compensation, which at that time required modules with dispersion-compensating fiber (DCF) and optical amplifiers to compensate for their loss.

The solution was to tweak the DCF, shifting wavelength of zero dispersion out of the C band (Figure 4.13), to provide just enough chromatic dispersion for operation of the DWDM link, but low enough to eliminate or minimize need for compensation. This created *non-zero dispersion shifted fiber* (NZDSF), covered by *ITU-T* Recommendation G.655 [4.14], and later also by G.656 [4.15]. Alas, manufacturers failed to agree common specifications, introducing 5 or 6 incompatible NZDSF types with different dispersion, 1–8 ps/nm·km at 1550 nm, dispersion slope and MFD, complicating network design and splicing. Despite that and high price, initially 3× in comparison to SMF, NZDSF captured most of the long-distance and metro market for almost a decade until circa 2009.

A new era began in 2008 with the development of adaptive digital dispersion compensation [4.17], and introduction of 40 Gbit/s, and later 100 Gbit/s transponders capable of dealing with almost any value of chromatic and polarization dispersion encountered in networks. This resulted in almost total return to SMF [4.10], which is the cheapest and easiest to splice of all fibers mentioned above, due to large MFD and simple step-index refractive profile. Moreover, the GeO_2 content is limited to 3.5–5%. Since 2000, SMF has been considerably improved, with new categories B/C/D added to ITU-T Recommendation G.652 (Table 4.3), including:

- reduction of PMD to 0.04–0.10 ps/$\sqrt{km}$
- narrowing of dimensional tolerances for easier splicing
- elimination of loss caused by OH-ion absorption around 1383 nm ("water peak")

Over 90% of current sales are G.652 C/D fibers, working at any wavelength between 1260 nm and 1625 nm, and with proper cabling of up to 1700 nm.

However, there is an application where the SMF was considered as deficient in one respect. This is the indoor wiring in Fiber To The Home (FTTH) access networks, often installed by workers without previous experience in fiber optics, using techniques forbidden in other fiber networks:

- routing thin cable around corners of doors and walls
- stapling it to wooden walls
- laying under the carpet (with the prospect of people walking and furniture standing on it)
- tight coiling of fibers in compact enclosures, etc.

This results in *fiber bending* with 5–10 mm curvature radius, despite introduction of special single fiber cables with 4.5–5 mm diameter as opposed to 1.7–2.8 mm in other indoor applications. Loss of indoor segment at 1550 nm is set at 1.2–1.5 dB, of which no more than 0.4–0.5 dB can be allocated to effects of bending.

Large-scale FTTH construction began in Japan in 2000 and in the US in 2004. Operators like NTT and Verizon wanted fibers tolerating sharp bending. SMF, however, had specified bending radius of 37.5 mm, later reduced to 30 mm [4.13]. While 125 μm fiber survives bending/coiling with 5–8 mm radius, depending on affected length and accepted probability of break [4.18], transmission is interrupted because weakly guided light escapes from the fiber. This effect is utilized for identification of fibers carrying traffic, making temporary non-reflective terminations, or in light injection and detection (LID) systems monitoring coupling of fibers during fusion splicing.

Suitable "bending-tolerant" and "bending-insensitive" fibers (BIF) were developed and standardized in three grades [4.19] (Table 4.4). Modifications with respect to SMF ranged from optimized multi-step refraction index profile to costly and complicated placement of special

Table 4.4 Categories and macrobending loss limits of bending-tolerant single-mode fibers [4.19]

ITU-T designation	G.657.A1		G.657.A2, G.657.B2			G.657.B3		
IEC designation	B6_a1		B6_a2, B6_b2			B6_b3		
Bending radius [mm]	15	10	15	10	7.5	10	7.5	5
Number of turns	10	1	10	1	1	1	1	1
Loss at 1550 nm [dB]	0.25	0.75	0.03	0.10	0.50	0.03	0.08	0.15
Loss at 1625 nm [dB]	1.00	1.50	0.10	0.20	1.00	0.10	0.25	0.45

micro- or nanostructures surrounding the core, blocking propagation in the radial direction. There is a compromise between bending tolerance and difficulties in fiber manufacturing and fusion splicing. G.657.A1/A2 fibers are compatible with SMF in transmission performance, dimensions and splicing; suppliers declare conformance to ITU-T G.652.D. However, G.657.B2/B3 fibers are compatible only on a "best effort" basis.

Many operators building FTTH networks do not use bending-tolerant fibers, opting for SMF, but bending-tolerant fibers found uses in other challenging environments, for example:

- in compact, high fiber count cables
- in cables subjected to wide variations of temperature and thermal shocks
- inside miniaturized equipment to reduce space for fiber storage

The history of multimode telecom fibers began with step-index designs, quickly abandoned because of limited modulation bandwidth, approximately 20 MHz km in a 50/125 μm fiber due to high modal dispersion. Improvement came with introduction of graded index multi-mode fiber (GI-MMF), where quasi-parabolic change of refraction index with distance from fiber axis compensates for variable length of propagation path. This provided adequate (in the early 1980s) bandwidth combined with easy splicing and low-cost connectors, thanks to a large core: 50–85 μm, 5–10× larger than in SMF.

However, optimum refraction index profile is wavelength dependent, and best performance is obtained in one band only – 850 nm or 1300 nm. In addition, a multimode fiber needs large amounts of costly dopant (GeO_2), with associated increase of Rayleigh scattering, geometry imperfections and fiber loss. GI-MMF is 2–4× more expensive than SMF and has higher attenuation: 0.5–1.0 dB/km at 1300 nm and 2–3 dB/km at 850 nm; operation at 1550 nm is not intended. Applications since 1990 are limited to local area networks (LAN), data centers, etc., which still constitute a large market.

Introduction of Gigabit Ethernet around 1999 revealed limitations of multi-mode fibers, as modal dispersion reduced link length to 150–550 m. The 50/125 μm fiber was upgraded, with focus at 850 nm operations, as length of multi-mode gigabit links is dispersion-limited. Modal dispersion was reduced by improvements in refraction index control and selective excitation of the central part of the fiber core, which required dedicated optics in transponders. New 50/125 μm fibers are suitable for short-distance transmissions of up to 40 Gbit/s. Despite this, steady proliferation of single-mode fibers in LANs is observed.

Apart from different core designs and transmission properties, all fibers listed above have identical cladding size, primary coating (applied during drawing for protection against scratching, humidity and microbending) and mechanical strength (Table 4.5). This is important for cabling, splicing and installation: the same cables, tools, fiber cleavers, fusion splicing machines, joint closures, fiber distribution frames, etc. can accommodate any fiber. An exception is the use of dedicated connector parts for termination, having tighter mechanical tolerances for single-mode fibers, and three options of endface shape:

- flat
- domed (PC)
- angled (APC)

Except for improvements and wider deployment of bending-tolerant fibers, which are otherwise compatible with regular SMF and can be considered as a new category of it, mass

Table 4.5 Multimode graded-index fibers for LANs

Acronym	Diameter (core/clad)	ITU-T designation	IEC designation	Main properties and applications
GI-MMF	50/125 μm	G.651 (withdrawn)	A1a.1	Fiber originally used in telecom networks in 1980s, later only in LANs. Works in 850 nm and 1300 nm bands, moderate bandwidth (200–800 MHz·km), easy splicing, 2–3× more costly than SMF. Obsolete.
GI-MMF (OM3, OM4)	50/125 μm	G.651.1	A1a.2, A1a.3	Fiber developed for gigabit LANs in 2000s. Works at 850 nm and 1300 nm, high bandwidth at 850 nm, up to 4.7 GHz·km with dedicated laser transmitters, easy splicing, more costly than SMF. Bending-tolerant versions recently.
GI-MMF	62.5/125 μm	(none)	A1b	Introduced in 1985 for LANs with LED transmitters. Easy splicing and assembly of low-cost connectors. Higher cost and lower bandwidth with respect to 50/125 μm fiber, ≈ 200 MHz·km. Not suitable for gigabit LANs, almost discontinued.

proliferation of new optical fiber is unlikely during this decade. Prospects for acceptance of new, revolutionary designs, for example, multi-core single-mode fibers for space division multiplexing systems and hollow-core microstructured fibers for high-capacity and low-latency DWDM networks, remain speculative.

A change already started is transition to fibers with the same 125 μm glass enclosed in a thin coating: 200 μm instead of 250 μm [4.13]. This increases fiber count in an identical cable by 46%; an example is a 9.6 mm cable with 288 fibers [4.20]. In current single-mode fiber, primary coating occupies 75% of the cross-section, while the light-guiding area with diameter of approximately 30 μm occupies just 1.5% (Figure 4.12). A thinner coating is suitable, mostly for BIF fibers, as it provides inferior protection against crush and microbending. Thin fibers are compatible with existing tools and accessories, but require modifications of the cable manufacturing process. Reduction of cladding diameter of general-purpose fibers is unlikely – reduced stiffness, strength and visibility would preclude good manual handling. However, 80 μm fibers are used in dispersion compensators, laser gyroscopes, etc., in protective enclosures.

Polymer optical fibers (POF) are restricted to niche applications due to high attenuation: 130–200 dB/km at 650 nm for most popular PMMA fibers. POF are predominantly of step-index multimode type with limited bandwidth, made in wide range of diameters up to 1 mm. Unlike silica fibers, PMMA fibers withstand severe bending and more than 15% strain without breaking. An important field of application is automobile and aircraft data wiring.

Table 4.6 Common properties of silica telecom fibers

Parameter	Value	Comments
Cladding diameter	125 μm	Narrow tolerance important for splicing and fitting connectors to single-mode fibers. Tolerance is ≤0.7 μm in good-quality single-mode fibers, ≤2 μm in multi-mode fibers.
Primary coating diameter	245 ± 10 μm 250 ± 15 μm	Data for uncolored (clear) and colored fiber. Coloring at cable plant requires adding new layer of polymer 2–3 μm thick.
Primary coating strip force (average)	1.0–5.0 N	Measured during mechanical stripping of coating. Excessive value results in fiber breaks during stripping.
Tight buffer diameter	900 μm	Hard polymer coating usually extruded over the primary one, mechanically strippable. Applied mostly on fibers for indoor cables and jumpers, where connector fitting is expected.
Proof stress level	0.69 GPa	Corresponds to 1% strain and 6.9 N force. Values up to 2.5% or 1.75 GPa available on request, with lower fiber yield and higher cost.
Tensile strength	3.80 GPa	Corresponds to 5.5% strain and 38 N force at break. Tested on 0.5 m samples, median value.
Stress corrosion susceptibility constant (n_d)	≥18	Higher value means lower probability of fiber break under strain. Special fibers have n_d up to 30.
Range of operating temperatures, with attenuation variations ±0.05 dB/km	–60 . . . +85°C (long-term)	Guaranteed for fiber in primary acrylate coating. Narrower for fiber in tight buffer (low temperature limit), much wider for fibers in special coating: polyimide, silicone, etc.

Attempts to introduce them to LANs and access networks failed, however. Low-loss (15–40 dB/km) graded-index POFs made of perfluorinated polymer (CYTOP) were developed, but are expensive and difficult to produce [4.21,4.22].

4.2.2 Cables: Applications, Operating Conditions and Requirements

Optical fibers in primary coating are fragile and unsuitable for use, except for controlled laboratory conditions or interior of equipment. *Cable* provides:

- protection against humidity
- mechanical forces
- rodents

- solar radiation
- efficient, error-free and safe installation
- identification of required number of fibers (2–1000s)
- additional application-specific features, e.g. lightning resistance

Fiber parameters, in particular attenuation and PMD, will remain within the specified range of operating conditions and during the network design lifetime of up to 40 years. Depending on application:

- indoor
- duct
- aerial
- power line
- field-deployable or other
- climate and other factors

Operating conditions and requirements vary considerably. This has led to the development of a wide variety of cable designs.

4.2.2.1   Indoor applications

Indoor cables are not subject to large pulling forces, extreme temperatures and humidity, but must be flexible, compact, flame retardant and almost always dielectric. Another requirement specific to indoor cables is termination with optical fiber connectors, to allow patching and connections to active equipment. For this, a rigid tight buffer is better:

- providing stiffness adequate for manual fiber handling
- having outside surface large enough for fixing with glue
- keeping rigidly the fiber inside

Fitting connector on fiber in compact *loose tubes* is difficult because of fiber movement back into the tube when pushing it into epoxy-filled ferrule. Simplex (1-fiber) and duplex (2-fiber) cables are used mainly for making patchcords (Figure 4.14). It allows connections within optical distribution frames for active devices and test instruments. Also pigtails (lengths of fiber or cable with a connector on one end), fusion spliced to fibers extracted from outdoor or indoor cable terminated in building, street cabinet, etc., can be produced.

Indoor cabling must be flame retardant, but fire codes, standards and practices differ worldwide, affecting choice of materials and selection of cable for a given application.

4.2.2.2   Duct applications

The most common is underground installation in ducts, usually at depths of 0.8–1.2 m to provide (some) safety against digging, traffic, and freezing. Cables laid underground (either in ducts, or directly buried) are not exposed to solar radiation and have a moderate range of operating temperatures, with a peak-to-peak yearly variability of approximately 20 K [4.23],

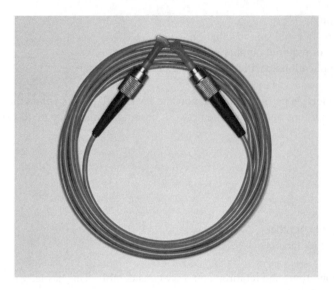

Figure 4.14   Fiber patch cord: 2.8 mm simplex cable terminated with FC/PC connectors

as opposed to 80–120 K encountered by aerial cables. Soil temperature at 1 m depth is usually within the 0 . . . 30 °C range, depending on climate. The operating temperature range specified for *duct cables* is wider, typically from –30 . . . –20°C up to 50 . . . 70°C, because cables can be exposed to above-ground temperatures in some places:

- on bridges or viaducts
- in shallow trenches

In cold climates, freezing of water surrounding the cable is possible. The cable must be able to withstand considerable crush forces and compressive deformation without damage.

During installation, a duct cable is either pulled or blown into a duct, and the typical installation length is 1–2 km, depending primarily on the condition of ducts and straightness of the cable route. Cable flotation [4.24] is also possible, with considerable reduction in friction forces, increased installation length, and fewer splices. This mode of installation, especially blowing, requires circular and smooth sheath surface.

Duct cables must withstand considerable pulling forces (500–3000 N) during pulling, but not during normal operation. Duct cables work in humid environment, as ducts and manholes are often filled with water or brine, and must resist water penetration after cut, or damage to joint closure.

The dominant threat to duct cables are cuts due to digging, drilling and other construction work, or worker errors in network maintenance [4.25,4.26], causing 65–80% of all outages. Average failure rates are several times higher in urban and suburban areas (metro networks) than in the countryside (long distance networks). Several countries, including UK, South Africa, Poland or India report severe problem of cable theft; thieves looking for copper cables routinely cut fiber cables as well. The damage described above cannot generally be prevented by cable design.

4.2.2.3 Buried applications

The main difference with respect to *duct cables* are more severe mechanical threats:

- cuts caused by rodents and other animals, sharp objects in the soil, digging, etc.
- crush forces generated by rocks or moving soil, bending, etc.
- crush and abrasion experienced during cable plowing
- impacts of rocks, tools, etc. during installation

On the other hand, the buried cable does not have to be as flexible, light and compact, as a typical duct cable. Methods ensuring required mechanical robustness are listed below, in order of severity of threats:

- external strength member of aramide or, better, glass yarns (termites, small rodents, impacts)
- outer sheath made of hard, slippery polyamide (termites, rodents, abrasion)
- armoring made of thin corrugated steel, approx. 0.2 mm (rodents, impacts)
- armoring made of galvanized steel wires and tapes or FRP rods (cuts, large rodents, crush forces, strong impacts)

The armoring should be separated from the cable core by an inner sheath, to avoid damage when it moves under crush force or during cable bending or twisting. It also considerably increases cable diameter and weight. Steel armoring, while low-cost and effective, is electrically conductive, and therefore not acceptable for operators requiring dielectric cables. Electrical safety codes require jointing of metallic cable members and their grounding at termination or splicing points.

4.2.2.4 Aerial networks

Self-supporting aerial cables operate in challenging conditions, substantially different from other cable networks, including:

- wide range of operating temperatures, from 0–50°C up to 50–85°C, with peak-to-peak band up to 60–120 K, depending on climate; both seasonal and daily variations are observed
- large, continuous, but variable tensile load due to stringing, wind and icing
- wind-induced vibrations
- solar radiation
- shot impact due to bird hunting (depends on location)

The key design issues are:

- provision of larger strength members
- handling of cable expansion/contraction with temperature
- preventing migration of optical fibers and gel to lower (middle) section of span

Vibrations are controlled by proper selection of accessories, including vibration dampers. Aerial cables are not directly threatened by digging and construction work, although support structures are often damaged by vehicles or construction machinery.

4.2.2.5   Aerial power lines

Optical fiber (unless metal- or carbon-coated) is dielectric and unaffected by electric and magnetic fields, except for polarization effects of no importance in transmission systems. Unlike metallic pairs, fibers can be routed in high voltage networks without interference, induction or stray currents. Fiber cables installed along overhead medium or high voltage power lines use right-of-way and infrastructure of power system (pylons, buildings, etc.) for added functions: communications, control and telemetry. While environmental, electrical and mechanical conditions are challenging, installation at elevation of 20–60 m in an electric power line eliminates threats familiar in underground cable networks: dig-ups, drilling, cable theft and vandalism, with approximately a 10-fold reduction in cable failure rate [4.24,4.25].

Several technologies have been tried for installing optical fibers on overhead power lines since 1980, including:

a) Optical Ground Wire (OPGW)
b) Optical Phase Conductor (OPPC)
c) All-Dielectric Self-Supported (ADSS) cable
d) Dielectric cable attached to or wrapped around power conductor or ground wire
e) Metallic Aerial Self-Supported (MASS) cable

Of those, OPGW (a) is most successful and deployed on the majority of high-voltage power lines in industrialized countries, with experience dating back to 1985 [4.27]. OPGW is the topmost conductor on the high voltage power line, suspended above phase conductors and electrically connected to each pylon for grounding. OPGW intercepts lightning strikes, and must withstand peak current and total charge transfer of 100–200 kA and 60–150°C, respectively. After a lightning strike, OPGW conducts a momentary fault current, whose typical amplitude and duration are 10–25 kA$_{RMS}$ and 0.5–1.0 s, respectively, and is subjected to thermal shock: rapid rise of temperature by 100–150 K and subsequent air cooling. Other challenges typical for aerial cables include aeolian vibrations. Thus, an installation of vibration dampers is often required. Tensile forces due to ice and wind loads can cause a span length of typically 250–500 m. OPGWs exhibit unique all-metal design with multiple steel wires as strength members. Nevertheless, the estimated service life of correctly installed OPGW is 25–40 years.

Before the introduction of single-mode fibers in high-temperature mechanically strippable acrylate coating around 2005 [4.28,4.29], the use of OPPC (b) was considered problematic due to continuous operation at temperatures of 80–100°C, leading to rapid deterioration of the acrylate fiber coating [4.30].

ADSS cables [c] installed on AC power lines with 220 kV and higher voltage frequently suffered sheath damage caused by dry-band surface arcing [4.31]. This problem is caused by induction of surface current of a few mA in wet cable by capacitive coupling to nearby phase conductors, while the ADSS is grounded at each pylon. During cable drying, arcing starts at the break of the conductive water film (actually a weak solution of acid or fertilizer), which leads to local burning of the surface of the polyethylene sheath. The punctured sheath allows water into the aramide strength member, which loses strength and breaks later on. Performance on 110 kV and lower voltage lines is better, but the cable must withstand large tensile forces

Table 4.7   Requirements for optical fiber cables. Scale: (none) ... * (mild) ... **** (severe)

	Indoor	Duct	Buried	Aerial	OPGW
Temperature range	*	**	**	***	****
Tensile strength	*	***	**	****	****
Crush/impact	*	**	****	**	****
Bending/flexibility	****	***	*	**	*
Flame retardant	****	–	–	–	–
Water protection	–	****	****	***	***
Cut/gnaw protection	–	**	****	**	****
Solar radiation resistance	–	–	–	****	****
Dielectric/metallic	D	D^a	D/M^2	D^1	M

^{1}Several network operators require dielectric cables, others do not
^{2}Requirements difficult to meet with dielectric design

and occasional gunshot damage by bird hunters. An additional problem is that ADSS adds extra load to pylons, because unlike OPGW it does not replace the existing earth wire.

Service record of light, low-cost and quick to install attached and wrapped cables (d) is the worst because of frequent mechanical damage by ice or gunshots. Also installation issues vary with location and supplier, from very good (UK) to very bad (Poland).

MASS cables, essentially thinner versions of OPGW, are strong and reliable, but heavy while having no function in power line; therefore it is not frequently used.

4.2.2.6   Comparisons

Table 4.7 provides a quick comparison of the most important operating conditions and requirements (or lack of them) for the cable applications reviewed above. They reflect the design of cable for particular application, the choice of materials and the product costs. It is impossible to make cables meeting all requirements, for example, those that are in direct conflict:

- metallic OPGW must conduct strong electric currents – ADSS must be dielectric
- buried cable shall be armored against rodents and digging – indoor must be flexible

Of particular importance for cable designer are:

- range of operating temperatures (min-max)
- mechanical specifications (axial load, crush, impact, twist, bending ...)
- required fire properties, dictating choice of materials
- required (or not) protection against ingress of water
- fiber count

4.2.2.7   Evolution of technology and applications

There are several trends which can be expected to last for at least another decade:

- increase of fiber count, particularly as prospects for further increase of system bit rates are limited [4.10]

- reduction of cable size and weight, taking advantage of more robust bending-tolerant fibers in thinner coating and tubes
- proliferation of fiber cables in access networks (Fiber To The X –FTTx)
- expansion of fiber backhaul and core networks for LTE and other types of wireless broadband access
- periodic upgrades of fiber interconnections in data centers and LANs

Directions of development after 2025 to 2030 are hard to predict, as carriage of steadily increasing traffic may need new disruptive fiber optic technology in core and metro networks. Therefore, radically different cables and accessories (connectors, splices, couplers, etc.) will be developed.

4.2.3   Fiber Protection and Identification in Cables

4.2.3.1   Protection of optical fibers

Optical fiber in primary coating (4.1) has several undesirable properties:

- Low strength to limit strain-induced crack development and random fiber breaks, medium-term (hours–days) tensile strain shall not exceed 33% of strain applied during proof testing (Table 4.6), equal to 0.33% and 2.3 N load. Conservative recommendation for long-term strain is 0.05%
- Macro-bending and micro-bending cause exponentially rising losses with wavelength and strain for fiber and polarization mode dispersion (PMD). Effective curvature radii of the whole fiber must exceed 50–80 mm for SMF [4.32], although stronger bending of short fiber lengths is acceptable
- Moisture penetrates standard acrylate coating in a matter of minutes [4.33], later on (months–years) this causes separation of the coating, bending of the fiber and drastic rise of the attenuation. Some solvents, e.g. gasoline, act much faster
- Sensitivity to hydrogen, in hours diffuse quickly through glass into fiber core and causes additional losses, primarily at the 1240 nm wavelength
- Exceptionally low temperature expansion coefficient, 2–3 orders of magnitude smaller than for polymers used in the cable industry (Table 4.8). Variations of temperature produce mismatch of dimensions, leading to strain and bending or buckling of the fiber with an increase of the attenuation and/or PMD

Unlike metal wires in telephone, data and power cables, optical fibers cannot serve as the load-bearing part of the cable. Even in a cable with 288 fibers, their total permitted load (a) is only 662 N. While forces encountered during pulling of 10–14 mm duct cable, weighing 90–150 kg/km, are between 1000 N and 2500 N. Self-supporting aerial cables are under permanent tensile load, unacceptable for glass fibers. Dedicated strength members must be included to minimize the transfer of load to the fibers.

Structures and materials in optical fiber cable must protect fibers against detrimental factors listed above, if present in given operation environment. The effects a) to d) appear with considerable, but variable delay and fiber breaks (a) occur in a random fashion. Effects of temperature are fast but may depend on the thermal history of the cable. Several polymers

Table 4.8 Thermal expansion coefficient (α) and Young modulus (E) of materials used in the production of optical fiber cables, at 20°C

Material	TEC/K^{-1}	E/MPa
Fused silica (optical fiber without coating)	0.55×10^{-6}	72 500
Optical fiber in 250 μm primary coating	$2.0–2.5 \times 10^{-6}$	–
High density polyethylene (HDPE)	5.9×10^{-5}	1400
Poly-butyl terephthalate (PBT)	$0.7–1.3 \times 10^{-4}$	2200–2800
Polyamide 6 (Nylon 6)	0.83×10^{-4}	1100
Polyamide 12 (Nylon 12)	$0.9–1.5 \times 10^{-4}$	900–1400
Plasticized poly-vinyl chloride (PVC)	$0.7–2.5 \times 10^{-4}$	20–500
Polypropylene (PP)	2.2×10^{-4}	1350
Hard UV-cured acrylate	$0.5–1.5 \times 10^{-4}$	500–1500
Aluminium (unalloyed)	2.4×10^{-5}	68 000
Carbon steel	1.3×10^{-5}	210 000
Stainless steel grade 304L	1.66×10^{-5}	200 000
S-glass	1.6×10^{-5}	83 000
Fiberglass/epoxy composite (FRP)	$4–7 \times 10^{-6}$	45 000–57 000
Aramide yarn (Twaron) – tested along fiber axis	-3.5×10^{-6}	90 000–110–000
Aramide yarn/epoxy composite	-2.0×10^{-6}	60 000

Note: properties of materials vary with product and method of processing; values in the table are typical. In additional, properties of polymers are strongly temperature-dependent.

are used to make tubes and tight buffers for fibers, for example polyamides and PBT, which have variable content of the crystalline phase, and shrink when slowly cooled after operation above their glass transition temperature (~60–80°C). This phenomenon in particular affects aerial cables working in continental climates [4.34]. The cable can protect fibers against mechanical forces, moisture, hydrogen and chemicals, but not against slow temperature variations; it can only accommodate resulting expansion/contraction of its parts without excessive strain and bending of fibers. An exception is behavior of OPGW during fault currents (Chapter 4.2.2).

Thermal expansion coefficient α_k of a cable comprising N parts, each with thermal expansion coefficient α_n, Young modulus E_n and cross-section A_n, can be calculated using the following formula [4.32]:

$$\alpha_k = \frac{\sum_{n=1}^{N} \alpha_n \cdot E_n \cdot A_n}{\sum_{n=1}^{N} E_n \cdot A_n} \tag{4.7}$$

Change of cable length ΔL_K resulting from change in temperature ΔT is:

$$\Delta L_k = \alpha_k \Delta T \cdot 100\% \tag{4.8}$$

The second effect to add is change of cable length ΔL_K due to tensile force F acting on the cable, which can be calculated as:

$$\Delta L_k = \frac{F}{\sum\limits_{n=1}^{N} E_n \cdot A_n} \cdot 100\% \qquad (4.9)$$

In both cases, considerable variability of polymer parameters with temperature must be taken into account. The second limitation is that the unglued fibrous materials, in particular loose aramid yarns, have almost zero modulus in compression mode and do not prevent cable contraction at low temperature.

Variations in cable length can be accommodated without fiber strain, if the fibers are put in a hollow structure such as in a tube, slot or groove. As a result, they are allowed to move freely. The length of the fiber L_F is longer than the tube (slot, etc.) length L_T, therefore the fiber undulates or forms a spiral. Fiber overlength ΔL_F is defined by this formula:

$$\Delta L_F = \frac{L_F - L_T}{L_T} \cdot 100\% \qquad (4.10)$$

indicating how much the cable can be elongated before fibers are strained. This is shown in Figure 4.15, where the ΔL_F is approximately 0.6% (value typical of large tube for aerial cable).

Fiber overlength is easily achieved during extrusion of the polymer tube, because cooling of the polymer (PBT, polyamide, HDPE) from processing temperatures of 220–280°C causes the tube to shrink, while fiber shrinkage is negligible. Shrinkage of semi-crystalline polymers like PBT depends considerably on speed of cooling [4.35]. To make welded metallic tubes, either controlled "overfeed" of fiber or mechanical longitudinal compression of tube after welding are used.

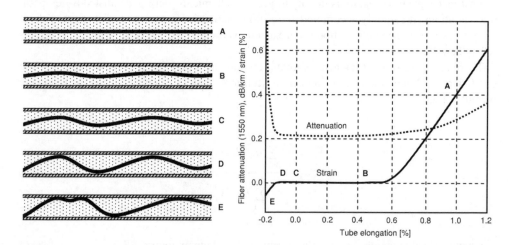

Figure 4.15 Fiber movement inside loose tubes in different operating conditions: (A) high tensile load, (B) moderate tensile load, (C) no load, room temperature, (D) low temperature, (E) extreme low temperature

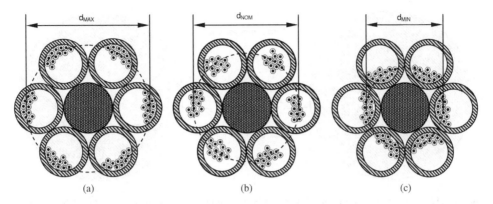

Figure 4.16 Movement of fibers in cable with stranded loose tubes: (a) contraction of cable at low temperature, (b) no tensile load, room temperature, (c) cable elongation due to tensile load and high temperature

Stranding introduces another kind of overlength of fiber with respect to the cable (ΔL_F, %), depending on stranding diameter d and pitch p:

$$\Delta L_F \approx 491 \left(\frac{d}{p}\right)^2 \tag{4.11}$$

For cable design, the difference ΔL between minimum and maximum ΔL_F, corresponding to d_{MIN} and d_{MAX} in Figure 4.16, respectively, called a "strain-free window", is important:

$$\Delta L \approx 491 \frac{d_{MAX}^2 - d_{MIN}^2}{p_S^2} \tag{4.12}$$

Formula 4.8 is valid for helical stranding. More detailed analysis is included in the book by Murata [4.32]. ΔL indicates how much contraction and/or elongation of cable can be accommodated without straining or excessively bending fibers. The range of fiber movement is reduced when more fibers are placed in a tube of given size, shrinking to zero if the tube is completely filled. Aerial cables designed for high tensile load and a wide range of temperatures have a small stranding pitch, 150 mm or less; values below 70–100 m are unfeasible because of induced fiber bending [4.32]. Stranding is used also in slotted-core cables [4.32].

Many cables incorporate so-called "tight-buffer" protection of fibers, which are encased in a structure without free space, hence providing no strain-free window. Both single-fiber (Figure 4.17) and multi-fiber (Figure 4.18) units exist, with fibers protected by an inner layer of soft polymer allowing movement during bending, crush, etc. The outer layer of hard polymer absorbs external forces with minimum deformation. Multi-fiber units usually have a strength member, providing stiffness and reducing strain due to variable tensile force or temperature. Fibers are usually stranded, which gives the benefit of reducing and stabilizing polarization mode dispersion under adverse operating conditions [4.36].

Fiber in a standardized 900 μm tight buffer (Figure 4.17), originally developed for missile guidance and other military applications, is now a default component of indoor and field

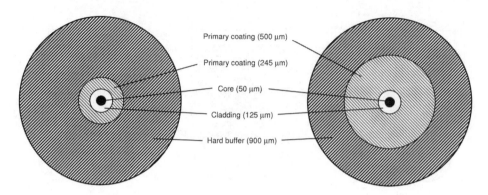

Figure 4.17 50/125 μm multimode fibers in 900 μm tight buffers, with 245 μm primary coating for indoor cables (left) and 500 μm primary coating for use in extended temperature range (right)

cables. Post-extrusion shrinkage of polymer introduces compressive strain in the fiber, typically 0.05–0.2%, and correspondingly reduces strain under tensile loads [4.35]. However, further shrinkage of the buffer in cold conditions ultimately causes fiber buckling and thus restricts lowest operating temperature [4.37]. This problem is eliminated by increasing the diameter of the fiber primary coating to 400–500 μm [4.38]. To facilitate stripping of the hard buffer, a thin "separation layer" of gel (in semi-tight buffer) or silicone polymer, is sometimes placed under the buffer.

4.2.3.2   Marking and identification of fibers

Multi-fiber cables must allow visual identification of sub-units and individual fibers, which is of particular importance for splicing and repair work. Several solutions are used, depending on cable component:

- fibers in tubes, ribbons and other modules are color-coded

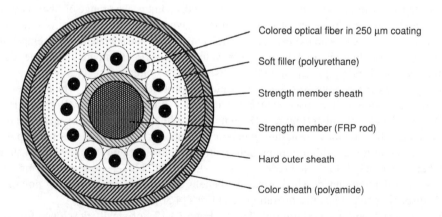

Figure 4.18   12-fiber unit of tight design (2.5 mm diameter) for OPGW cables

Table 4.9   Standard colors for marking optical fibers

IEC 60304[1] Color	TIA/EIA-598C (US)		
	No.	Color	Abbreviation
Blue	1	Blue	BL
Orange	2	Orange	OR
Green	3	Green	GR
Brown	4	Brown	BR
Grey	5	Slate	SL
White	6	White	WH
Red	7	Red	RD
Black	8	Black	BL
Yellow	9	Yellow	YL
Violet	10	Violet	VI
Rose	11	Rose	RS
Turquoise	12	Aqua	AQ

[1]IEC 60304 published in 1982 (Ed. 3) does not dictate order of colors

- multiple tubes are color-coded, with distinct colors of all tubes, or only of No. 1 and No. 2, the rest being of the same color, most often white [4.39]
- slots in slotted-core cables are identified with markers (Figures 4.28 and 4.29)
- ink-jet printing on ribbons and sub-units in breakout cables (Figure 4.20)
- fibers in individual tight buffers are marked by color of buffer

IEC 60304 [4.40] and *TIA/EIA-598C* [4.41] standards define 12 colors for fiber marking (Table 4.9). Some suppliers add a few more, but this makes recognition harder. When the number of fibers in a cable unit exceeds 12, they are divided into bundles of up to 12, marked by wrapping with cords in colors selected from the same set. They can also be marked by printing periodic black rings (Figure 4.19). Recognition of some colors, for example blue vs. violet, is negatively affected by using artificial light sources with a poor color rendering index (CRI), such as compact fluorescent lamps; halogen incandescent lamps are best for this purpose.

The type of fiber in an indoor cable is indicated by the color of the jacket, a separate color being traditionally assigned to each, for example yellow – SMF, red – NZDSF, green – 50/125 μm, orange – 62.5/125 μm, etc. [4.39]. The US standard TIA-568C [4.41], published in 2005, reduced the set of colors for typical indoor civilian cables to only three (Table 4.10), and this convention is gradually being adopted worldwide.

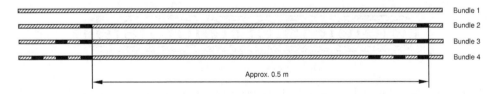

Figure 4.19   Marking of fibers belonging to different bundles with black rings

Table 4.10 Jacket colors of indoor optical fiber cables according to TIA-568C

Fiber type	Jacket color
Polarization maintaining single mode[1]	Blue
Single-mode (any other type)	Yellow
Multimode 50/125 μm (old)	Orange
Multimode 50/125 μm (new)	Aqua
Multimode 62.5/125 μm	Orange
Multimode 100/140 μm	Orange

[1]Special product, used mainly in test instruments, not as transmission medium in networks

Markings must remain legible during the lifetime of the cable, to enable repairs and modifications. This was initially problematic due to discoloration of fibers caused by dark fading of dyes and/or their dissolution in filling gel; both processes progress faster with increasing temperature. Fibers having a dyed outer (hard) layer of a primary coating, applied during drawing, are superior in this respect to fibers colored by adding a thin color coating to the clear fiber later. Identification of tubes and fibers in tight buffers is ensured by extrusion of these parts from dyed plastics; stability of color is good, as those parts of cable are not exposed to light during operation.

4.2.3.3 Fiber ribbons

Another way of fiber bundling is to glue together several, most often 4–12 parallel fibers (Figure 4.20) with soft, transparent material(s), forming a flat, smooth unit. Several ribbons can be stacked together and placed in a slot or *loose tube*, offering the highest possible density of fibers in the cable.

Another important advantage of ribbons is the possibility of cleaving and fusion splicing a complete ribbon in one step, making splicing work much faster. A thin ribbon, however, does not provide much protection to fibers (special versions with hard buffer exist), and contrary to a single fiber, is sensitive to twisting.

Table 4.11 lists basic parts of optical fiber cables, their functions and typical materials. It is not exhaustive

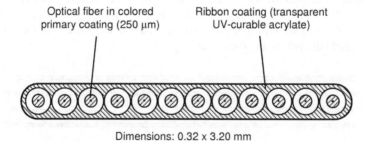

Optical fiber in colored primary coating (250 μm)   Ribbon coating (transparent UV-curable acrylate)

Dimensions: 0.32 x 3.20 mm

Figure 4.20 Optical fiber ribbon with 12 fibers. Fiber count varies from 4 to 24 fibers. Other coating ("matrix") designs and materials are sometimes used

Table 4.11   Basic components of optical cables and their functions

Element	Function	Form and material(s)
Core	Holds optical fibers, provides their grouping and identification. Fibers and units are identified by color, position or printed number. Protects fibers against strain, water and direct mechanical damage.	Loose tubes, slotted-core, multi-fiber tight buffered units (outdoor cables), tight buffers (indoor, jumper cables, flexible field cables), located inside cable. Longitudinal water protection with filling compounds or swellable materials. Fibers may be formed in ribbons. Tube materials: PBT, PA, PP, HDPE, other plastics, stainless steel, aluminum, copper. Slotted-core materials: PP, HDPE, aluminum (in OPGW).
Moisture barrier	Protects the core against radial migration of moisture through polymer sheath. Used in outdoor cables operating in humid environments.	Aluminium foil laminated with polyethylene, wrapped around the core, with polymer fused. Alternatively, a hermetic, welded aluminum or copper tube holding the core of cable. Metallic tubes provide equivalent functionality.
Strength member	Bears tensile loads, provides stiffness, reduces variations of cable length and attenuation with temperature.	Rigid members: steel wire(s) or strand(s), glass-epoxy or aramide-epoxy rods (FRP). Fibrous members: aramide or glass yarn. Cable may include several strength members made of different materials.
Sheath (in outdoor cables). Jacket (in indoor cables)	Prevents ingress of water, solar radiation and contaminants into the core. Protects core against abrasion, crush and impacts.	Continuous, tight layer of flexible material surrounding the core. Materials: polyethylene (HDPE, MDPE, LLDPE) or polypropylene (PP) stabilized with 2.5% carbon black (outdoor cables), plasticized poly-vinyl chloride (PVC), polyurethane (PU), fluoropolymers, LSZH compounds (indoor cables). Absent in OPGW and similar metallic cables.
Ripcord	Quick longitudinal cut and removal of sheath without dedicated tools.	Thick cord of polyester or aramide, strong enough to cut through the sheath or jacket.

Table 4.11   (*continued*)

Element	Function	Form and material(s)
Armor (in outdoor cables)	Protects cable core and inner sheath against mechanical damage: impacts, cuts, gnawing, crush, etc.	Steel tapes or wires, or fiberglass (FRP) rods wrapped over inner sheath and core. Corrugated steel tape surrounding inner sheath and core. Against mild attacks of rodents and termites: glass or aramide yarns or tapes, outside sheath of hard polyamide. Steel or FRP armor is also a strength member.
Outer sheath (in outdoor cables)	Protects core, external strength member and armor against water and solar radiation. Allows identification marking.	Continuous, tight layer of strong and (relatively) flexible material surrounding the core. Materials: polyethylene (HDPE, MDPE) or polypropylene (PP) stabilized with 2.5% carbon black Sometimes made of polyamide, providing protection against termites, rodents, cuts, etc. and reducing friction during pulling into duct. Absent in OPGW and similar metallic cables.

4.2.4   Indoor Cables

Fiber counts in *indoor cables* are relatively low; 1-fiber, or "simplex" (Figure 4.21) and 2-fiber, or "duplex" (Figure 4.22) cables constitute most of production mileage. Although manufacturers offer up to 72–144 fibers [4.38] in many designs, including features like:

- central strength member

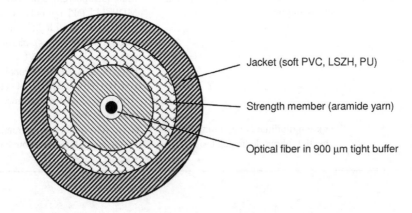

Figure 4.21   Simplex indoor cable (2 mm) with fiber in 900 μm tight buffer

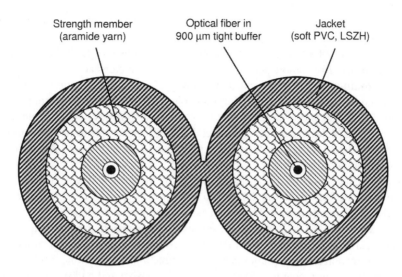

Figure 4.22 Duplex "zipcord" cable with fibers in 900 μm tight buffers, composed of two 2.8 mm units, identified with different colors of buffers

- breakout: cable can be separated into several simplex cables at the end (Figure 4.23)
- fibers with equalized optical length for parallel data transmission
- additional mechanical protection/armoring

The default strength member in indoor cables is aramide yarn, consisting of 5–10 μm fibers. It has excellent strength when pulled (Table 4.8) and provides cushioning against crush or impact, but does not prevent cable contraction at low temperature or excessive bending. Larger and heavier cables have also rigid central strength member of fiberglass-epoxy or aramide-epoxy composite (FRP) (Figure 4.23). Examples are presented in the figures below, with all parts shown to scale wherever possible.

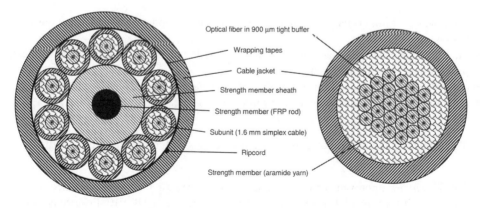

Figure 4.23 10-fiber breakout indoor cable (left) and 24-fiber distribution cable with fibers in tight buffers (right) shown at the same scale

Several suppliers make factory-terminated multi-fiber cables of almost any type and fiber count to customer specifications, primarily for LANs and data centers.

Rising numbers of fibers installed and patched, in particular in DWDM networks, large distribution frames (1200 or more per rack) and data centers, created acute problems of putting more fibers into limited spaces in racks and raceways. The simplest solution is reduction of cable diameter, making it more fragile. Simplex cables have been downsized from 3.0–4.5 mm in the 1980s to 1.6 mm and even less today [4.38,4.39]; also fibers in the 0.9 mm buffer are used to save space. Major exceptions are special 4–5 mm cables for in-house connections in FTTH networks, which must limit the fiber bend radius to about 5 mm (Table 4.4) when the cable is fixed to a 90° corner of a wall or a door.

Two-way optical connection to transponder requires two fibers and two connectors, usually of the SC or LC type. It is made simpler by duplex "zipcord" cabling, comprising two simplex units joined side by side (Figure 4.22), which can be easily separated.

Breakout cables (Figure 4.23) are useful for tasks like connecting multiple active devices within a certain area, as the subunits constitute regular simplex cables. There is no need to splice pigtails or use furcation tubing (simplex cable with empty *loose tube* for mechanical protection of the fiber in the primary coating). Such a cable is expensive and bulky in comparison to a "distribution" cable with a bundle of buffered fibers and (a) common jacket.

Distribution cables with more than 24 fibers comprise multiple bundles usually including no more than 12 fibers each, due to the set of fiber colors, and the central strength member; the advantage of compactness is partly lost. For dense cabling in large data centers, etc., distribution cables with bundles of primary coated fibers have been introduced, with up to 12 fibers in a 3 mm cable [4.38]. These are rather fragile, requiring careful handling.

Rather surprisingly, cables similar to those shown in Figure 4.23 are used in challenging conditions as field-deployable and recoverable cables intended for military communications, temporary fiber links for TV cameras, ad-hoc temporary networks for events, etc. Such cables must be strong, flexible, work in the temperature range known from aerial networks, say – 40 ... +85°C, resist solar radiation, abrasion, humidity, petroleum products, ionizing radiation (military cables), etc. and are most often factory-terminated with connectors of various styles. Modifications with respect to *indoor cables* include:

- fibers with 500 μm primary coating (Figure 4.17), accommodating more buffer contraction without fiber buckling
- thicker jacket of special materials, like military-grade polyurethane, sometimes with extra protection against rodents or cuts
- colors: orange in broadcasting (visibility), black (solar radiation), olive (camouflage)
- radiation hardened fibers and plastics (military use, nuclear facilities, satellites)

There are three most common classes of indoor cables with respect to fire properties:

- *Plenum cables*: approved for use in spaces with forced air flow, in accordance with US standards NFPA 90A [4.42], and NFPA 262 [4.43], formerly UL 910. Priority is given to flame-retardant and self-extinguishing properties of the cable, while opacity and toxicity of smoke are not important.

- *Riser cables*: approved for use in rooms and in air spaces without forced air flow, in accordance with US standard ANSI/UL 1666 [4.44]. Again, emission and properties of smoke are not important.
- *Low-smoke zero halogen (LSZH)*: flame retardant cables emit a small amount of smoke and corrosive/toxic gas during fire – conforming to European standard EN 50399 [4.45], which evolved from previous documents EN 50266 and IEC 60332-3.

Flame retardancy is highest in plenum cables, which automatically meet the requirements for riser cables. This, however, comes at the expense of inferior flexibility and higher cost. Plenum and riser cables have traditionally been made with jackets, and often fiber buffers. They are made of plasticized PVC or fluoropolymers like PVDF, which contain chloride or fluorine. These materials burn little and slowly, but emit dense, corrosive and toxic smoke under fire.

The jacket, and sometimes other parts of the LSZH cable, is made of a polyethylene compound filled with hydrated aluminum or magnesium oxide ($Al_2O_3{}^*nH_2O$, $MgO_2{}^*nH_2O$) powder. The filler emits large amounts of water vapor, but no dense or toxic smoke when heated. It does not prevent the combustion of polymer but delays it instead. For design of a cabling system, an important parameter of the LSZH cable is combustion energy per unit length [4.45]. LSZH cables do not meet plenum requirements, but can meet riser specifications.

4.2.5 Duct Cables

Fiber counts vary from 8–12 in a cable laid to a street cabinet or BTS, up to 144, 240 or more on major routes in metro networks; and steady increase is observed. Extreme fiber counts are required in FTTH networks, in particular when passive optical splitting is not used. Cables used in Japan reportedly have 4000 and more fibers [4.46]. Duct space is generally rare and cables shall be as compact as possible.

Dominant designs of *duct cables* are of the loose type, able to accommodate extension without excessive fiber strain – with stranded loose tubes (Figure 4.24), central tube

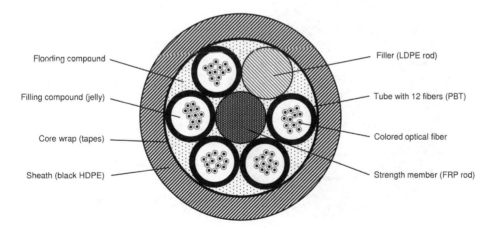

Figure 4.24   Dielectric, gel-filled duct cable with stranded loose tubes

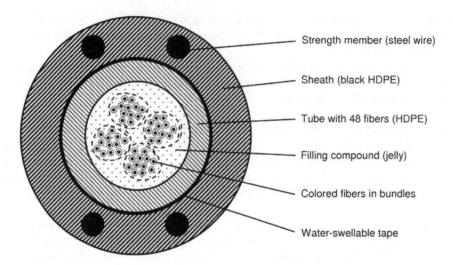

Figure 4.25   Duct cable with gel-filled central loose tube and fibers in bundles

(Figures 4.25 and 4.26), or slotted core with single fibers (Figure 4.27) or ribbons (Figures 4.28 and 4.29). Protection against longitudinal water penetration is ensured by filling the tubes and interstices with hydrophobic petroleum or synthetic gel, or inclusion of water-swellable elements: tapes, powders or strings in "dry" cables. Hybrid designs with tube(s) filled with gel, and interstices protected by swellable elements, are frequently used.

A wide variety of cable designs and sizes exists. Below are only examples of typical products. Selection of the cable for a particular project depends on multiple factors, including:

- cost
- diameter and weight

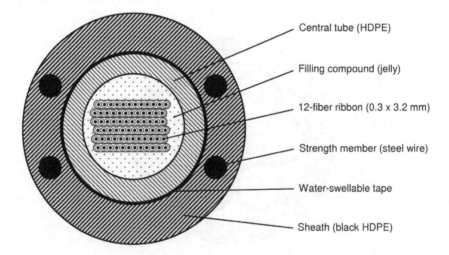

Figure 4.26   Duct cable with gel-filled central loose tube and 84 fibers in stacked ribbons

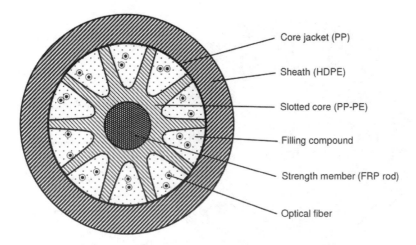

Figure 4.27   Dielectric slotted-core cable with 20 fibers

- ease of splicing
- compatibility with available tools and accessories
- requirements, experience and practices of network operator
- local regulations

Most operators in Europe require dielectric cables, avoiding onerous requirements (permits, safety grounding, etc.) when crossing power lines, laying cables along electrified railway, etc. This requirement is less common in Asia and the United States.

Particular cable designs are not equal even if they meet the same specifications, with specific relative advantages and disadvantages.

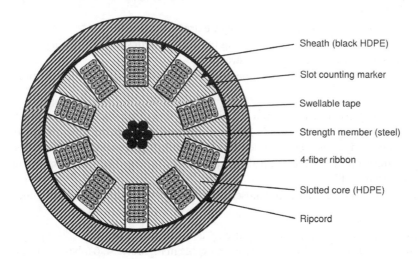

Figure 4.28   200-fiber slotted-core cable with fiber ribbons in rectangular slots. Water blocking provided by swellable tape wrapped around "dry" core

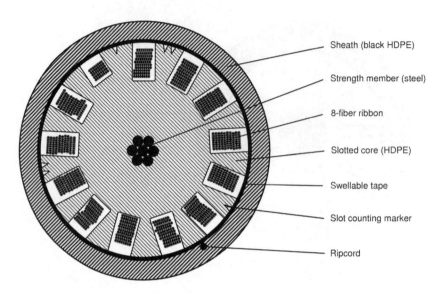

Figure 4.29   1000-fiber slotted-core cable with ribbons used in the feeder segments of FTTH networks in Japan. Diameter: 29.5 mm

A cable with multiple tubes (Figure 4.24), stranded in a helical, or more often, reversible (SZ) pattern has, since its introduction by Northern Telecom around 1979, been the "gold standard" in outdoor cable design, displaying multiple advantages:

- easy to adjust and accurately controlled fiber overlength, by tube stranding pitch
- convenient location of rigid strength member in the center, good flexibility independent of bending direction
- excellent behavior when subject to strain or extreme temperatures
- convenient fiber identification and routing to splice cassettes during splicing
- prevention of fiber migration by tube stranding
- selective access to fibers from one tube without disturbing others (SZ designs only)

A single tube preferably holds up to 12 fibers (number of standard colors), sometimes 24. Most versions have 5–12 tubes, holding up to 144 or 288 fibers, including a plastic sheath over a central strength member if necessary. Fillers allow placement of fewer tubes without redesign of the cable. To increase capacity, tubes may be arranged in two layers separated with tapes, up to 18 (6 + 12).

Cables of this type have an excellent service record, but also drawbacks including:

- large diameter and weight; especially if only a few fibers are needed
- high manufacturing cost: 7 extrusion runs for cable in Figure 4.24: 6 tubes/fillers and sheath, more materials needed in comparison to central tube cable (Figure 4.25)
- large total fiber overlength (1.5–4%) and error in locating cable cut with OTDR

Central tube design (Figure 4.25) is preferred for low and medium fiber counts, up to approximately 48, and much more when fiber ribbons are used (Figure 4.26). The reasons are:

- low diameter and weight
- low cost: 2 extrusion runs: tube and sheath, less materials used

List of disadvantages is longer, however:

- limited fiber overlength in small diameter tube
- difficult splicing when fiber count exceeds 12–24: bundle identification by means of color threads or black rings (Figure 4.16) is inconvenient, and routing of fibers to multiple splice cassettes, holding 12 or 24 splices each requires protective tubing
- inconvenient location of strength members over the core; at least 2 rigid strength members are needed, and cable stiffness depends on direction
- possible fiber migration, especially in aerial cables

Slotted-core cable with fibers in primary coatings (Figure 4.27), introduced around 1980 were once made in many countries (France, Poland, Denmark, Sweden . . .) as a cheaper substitute to stranded-tube cables when a small number of fibers was needed. Manufacturing required only two extrusion operations. The slotted core was made of low-cost plastic, like PP/PE copolymer, and the rigid strength member was centrally located. Fiber overlength was controlled by adjusting pitch of slots, as in a cable with stranded tubes.

Unfortunately:

- splicing and termination is difficult: loose fibers extracted from a cable require protective tubes. This work is time-consuming for higher fiber counts
- placing multiple fibers in one slot was problematic, limiting fiber capacity

Slotted core *must* have extruded tight jacket, fused with it. If foil wrap is used, fibers get squeezed between the slotted core and sheath when the cable is bent and damaged. Other errors are narrow ridges between slots, resulting in poor impact resistance, and use of difficult to remove silicone gel. In the 1990s, such cables disappeared.

However, another kind of slotted-core cable (Figure 4.28) still enjoys wide use, particularly in Japan [Furukawa, etc.]. Its key features are:

- rectangular slots holding stacked fiber ribbons
- wide ridges between slots, resistant to crush or impact

Again, core is extruded from low-cost polymer like HDPE in a single run, but with ribbons:

- any number of fibers can be accommodated and identified
- cable with a high number of fibers is quite compact (Figure 4.29)
- fusion splicing (of complete ribbons) is very fast

Cables of this type with up to 4000 fibers are used in Japan [4.46–4.48]. However, ribbon splicing requires dedicated tools, equipment and training of personnel. This creates an entry

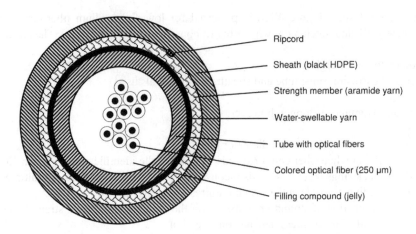

- Ripcord
- Sheath (black HDPE)
- Strength member (aramide yarn)
- Water-swellable yarn
- Tube with optical fibers
- Colored optical fiber (250 μm)
- Filling compound (jelly)

Figure 4.30   3.5 mm micro-cable with 12 fibers in central polymer tube

barrier due to cost and disruption. The acceptance of ribbon cables is selective. Many countries and in particular local installers reject them [4.46].

Duct cables can have a second strength member, a layer of aramide or glass yarns, placed between the core (the structure comprising all optical fibers) and the sheath. Besides an increase in tensile strength, this considerably improves cable resistance to impacts, cuts and rodents, with glass yarn being particularly effective in the latter role.

There is much interest in reducing cable diameter, or putting more fibers in a cable of the same size. Duct space is at a premium in most cities, and proliferation of fibers in access networks forces operators to lay fiber cables in new areas. 100 mm cable ducts built for copper cables are sub-divided with 40 mm or similar HDPE tubes serving as secondary ducts; 3 or 4 HDPE tubes are placed in a single 100 mm hole. Plastic tubes are also laid as new ducts. This idea is extended to using micro-ducts of 2.5–10 mm inner diameter and micro-cables with relatively few fibers [4.49]. The amount of fibers reaches from 2 up to 96 and even more, with diameters from 1.2–2 mm to approximately 9 mm, respectively. Methods used to reduce cable diameter include:

- bundled fibers in thin-walled "micro-tubes", with almost no space for movement
- thin sheath and drastic downsizing of strength members
- bending-tolerant fibers, also in 200 μm coating

An example of a small micro-cable is shown in Figure 4.30.

4.2.6   Aerial Cables

Compared to duct cables, self-supporting *aerial cables* must withstand higher, long-term tensile loads. The cross-section of strength members is increased, usually by adding an external member to the structure adopted from the duct cable. A thick FRP rod or steel wire in the center would make the cable rigid and so is normally avoided. The "Figure 8" cable shown in Figure 4.31 is suitable for relatively short, approximately 25–100 m spans.

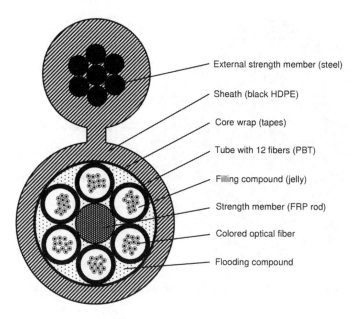

External strength member (steel)

Sheath (black HDPE)

Core wrap (tapes)

Tube with 12 fibers (PBT)

Filling compound (jelly)

Strength member (FRP rod)

Colored optical fiber

Flooding compound

Figure 4.31   Figure 8 aerial cable

The external strength member is easily separated from the rest of the cable at the splicing location. It is usually made of galvanized steel wires; FRP rod can break when the cable is bent in support clamps. The optical core may be of any type presented in the previous section, but designed for a wider range of length variations due to temperature and strain. "Figure 8" fiber cable, with a relatively light lower part (unlike similar copper cable), is not suitable for long spans and windy areas, due to wind-induced "dancing", causing loosening and damage of the cable and hardware. For such purposes, round aerial dielectric self-supporting (ADSS) cable, incorporating aramide yarn over a dielectric core (Figure 4.32) is better.

Aramide yarns protect cable core against impacts, including shot from shotguns. Unique properties of aramide fibers include negative thermal expansion coefficient (Table 4.8). With proper selection of relative cross-sections of parts, thermal expansion coefficient of the cable is matched to one of the optical fibers, minimizing effects of temperature. In ADSS for long spans, the aramide layer is separated from the core by an inner sheath to reduce crush forces acting on the core. Cables of this type can be strung between pylons of overhead power lines, with 400 m and longer spans.

Micro-cables are also used in aerial networks, having low tensile strength. They are attached to overhead metallic conductor or cable serving as support, by wrapping around, lashing, etc.

4.2.7   Optical Ground Wires

Most OPGWs include a combination of aluminium alloy (AlMgSi) wires, serving as current conductors and steel wires providing mechanical strength. To prevent corrosion, steel wires are clad with aluminium; protection by zinc coating is ineffective in industrial or coastal areas. Optical fibers are most often enclosed in hermetic metal tube(s), made of aluminium, stainless

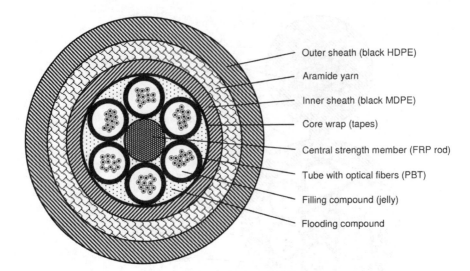

Outer sheath (black HDPE)

Aramide yarn

Inner sheath (black MDPE)

Core wrap (tapes)

Central strength member (FRP rod)

Tube with optical fibers (PBT)

Filling compound (jelly)

Flooding compound

Figure 4.32   ADSS cable with double sheath and stranded loose tubes

steel (Figure 4.33), or a combination of both (Figure 4.34). However, designs incorporating plastic loose tubes or tight-buffered units, like the one shown in Figure 4.30 and Figure 4.17, are still encountered, the service record of the latter being good.

OPGWs are manufactured in a wide range of sizes, depending on requirements:

- rated tensile strength (RTS, kN)
- fault current rating, aka heat integral (kA^2s)
- fiber count

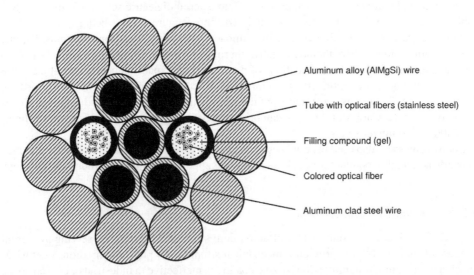

Aluminum alloy (AlMgSi) wire

Tube with optical fibers (stainless steel)

Filling compound (gel)

Colored optical fiber

Aluminum clad steel wire

Figure 4.33   OPGW with stranded steel tubes

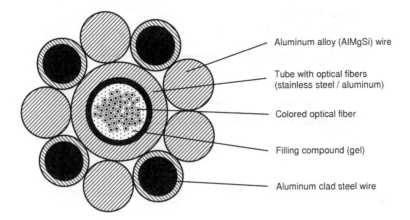

Aluminum alloy (AlMgSi) wire

Tube with optical fibers
(stainless steel / aluminum)

Colored optical fiber

Filling compound (gel)

Aluminum clad steel wire

Figure 4.34   OPGW with central aluminium/steel tube and single layer of armoring

A fiber strain-free window suitable for a wide range of tensile forces (including wind and ice loads) and operating temperatures, for example $-40\ldots +80°$C, is best provided by loose tube designs with central or stranded tubes. The first option is default for compact OPGWs with only one layer of wires (Figure 4.34).

Several alternative OPGW designs exist, for example with tight-buffered unit(s) similar to the one shown in Figure 4.20 or a plastic tube. They can be enclosed in a central tube made of aluminium, or with a central slotted core, also made of aluminium, holding multiple plastic tubes with optical fibers. Because of the need to use very compact tubes, ribbons are not used in OPGW.

4.2.8   Fiber Cabling Summary

After almost 40 years of development, optical fiber cable technology has reached a degree of maturity, with successful deployment in most sectors of telecom and data networks. The only telecom segment with limited fiber penetration so far is the access network, which is still preserved by copper cabling. This is a consequence of missing huge investments and unresolved regulatory issues, rather than technology.

In the data sector, twisted-pair copper cables hold as the dominant transmission medium for the "horizontal" segment of LANs; even 40 Gbit/s bit rates are to be supported by Category 8 cabling. Besides the issue of cost (of active equipment, not cabling), this is due to short distances, between 30 m and 100 m, where superior fiber performance is not essential.

4.3   Coupling Elements for Fiber-Optic Systems

As with electrical systems, there is also a desire for a modular construction of optical systems. This creates special demands for optical coupling elements. Compared to their electrical equivalents, these requirements are more complex in most cases. The following individual tasks need to be completed (Figure 4.35):

- coupling a light source to a waveguide

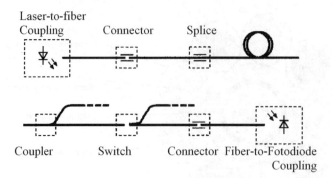

Figure 4.35   Coupling elements for optical transmission systems

- detachable connections between waveguides (optical connectors)
- permanent connections between waveguides (optical splices)
- optical junctions (coupler)
- optical switches
- fiber-to-photodiode coupling

4.3.1   Light Source-to-Fiber Coupling

The quality of such a coupling and the necessary efforts largely depend on the type of light source used, as well as the waveguide. The quality is first and foremost derived from the light source-to-fiber coupling efficiency which can be obtained. The coupling attenuation should therefore be as small as possible:

$$A_C = -10\text{dB} \lg \frac{P_F}{P_L} \tag{4.13}$$

where:

P_L light source output power
P_F power coupled to the fiber

If lasers are used as a light source, the feedback from the coupling system to the laser also plays a role [4.50–4.52]. Concerning coherent systems for optical transmission technology (Chapter 9.5), unwanted feedback must be as small as possible.

Concerning couplings, it is usually the case that the field distribution emitted by a light source does not coincide with the field distribution which can be accepted by the fiber. An interconnecting element, the coupling system, will consequently have to manage the necessary adaptation. With multi-mode fibers and their relatively large core diameters and large numeric apertures, coupling is easy to put into effect, and the attainable coupling efficiencies are relatively large. A good approach to the examination of these cases can also be achieved by applying geometrical or ray optics [4.53]. Geometrical optics is concerned with optical images. Beams coming from an object point reunite in its respective image point. In our case,

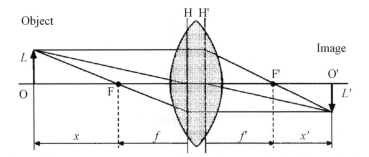

Figure 4.36 Image formation for $\beta' < 0$ of an object O sized L to an image O' sized L', object distance x (image distance width x' respectively), focal lengths $f' = -f$, object and image focal points F and F' respectively, and main planes H and H' respectively

the illuminating area of a light source is projected onto the core of the fiber. Figure 4.36 shows the according *Listing's construction*. This construction is a good help to determine the optical image formation graphically. Here, L is the size of the object, in our case the radius of the light source spot size (simplification, Chapter 6.1), and L' is the image size, so it should fit with the core radius of the fiber. The optical system comports the focal points F and F' respectively, and the planes H and H' respectively. If the system is considered to be a thin lens, the mains planes H and H' coincide. For the magnification ratio β' holds:

$$\beta' = \frac{L'}{L} \tag{4.14}$$

The object size L and the image size L' have opposite signs in Figure 4.36, because they are calculated from the optical axis (upwards positive, downwards negative). Such an image formation produces a real image; if β' is positive we talk about a virtual image. The light direction (generally from left to right) is defined as a positive direction.

Hence: $x < 0$ and $x' > 0$, x depicting the distance from focal point F to object point O, and x' describing the distance from focus point F to image O'. The focal width f is the distance between focal point and main plane. Given Figure 4.36, the Newtonian equations are as follows:

$$\beta' = \frac{-x'}{f} = \frac{f}{x} \tag{4.15}$$

and

$$xx' = -f^2 \tag{4.16}$$

If f' is positive, the lens is a collective lens (Figure 4.36); if the focal width is negative, it is a divergent lens. A diaphragm limits the maximum angle of incidence, the aperture. This can be, for example, the lens boundary of the optical system (Figure 4.37). The transformation of

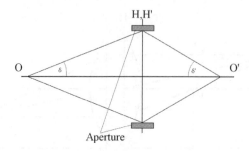

H,H'

O δ δ' O'

Aperture

Figure 4.37 Numerical aperture of an optical image formation system, δ and δ', are respectively the angles of the incoming or outgoing light beam with the optical axis.

angles in front of, as well as behind, the optical system is equally defined by the reproduction magnification ratio β'. The optical sine condition applies:

$$\beta' = \frac{\sin \delta'}{\sin \delta'} \tag{4.17}$$

From Eq. (4.14), it follows:

$$L \sin \delta = L' \sin \delta' \tag{4.18}$$

If the system is airborne, $\sin\delta$ respectively $\sin\delta'$ are exactly identical with the numerical aperture A_N, in front of $A\prime_N$ or behind the system:

$$LA_N = L'A'_N \tag{4.19}$$

The goal of the coupling tasks is to project the illuminating area $A = L^2\pi$ onto the area of the fiber core $A' = L'^2\pi$. Thus, the following linkage between areas and numerical apertures results from Eq. (4.15):

$$AA_N^2 = A'A_N'^2 \tag{4.20}$$

We often encounter the problem of diminution (except with laser diodes (Chapter 6.1.2)), as the illuminating area is usually larger than the area of the fiber core. Area and beam angle are given by the light source. As δ is always <90°, the numerical aperture of the fiber is $A'_N < 1$ and thus $A' \geq AA_N^2$.

If we desire a coupling without losses, one condition is that the fiber core area must be at least as large as the product of the area and the numerical aperture squared on the light source side. With multi-mode fibers, we try to deal with tolerable coupling losses, even without optical image formation because of the large fiber core surfaces and the large numerical apertures. If the light source is a Lambert radiator (Chapter 6.1.1, $A_N = 1$), and if a step index fiber is used, the coupling attenuation A_C according to Eq. (4.20) is:

$$A_C = -10 \text{ dB } \lg \left(A'/A \right) - 20\text{dB } \lg A'_N \qquad \text{for } A > A' \tag{4.21}$$

$$A_C = -20\text{dB } \lg A'_N \qquad \text{for } A < A' \tag{4.22}$$

The coupling considerations are considerably more complex with graded index fibers. In contrast to step index fibers, the numerical aperture depends in this case on the radial position of the light cone (Figure 4.42) according to the Eq. (4.7). Following Eqs (3.8) and (4.9):

$$A_N(0) = A_{Nmax} \quad \text{and} \quad A_N(a) = 0 \tag{4.23}$$

In this case, the following results for coupling attenuation A_C can be found [4.54]:

$$A_C = -10 \text{ dB } \lg(A'/A) - 20\text{dB} \lg A'_N + 3\text{dB} \qquad \text{for } A > A' \tag{4.24}$$

$$A_C = -10\text{dB} \lg(A'^2_N - A'^2_N A/2A') \qquad \text{for } A' < A \tag{4.25}$$

If the illuminating area of the light source is larger than the fiber core area, twice as much light can be coupled into a step index fiber as compared to a graded index fiber. This follows Eqs (4.21) and (4.24). The equations for coupling attenuation are only an estimate with regard to a possible maximum. Further losses due to a false adaptation of the field distribution via reflection losses and through faulty adjustment can also occur (see below).

For single-mode fibers, the description of coupling tasks with the help of geometrical optics fails. Instead, the Gaussian beam formation (Figure 4.38) of the light source must be transformed into the Gaussian beam formation of the fiber.

As light sources, semiconductor lasers are almost exclusively used for application with single-mode fibers (see below). Therefore, ideally we assume the emission characteristics of the semiconductor laser can be described theoretically by a rotation-symmetrical Gaussian beam, and the same holds for the acceptance characteristics of the single-mode fiber. In reality this is not true, thus additional losses occur. Moreover, the coupling efficiency also depends on the alignment behavior of the two experimental real beams.

Figure 4.38 shows the Gaussian beam formation, in this case the image formation of the beam waist (spot size) of a semiconductor laser into the beam waist of a single-mode fiber. The delimitation lines show the $1/e^2$ values of the beam width in propagation (Eq. (3.21)) in front of and behind the formation lens.

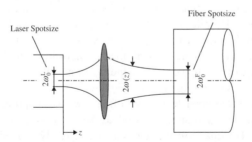

Figure 4.38 Gaussian Beam formation

where:

w^L_0 laser spot size
w^F_0 fiber spot size
$w_0(z)$ spot size at the local coordinate z

Figure 4.39   Plug connectors and LED sockets for LED-to-polymer fiber coupling as well as for receiver diodes

An LED surface emitter is a *Lambertian emitter* (Chapter 6.1.1). It comports a very large numerical aperture. The coupling attenuation of such an LED into a multi-mode fiber is relatively large in comparison to laser diodes, which have a considerably more favorable dispersion characteristic. However, the coupling into a single-mode fiber is even more problematic. There are only a few µW of optical power that can be coupled into the fiber – working with some mW of an LED. At this point it has to be mentioned that we are talking of a typical commercially available LED. There are also much better results achievable with more sophistication but significantly higher costs. Despite this fact, there are special cases of applications, in particular demanding LED light and not laser light, for example if an especially low coherence is required (Chapter 6.1.1). For polymer optical fibers comporting core diameters of up to 1mm and large numerical apertures, however, eligible coupling effects can be achieved without having to resort to complex coupling systems (Figure 4.39).

In lab experiments, microscope lenses are often used as coupling systems. Conversely, micro-lenses, for example ball lenses, as well as Selfoc lenses, are used in light source modules (Figure 4.40). The image-forming lens is often directly applied to the end of the fiber or even created by the end of the fiber. Figure 4.40 shows a variety of these coupling types [4.50,4.52, 4.56,4.57]. Provided semiconductor lasers are used as light sources, the coupling attenuations are about 1 dB with multi-mode fibers; with single-mode fibers, best values of about 2 dB [4.51] can be achieved. These values often are downgraded, mostly due to the real beam characteristics of the light sources used. Thus, often not only the numerical aperture is crucial but also the form of the field distribution, which does not correspond to the necessary rotation-symmetrical Gaussian beam formation (Figures 6.23 and 6.24 in Chapter 6.1.2).

Laser Diode   Launching FIber

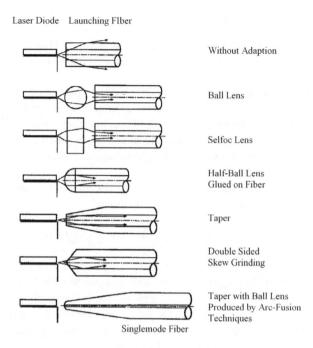

Without Adaption

Ball Lens

Selfoc Lens

Half-Ball Lens
Glued on Fiber

Taper

Double Sided
Skew Grinding

Taper with Ball Lens
Produced by Arc-Fusion
Techniques

Singlemode Fiber

Figure 4.40   Different coupling techniques for light source-to-fiber coupling

With the help of a double-sided skew grinding of the end of a multi-mode fiber, a higher symmetry can be achieved, compensating partly for the asymmetry of the beam characteristics of typical semiconductor lasers (Figures 6.23 and 6.24) [4.59]. A *taper* at the end of a fiber is realized by reduction of the fiber diameter and consequently of the fiber core. Afterwards the tapered end will be melted by an arc fusion technique and thus due to the surface tension a half-ball lens is created. In this way, the numerical aperture of the fiber can be enlarged and an adaption to the also very large numerical aperture of the light source can be achieved. This technique applied to single-mode fibers allows very small outer diameters, respectively radii, enabling a high refractive power leading to a significant coupling efficiency improvement.

When constructing a light source module, there is also the problem of adjustment and fixation. For single-mode fibers, adjustment errors to the extent of 1 μm in the propagation direction are already large enough to cause drastic coupling losses [4.58]. If these errors occur vertically to the propagation direction, this is already the case when we are dealing within a domain of sub-micrometers. Once the optimum adjustment has been found, the fiber needs to be attached in relation to the light source. Techniques are carried out with the help of glue, welding, brazing, etc. This process must not change its spatial position. It poses high challenges to the fixation technology. Moreover, the said fixation also ought to be stable over a long period of time, as well as regarding fluctuations in temperature. Hence, during the fixation, we must especially ensure that no mechanical tensions occur, which would later on change the coupling efficiency, in particular when the temperature alters. Symmetrical alignments are advantageous in such cases.

4.3.2   Fiber-to-Fiber Coupling

In practice, fibers must very often be connected to one another. This is done via detachable connections (optical connectors) and permanent connections (optical splices). For an extensive discussion of this area of interest, see [4.60].

Instead of the illuminating area of a light source we can also consider an area A to be the core area of an arriving step index fiber, and likewise the numerical aperture the arriving fiber. Then, Eq. (4.20) also applies for butt coupling of fibers amongst each other. Coupling without any losses between the arriving and the leaving fiber is only possible if the product of the areas and the numerical apertures of the leaving fiber squared are at least as large as that of the arriving ones.

If this is not the case, the smallest possible coupling attenuation with the data A, A_N for the arriving fiber and A', A'_N for the leaving one, can be found:

$$A_C = -10 \text{ dB } \lg(A'/A) - 20\text{dB} \lg \left(A'_N/A_N \right) \tag{4.26}$$

If we attach a projection lens between the fibers, the image size of the fiber core area can indeed be varied; however, the angle of the beam will then also be varied inversely proportional. Consequently, the product of the illuminating area and the corresponding numerical aperture squared remains consistent. This means that the lowest possible coupling attenuation following Eq. (4.26) cannot be further improved upon with the help of image formation.

Considering Eqs (4.27) and (4.28), Eq. (4.20) we have:

$$AA_N^2 = N\frac{\lambda^2}{2\pi} \quad \text{resp.} \quad A'A_N'^2 = N'\frac{\lambda^2}{2\pi} \tag{4.27}$$

Here, N and N' respectively name the number of modes capable to propagate in the arriving and the leaving step index fiber (Eq. (3.17)). The minimum coupling attenuation therefore can also be calculated by comparing the mode numbers:

$$A_C = -10 \text{ dB } \lg(N'/N) \quad \text{for} \quad N' < N \tag{4.28}$$

From Eq. (4.28) it follows that it is impossible to couple from a fiber guiding a certain number of modes into a fiber guiding a smaller number of modes without losses. There is no mode cone. Coupling without losses is possible in the reverse direction (here ignoring any other effects). Equation (4.28) shows the lowest possible coupling attenuation between fibers; it is also valid for graded index fibers. Two-step index fibers that are meant to be connected may have different areas or suffer from an axial misalignment at the fiber ends. In this case, for butt coupling, a misalignment of the areas determining coupling efficiencies occurs. A coupling attenuation A_{CA} can be found (Figure 4.42a):

$$A_{CA} = -10 \text{ dB } \lg \left(A''_N/A_N \right) \tag{4.29}$$

where:

   A_N fiber core area of the arriving fiber
A''_N common area of arriving and leaving fibers

If the two numerical apertures do not align, the solid angle overlap determines the reduced coupling efficiency. Coupling without any losses is only possible if the numerical aperture of the leaving fiber (A'_N) is larger than the one of the arriving one (A_N). If this is not the case, there is an additional coupling attenuation A_{CAN}:

$$A_{CAN} = -20 \text{ dB } \lg(A''/A) \tag{4.30}$$

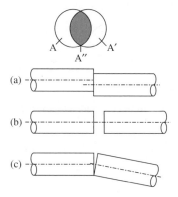

Figure 4.41   Misalignment of fiber-to-fiber coupling

where:

a) axial misalignment
b) distance between fiber ends
c) angular misalignment

The overall minimum coupling attenuation results from the sum of A_{CA} and A_{CAN}. Further losses additionally occur due to necessary distances between fiber ends and angular misalignment [4.60,4.61] (Figures 4.41b and c).

For butt coupling of graded index fibers, the coupling considerations are, similar to light source-to-fiber coupling, considerably more complex because of the radii-dependent numerical apertures (Eq. (4.23)).

A practical tool for the determination of the coupling efficiency between graded index fibers is the *phase space diagram* [4.62]. This refers to the graphical depiction of the numerical aperture squared as a function of the radial coordinates squared (Figure 4.42). For butt coupling, the coupling attenuation of graded index fibers can easily be determined with the help of such phase space diagrams in a graphic depiction – numerical aperture squared versus local coordinate r squared. The graphs are triangles, thus making area calculations very simple. Figure 4.43 shows the phase space diagrams of two graded index fibers with different core radii and different numerical apertures. For the minimum coupling attenuation from fiber 1 to fiber 2, it holds that [4.62]:

$$A_C (1 \rightarrow 2) = -10 \text{dB} \lg(A_3/A_1) \tag{4.31}$$

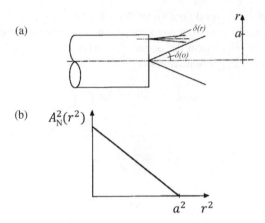

Figure 4.42   Numerical aperture of a graded index fiber versus radial coordinate r

where:

A_N Numerical aperture
α Fiber core radius

a)   divergence angle
b)   corresponding phase space diagram

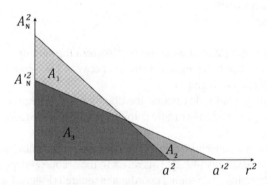

Figure 4.43   Phase space diagrams of two coupled graded index fibers for determination of coupling attenuation

where:

r Radial coordinates
$A_N \, A\prime_N$ Numerical apertures
$a, a\prime$ Fiber core radii of both fibers
A1, A2, A3   Areas in the phase space diagram

Inversely, it follows from fiber 2 to fiber 1:

$$A_C (1 \rightarrow 2) = -10\text{dB lg}(A_3/A_2) \tag{4.32}$$

$(A_1, A_2$ and A_3, Figure 4.43)

For single-mode fibers, the calculation needs to be done with the help of the Gaussian beam formation, as it was done with the laser-to-fiber coupling. The coupling efficiency again depends on how well the Gaussian fields of the two fibers align. It can be calculated using the corresponding alignment integral [4.55]. For the butt coupling of two single-mode fibers with the corresponding beam waists w_1, and w_2, it follows that:

$$A_C = -10\text{dB lg} \frac{2w_1 w_2}{w_1^2 + w_2^2} \tag{4.33}$$

With additional axial misalignment and occurring distances between both fiber ends, further losses emerge [4.63].

Apart from the losses described above, there are more losses due to Fresnel reflection (Chapter 2.5). This means a loss of at least 4% for the frequently occurring glass–air transition per interface (for vertical incidence of light) – consequently, at least 8% with the butt coupling of two fibers. Moreover, if further lens systems with multiple interfaces in between the fibers are applied, a considerable total loss can add up because of this effect. This can be partly avoided if anti-reflection coatings are attached to the interfaces [4.64]. The result is a destructive interference of the undesired reflection (Chapter 2.2). It is, however, not possible to entirely suppress the reflection; in practice, we can achieve remaining reflection coefficients of about 1% per interface with reasonable efforts.

Another option to lessen the problem is to apply a liquid with a rather well-adapted refraction index ($n \approx 1,457$) between two adjacent interfaces, that is, for example between two fiber end faces. This is easy to handle in a laboratory but hardly used in practical systems, as the fiber ends need to be cleaned after a certain time, or because the liquid is volatilized and needs to be renewed. If two waveguides, that are meant to be connected, consist of materials with different refraction indices, the optimum refraction index of the liquid is calculated as follows:

$$n_L = \sqrt{n_1 n_2} \tag{4.34}$$

where:

n_1 Refraction index of the arriving waveguide
n_2 Refraction index of the leaving waveguide

A third possibility is to realize an optical coupling distance $d < \lambda/2$ between both core surfaces. In this case, there is no reflection at the two interfaces, the light tunnels without loss to the second fiber via the narrow gap [4.65]. Yet, this technology poses high challenges to the evenness and the 90° angling of the fiber end surfaces.

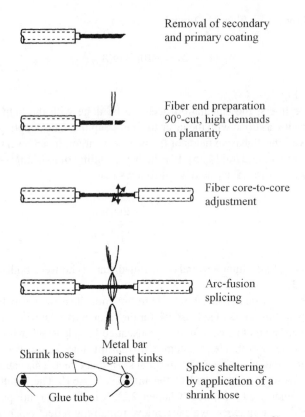

Removal of secondary
and primary coating

Fiber end preparation
90°-cut, high demands
on planarity

Fiber core-to-core
adjustment

Arc-fusion
splicing

Shrink hose

Metal bar
against kinks

Glue tube

Splice sheltering
by application of a
shrink hose

Figure 4.44   Arc-fusion splicing technology for glass fibers

4.3.3   Fiber-Optic Splices

The most frequent way of coupling two waveguides, mainly fibers, is a permanent optical
connection, the *splice* [4.60]. Here, we have the option of using gluing techniques. A transparent
glue is used, the refraction index of which is supposed to come as close as possible to the
refraction indices of the waveguides (Eq. 4.34)). This technology is applied in particular for
polymer fibers, but can also be applied to glass fibers. The most common method for fiber-optic
connections is *arc-fusion splicing*. For this purpose, fibers are first freed from their secondary
coating (only necessary with tight buffered fibers) with the help of a cable stripper. Then, the
protective layer directly on the glass fiber surface, the primary coating, is taken off. The latter
is achieved by pulling the layer off with the help of a nylon-fiber loop in the case of a soft
coating. With the nowadays more frequent hard coatings, solvents must be used [4.60]. In
order to achieve low attenuations at the coupling point, even 90° end surfaces are necessary.
For production of such fiber end surfaces, special cutting machines are used. A simple way
of preparing such a fiber end is to cut the fiber with a diamond and to break it under defined
tensile stress without any torsion (Figure 4.44).

As a result, fractures with angular errors of less than 1 degree and high surface qualities are
created [4.60]. Grinding and polishing procedures are considered alternatives to the described
method.

Figure 4.45 Arc-fusion splicing set-up of a glass fiber

For all the coupling techniques mentioned above, manipulation equipment with high requirements is needed in laboratories. If a propagation direction of about 1 μm is necessary, vertically to the fiber end, the requirements range is in the sub-micrometer area.

The fiber ends, which have been stripped of their primary coating, prepared and aligned, are welded together at about 2300°C without any other tools (Figure 4.44). Figure 4.45 shows a splicing set-up working with micrometer screws (left-hand side). The adjustment procedure and the fusion can be regarded with a microscope at the same time in two perpendicularly oriented directions (right-hand side of the figure). To protect the splices, afterwards they are covered with a splice sheltering along with the free fibers around them, for example by a shrink tubing (Figure 4.44). The shrink tubing is filled with glue to avoid direct fiber contact with air. Air always contains water and glass fibers are hygroscopic. This leads to the danger of fiber brakeage. Splices cause a heightened attenuation of a fiber link compared to a splice-free link, and they also reduce the tensile strength. With this connection technology, the reflection losses of fibers with equal refraction indices are omitted, because there is no more butt coupling after the splicing procedure. For a good splice, the splice is not recognizable anymore by use of a normal microscope. With the help of such splicing procedures in a laboratory, coupling attenuations of less than 0.01 dB can be achieved, even for single-mode fibers. Typical values gained, for example in a cable duct, are below 0.2 dB [4.58,4.60].

4.3.4   Fiber-Optic Connectors

Especially at the beginning and at the end of an optical transmission link, a detachable optical connection, an optical connector [4.60], is required.

We distinguish between connectors with direct end face couplings (Figure 4.46) and lens connectors, in which an optical image system is applied in between the fibers (Figures 4.47

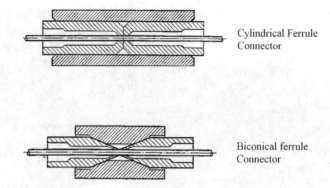

Figure 4.46   Examples for optical connector execution

and 4.48). The requirement for an optical connector first is a low coupling attenuation similar to the splice [4.60]. But the possibility to detach and connect as often as possible is demanded. Furthermore, the connectors are supposed to be interchangeable without any grave fluctuations in attenuation.

Even after a multitude of connection cycles, the attenuation values must be reproducible. In the case of large-scale temperature changes, the changes in attenuation ought to be limited to a tolerable value.

When constructing a connector, first the fibers (or the lenses) are aligned. The optimum adjustment is achieved by measurement of the transmitted power or centering the fibers according to the core positions. Sometimes, centering the fibers according to the outer diameters of the fibers is enough. Once the optimum position has been found, the fibers are fixed by a corresponding holding device; this is usually done by using an adequate glue. Sometimes, fiber ends are metalized to allow brazing techniques. For multi-mode, as well as for single-mode fibers, attenuation values below 0.2 dB [4.60] can be achieved.

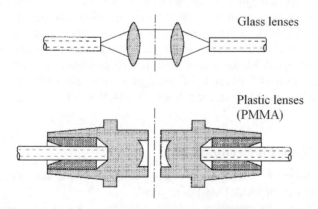

Figure 4.47   Optical connectors applying beam expansion techniques

Figure 4.48   Optical fiber-lens connector

4.3.5   Fiber-Optic Couplers

For bidirectional transmission systems (Chapter 9.4), in which only one fiber is supposed to be used, optical couplers are necessary [4.66,4.67]. There are *end face* and *surface couplers*. End face couplings are accomplished through wave front division or power division (Figure 4.49). Applying wave front splitting, the division is caused by the spatial alignment of the output fibers. With power division, a semipermeable mirror, a beam splitter, guarantees that

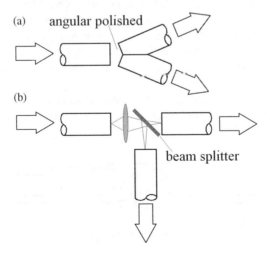

Figure 4.49   End face coupling through light wave division

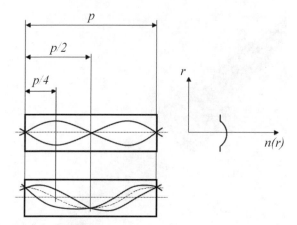

Figure 4.50   Selfoc lens, p pitch, and refraction index *n* versus radial coordinate *r*

only a part is transmitted through the mirror and leads to a subsequent fiber, whereas a second part is reflected in the mirror and then coupled into a second fiber. The distribution relation can also be varied.

where:

a) Division of power through front wave division
b) Division of power through amplitude division

The fibers need to be adjusted and fixed accordingly for this coupling type. If the coupling is to be efficient, an optical image formation system must be added. These image systems are often realized with simple *selfoc lenses* [4.68] (Figure 4.50). In this case, this is a piece of an optical stick with a parabolic refraction index comparable to the profile of a graded index fiber. Such an element operates like a typical optical lens made of one homogenous material, but with varying thickness (Figure 4.36). In contrast to a typical lens, with a selfoc lens, the geometrical thickness is constant, but the refraction index changes, however. In both cases, the product of the thickness and the refraction index is constant. Given an adequate length of the selfoc lens, an object is projected from a position A onto (Figure 4.50) a position B with an image magnification of $\beta' = -1$ (distance AB = $p/2$). The length p is called the pitch of the selfoc lens.

With the help of such selfoc lenses, 4-port arrangements can be realized (Figure 4.51) if a partly permeable mirror layer is applied in position C. This layer could also be wavelength

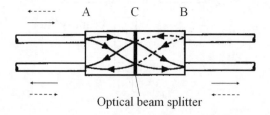

Figure 4.51   4-Port arrangement and power division via selfoc lens

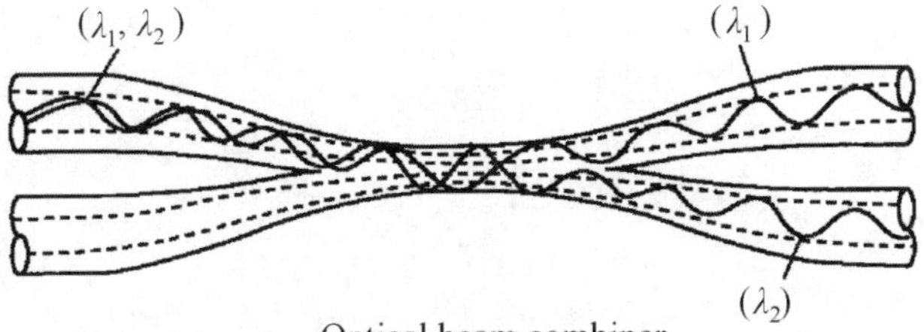

Figure 4.52   4-port arrangement by surface couplers. λ_1, λ_2: different wavelengths

selective, such that the light could be transmitted or reflected, depending on the wavelength. Thus, a *wavelength selective coupler* is achieved.

Fiber surface couplers can avoid the need of an extra mirror. In this case, the optical fiber cladding of two fibers is partly removed over a certain length. This is done with grinding and corroding techniques. The fiber surfaces are then directly adjusted and attached [4.69,4.70].

All of the coupling techniques mentioned above require complex adjustment and fixation techniques. In such cases, temperature and long-term stability pose a special problem. The creation of an *optical fusion coupler* promises better results [4.67,4.71,4.72] (Figures 4.52 and 4.54). Here, the two fibers that are to be coupled are first melted together and tapered afterwards. Due to the reduction of the optical fiber diameter and thus also the fiber core, the optical field reaches higher into the cladding area and is then able to couple into the core of the second fiber via its cladding.

When choosing the fibers, it is, however, vital to make sure that there is no other heightening of the refraction index, as for example, with a dispersion flattened [3.2], because this increase in the refraction index is considered by the optical field to be a third waveguide. Hence, there is coupling from the arriving fiber into its own cladding area limited by the next area with the higher refraction index, but further coupling into the second fiber is not possible. Thus, fibers with matched cladding are needed. After the melting procedure, a durable connection such as with a splice is created. Such couplers consequently display an excellent temperature and long-term behavior. With this technology, couplers for multi-mode fibers, as well as for single-mode fibers, can be created. Coupling losses below 0.5 dB are possible; the distribution on both fiber is variable within a large range, depending on how much the fiber is tapered or how closely the two cores are connected.

With single-mode fibers, there are further interesting opportunities: The coupling between the two fibers can be made different for different wavelength, thus again a *wavelength-selective optical fusion coupler* is attained:

Attaching and then melting the two fibers together of such surface couplers creates a certain length of touching fiber cores. If this length, the interaction length, is large enough, the light tunnels back and forth between those two fibers. This process is wavelength-dependent. For large tunneling numbers, there is -wavelength-selectivity [4.10]. An optical wave with the wavelength λ_1, leaves the coupler in the case of adequate dimensioning at the output port 3 (Figure 4.52), whereas another wave with the wavelength λ_2, is guided to the output port 4.

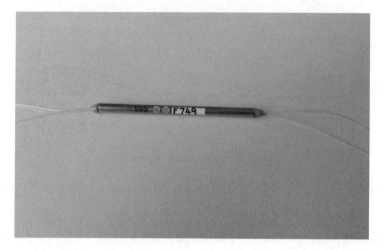

Figure 4.53   Wavelength-selective optical fusion coupler made of single-mode fibers

If both light waves come from the arriving fiber, in our case from the left-hand side, they are then spatially separated, leaving the coupler on the right-hand side. Conversely, such a coupler allows to couple light with wavelengths λ_1 and λ_2. In our case, if coming from the right-hand side into ports 3 and 4, they are unified together, leaving the coupler at port 1 on the left-hand side.

The coupler which propagates light in a forward direction, in our case from left to right, is also referred to as a *demultiplexer*, vice versa if light propagates in a backward direction it is called a *multiplexer*. Thus, such couplers are used in multiplex systems (Chapter 9.4). With selfoc lens image systems, a wavelength-selective mirror can also be realized by use of *dielectric mirrors*. Thus, multiplexers and demultiplexers are equally created. Such arrangements of multiplexers and demultiplexers are also referred to as bulk-optical (macro-optical) devices, as they are not constructed by fiber or integrated optics (Chapter 5.2). Moreover, by choosing the correct parameters, such couplers can also be *polarization-selective* ones [4.24]. If general, polarized light is propagating in the arriving fiber, there is different behavior for the parallel and perpendicular polarization component (Chapter 2.4). The parallel polarized part is then fed to output port 3 for example, whereas the perpendicular polarized part is then fed to output port 4. With the help of a drilled fusion coupling, also *polarization-independent* couplers can be realized [4.75]. Polarization-independent polarization selective couplers are essential for coherent use in transmission and sensor technologies.

In communications engineering, not only two-point connections but also networks are necessary (Chapter 9.4). For this usage, also couplers are needed which distribute the light from an arriving fiber onto a multitude of output fibers. These couplers are called *star couplers*. For this purpose, a large number of fibers are melted together, thus a transmission star coupler is attained. We can also make a cut in the melted area and apply a mirror layer in order to receive a reflection star coupler (Figure 4.54). By positioning several 2×2 couplers next to each other, a cascaded arrangement can also perform a split into a multitude of fibers. This particular technique is used in particular with single-mode fibers [4.76].

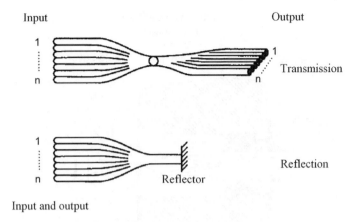

Input and output

Figure 4.54   Star coupler

where:

Transmission mode, upper part: inputs left outputs right

Reflection mode, lower part: inputs and outputs left

4.3.6   Fiber-Optic Switches

Optical fibers also allow the creation of optical switches (Figure 4.55) [4.77]. The movement of an arriving multi-mode fiber is electromechanically carried out via modification of an electrical relay [4.77]. The switching times, however, are a few milliseconds due to the mechanical alignment. Techniques of this kind are hence not suitable for short duty cycles. They can only be used for purposes of monitoring. If fast optical switches are necessary, integrated optical techniques are called for.

4.3.7   Fiber-to-Detector Coupling

The final optical element in an optical link is a photodiode. Coupling of light emerging from a fiber to the diode is mostly easy. Usually the diode area is larger than the fiber core area, except for high-speed photodiodes. For high bit rate transmission systems, we need a low diode junction capacity which determines the cut-off frequency together with the load resistor

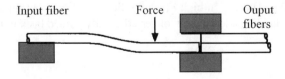

Figure 4.55   Fiber-optic alignment in a fiber-optic switch

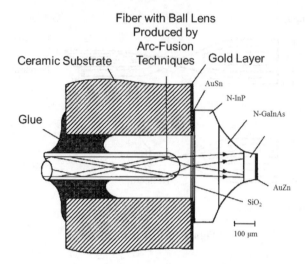

Figure 4.56   Fiber-to-detector coupling applying a mesa PIN photodiode with an active reception zone area of 100 μm

of the following front end amplifier (Chapter 7.2.1). Therefore, the diode area should be very small. This then raises higher demands on the coupling procedure.

For middle-speed systems having data rates with some hundred Mbit/s, multimode fibers n can still be used. In this case, for certain constellations, problems could occur. Figure 4.56 shows a fiber-to-photodiode coupling applying a mesa photodiode.

The reception zone is at the rear end of the diode, thus there is a large obligatory gap between the fiber end and the diode reception zone. Due to the fiber numerical aperture resulting typically in divergent angles of 30 degrees or more, the light spot size would be much larger than the diode reception zone area, leading to high coupling losses. A simple solution to this problem is melting the multi-mode fiber end by an arc fusion technique. Due to the surface tension, a half-ball lens is created reducing the beam divergence significantly. Thus the coupling efficiency is also improved considerably.

For very high-speed photodiodes used in the 10 G region, very small diode areas are necessary: 50 μm to 30 μm or even smaller. In this case, as a coupling fiber, exclusively single-mode fibers can be applied which solves the problem due to their ability of data transmission over 100 G. Often it is possible just to cut and brake the fiber (Chapter 4.3.3). To avoid Fresnel-induced back reflection to the laser, afterwards the fiber will be glued directly onto the photo diode by applying transparent glue. Further techniques to avoid back reflections towards the laser can be attained by producing skew fiber ends. This can be done as in the fiber cutting and braking process for fiber splicing. In contrast, in this case we apply a high torsion before braking. Thus, even high Fresnel reflections occur, but the reflected light will be coupled into the cladding and not any longer efficiently guided back to the laser [4.78]

5

Integrated-Optic Components

Integrated-optic devices [5.1,5.2] are often subdivided into active and passive components. Active devices are concerned with electro-optic transformers, such as optical light sources and drains or semiconductor-based optical amplifiers. These active devices are described in Chapter 6. This chapter is primarily limited to the passive components. Here, the passivity refers to the creation and recombination of charge carriers in connection with photon generation and absorption, respectively. However, they can be active with regard to influencing the optical properties of a light wave.

Amplitude, frequency, phase and polarization of a light wave can be influenced. The aim is also to achieve integrated optics for large-scale integration. This is achieved as in the electric analog, to arrange a very high number of optical components with different functions in a very small area. However, such integration is not yet very advanced. Many components already come as single parts, but combinations are hard to find. In particular, this applies to a combination of active and passive components. Concerning active components, quaternary compounds from III-V-elements are used as substrate materials: Gallium, Aluminium, Arsenic, Indium, Phosphor (Chapter 6.1). For passive components [5.3,5.4], glass, silica, lithium tantalate and lithium niobate ($LiNbO_3$) are used. The latter is the most common substrate material for passive optical components. Moreover, regarding today's state-of-the-art technology, it is the best controllable material. Lithium niobate as a substrate is suitable for a multiple of operations. This is due to the fact that $LiNbO_3$ features a variety of different physical effects, such as electro-optic effects (e.g. the Kerr and Pockeks effect), magneto-optic effects (e.g. the Faraday effect) and acousto-optic effects. Lithium niobate is strongly birefringent. Among the electro-optic effects, the Pockeks effect is the dominant one. Moreover, many effects are temperature-dependent, which causes serious problems for certain applications.

Optical and Microwave Technologies for Telecommunication Networks, First Edition. Otto Strobel.
© 2016 John Wiley & Sons, Ltd. Published 2016 by John Wiley & Sons, Ltd.

5.1   Integrated-Optic Waveguides

An integrated-optic waveguide is the simplest device [5.5–5.8]. Figure 5.1 depicts a simplified presentation of the production process. Here, a titanium (Ti) doped coating is added to a lithium niobate substrate by means of evaporating or sputtering techniques.

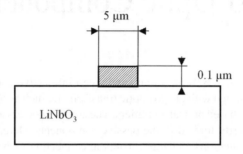

Figure 5.1   Production process of an integrated-optic waveguide

where:

Ti   Titanium
$LiNbO_3$ Lithium niobate

The desired structure of the waveguide is gained with the aid of photo-lithographic techniques – in this case a straight photo strip in the middle of the substrate surface (Figure 5.2). Afterwards, the device is placed into a diffusion furnace at about 1000°C for several hours. The strip consisting of titanium diffuses into the lithium niobate-substrate. This causes an increase in the refraction index of this strip compared to the environment. This implies the development of an optical waveguide similar to a glass fiber (Chapter 3). Choosing the diffusion depth and the strip width that is small enough, taking into account the wavelength of the light, the structural parameter becomes $V \leq 2.405$, so that the waveguide is single-moded. The waveguide actually appears two-moded under polarization.

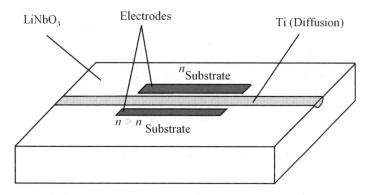

Figure 5.2   Integrated-optic phase modulator

where:

n Refraction index

5.2   Integrated-Optic Modulators

Attaching electrodes to both sides of the waveguide by means of applying a voltage application (Figure 5.2) generates an electrical field. That causes a change of the refraction index (Δn) in the waveguide and thus a phase change $\Delta\Phi$ of the guided light due to the electro-optic effect:

$$\Delta\Phi = 2\pi L \Delta n / \lambda \tag{5.1}$$

Thus, the phase of the light can be modulated by applying a voltage. An *integrated-optic phase modulator* is achieved [5.9]. By means of vapor deposition, gold or aluminium electrodes are attached to the substrate. This procedure is carried out after diffusion. The structuring again takes place with the aid of photo-lithographic processes. The electrical contact of the electrodes occurs by placing gold wires using bonding techniques. By dividing an arriving waveguide into two arms and reunifying them again, interference of guided light waves in both arms occurs (Chapter 2.2). An integrated-optic Mach-Zehnder interferometer is obtained (Figure 5.3).

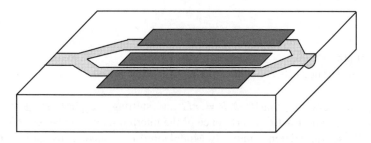

Figure 5.3   Integrated-optic Mach-Zender interferometer

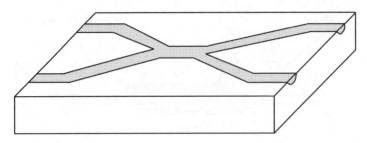

Figure 5.4   Integrated-optic 2 × 2 Coupler

Therefore, two optical beam splitters need to be manufactured as Y-couplers from waveg-uides. Bent waveguides are developed featuring a greater attenuation as compared to straight ones. Therefore, the angles of inclination of bent waveguides should be below one degree. Thus, the increase of attenuation is tolerable. By means of this technique, 2 × 2- or 3 × 3-couplers can be achieved (Figure 5.4).

By placing electrodes on both sides of the waveguide in the central part, a tunable integrated-optic Mach-Zehnder interferometer is gained. By means of an applied voltage, the phase dif-ference between the two guided waves in both arms can be varied [5.10]. Thus, corresponding to Eq. (2.17), a cosine-characteristic of the optical measure by a photodetector at the output gate is achieved (Figure 5.5).

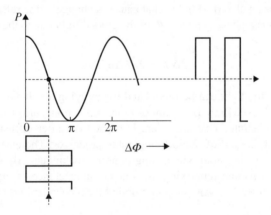

Figure 5.5   PM/AM (Phase Modulation – Intensity Modulation) –conversion

where:

P Optical power
$\Delta\Phi$ Phased difference between both superimposed light waves

Choosing an operation point at $\Delta\Phi = \pi/2$ and shifting the phase by $\pm\pi/2$, a change from constructive to destructive interference of the interferometer can be realized. A PM/IM conversion (Phase Modulation – Intensity Modulation) is yielded. Thus, an integrated-optic intensity modulator, moreover, an *integrated-optic switch*, is received. Adjusting the central

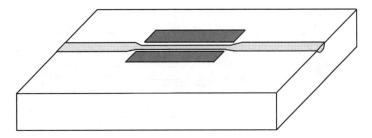

Figure 5.6   Integrated-optic cut-off modulator

electrode on the electric potential of the ground and both outside electrodes on potentials with opposite sign, we achieve a so-called push-pull operation. Thereby, the necessary voltage is halved. This is particularly favorable concerning high-frequency modulation, considering the difficulties in the construction of HF-drivers with high electric power. Due to the nearly vanishing rise- and fall-time of the electro-optic effect, very fast optical switches can be built by means of such components. Modulators with several GHz are already commercially available. In laboratory experiments, frequencies of about 100 GHz have already been measured [5.11].

Again, based on the phase modulator, an optical switch can also be built in other ways. In the central part, the waveguide width is strongly reduced (Figure 5.6), so that the fundamental mode is still guided but at the limit ($V \approx 1.5$, Chapter 3). By placing electrodes in this range, the refraction index and the structural parameter can be varied by the voltage applied. Thus, the fundamental mode is no longer guided.

The optical wave propagates in the substrate. Such an intensity modulator is named the *cut-off modulator*.

An *integrated-optic directional coupler* (Figure 5.7) can be considered as a further opportunity to build an integrated-optic switch; in this case it works as an optical toggle switch.

Similar to a fiber coupler, a multiple seesaw of the light between the closely neighboring waveguides can be observed. Whether light exits at gate 3 or at gate 4 depends on the waveguide length in the over-coupling range. The link can again be influenced by a change of the refraction index with the aid of an electric field:

$$\Delta(nL) = \Delta n \cdot L \tag{5.2}$$

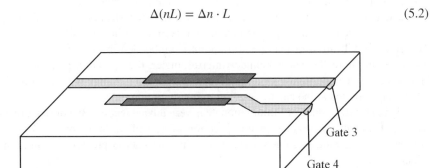

Figure 5.7   Integrated-optic directional coupler with modulation opportunity as an optical switch

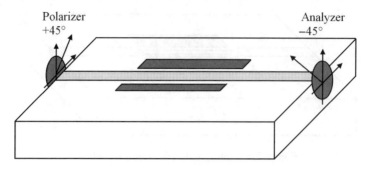

Figure 5.8   Integrated-optic polarization modulator as an optical switch

By exploiting the polarization properties, lithium niobate provides a further opportunity to obtain an optical switch. Lithium niobate features a strong birefringence, which also depends on an applied electric field. Regarding a suitable cut (Y-Cut, Z-Cut [5.12]), the refraction index parallel to the plane of a waveguide displays a different field dependence than if it was perpendicular to this plane [5.13]. Both distinguished refraction indices comply with the position of the principal axes of polarization.

A development of polarization is gained according to Figure 3.14, by coupling light into the lithium niobate waveguide under 45 degrees with respect to both principal axes. First, no electric field is applied (Figure 5.8). Concerning an appropriate optical path length, linearly polarized light with the same direction is again achieved at the end of the waveguide. By attaching polarizers at the input and output of the waveguide in this direction, the incident optical wave described above is transported unhampered. The refractive index difference can be varied by voltage application, with the aid of electrodes attached alongside the birefringent waveguide. Thus, the path difference of both components parallel to the principal axes can be controlled. The incident optical wave can be decomposed into both of these components. In the case of transmission, the phase difference must be an integer multiple of 2π corresponding to the beat wavelength Λ (Chapter 3).

If the phase difference is $\pi/2$, or an uneven-numbered multiple of $\pi/2$, linearly polarized light at the end of the waveguide with an orientation change of 90 degrees to the transmission case is achieved. Thus, relating to the position of the polarizers described above, a suppression of light transmission happens. A phase modulation-/intensity modulation conversion is obtained, that is, an intensity modulator being based on polarization effects. Thereby, the temperature dependency is also a problem, as in an integrated-optic Mach-Zehnder interferometer. Therefore, stabilization of the operating point is embedded in the systems [5.10]. Concerning the example of the Mach-Zehnder interferometer, the operation point A (Figure 5.5) is held steady by compensation of an undesired temperature change with the aid of additional DC-voltage applied at the electrodes.

The high birefringence of lithium niobate is very advantageous. By coupling linearly polarized light into the waveguide in the direction of one of the principal axes, a polarization preserving waveguide is gained, as achieved in polarization preserving single-mode fibers (Chapter 3).

In the case of an acoustic wave crossing an optical wave, a spatial deflection of the propagation direction and a frequency change of the light take place (Figure 5.9). This acousto-optical

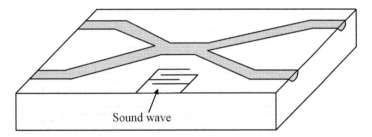

Figure 5.9   Integrated-optic Bragg cell as a frequency modulator and as an optical switch

effect is called the *Bragg effect* [5.14]. The increase of the frequency is equal to the frequency of the acoustic wave. The angle of inclination α is proportional to the power of the acoustic wave (Figure 5.10). An *integrated-optic frequency modulator* is achieved. Crossing two waveguides in a lithium niobate substrate under a suitable angle and attaching a generator of an acoustic wave transverse to the direction of light, the distribution of an optical wave arriving from gate 1 to gates 3 and 4 can be influenced [5.15] (Figure 5.9).

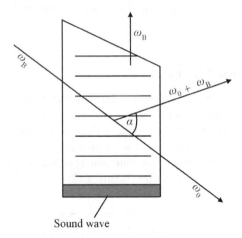

Figure 5.10   Interaction between light and sound wave

where:

ω_0 Angular frequency of the light wave
ω_B Angular frequency of the sound wave

5.3   Integrated-Optic Polarizers

It is also possible to produce integrated-optic polarizers. Therefore, a deposition on the waveguide of a buffer layer consisting of calcium fluoride (CaF_2) or silicon nitrite (Si_3N_4) will be carried tangentally to the core area – followed by a further layer of aluminium (Figure 5.11).

An oscillating wave vector (TE-wave) parallel to the plane of the substrate does not penetrate into the buffer or aluminium layer. Therefore, the TE-wave remains nearly unattenuated. In contrast, the wave vector (TM-wave) oscillates perpendicular to the substrate and is strongly

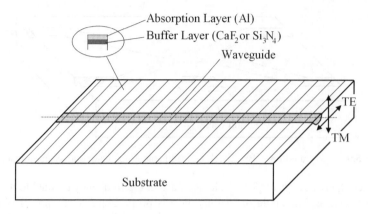

Figure 5.11   Integrated-optic polarizer

attenuated by the aluminium layer. Values of attenuation of more than 100 dB/cm can be realized [5.16,5.17]. By means of this technique, fiber polarizers can also be realized. Therefore, such attenuating layers are deposited on a glass fiber grinded until the fiber core is reached (Chapter 4.3.5).

5.4   Integrated-Optic Filters

By means of deposition of a partial reflecting silver layer on both abutting sides of a lithium niobate waveguide, an integrated-optic *Fabry-Perot Interferometer* is obtained. The optical waves are multiply reflected between the end faces. Constructive and destructive superposition appears, depending on phase differences of the waves propagating through this multi-beam interferometer. The following transmission function $\tau(\lambda)$ is a function of the wavelength if refraction index n and length L are constant (Figure 5.12):

$$\tau(\lambda) = \frac{\tau_{max}}{1 + \left(\frac{2F}{\pi}\right)^2 \sin^2\left(\frac{2\pi}{\lambda}nL\right)} \tag{5.3}$$

where:

τ_{max} Maximum transmission
F Finesse (see below)
L Length of the resonator

where:

FSR   Free spectral range
λ Wavelength
$\Delta\lambda^F$ FWHM width of the transmission area
$\bar{\lambda}^F$ Center wavelength

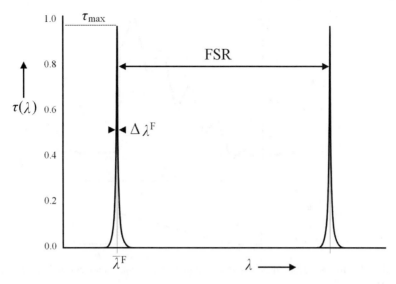

Figure 5.12   Transmission function $\tau(\lambda)$ vs. wavelength λ of a Fabry-Perot interferometer with high finesse, F

An *integrated-optic filter* is achieved. Light is transmitted for definite wavelengths only. The distance of the wavelengths $\Delta\lambda$ and the frequencies Δf is given by the optical length of the Fabry-Perot interferometer:

$$\Delta\lambda = \frac{\lambda^2}{2nL} \quad \text{and accordingly} \quad \Delta f = \frac{c}{2nL} \tag{5.4}$$

$\Delta\lambda$ and Δf are called the *free spectral region* (FSR). The quality factor of the Fabry-Perot interferometer is described by the *finesse, F*, where $F = \Delta\lambda / \Delta\lambda^F$.
where:

$\Delta\lambda^F$ Full width at half maximum of the filter peak (Figure 5.12)

The quality factor depends on the reflectivity at the end faces of the waveguide, on the surfaces, evenness and the attenuation in the waveguide. Due to the different refraction indices from air and lithium niobate ($n_{LiNbO3} \approx 2.2$), a Fresnel-reflection is gained of about 14% and thus a Fabry-Perot interferometer also. It has to be pointed out that the Fabry-Perot effect is achieved without any extra layer deposition on the end faces.

Using a phase modulator instead of a simple waveguide, the value of the wavelength at maximum transmission of the filter function $\bar{\lambda}^F$ in Figure 5.12 can be varied by voltage application. A controllable integrated-optic Fabry-Perot interferometer is achieved [5.18]. A simple unmodified lithium niobate-phase modulator is applied as a Fabry-Perot interferometer in the following experiment [5.34,1.13]. By means of a slow linear electric current rise, the wavelength of a thermally stabilized semiconductor is varied (Chapter 6.1.2) and the signal of the detector recorded (Figure 5.13, upper curve).

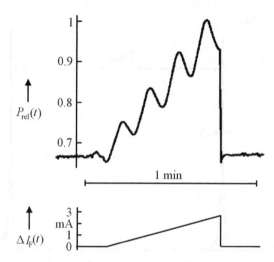

Figure 5.13   Relative transmitted power $P_{rel}(t)$, (upper curve) and the according current rise $\Delta I_F(t)$ (lower curve)

where:

$\Delta I_F(t)$ Current change
t Time

In the lower part of the diagram, the resulting current change ΔI_F is shown. Several Fabry-Perot modes can be distinguished, superimposed by a continuous rise of the optical power due to the current increase. The modulation depth due to the Fabry-Perot effect is about 10%. Thus, the controllable integrated-optic Fabry-Perot interferometer can be used for purposes of filtering, measuring problems and tasks concerning stabilization. It has to be mentioned that the problem of temperature changes has to be taken into account. A remedy has to be found by use of an operating point control.

5.5   Losses in Integrated-Optic Devices

Today's achievable attenuation in integrated-optic lithium niobate devices is negligibly low concerning straight waveguides. Values have been measured of 0.03 dB/cm [5.19]. The actual values of attenuation arise from required bending of the waveguide and the coupling of the lithium niobate chip to fibers. The attenuation caused by bending is about 1 dB by large radii. The losses per coupling point are also about 1 dB, provided that an adaption of the refraction index is carried out [5.9]. This is gained by gluing techniques achieving a long-term thermal stability. Due to the high reflectivity of about 14%, refraction index matching is required in order to avoid optical feedbacks. Moreover, the ends of the chips can be skew grinded for this purpose.

6

Optical Light Sources and Drains

The optical *transmitter* and *receiver diodes*, also called *electro-optic* and *opto-electric trans-ducers*, convert an electrical current into optical power and vice versa. They are also known as *light sources* and *drains*.

Regarding optical communication engineering and fiber sensor techniques, semiconductor devices are predominantly used as light sources and light drains. Due to their size, semiconductor devices allow a good geometric adjustment to the fiber. Moreover, a high efficiency of conversion is gained. After overcoming primary difficulties, these components now feature good operational reliability. Sometimes gas and solid state lasers are applied as light sources for measurement problems in the laboratory. In this case their size and weight can be tolerated.

Relating to semiconductor light sources and drains, generation and absorption of light can be demonstrated by means of the *energy-band model* in semiconductors [6.1,6.2]. The description of energy states of electrons in semiconductors is carried out by conduction and valence bands, separated from each other by a band gap, where no energy states exist. The conduction band lies above this band gap, with the highest valence band beneath it. The absolute energy maximum of this valence band is called the valence band edge and the absolute energy minimum of the conduction band is called the conduction band edge.

Regarding n-doped semiconductors, there is an excess of electrons in the conduction band, whereas in the p-doped conductors an excess of holes in the valence band is registered (Figure 6.5a). The electrons comport a higher energy level, such that they are excited and leave holes in the valence band. Figure 6.1 depicts the energy-band scheme with the narrowest neighboring conduction and valence band. If an electron is falling from the conduction band with an energy level E_2 into the valence band with a lower energy level E_1, the energy difference ΔE will be released. Vice versa, to lift an electron from the valence band into the conduction band, the energy difference ΔE has to be supplied. Referring to Einstein's formula, the energy difference ΔE is defined as:

$$\Delta E = hf = hc/\lambda \tag{6.1}$$

where:

h Planck's constant ($h = 6.626 \cdot 10^{-34}$ Js)

Optical and Microwave Technologies for Telecommunication Networks, First Edition. Otto Strobel.
© 2016 John Wiley & Sons, Ltd. Published 2016 by John Wiley & Sons, Ltd.

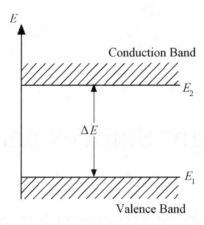

Figure 6.1   Energy band scheme of a semiconductor

A photon can be emitted during the transition of an electron from E_2 to E_1. Conversely, the absorption of a photon can induce the transition of an electron from E_1 to E_2. The most important electron transitions for light generation and absorption are presented in Figure 6.2. Figure 6.2a shows the process taking place in a photodiode, the *absorption* of a photon. The electron in this case is excited from the energy level E_1 to the higher energy level E_2. Conversely, if an electron spontaneously falls from the energy level E_2 to the energy level E_1, it may lead to *spontaneous emission* of a photon. The spontaneous transition happens without any influence of a further cause. This process is found in an LED, but partially occurs in lasers too. An already existing photon interacting with a semiconductor can cause the transition of an electron from E_2 to E_1. This may lead once again to the emission of a photon (Figure 6.2c). Consequently, this process has been activated by the already existing photon. Therefore, it is called *stimulated* or *induced emission*.

The development of a stimulated emission absolutely presupposes the existence of a *population inversion*. In both phenomena, the stimulated emission, as well as the absorption of

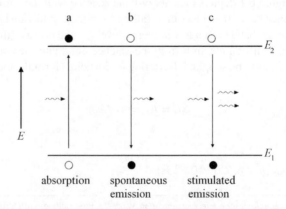

Figure 6.2   Photon emission and absorption at electron transitions between valence and conduction band

photons in the semiconductor material, possesses a certain probability. The absorption probability of a photon and thus the transition probability of an electron from E_1 to E_2 is proportional to the number of the existing electrons, the population number N_1 and the lower energy level E_1. Vice versa, the emission probability and thus the transition probability of an electron from E_2 to E_1 is proportional to the number of the existing electrons, the population number N_2 and the higher energy level E_2. In nature, at thermodynamic equilibrium, it always holds that $N_1 > N_2$.

The rate of absorption always prevails over the stimulated emission rate. Just as in the case of population inversion $N_2 > N_1$, a measurable stimulated emission can be achieved. Such a state does not occur in "natural" light sources (sun, candle, light bulb and LED), therefore it has to be produced artificially. The population inversion is achieved by supplying energy, a process called *pumping*. In general, this can happen by supplying heat (thermal pumping), light (optical pumping) or by injection of a current (electrical pumping). The latter is used in semiconductor light sources.

The stimulated photon possesses the same wavelength and phase as the stimulating photon. This process appearing in a laser is also termed the *coherent light generation*. A light source with a narrow spectral width is obtained (Chapter 6.1.2).

Semiconductor light sources and drains are basically designed as PN diodes. A barrier layer of P- and N-doped semiconductor material is achieved. Light source devices are activated in the forward direction. This leads to the *recombination of charge carriers* and the generation of photons. Accordingly, light drain devices are activated in the reverse direction. This leads to the *absorption of photons* and thus to the *separation of charge carriers*. The latter are accelerated in the occurring electrical field, which implies a drift current. The particle character of light arises from the point of view as described above. The photon is also called a light quantum.

Concerning Eq. (6.1), a certain wavelength λ is achieved from the energy difference. Thus, it follows a limited condition regarding light sources and light drains, respectively. Light sources can be activated above this wavelength, while light drains are activated below it. When considering the particle character, it is possible to assign the photon a momentum p. Concerning the energy difference, Einstein's relation from his special relativity theory can be used:

$$\Delta E = mc^2 \tag{6.2}$$

where:

m Photon mass

For the absolute value of a particle momentum, it follows that:

$$p = mc = \Delta E/c \tag{6.3}$$

and with Eq. (6.1), $k = 2\pi/\lambda$ and $\hbar = h/(2\pi)$, it can be written as:

$$p = hf/c = h/\lambda = \hbar k \tag{6.4}$$

According to *de Broglie*, physical matter is assigned to wave characteristics too. Thus, in a solid state, an electron also possesses a wave number k corresponding to Eq. (6.4). This simple idea, born in 1924, was a scientific revolution. It says light can be considered as a particle and as a wave! This is often named light dualism. Equation (6.4) can be written from left to right as well as from right to left. From left to right we give a wave a particle character to understand the absorption process. From right to left an electron beam can be seen as wave-like under the electron microscope.

In the valence band, as well as in the conduction band, the electron energy depends on the momentum and therefore on the wave number k. For transition of electrons between the valence band and the conduction band energy as well as momentum, conservation must be retained. For energy conservation it follows that:

$$E_2 = E_1 - E_p \tag{6.5}$$

where:

E_p Photon energy

For momentum conservation and thus for the conservation of the wave number and the wave vector it follows that:

$$\vec{k}_2 = \vec{k}_1 + \vec{k}_p \tag{6.6}$$

where:

$\vec{k}_p$ Wave vector of the photon

The momentum of an electron and its wave number is much greater than that of a photon, thus $\vec{k}_p$ is negligible in Eq. (6.6), assuming no further particles participate in transition, expect photons and electrons. From Eq. (6.6), it follows that such transitions are exclusively possible if the wave number (absolute value of the wave vector (Chapter 2.1)) does not change: $k_2 \approx k_1$. In a diagram of the electron energy E versus wave number k, the transition point in the valence band has to face exactly the transition point in the conduction band (Figure 6.3). Hence, such a transition is called a *direct transition*. A corresponding semiconductor is described as a *direct*

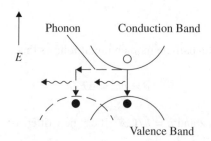

Figure 6.3   Direct and indirect electron transitions from the conduction to the valence band

semiconductor. In this case, the absolute maximum of the energy in the valence band has to match exactly the absolute minimum of the energy in the conduction band, that is at the same wave number (Figure 6.3 continuous line). If both extrema do not match each other, it becomes known as an *indirect semiconductor* (Figure 6.3 dashed line). The generation of a photon in the direct way in indirect semiconductors is not possible if energy and momentum need to be obtained. This is only possible through involvement of a third particle, a *phonon*.

A phonon represents lattice vibrations and has to possess a corresponding wave number to conserve momentum. This three-particle process rarely appears in comparison with the two particle processes at the direct transition. A photon generating transition in indirect semiconductors is much more improbable than in direct ones. Therefore, in direct semiconductors *radiant recombinations* are primarily achieved, whereas in indirect semiconductors *non-radiant recombinations* occur.

Under certain circumstances, there are *Auger processes* [6.2,6.3]. A recombined electron does not cause the generation of a photon. Rather it transmits its energy to a further free electron in the conduction band. The latter releases this energy difference by executing impacts with the crystal lattice and heating it up. Energy can be transmitted to one of several phonons. The energy is essentially transformed into thermal energy and is lost from the radiant component. Therefore, direct semiconductor materials are only deployed in semiconductor-based light sources. For light drains, direct as well as indirect semiconductors can be used. The most important direct semiconductor materials for optical transmitters and receivers are Gallium-Arsenide (GaAs), Gallium-Aluminium-Arsenide (GaAlAs) and Indium-Gallium-Arsenide-Phosphide (InGaAsP). These are chemical compounds from the element group III and V. The most important indirect semiconductor materials for receivers are Silicon (Si) and Germanium (Ge). Silicon features a band gap of about 1.1 eV, which corresponds to a wavelength of about 1.1 μm. Concerning Germanium, the values lie at about 0.7 eV and accordingly at about 1.6 μm, respectively.

Thus, *Silicon* is suitable as a semiconductor material for photodiodes *below* a wavelength of 1.1 μm, and Germanium *below* 1.7 μm. Silicon and Germanium diodes are built up of one uniform semiconductor material. In contrast, ternary and quaternary diodes consist of three and accordingly four chemical elements of different materials. Several layers are often deposited on a substrate. The substrate material GaAs is used for GaAlAs and InP for InGaAsP respectively.

For the production process it is very important that the lattice constants of the substrate material and those of ternary, respectively, quaternary mixed crystals match. Relating to GaAs and GaAlAs, this is fulfilled in a good approximation [6.4,6.5]. Vegard's law describes the linear dependency between different lattice constants of mixed crystals. *GaAlAs/GaAs semiconductor light sources* can be realized by substitution of Ga with Al. Their energy gap can be varied in such a way that wavelengths of about 0.65 μm up to 0.87 μm can be achieved. This is possible by replacing a certain amount of gallium with the same amount of aluminium. The nomenclature defines $Ga_{1-x}Al_xAs$, where x represents the amount of gallium which is substituted by aluminium. Active band gaps can only be produced up to about 45% of Al in the GaAlAs mixed semiconductor. Increasing the amount of Al leads to an indirect semiconductor material.

Due to the realization of fiber-optic amplifiers (Chapter 8), further developments using indium allow the production of InGaAs/GaAs light sources [6.6]. Thus, wavelengths of above 900 nm can also be achieved. Exclusively, special compounds of InP and InGaAsP should match in order that the lattice constants fit together. This case holds for $In_{1-x}Ga_xAs_yP_{1-y}$, where x specifies the substituted part of indium by gallium and y that of phosphor by arsenic [6.4,6.5].

Realizable *InGaAs/InP semiconductor light sources* can be operated in a wavelength range of about 0.9 μm up to 1.6 μm. *InGaAs/InP photodiodes* can be used in the wavelength range of about 0.88 μm up to 1.7 μm (Figure 6.39).

6.1 Semiconductor Light Sources

Concerning the used light sources for optical communication and sensor techniques, *light emitting diodes* (*LEDs*) and *semiconductor* lasers are primarily applied. Electrically considered, these semiconductor light sources can be represented as PN diodes, which are operated in the forward direction. A typical voltage-current characteristic is shown in Figure 6.11 (dashed line). It follows that [6.7]:

$$I \sim e^{\frac{eU}{kT}} - 1 \qquad (6.7)$$

where:

I   Current in forward direction
U   Voltage between anode and cathode
K   Boltzmann's constant ($k = 1.381 \cdot 10^{-23}$ J/K)
T   Temperature
e   Unit charge ($e = 1.602 \cdot 10^{-19}$ As)

Typical styles of semiconductor light sources for fiber-optic applications are presented in [6.8]. Figure 6.4 shows a schematic set-up of semiconductor layers. These semiconductor light sources are often arranged as diodes in a *double heterostructure* [6.9]. The active layer in which the charge carrier radiants recombine is surrounded by layers with higher band gaps. The corresponding energy-band scheme of a forward current activating a double heterostructure diode is depicted in Figure 6.5.

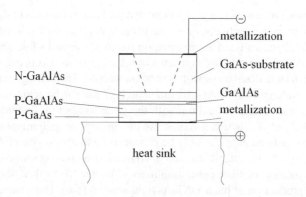

Figure 6.4   Structure and set-up of the different doped semiconductor layers of a GaAs light source (front view)

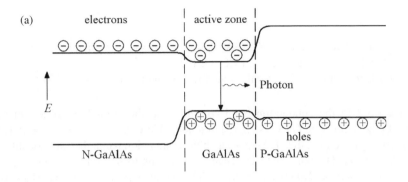

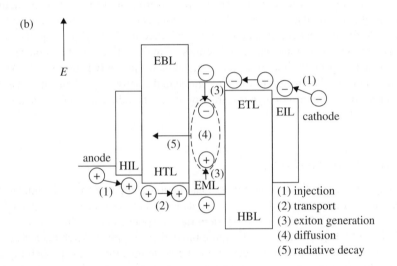

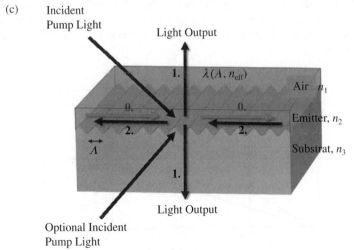

Figure 6.5 LEDs, LASERs and OLASERS

where:

a) Forward current actuated GaAs/GaAlAs double heterostructure diode
b) Energy-band scheme of an organic LED, OLED
c) Organic Light Amplification by Stimulated Emission of Radiation, OLASER

Free electrons in the conduction band coming from the N-GaAlAs zone can pass the boundary to the non-doped GaAlAs conduction layer. However, due to the appearance of the energy barrier at the end of the GaAlAs zone in the conduction band, they are not able to enter the P-GaAlAs layer. The holes in the valence band coming from the P-GaAlAs zone can also flow through the GaAlAs/N-GaAlAs interface. The energy barrier in the valence band prevents holes to flow into the N-GaAlAs layer. Therefore, electrons as well as holes remain in the GaAlAs zone, the so-called active zone. This active zone can be represented with the model of a potential well for the conduction and valence band; here the band gap comports the lowest value. Charge carriers recombine and photon emission in the active zone takes place. Limiting the active layer is especially important for semiconductor lasers (Chapter 6.1.2). Moreover, this process is improved by means of the wave guide in the active layer. The latter features not only a lower band gap but also a higher refraction index than the surrounding layers. Thus, a waveguide is created (Chapter 3.1).

6.1.1   Light Emitting Diodes

The LED (light emitting diode) is a simple construction compared to semiconductor lasers. The generation of radiation is achieved by spontaneous emission of photons (Chapter 6.1). The *P-I characteristic* is one of the most important response curves. In this case, P is the optical power and I the forward current. Figure 6.6 depicts a typical *P-I* characteristic.

The temperature occurs as a parameter. With higher temperatures, the probability of non-radiant recombination increases (e.g. by means of Auger processes). Therefore, the *P-I* characteristic shows a decreased slope for higher temperatures. Relating to a given current, the radiant power decreases with increasing temperature. Furthermore, an effect of saturation is recognizable (Figure 6.6).

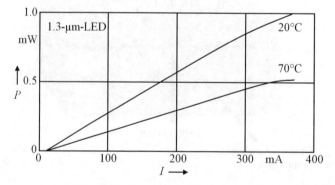

Figure 6.6   *P-I* characteristic of a quaternary LED at two different temperatures

Higher currents create higher dissipation power. Therefore, the diode heats itself up and the power consumption decreases. In order to minimize such heating, good heat dissipation has to be provided. The heat sink has to be placed as close as possible to the recombination zone. Therefore, the GaAs substrate (bottom layer) with its superimposed active layers has to be mounted on the top. The diode is turned upside down, which is called *Upside-Down mounting* (Figure 6.4).

To make use of the LED light for the applications described above, it is necessary to decouple it from the diode. For example, a certain part of the GaAs substrate can be etched off (see broken lines in Figure 6.4). Thus, the light can be emitted from the surface of the diode (in Figure 6.4 upwards). That is called a *surface emitter* or *Burrus LED*, named by its inventor, C. Burrus [6.10]. Due to the transparency of the substrate material InP regarding wavelengths higher than 1 µm, the etching off is not required for a long-wave InGaAsP/InP LED. Such a surface-emitting diode is also called a *Lambertian radiator*. This radiator possesses a strong divergent emission characteristic.

The radiant intensity is proportional to the radiant surface A. On closer examination of this surface under the angle δ, only the projection $A \cdot \cos\delta$ of the radiant surface is effective (Figure 6.7). The luminous flux and thus the optical power P is proportional to the radiant intensity. Thus, it follows that:

$$P(\delta) = P(0) \cos \delta \qquad (6.8)$$

where:

δ Angle between surface normal and viewing direction

Figure 6.7 represents a sectional view through the rotationally symmetrical radiation characteristic of the Lambertian radiator. The decrease of performance is represented by the lengths of the arrows. All arrow peaks lie in the Thales circle. Due to the strongly divergent radiation characteristic, the coupling into a fiber can be sufficient for multimode fibers using butt coupling or taking into account a lens-coupling design. In particular, for single-mode fibers, it is not effective.

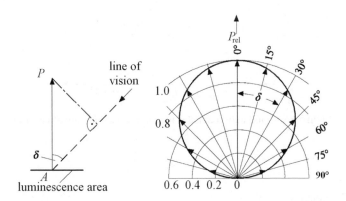

Figure 6.7   Emission characteristic, the *farfield* of an LED

where:

P Optical power
P_{rel} Optical power referred to the optical power for perpendicular emission

From this point of view, it is more favorable to use an *edge-emitting LED* [6.11,6.12]. For this type of LED, the emission does not occur on the surface like the Burrus LED but on the side (front in Figure 6.4). The build-up of this LED type already closely resembles a semiconductor laser and requires a similar complexity during manufacturing. Therefore, usually semiconductor lasers are preferred. Edge-emitting LEDs are only used in special applications.

To obtain a sufficiently high optical power (about 1 mW), the charge carrier lifetime τ_e is not allowed for decrease by very much, because in this case the quantum efficiency, that is the optical output ratio, will diminish severely (Eqs (6.11,6.12)). τ_e represents an effective charge carrier lifetime arising from the durability of the spontaneous emission and that of the concerned non-radiant recombination. However, a high charge carrier lifetime τ_e causes a narrow modulation bandwidth of the LED [6.13]. This effect determines the cut-off frequency f_c, and can be written as:

$$f_c \sim 1/\tau_e \tag{6.9}$$

Achieved cut-off frequencies of such an LED lie at about 100 MHz. Figure 6.8 shows an example of the transfer function of an injection current modulated LED. It represents the so-called Bode diagram of a transmitted signal, the modulation transfer function (MTF). The cut-off frequency is defined as the frequency at which the absolute value of the transfer function is reduced to the half (i.e. 3 dB for powers, Chapter 3.2.4). The bandwidth of an LED is determined by the frequency range from 0 Hz up to the cut-off frequency f_c.

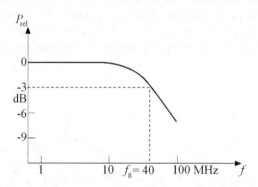

Figure 6.8   Modulation transfer function, Bode diagram of an LED

where:

f Frequency
f_c Cut-off frequency
P_{rel} Modulated optical power related to modulation frequencies tending to 0 Hz

Figure 6.9 presents the *optical spectrum* of an LED. Regarding GaAs and InP LEDs, the full widths at half maximum (FWHM) $\Delta\lambda$ comport about 30 nm up to 40 nm, and 80 nm to 130 nm, respectively. In the context of chromatic dispersion of a fiber, these large FWHMs are a problem for application in optical telecommunication engineering. Hence, applications for LEDs with shorter transfer links and lower bit rates are possible. The lower output power (in comparison to the laser) also raises problems. In contrast, the wide-range linear *P-I* characteristic is favorable for analog modulation. Therefore, formation of harmonics is prevented. All in all, the LED is particularly used in analog techniques with high linearity requirements.

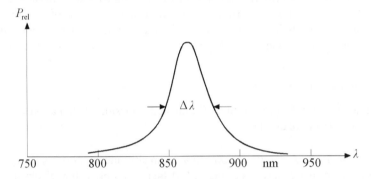

Figure 6.9   Relative optical power of an LED vs. wavelength λ

where:

$\Delta\lambda$ Full width at half maximum

A further type of LED is the organic light emitting diode (OLED). Their commercial use in general is not in optical communication or sensor applications, their use being in consumer electronics such as display application. OLEDs are built up with planar techniques. Basically, a glass substrate is needed. Besides the glass substrate, flexible substrates are possible. Between the anode and cathode the following layers can be set up: electron injection layer (EIL), electron transport (ETL) and hole blocking layer (HBL), emission layer (EML), hole transport layer (HTL) and electron blocking layer (EBL) and hole injection layer (HIL). Figure 6.5b depicts a set-up of an OLED. There are materials which combine the properties of two or three layers, thereby reducing the amount of layers.

OLEDs possess a transparent and a non-transparent contact. Therefore, either the anode or the cathode is transparent. This defines whether the OLED is a Bottom Emitter or a Top Emitter. The transparent material is a widely transparent semiconductor called ITO (indium tin oxide), but also completely transparent OLEDs are possible.

The electron conduction in organic materials is caused by delocalized π electrons in the molecule, but also hopping is a possibility in electron transport. These materials can have conductivity levels ranging from insulators to conductors. Therefore, they are counted among semiconductors, more precisely named organic semiconductors. The highest occupied molecular orbital (HOMO) can be interpreted as the valance band edge of an inorganic semiconductor. Furthermore, the lowest unoccupied molecular orbital (LUMO) can be interpreted as the conduction band edge.

During operation, the OLEDs are connected in the forward current direction. Therefore electrons are injected into the LUMO of the injection layer, respectively, holes in the HOMO of the transport layer. The injection layer possesses a lower energy gap compared to the transport layer. The transport layer therefore possesses high charge carrier mobility. Reaching the electron and hole blocking layer, which forms a barrier to the charge carriers, electrostatic forces recombine them forming a Frenkel exciton. This is a bound state for the electron and hole. They do not recombine completely but are bonded to each other. Generally the exciton state is energetically lower as in the transport layer. The decay of this excited state results in emission of radiation. Furthermore, dye molecules can be doped in the emission layer forming traps for the holes and electrons. They recombine at the band gap of this dye molecule. Therefore, doping the emission layer can adjust the wavelength of the radiation. HOMOs and LUMOs of all materials are Gaussian-shaped, the wavelength spectrum is widened [6.71].

6.1.2   Semiconductor Lasers

To realize a laser (Light Amplification by Stimulated Emission of Radiation = LASER), we have to fulfill two laser conditions:

- *First Laser condition*: A *population inversion* as an absolute supposition for the stimulated emission (Chapter 6.1.2) has to be realized. If there are only a few electrons in the conduction band, as in natural light sources, consequently only a few stimulated photons can be produced. Therefore, artificially, the conduction band must be filled up with numerous electrons.
- *Second Laser condition*: Moreover, the semiconductor laser as any laser is characterized by an *optical resonator* (Figure 6.10). Partially transmitting reflectors are arranged at the front and the end of the lasing area. They are created by cutting and breaking the semiconductor crystal media.

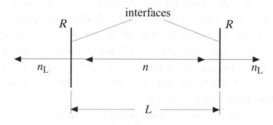

Figure 6.10   Optical resonator

where:

n_{air} Refraction index of the ambient air
n Refraction index of the semiconductor material
R Reflection coefficient

Regarding the transition of light from the semiconductor materials GaAs and InP, respectively, to the environment, such as to the air, Fresnel reflection occurs (Chapter 2.5). The refraction indices of GaAs and InP are about $n \approx 3.5$. Thus, a power reflection coefficient of about 31% is gained. Semiconductor lasers working with the above-described principle are also called *Fabry-Perot lasers*. They are built up as a Fabry-Perot interferometer (Chapter 5.4). An incident light wave at the interface layer GaAs/air (respectively InP/air) is split up into a transmitted part and a reflected part. The reflection recurs at the opposed laser mirror, thus multiple reflection occurs. Every circulation in the laser resonator amplifies the light wave through the occurring laser process

By means of current injection (pumping), the electrons are lifted from the valence band to the conduction band with higher energy level (Chapter 6.1). At low currents, first spontaneous emission occurs by recombination of electrons with holes, like those in an LED. Afterwards these existing photons stimulate further electron transitions, thus a stimulated or induced emission is generated (Figure 6.2). For the laser power, it can be written as:

$$P \sim e^{gz - \alpha z} \tag{6.10}$$

where:

g Gain exponent
α Resonator loss
z Spatial coordinate in propagation direction

The losses of the resonator arise from scattering in the laser and the mirror losses. From the viewpoint of an observer inside the laser, the latter is defined as losses. However, relating to the user, this presents no loss but the effective power which, for example, can be coupled into a fiber.

Above a certain current, the amplification compensates all losses of the resonator, $g = \alpha$. In this case, the laser oscillation condition is fulfilled. The output power of the laser increases drastically [6.14], so the *P-I* characteristic shows a drastic kink. The current at this transition is called the *threshold current*. Such a drastic kink is not observed in practice. Depending on the type of laser, the characteristic shows a more or less rounded curve (Figure 6.11). Therefore, the exact position of the threshold current is defined in the intersection of the downwards extended straight line in Figure 6.11 with the abscissa (*I*-axis).
where:

P Optical power
I Current in forward direction
I_{th} Threshold current
U Applied voltage

Appling a further increase of the injection current, a corresponding higher optical power part is emitted, and $g = \alpha$ still applies. The output power increases approximately linearly. Assuming $g > \alpha$, the power would increase ad infinitum and destroy the laser. The increase

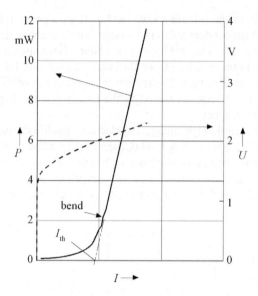

Figure 6.11   *P-I* and *U-I* characteristics

of the optical output power versus the injection current is characterized by the *differential quantum efficiency*:

$$\eta_d = \frac{dP}{dI} \cdot \frac{e}{h \cdot f} \tag{6.11}$$

Including dissipation losses, it follows for the quantum efficiency that:

$$\eta_d = \frac{P}{I} \cdot \frac{e}{h \cdot f} \tag{6.12}$$

The power losses lead to an intrinsic heating of the laser. Significant effects on the *P-I* characteristic appear. This also applies to a change in ambient temperature. Regarding the semiconductor laser, the temperature dependency is much higher as compared to an LED. The threshold current increases with increasing temperature (Figure 6.12).

Therefore, at a fixed current, it is possible that very little optical power will no longer be emitted. In contrast, regarding an intense decrease of temperature, the output power can increase in such a way that a very high power density is achieved. Thus, self-destruction of the laser can occur. The following equation applies to the temperature dependency of the threshold current I_{th} [6.15]:

$$I_{th}(T) \sim e^{\frac{T}{T_0}} \tag{6.13}$$

where:

T_0 Characteristic temperature

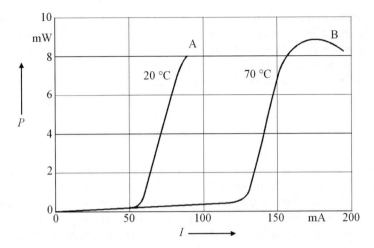

Figure 6.12   *P-I* characteristic of a semiconductor laser at two different temperatures

T_0 is a measure of the temperature threshold current dependency. The higher T_0, the lower the influence of the temperature. Therefore, we try to gain T_0 values as high as possible in the manufacturing process. Concerning GaAs-laser diodes, the achieved values lie between 150 K and 250 K [6.15]. For InP they are more unfavorable at about 50 K to 100 K [6.16]. Regarding a very strong increase in the injection current, a corresponding very high temperature arises. Thus, the differential quantum efficiency decreases again. The *P-I* characteristic becomes non-linear again (see zone A in Figure 6.12). In this case it is also possible that non-radiant recombinations occur and thus the output power decreases with increasing current (zone B in Figure 6.12).

Moreover, non-radiant recombinations are enhanced with increasing operating lifetime. This is called *aging* of the laser [6.17,6.18]. Figure 6.13 presents the aging characteristic of a

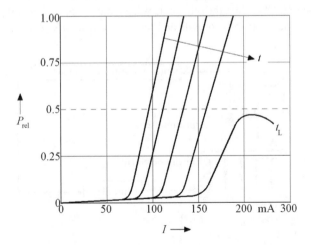

Figure 6.13   Aging characteristic of a semiconductor laser, *P-I* characteristic versus increasing operating time *t*

GaAs laser. A shift of the threshold current to higher values and a reduction in the differential quantum efficiency appear (decrease of the slope in the linear range of the curve). Therefore, the dissipation power rises. In particular, this adversely affects the modulation capability. Due to these circumstances, it is difficult to generate high amplitudes of the injection current, in particular at high frequencies by an electronic circuit.

where:

t_L Life-time

Moreover, the decrease of performance appears with increasing injection current at aging. The operating time at which the power in the reversal point is falling below half of its original maximum output is called the fatigue life of the semiconductor laser. Today's manufacturers of lasers specify fatigue lives of more than 10^5 hours. These are no real-time measurements (waiting 11 years is not desirable), but equivalent times can be generated by artificial aging due to operation at high temperatures.

A defined *P-I* characteristics is of particular importance for laser modulation. In order to overcome the temperature problem, two approaches are taken: a thermal stabilization or a control of the operating point in the *P-I* curve. Regarding *temperature control*, the threshold current is kept constant. Figure 6.14 depicts a schematic diagram of such a control. The actual temperature value is measured as close as possible to the active zone of the laser chip by means of a temperature sensor. This can be done by an NTC resistor or a semiconductor temperature sensor. A suitable electronic compares this actual value with an indicated nominal value and feeds the difference to a control unit. Mainly *P-I* controllers are applied (Proportional-Integral controller [6.19]). The *P-I* controller drives a certain current through a *Pelletier element* [7.20] (Pelletier cooler). A Pelletier element is a semiconductor component composed of Bismuth-Telluride.

Between its ends, a temperature difference is generated in the case of a current flow. This behavior is caused by the phonon-drag effect. One side of the Pelletier element is cooled down (e.g. side A in Figure 6.14), while the other side is heated up (e.g. side B in Figure 6.14). Thus, thermal energy is conducted from A to B and led into a heat sink. A reversal of the current

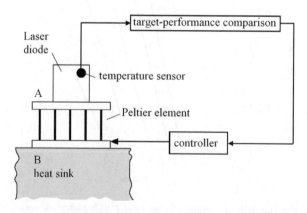

Figure 6.14   Temperature control scheme of a laser diode

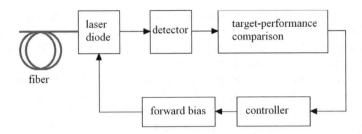

Figure 6.15   Operating point control of a laser diode with aid of a monitor diode control

direction also yields to an inversion of the thermal flux direction. Therefore, the laser diode can be cooled as well as heated. Relating to a temperature control, an almost ideal component is achieved. The *P-I* controller drives a correspondent current, consequently to sign and absolute value through the element until the deviation of the actual thermal value from the nominal value is controlled to zero.

Appling an operating point, or *monitor diode control,* provides a further opportunity to solve the temperature problem. Figure 6.15 shows the appropriate block diagram. As a Fabry-Perot laser has two outputs (Figure 6.10), the rear output can be applied to monitor the operation. Applying a monitor diode at this point, the power output of the laser can be measured by means of the current flow through the monitor diode (Chapter 6.2). As with the temperature control, a set/actual comparison (target-performance comparison) and a control unit are again used. The control unit directly affects the bias current (Figure 6.16) of the semiconductor laser. A decrease of the power output of the laser could occur, for example by means of enhancement of the ambient temperature. In this case, the bias current is increased in such a way that the primal power will be achieved again. This control only works as long as $dP/dI > 0$ is fulfilled, but does not occur in zone B of Figure 6.12.

Applying a modulation means that the mean power has to be controlled. Thus, the control has to be slow in comparison to the modulation frequency. In general, this is no problem. However, attention has to be paid in such a way that the high and low power level of the binary signal statistically appear with the same probability (Figure 9.4), otherwise a variation of the mean power is produced. The desired operating point in the *P-I* curve cannot be redeemed. Regarding a digital signal, this means an equal distribution of logic zeros and logic ones, respectively.

Modulation of semiconductor lasers in particular plays an important role in optical data transmission. However, it is also needed for signal analysis in the sensor systems. But data rate requirements are very different. While requirements for data transmission demand signals up to the GHz range, sensor techniques often handle MHz signals and lower frequencies (simple sensors also work without modulation). Typically light modulation is performed directly by means of injection current modulation (see exceptions, integrated-optical applications, Chapter 9.4.3). This also holds for analog and digital information. In optical communication, digital data processing is preferred. One reason therefore is the non-linear behavior of the applied devices, in particular on the *P-I* characteristic of a semiconductor laser. This component shows strong non-linearities at the threshold and in the saturation region. Moreover, also kinks (bends) exist (Figure 6.11). Kinks in the linear part of the *P-I* curve are often related to mode jumps (see below).

Together with analog modulation, these non-linearities lead to harmonics disturbing the system performance [6.21,6.22], in particular, in the case of the use of several optical time-division-modulated (TDM) emitters coupled into a transmission path. Modulation for data transmission is mainly carried out digitally, except in low-cost systems. In some cases, non-linearities can be compensated for by pre-equalization networks [9.5,9.6]. Due to reasons of economy, the rollout strategy of an optical fiber is forced by such developments. For example, last-mile communication, fiber-to-the-home systems are aimed for, that is to bring the optical fiber to a huge number of participants, even to one-family homes (Chapter 10.1).

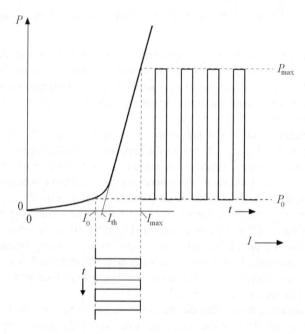

Figure 6.16   Direct modulation of a semiconductor laser by the injection current

where:

I_0 Bias current
I_{th} Threshold current
I_{max} Maximum current
P_{max} Maximum power
t Time

Regarding optical signals as small as possible, not only for an optical transmission system but also for a sensor system, should be able to process data exactly. Therefore a high *signal-to-noise ratio S/N* is required (Chapter 9.1). Concerning laser diode modulation, this implies that the threshold power P_0 (Figure 6.16) should be as low as possible, whereas the maximum output P_{max} should be chosen to be as high as possible. Moreover, a high maximum output power enables a high modulation capability. For very fast modulation, the rounding of the *P-I* curve in the region A (Figure 6.12) does not occur. Due to large thermal time constants, heating is not achieved. However, the maximum output is restricted by reason of laser chip destruction hazards, thus the possible maximum output is restricted. Moreover, the minimum output must not be too small, otherwise the modulation capability would be slowed down again; in this case we are talking about LED behavior of the laser. Regarding an ideal current jump of the injection current from I_0 to I_{max}, the optical power does not directly follow this jump in time, but increases later after a delay t_d (Figure 6.17, [6.23] photon density $S(t)$):

$$t_d \sim \ln \left(\frac{I - I_0}{I - I_{th}} \right)$$

(6.14)

where:

I_0 Forward current
I_{th} Threshold current

For $I_0 \rightarrow I_{th}$, it follows that $t_d \rightarrow 0$. The minimum output power has to be adjusted in such a way that the power approximately corresponds to the threshold power at the logical zero. That causes an optical background leading to undesired noise (Chapter 9.1). Thus we have to make a compromise to achieve modulation frequencies above 10 GHz. Therefore, appropriate parameters have to be chosen, an unavoidable agreement. Figure 6.18 depicts the absolute value of the transfer function of a semiconductor laser modulated by the injection current. In this example, a 3-dB cut-off frequency of 1.9 GHz is gained.

Figure 6.17 shows that a transient phenomenon takes place. The decaying oscillations are called *relaxation oscillations* [6.23,6.24]. These are generated by interaction between the dynamic charge carriers and the photons in the semiconductor material [6.24]. After the current jump, the density of the charge carriers $n(t)$ is very high at the beginning due to the lack of radiant recombinations during time t_d. The photon density $S(t)$ features a delayed increase until the steady state charge carrier density is obtained after the transient effect ($S(t)$ at an instant of time A (Figure 6.17).

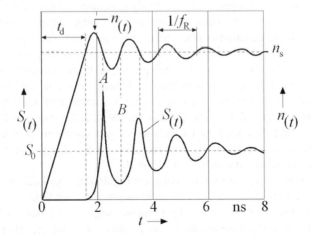

Figure 6.17   Relaxation oscillations following current pulse activation

where:

t_d Time delay
f_R Relaxations frequency
$n(t)$ Charge carrier density
$S(t)$ Photon density

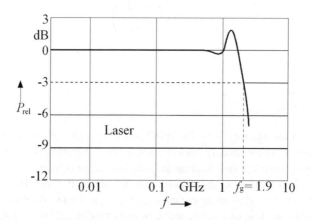

Figure 6.18   Modulation transfer function of a semiconductor laser

where:

f Frequency
f_c Cut-off frequency
P_{rel} Amplitude of the modulated optical power related to values for modulation frequencies tending to zero

The photon density is now higher as in the steady state after the transient phenomenon (S_0), because of the whole developed reservoir, whereas t_d of charge carriers is now available. The output power exceeds the maximum value after the transient state. The existing high number of photons is able to activate many stimulated emissions. Many more charge carriers recombine as compared to the steady state, more than being filled up by the injection current. Thus, the charge carrier density drops down below the steady state situation, and the photon density decreases again. The output power decreases below the value after the transient procedure (instant of time B in Figure 6.17). The process recurs until the transient effect is decayed. The exact relations are described by the balance or rate equations. It deals with a set of differential equations. Both differential equations, one for the charge carrier and another for photon density, are coupled with each other [3.5,6.26].

Wave guiding in semiconductor lasers is of great importance [6.14,6.27,6.28]. To make the amplification of the optical radiation as efficient as possible, the light wave and the region of recombinations have to be restricted to a narrow range. The guiding of the light wave has to be limited in two dimensions, as in optical fibers or in integrated-optic waveguides.

Concerning the configuration and the layer sequence of a semiconductor laser, these are similar to those relating to the LED (Figure 6.4). Thus, a limitation perpendicular to the direction of the current flow is gained (Figure 6.19). The band gap ΔE plays an important role in this purpose. Due to the double heterostructure (Chapter 6.1.2), the recombinations are restricted to layer I. Moreover, by reason of doping, the refraction index of the layer I, the active layer, is higher than that of the ambient layers: $n_1 > n_2, n_3$. Thus, a slab waveguide is developed.

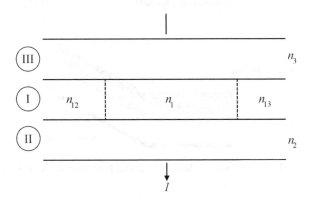

Figure 6.19   Waveguiding in a semiconductor laser

where:

I	Active layer
II and III	Surrounding layers and their refraction indices n_2, n_3
n_{12}, n_{13}	Refraction index fine structure for index guiding lasers

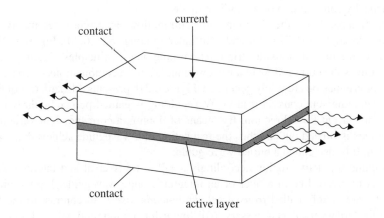

Figure 6.20   Semiconductor laser without lateral limitation of the active layer

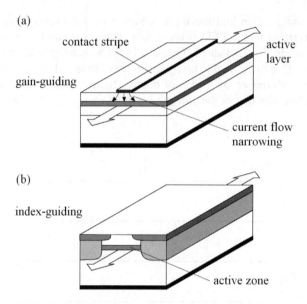

Figure 6.21   Semiconductor laser with lateral limitation of the active layer

where:

a) With gain guiding
b) With index guiding

The first semiconductor lasers were connected onto the whole surface. Thus, the current flow took place in the total semiconductor material (Figure 6.20), which implicated high threshold currents. Therefore only pulsed operation was completed; otherwise that would have led to destruction of the laser. CW operation (continuous wave) has been allowed by means of further advancements by compression of the radiant range.

The limitation parallel to the direction of the current flow can be classified into two groups, lasers with gain guiding [6.27,6.29] and with index guiding [6.27,6.30]. Figure 6.21a shows the schematic structure of an oxide strip laser working on the principles of gain guiding. The contacting affects the current flow in a narrow range. Here, recombinations of charge carriers and thus photon emission are solely generated. Figure 6.21b presents a schematic configuration of an index guiding semiconductor laser. Waveguiding is gained perpendicularly as well as parallel to the current flow direction. By means of doping, a correspondent variation of the refraction index is carried out and thus the realization of a waveguide achieved: $n_1 > n_{12}$, n_{13} (Figure 6.19). It is also called a buried wave guide.

Dimensioning the active region accordingly small, a transversal and lateral single-mode waveguide is received, like in a fiber or an integrated-optic waveguide. Lasers with such a characteristic are simply called *transversal single-mode lasers*. In contrast to the fiber, the addition of the "transversal" is necessary to distinguish from longitudinal modes (see below). The transversal single-mode characteristic represents a basic requirement for a semiconductor laser, especially in terms of coupling light into the fiber.

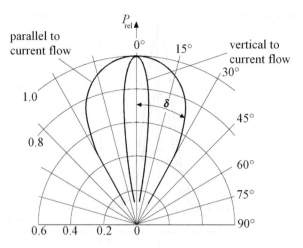

Figure 6.22   Radiation characteristics of a typical Fabry-Perot semiconductor laser

where:

P_{rel} Optical output power related to the optical output power for perpendicular radiation

Figure 6.22 depicts the *far field characteristics*, the spatial divergence of the most common semiconductor laser type, the Fabry-Perot laser [6.31]. It is narrower than that of an LED (Figure 6.7). However, due to the asymmetrical radiating surface, an asymmetrical radiating characteristic is usually gained. That is clarified with the aid of the correlations between the angle of radiation δ and the waist of a Gaussian ray (Eq. (3.20)). The radiating surface of a typical semiconductor laser has a width of about 4 µm up to 10 µm and a thickness of less than 1 µm. Figure 6.23 shows the far field of an index guided semiconductor laser in two orthogonal directions. The optical absolute power appears as a parameter.
where:

P Optical power as a function of the angle of radiation δ in two orthogonal directions at three different absolute powers

In general, gain guided lasers feature a higher divergence of the radiated light and a stronger astigmatic behavior than index guided lasers. Competing with the waveguide, a decrease of the refraction index arises from the current injection. This also causes a significant deviation from the Gaussian form of the far field. The so-called ears in the far field, perpendicular to the current flow direction, are developed.
where:

P Optical power P as a function of the angle of radiation δ in two orthogonal directions

Figure 6.24 shows a measurement curve of the far field from a gain guided laser in both directions perpendicular to each other. Thus, the propagating fundamental laser modus is not

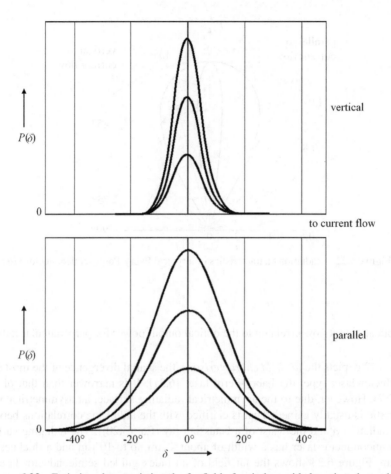

Figure 6.23   Emission characteristics, the far field of an index guided semiconductor laser

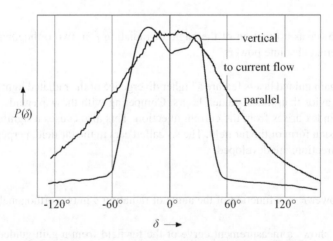

Figure 6.24   Far field of a gain guided semiconductor laser

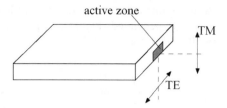

Figure 6.25   TE- and TM-direction related to the laser chip and its active zone

a plane wave but a divergent radiation occurs, simplified considering there are no vertical reflections at the laser mirror. Moreover, concerning both orthogonal directions the divergence and thus also the reflection angles are different. Thus, different power reflection coefficients for a TE- and a TM-wave [6.32] (Figure 6.25) arises due to Fresnel reflection:

$$R_{\mathrm{TE}} > R_{\mathrm{per}} \text{ and respectively } R_{\mathrm{TM}} < R_{\mathrm{per}} \text{ and thus } R_{\mathrm{TE}} > R_{\mathrm{TM}}, \quad (\text{Chapter 2.5})$$

where:

R_{per} Power-reflection coefficient for the reflection perpendicular to the front surface of the semiconductor laser

The TE wave is much more reflected than the TM wave, and thus more amplified in the laser material. Due to the fact that the amplification process is subject to an exponential function (Eq. (6.10)), the emission is mainly linearly polarized in the TE direction. Due to spontaneous emission of a small but not negligible amount in a semiconductor laser, the rest of the unpolarized light remains. Regarding index guided lasers, the ratio of polarized and unpolarized light is about 10^{-3} up to 10^{-2}. Concerning gain guided lasers, the unpolarized part can amount up to 10 %. This is due to a significantly higher part of the spontaneous emission. This also leads to a stronger rounding of the *P-I* curve in the threshold region [6.21] (Figure 6.26).

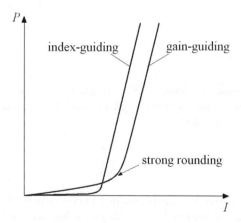

Figure 6.26   *P-I* characteristic of an index and gain guided semiconductor laser in comparison

Besides transversal modes in a semiconductor laser, there are longitudinal modes [6.33]. Standing waves in the resonator are generated (Figure 6.27). Therefore, at the mirrors, the existence of nodes is required. It follows that:

$$L = \frac{\mu \lambda}{2n} \tag{6.15}$$

where:

μ Integer

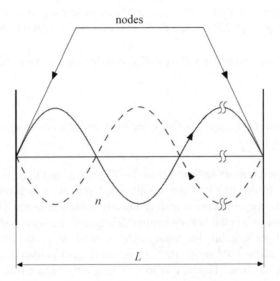

Figure 6.27   Standing wave in a laser resonator

where:

n Reflection index
L Length of the resonator

Approximately 3000 nodes are obtained from a typical GaAs laser with a wavelength of 850 nm, a refraction index of 3.5 and a typical length of about 300 μm. Neglecting the dispersion, it applies to the *mode distance* $\Delta\lambda$:

$$\Delta\lambda = \frac{\lambda^2}{2nL} \tag{6.16}$$

That corresponds to the free spectral region of a Fabry-Perot interferometer (Chapter 5.4). Regarding the type of lasers described above, the mode distance is about 0.3 nm. Figure 6.28 shows measurement curves of the according laser mode spectrum.

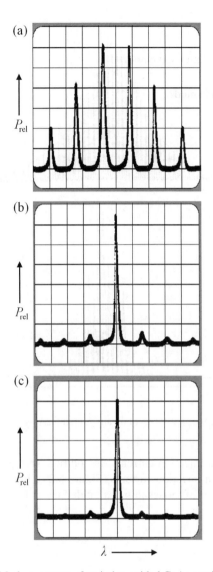

Figure 6.28   Mode spectrum of an index guided GaAs semiconductor laser

a) Near threshold current
b) At intermediate current
c) At high current, mode distance $\Delta\lambda \approx 0.3$ nm center wave length $\lambda \approx 850$nm [6.13,6.34]

Figure 6.28 (a) was measured applying a forward current below the threshold current. The typical spectrum of a *multi-mode laser* is produced. Figure part (b) is recorded at about 1.4 times the threshold current and (c) represents the last third in the *P-I* curve. In the last case, the side mode suppression approximately amounts to 12 dB for next-neighbor modes. The investigated laser type was an index guided laser. Index guided lasers are more like few-mode

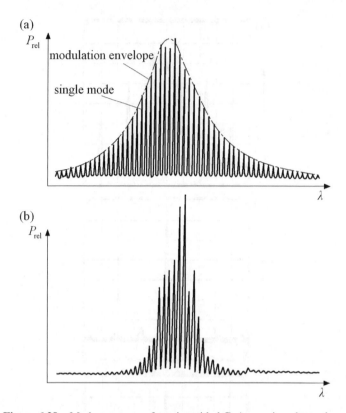

(a)

P_{rel}

modulation envelope

single mode

λ

(b)

P_{rel}

λ

Figure 6.29 Mode spectrum of a gain guided GaAs-semiconductor laser

lasers, in particular at high injection currents. In contrast, gain guided lasers show a multi-mode behavior, also at high injection currents (Figure 6.29). Injection current modulation in general increases the number of modes in index guided lasers [6.71 and 1.13].

a) At low currents
b) At high currents

For optical communications systems, *single-mode lasers* are preferred for use with single-mode fibers. Thus, a small signal distortion due to chromatic dispersion (Chapter 3.2.1) is achieved. In contrast, using multi-mode fibers, multi-mode lasers are required. The small spectral width of a single-mode laser yields a high coherence. That leads to a problem concerning the connection of single-mode lasers with multi-mode fibers. The high coherence implies a high capability of interference between different fiber modes, and for even longer time delays (Chapters 2.2 and 2.3). When investigating the face surface of a fiber at a random point of the fiber link, a speckled pattern is perceived (Figure 6.30). Due to external influences, such as temperature and vibration, the speckled pattern is time-variant. That is, a less problematic fact if the field of the whole speckled pattern is absolutely fed to a detector. However, at certain points in the fiber link, in particular at connectors and splices, part of the speckled

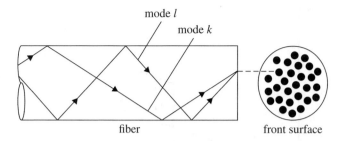

mode *l*

mode *k*

fiber   front surface

Figure 6.30   Interference of transversal modes of a multi-mode fiber by use of a coherent light source

pattern is lost. A so-called mode selectivity (Figure 4.41) is produced, leading to *modal noise* [6.35,6.36].

Distinct modes have different strengths attenuated by the fiber. Therefore, this effect of modal noise appears not only at points of impact, but also by the fiber itself. A remedy is found by the use of multi-mode lasers. On the basis of the high spectral width involving low coherence, the capability of interference due to the dispersion of the fiber is much more reduced after a short link so that practically no modal noise is developed. However, signal distortions caused by chromatic dispersion are created and cannot be eliminated, respectively. Therefore, this procedure can only be applied to low transmission capacities (bit rate times length). Regarding high standards for transmission characteristics, the use of single-mode fibers in connection with single-mode lasers is required.

Concerning sensor systems, the acceptable laser is a matter of application. Single-mode lasers are required when high coherence is needed. Relating to many application systems, especially with fibers, a too high coherence is interfering [6.37,6.38]: the high coherence may lead to undesired interferences by reflections at impact points affecting the measurement signal as disturbance. This happens, even if the impact points are widely spread.

The temperature dependence of the spectral properties of semiconductor lasers can pose a problem regarding optical communication systems as well as sensor systems [6.39,6.40]. Considering the spectrum of a multi-mode laser, with increasing temperature the gravity center of its envelope (Figure 6.29a) is shifted to higher wavelengths. This is caused by the temperature dependence of the amplification curve. Relating to GaAs diodes, their values lie at about 0.3 nm/K, for InP diodes the values are at approximately 0.5 nm/K. The displacement of a single-mode (also with increasing temperature to higher wavelengths [6.41]) is determined by the change of the length of the resonator (refraction index times length).

The chief cause is to be found in a change in the refraction index. The dependence for both semiconductor laser materials lies at about 0.1 nm/K. Regarding a single-mode laser, a continuous displacement of the longitudinal modus (Figure 6.31) is first developed until the amplification curve is more favorable for the next higher mode. Then a wavelength jump is generated and the original modus disappears. The next higher modus begins to oscillate and changes its wavelength continuously with higher temperatures. Regarding the transition, it is possible that an unstable state arises, that is the wavelength permanently jumps between two adjacent modes. An additional increase of the injection current affects an earlier appearance of the saltus (Figure 6.31).

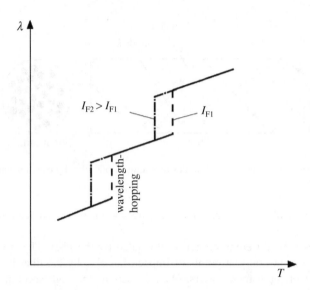

Figure 6.31   Wavelength λ of a single-modus (for single-mode lasers the only modus) as a function of the temperature T with the injection current I_F as parameter

In particular, the tendency to jump is enlarged by injection current modulation. Concerning multi-mode lasers, the modulation causes time-fluctuant power distribution between the distinct modes. The temporal variation can be very fast and therefore cannot be observed with the aid of slow measurement equipment, for example with a mechanically operated spectrometer. Time-resolved measurements have to be accomplished. Due to this effect in connection with the dispersion of the fiber, a further noise, the so-called *mode partition noise*, develops [6.21,6.42].

Exact analyses indicate that the spectrum of a single-mode laser or of a single mode from a multi-mode laser does not have the form of a Dirac delta function (Chapter 2.3). In fact, the power density spectrum shows a *Lorentz-function p(f)* with a finite spectral width Δf:

$$p(f) = \frac{p_0}{1 + c' \left(f - f_0\right)^2} \tag{6.17}$$

where:

f_0 Medium frequency
c' Constant

The constant c' describes the level of the decreasing power density part with increasing deviation from the center frequency. The spectral half width Δf depends on the optical laser power P (*Schawlow-Townes formula* [6.43]) and the *line broadening factor* α from Henry [6.44]:

$$\Delta f \sim \frac{1 + \alpha^2}{P} \tag{6.18}$$

Thus, not only the envelope of the mode spectrum (Figures 6.28 and 6.29), but also the spectral width of the single mode, becomes narrower with increasing injection current. Changes of charge carriers affect variations of the imaginary part of the refraction index n'' by means of modulation and noise, respectively (Chapter 3.1). That also causes a change of the real part n' and leads to line broadening (chirping) [6.45]:

$$\alpha = \Delta n' / \Delta n'' \tag{6.19}$$

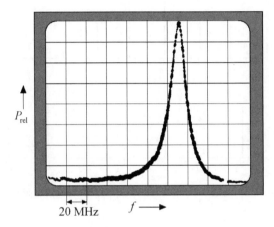

Figure 6.32 High-resolution measurement of the central modus of a semiconductor laser related to Figure 6.28c

where:

f Laser frequency

Figure 6.32 shows the measurement of the central modus of a GaAs laser at a high injection current with high resolution by means of a Fabry-Perot interferometer. The FWHM is aboutbrk 10 MHz, determined after elimination of influences from the instrument [6.34,1.13]. Generally, quaternary laser diodes still feature larger line widths. In certain circumstances, these high values cannot be tolerated, so arrangements against line broadening have to be make. Often the operation of a system presupposes a single-mode laser with high demands on side mode suppression. In particular, that is not fulfilled for long-wavelength Fabry-Perot lasers. By attaching such a laser to an *external resonator*, the side mode suppression can be drastically enhanced.

Figure 6.33 shows the action principle. Therefore, the optical path length of an external resonator has to lie about 10 to 20% above or below the optical path length of the laser. Longitudinal modes of the laser as well as of the external resonator are generated. Regarding a constellation (Figure 6.33a), coincidence of the spectral position of the laser modus and a modus of the external resonator is yielded [6.46]. However, concerning the central mode, the coincidence appears again just modus #5 of the laser (in accordance with modus #4 of the external resonator).

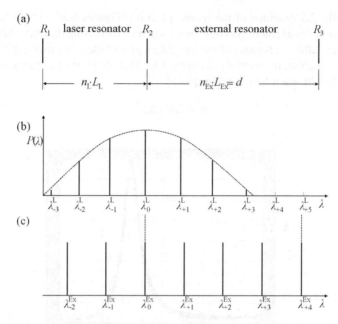

Figure 6.33   Schematic illustration of a laser with external resonator

where:

a) Constellation of a laser with an external resonator
b) Mode structure of the laser resonator
c) Mode structure of the external resonator

$R_1 = R_2$ Reflection coefficient of the Fabry-Perot laser
R_3 Reflection coefficient of the external resonator
n_i, L_i, λ_i Refraction index, length and wave length of the laser and external resonator

But this modus already lies outside of the laser gain curve. Thus, a significant increase of the central mode takes place, whereas the other modes are suppressed [6.46,1.13]. Figure 6.34a depicts the spectrum of an InP laser without external resonator. In Figure 6.34b, an external resonator is placed. In both cases, the same injection current and the same temperature have been adjusted. The achieved suppression of secondary modes is higher than 20 dB. Moreover, by changing the position of the external mirror, it is also possible to vary the wavelength of the laser [6.46].
where:

a) Multi-mode state without external resonator
b) Mono-mode state with external resonator

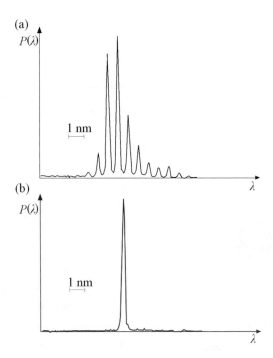

Figure 6.34 Mode spectrum of an InGaAsP Fabry-Perot semiconductor laser, center wavelength $\lambda \approx$ 1290 nm

Modern concepts concerning the suppression of secondary modes are based on the characteristics of the semiconductor itself. The C^3 laser (cleaved-coupled cavity) [6.47] similarly works as described above. It consists of two coupled active resonators being adjusted by the injection current (Figure 6.35a). The *DBR laser* (distributed Bragg reflection) [6.48] eliminates the Fabry-Perot reflection at the ends of the laser by means of two grating reflectors outside the active layer (Figure 6.35b). The *DFB laser* (distributed feedback, [6.48,6.49]) represents the most developed laser type. In this case, the feedback procedure is distributed over the total active region by an applied grating (Figure 6.35c).

Compared to the Fabry-Perot laser, the DFB laser shows further favorable properties. It is dynamically single-mode, such that during modulation a single longitudinal modus is exclusively emitted. Because of the diffraction at the gratings, there are modes transferred into guided and radiant modes. However, a guided mode can be traced back forming a standing wave, therefore a guided mode m can induce another mode m (grating orders), which is traced exactly in the opposite way [5.15]. This leads to a mutual energy exchange between these modes. Thus, the diffraction gratings work like reflectors. Radiation or leaky modes can be counted as losses. In addition, it features a smaller line broadening in comparison to the Fabry-Perot laser. Moreover, its *P-I* characteristic is linear to a high degree.

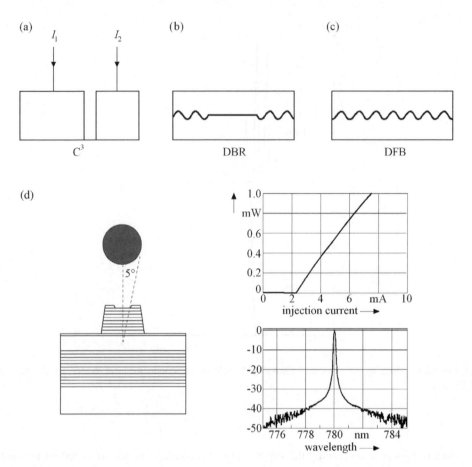

Figure 6.35 (a,b,c) Further developments of semiconductor lasers; (d) Vertical Cavity Surface Emitting Laser (VCSEL)

where:

a) Cleaved Coupled Cavity laser (C^3)
   I_1 Modulation current
   I_2 Current for adjusting the second resonator
b) Distributed Bragg Reflection laser (DBR)
c) Distributed Feedback laser (DFB)

The Vertical Cavity Surface Emitting Laser (VCSEL) seems a promising light source for use in optical communications engineering and in sensor technology [1.12,6.70]. In contrast to the above described laser types, the VCSEL is not an edge emitter but a surface emitting laser diode. Therefore, the construction of the resonator vertically works with multi quantum well structures (Multiple Quantum Well Laser, MQW) [6.70].

The surface emission facilitates a favorable coupling of the fiber. That can be considered as an important advantage of this type of laser. In addition, coupling to a fiber is made easy by low and symmetrical emission characteristics (Figure 6.35d). Moreover, very low threshold currents and longitudinal single-mode structures are gained. Regarding optical transmission and sensor systems, the stability of the wavelength and the frequency respectively play an important role in certain application [6.39]. With the aid of employed external reference elements, the wavelength can be measured. For example, external Fabry-Perot interferometers or ring resonators can be applied as reference elements. By use of control set-ups, deviations from an indicated nominal value can be compensated for by an injection current variation [6.18] (Figure 6.36a). Figure 6.36b shows the stabilized frequency of a GaAs laser. The achieved stability is about 1 MHz ($\sim 10^{-9}$ of the nominal value).

Fluctuations of wavelength and frequency can be interpreted as the quantity of noise. For slow changes with time constants, in the range of minutes, they can be counted as drifts (Figure 6.36b). High-frequency variations are called *frequency noise* [6.51]. Including the fluctuations detecting bandwidth, this noise is indicated in measurement units of light frequency/$Hz^{0.5}$. Thereby, the fluctuations of the frequency with finite half width are considered. These fluctuations can be detected by means of FM-AM converters, for example a Michelson interferometer. A Fabry-Perot interferometer can also be used. Here, the central

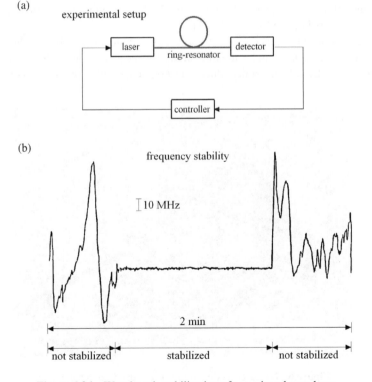

Figure 6.36   Wavelength stabilization of a semiconductor laser

wavelength of the laser is to be found at an operating point on the edge of the transfer function (Chapter 5.4).
where:

a) Experimental set-up
b) Wavelength in the stabilized and non-stabilized state

The finite line width of the laser can be regarded as a process of noise, that is, as an ideal oscillator whose relative spectral power part is different from zero in an infinitesimal range (Dirac delta function, Chapter 2.3). Its statistical temporal interval complies with the Lorentz function [6.43,6.44]. Fluctuations of power, also called intensity noise, can be considered as the square of the *amplitude noise*. Their lower limit is defined by the quantum noise of the photons [6.52]. Regarding the semiconductor laser, the amplification is also subject to statistical processes. Thus, further noise has to be taken into account [6.53]. Slow changes of the amplitude in the Hz range are caused by slow changes of temperature of the laser heat sink or by slow fluctuation of the injection current. Higher-frequency noise parts are primarily affected by the injection current. Fluctuations of the charge carrier density in the active zone of the semiconductor laser induce the change of the real part of the refraction index as well as the imaginary one [6.44]. Thus, they effect variations of the amplification factor of the laser. The maximum value of the amplitude noise is achieved close to the threshold current and decreases at higher injection currents again [6.54]. Considered spectrally, the maximum appears at the relaxation frequency f_R (Figure 6.17). A detailed representation about noise of semiconductor lasers is to be found in [6.55].

Figure 6.37 shows an opened laser module with a coupled pigtail. The Pelletier cooler for temperature stabilization and a monitor diode for measurement of the power of the rear light can be seen. Based on semiconductor lasers, semiconductor laser amplifiers can be

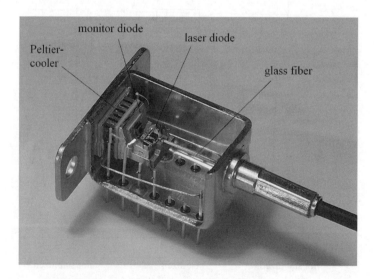

Figure 6.37   Laser module for optical transmission technique

realized. Like the semiconductor laser, these components have a low power loss. Relating to the geometry, they are well aligned to the glass fiber [1.12,6.56,6.57]. Two types are distinguished, the *Fabry-Perot amplifier* and the *traveling wave amplifier*. The Fabry-Perot amplifier only works in a narrow spectral range. In contrast, the traveling wave amplifier can be used for broadband applications. Thus, simultaneous amplification of many light waves from different transmitters with adjacent optical frequency is possible. Additional noise by amplification of the spontaneous emission and the occurrence of signal saturation due to the multitude of channels can be problematic. Moreover, due to the high coherence, these components are very sensitive to back reflected light.

6.1.3   Organic Lasers

Organic thin film lasers are very special light sources. They combine spectral properties of organic emitter materials with compact device designs. There are plenty of organic dyes and semiconducting materials that can be used as laser active materials. The assortment of emitters allows the production of lasers with emission wavelengths starting in the near UV, covering the whole visible part of the electromagnetic spectrum, and ending in the near infrared [6.77–6.80]. Organic lasers (OLasers) with a liquid or solid state active medium of macroscopic scale, are very common [6.81,6.82]. But it is also possible and more elaborate to fabricate organic thin film lasers. Figure 6.5c shows an OLASER. The incident pump light can be coupled into the active zone from the top as well as from the bottom.

They can be based on optical feedback provided by Fabry-Perot resonators, for example in slab waveguides or as vertical cavity surface emitting lasers (VCSEL), by random scattering [6.83, by ring resonators [6.84] or by distributed feedback (DFB) caused by periodic structures [6.85]. The latter can be in the form of sub-micrometer surface or volume gratings within a thin film waveguide leading to DFB laser emission. Since it is easy to replicate surface relief gratings, for example by nano-imprint lithography, these lasers can be implemented on a variety of surfaces, even on curved, flexible and transparent surfaces [6.86].

Organic thin film lasers are not limited in size but a typical dimension lies in the range of 100 μm × 100 μm × 100 nm. This allows for high degrees of miniaturization and integration. The production of these lasers benefits from the fact that organic emitter materials can be processed by polymer typical casting technologies on a variety of substrate materials, such as silicon, glass or polymers. This enables high production rates and low costs.

Actually there has been no demonstration of an electrically driven organic laser. Besides solid state lasers, it has been shown that laser diodes [6.87] and even LEDs [6.88] can be used as optical pump light sources. This means a huge step forward for the ability of being integrated. In general, the lasers are featured by low laser threshold values, high efficiency, and their ability of being adjusted to a wide range of emission wavelengths, single mode behavior and narrow laser lines.

6.2   Semiconductor Light Drains

By illumination with photons and absorption in semiconductor drains, charge carrier release happens (Chapter 6.1) [6.58,6.59]. In general, photodiodes are operated in a reverse biased direction. Photons, which possess a higher energy hf than the band gap (ΔE), are able to generate electron-hole pairs in semiconductor materials. Due to the reverse applied electrical field, the charge carriers drift towards the anode and cathode, respectively. The transport of charge carriers implies the existence of electrical current via an externally closed circuit (Figure 6.38).

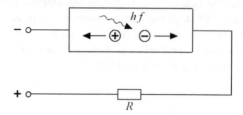

Figure 6.38 Photon absorption and charge carrier generation in semiconductor materials

where:

R External resistor

The probability of electron-hole pair generation by incident photons depends on the *penetration depth* of the photons [6.60–6.62]. The penetration depth is also called the *absorption length l*. It is defined to be the depth z at which the number of photons is declined to the value $1/e$:

$$P = P_0 e^{-\alpha z} \tag{6.20}$$

$$l = 1/\alpha \tag{6.21}$$

where:

α Absorption coefficient
P Reduced power at the depth z

The penetration depth and the absorption coefficient, respectively, depend on the semiconductor material and the wavelength of the applied light source (Figure 6.39).

The long-wave limit of the function (Chapter 6.1) is located slightly above the intersection points of the curve with the λ-axis. Above this value, the respective semiconductor material is nearly transparent. The absorption characteristics of InGaAs can be modified by means of variation of the percentage of In and Ga and accordingly by addition of power P. Elemental semiconductors show a smaller decrease of the absorption coefficient with increasing wavelength, compared to compound semiconductors [6.65]. This is due to direct and indirect band gaps of the semiconductor materials (Figure 6.3). For example, besides the minimum energy of an incoming photon, silicon needs an interaction with an additional phonon to overcome the indirect band gap.

However, not every incident photon leads to the generation of electron-hole pairs resulting in an electrical current via the external circuit. Some photons are lost due to reflection and some generated electrons and holes are captured by acceptors and donators. The probability of

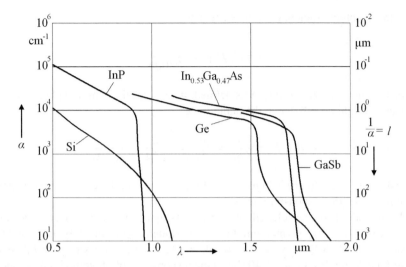

Figure 6.39 Absorption coefficient α and the penetration depth $l = 1/\alpha$ as a function of the wavelength λ for certain semiconductor materials

the emergence of useable charge carriers is expressed by the *quantum efficiency η*. It describes the ratio between the number of usable generated electron-hole pairs and the number of the incident photons. It follows that:

$$\eta = (1 - R)(1 - e^{-\alpha z}) \qquad (6.22)$$

where:

R Fresnel reflection coefficient

In order to gain a quantum efficiency as high as possible, the thickness d of the absorption layer ($z = d$) also has to be large. Indeed, this implies that the charge carriers remain in this region for a very long time. Therefore, the diode slows down. The reduction of the Fresnel reflection coefficient R represents a further measure to increase the quantum efficiency (Chapter 2.5). Concerning InP, the Fresnel coefficient is about 31%. The reflectance can be degraded to about 1% by applying antireflection coatings.

Typical achieved quantum efficiencies are about 0.8. The achievable photocurrent I_P is normalized to the elementary charge e: I_P/e and the incident optical power P to the energy of a light quantum hf: $P/(hf)$. Thus, the number of the generated electron-hole pairs N_L and accordingly the number N_{Ph} of the incident photons per time unit Δt can be determined. It follows that:

$$N_L = (I_P/e)\Delta t \qquad (6.23)$$

$$N_{Ph} = (P/(hf))\Delta t \qquad (6.24)$$

Hence, the quantum efficiency can be written as:

$$\eta = N_L/N_{Ph} = \frac{I_P/e}{P/(hf)} \tag{6.25}$$

For practical use, the *responsivity* S^* is defined. It specifies the achieved photo current I_P at a given incident light power P onto the photodiode:

$$S^* = I_P/P = \eta\frac{e}{hf} = \eta\frac{e\lambda}{hc} = \eta\frac{\lambda}{1.24} \cdot \frac{A}{\mu m \cdot W} \tag{6.26}$$

The final result of Eq. (6.26) is in particular for practical use; if we insert the wavelength in μm, we directly get the responsivity in A/W.

For silicon diodes and a wavelength of 0.85 μm (1st window, Figure 9.1), typical values lie at $S^* = 0.5$ A/W [6.63]. For germanium diodes, the value is approximately 0.7 A/W (1.3 μm, 2nd window) and 0.9 A/W (1.5 μm, 3rd window). At 1.5 μm, values up to 0.9 A/W can be gained for InGaAs diodes. Figure 6.40 shows the responsivity of photodiodes as a function of the wavelength for the corresponding semiconductor materials.

6.2.1   Types of Photodiodes

After the generation of electron-hole pairs by incident photons onto a pure undoped semiconductor material, there is a risk that electrons and holes will partially recombine again. Thus, they are unusable and cannot contribute to the photocurrent. Therefore, a PN junction is embedded into the semiconductor material. Moreover, a reverse voltage application is carried out to accelerate the charge carriers. A simple *PN diode* is gained (Figure 6.41).

The highest electrical field strength due to the applied reverse voltage is developed at the PN junction. At this zone the charge carriers are strongly accelerated and removed from this region. They drift in the direction of the anode and the cathode, respectively, with relatively

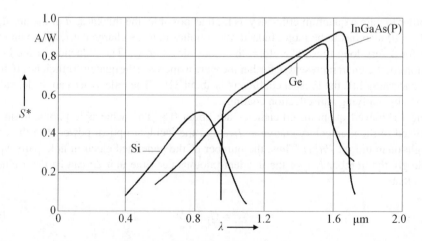

Figure 6.40   Responsivity S^* of Si, Ge and InGaAs/InP photodiodes vs. wavelength λ

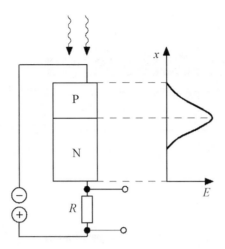

Figure 6.41 Photodiode with a simple PN junction and the according electrical field distribution *E(x)*

high drift velocities. In this region a depletion of charge carriers occurs, but outside of this zone the field strength is small. There, the electron-hole pairs move with relative low diffusion velocities. Due to the fact that the absorption of the photons mainly happens outside the drift zone, such photodiodes are relative slow.

Therefore, the use of *PIN diodes* is more favorable [6.64]. An intrinsic zone I is introduced into a PN diode, that is between the P- and the N-doped region. In this non-doped or respectively just slightly-doped region, a constant electric field over the whole I-zone is guaranteed (Figure 6.42)

The region of depletion basically coincides with the absorption zone. Therefore, slow diffusion currents are avoided [6.66]. A relative high-speed photodiode is gained. Instead of

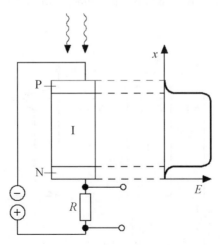

Figure 6.42 PIN photodiode and the according electrical field distribution *E(x)*, where I is the intrinsic layer

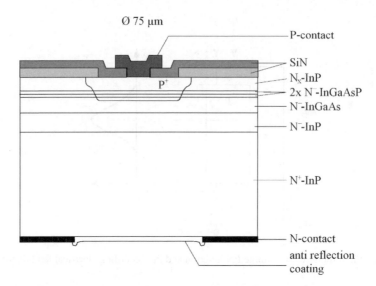

Figure 6.43   Construction set-up of an InGaAs-PIN-photodiode

the I-zone, slightly P-doped zones (π-zones) or N-doped zones (v-zones) can be introduced. The word "slightly" should be understood relatively compared to the doping of the surrounding layers. Figure 6.43 depicts a typical arrangement of a PIN diode.

The *Avalanche Photodiode (APD)* represents another often used component [6.67,6.68]. Here, the I-zone is located between two different doped P-zones (P$^+$: highly doped). The applied reverse voltage is significantly higher compared to PIN diodes. At the PN junction, a very high electric field is developed (Figure 6.44).

This leads to a high acceleration of charge carriers in this zone. They move with a very high speed. Collisions occur between the electrons and the atoms of the semiconductor material. Thus, further electrons will be released from the atoms, so that they are lifted from the valence

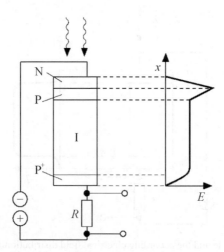

Figure 6.44   Avalanche Photodiode (APD) and the according electrical field distribution *E(x)*

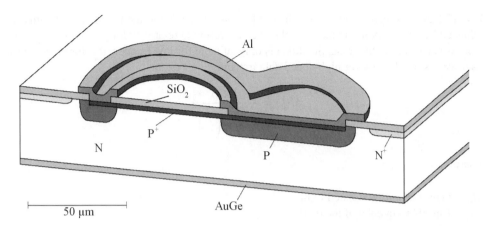

Figure 6.45 Set-up of a Ge-avalanche photodiode

band into the conduction band. Thus, secondary electrons are received, an effect which is called *impact ionization*. Afterwards, the achieved secondary electrons are also accelerated and generate tertiary electrons and so on. The electrons multiply themselves. This process leads to an avalanche of charge carriers, therefore the photodiode is given the name *avalanche photodiode, APD*. The diode has an internal gain. For Si diodes the required reverse voltages for avalanche photodiodes amount to about 160 V, and for Ge diodes approximately 30 V are required [6.69]. Figure 6.45 depicts a possible arrangement of an avalanche photodiode. Figure 6.46 shows the multiplication factor, the *gain* as a function of the applied voltage of a silicon diode.

The temperature influences the impact ionization and thus affects the gain. A multiplicity of the photocurrent appears. Gains of up to $M = 10^4$ are possible. However, only values below

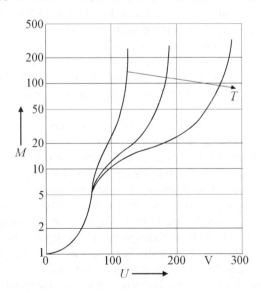

Figure 6.46 Gain M of a Si-avalanche photodiode versus applied voltage U at different temperatures

$M \approx 100$ are reasonable. Otherwise, the diode slows down due to the fact that high gains are achieved by a corresponding major avalanche. Therefore, a relatively long continuance of the process is needed. In this case, the diode is only suitable for low frequencies, for example for a sensor system, but not for high-speed data transmission systems:

$$M(f) = \frac{M(0)}{\sqrt{1 + \left(\frac{f}{f_c}\right)^2}} \tag{6.27}$$

where:

M_0 Gain for non-modulated light
f_c Cut-off-frequency of the diode

$$f_c = \frac{1}{2\pi M(0) t_M} \tag{6.28}$$

where:

t_M Delay in the gain zone

Therefore, concerning an APD, the *gain-bandwidth product* $M(0)$ times f_c is defined as the figure of merit. For silicon diodes, typical values of 300 GHz are gained, for germanium diodes 60 GHz and for quaternary diodes about 80 GHz.

An advanced operating mode of an APD represents the *Geiger Mode*. In this case, the applied voltage is biased above the breakdown voltage. Geiger-Mode APDs (G-APD) are in their set-up similar to conventional APDs. The n-doped zone is highly doped and therefore is marked with n^+ (Figure 6.44). Usually these kinds of photodiodes are built up as a composite of many Geiger-Mode APDs connected in parallel. They form an optical sensor, respectively a photomultiplier or photon detector. The parallel G-APDs can be seen as pixels in the detector. The detector is also called the Pixelized Photon Detector or Multi Pixel Photon Counter. Regarding the Si-photomultiplier, the reverse voltages amounts to about 20–70 V, depending on the manufacturer.

Due to the high gain, single photons can be detected. Gain values can be measured in the range of up to 10^6. More created charge carriers are located in the avalanche. A higher photocurrent can be measured. In this case, the electric field in the p-n$^+$ junction is now high enough to accelerate not exclusively electrons, but also holes. Thus, even holes get a sufficiently high kinetic energy to enhance the impact ionization process. In silicon, impact ionization coefficients for electrons α_n are higher than for holes α_p ($\alpha_n > \alpha_p$). Therefore, electrons contribute more strongly than holes in the creation of a charge carrier avalanche. However, due to the Geiger mode, the charge carrier avalanche spread out in two ways, which finally leads to a higher number of charge carriers and a higher gain. Due to the breakthrough behavior and the self-preservation of the avalanche, the current continues, even if there is no additional photon. Therefore, this light-independent current has to be quenched. Hence, a quenching resistor is integrated in series with each G-APD. Thus, the

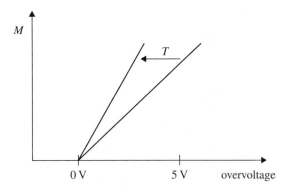

Figure 6.47   Achieved gain M in one pixel as a function of the overvoltage U_{br}

current is reduced by producing a voltage drop and thus decreasing the reverse voltage at the G-APDs.

The breakdown voltage U_{br} determines the Geiger mode and the gain value. Figure 6.47 depicts the gain as a function of the overvoltage $(U - U_{br})$. It can be seen that with $U < U_{br}$ the gain values are relatively small compared to the gain values at $U > U_{br}$. The latter condition determines the Geiger mode. The first condition is the normal mode of operation of a conventional APD. Here also the temperature influences the process of impact ionization and therefore the gain (APD) [6.72,6.73].

$$M(U - U_{br}) = \frac{C_{Pixel}\left(U - U_{br}\right)}{e} \tag{6.29}$$

where:

C_{Pixel} Pixel capacitance of a G-APD
e Elementary charge

It is necessary to keep the overvoltage $U - U_{br}$ = const. For stable measuring conditions, a constant ambient temperature is required. However, operating the device at lower temperatures, for example cooling it down with a Peltier Element, reduces thermal breakthroughs at the pixels. This leads to lower noise.

Photomultipliers or photon detectors seem to be complicated devices [6.74]. To realize their function and their time behavior when detecting a photon, an equivalent circuit diagram can help (Figure 6.48). Figure 6.49 shows the measured photocurrent by detecting an incident photon.

where:

U Reverse applied voltage
R_Q Quenching resistor
C_{Pixel} Pixel capacitance of a G-APD
U_{br} Breakdown voltage
R_B Series resistance of the microplasma in the avalanche

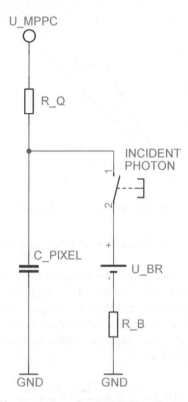

Figure 6.48   Equivalent circuit diagram of a pixel (G-APD)

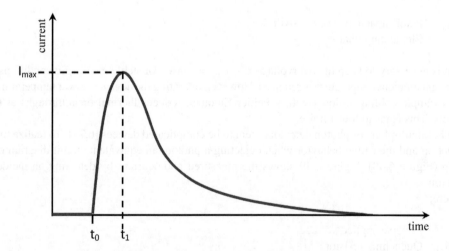

Figure 6.49   Photocurrent as a function of time

where:

I_{max} Maximum current
t_0, t_1 Moments in time

The current profile can be divided into two sections: the rising and the falling edge. Regarding the equivalent circuit diagram, the rising and falling edge can be described with the following formulas:

$$I_{max} \approx \frac{U - U_{Br}}{R_Q + R_B} \qquad (6.30)$$

$$\text{rising edge} \approx I_{max} \left(1 - \exp \left(\frac{t - t_0}{R_B \cdot C_{Pixel}} \right) \right) \qquad (6.31)$$

$$\text{falling edge} \approx I_{max} \cdot \exp \left(\frac{t - t_1}{R_Q \cdot C_{Pixel}} \right) \qquad (6.32)$$

The time constant τ_{rising} is relatively small and the entire rising edge amounts some ps to about 1 ns, therefore R_B and C_{Pixel} are very small. However, the falling edge takes longer. Thus, the time constant $\tau_{falling}$ is higher. The entire falling edge amounts to about 10 ns. During this time no further incident photon can be detected, so it is called the *recovery time* [6.75, 6.76]. The recovery time is voltage independent, but depends on the quenching resistance R_Q and on the pixel capacitance C_{Pixel}. The quenching resistance can be up to some hundreds of kΩ. Devices with larger pixels show a longer recovery time due to higher pixel capacitance.

7

Optical Transmitter and Receiver Circuit Design

A light source with a driver is called an optical transmitter. By completing the photodiode with a following preamplifier, an optical receiver is obtained.

7.1 Optical Transmitter Circuit Design

In optical transmitters, laser diodes and LEDs are applied. These components have been described in detail in Chapter 6. Both components work electrically, much like PN diodes actuated in flow directions. Therefore, a forward current is achieved. The optical power of both light sources depends on the injection current I_F. The electrical control is called the *driver*, which has to be arranged in such a way that an impressed current is gained. The current supplied by the driver should be nearly independent of the load resistance. This implies that the small-signal source resistance of the driver has to be much higher than the dynamic resistance of the laser diode or the LED, respectively. The dynamic resistance amounts to about 3 Ω. The semiconductor laser and the APD show a strong non-linear behavior (Chapters 6.1.2 and 6.2.1). Therefore, in optical communication engineering, digital modulation is often carried out, as non-linear distortions and intermodulations could cause problems [6.21,6.22].

For simple drivers, a transistor can be used as a switch (Figure 7.1). Corresponding to the digital signal, the input voltage (U_E) lies between 0 V and a maximum value (e.g. 5 V). At high levels of U_E, a high base-emitter current I_{BE} flows. Therefore, the collector-emitter voltage

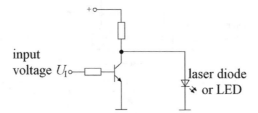

Figure 7.1 Driver circuit with a transistor as a switch for semiconductor light sources

Optical and Microwave Technologies for Telecommunication Networks, First Edition. Otto Strobel.
© 2016 John Wiley & Sons, Ltd. Published 2016 by John Wiley & Sons, Ltd.

Figure 7.2   10-Gbit/s laser module with a 1.55 μm DFB laser

U_{CE} becomes very small (e.g. < 0.6 V or < 0.2 V for silicon and germanium, respectively), depending on the applied semiconductor material.

where:

1   50 Ω strip conductor submount
2   Laser diode with coupled glass fiber
3   Monitor diode

The transistor has a fast switch and the current primarily flows through the transistor and not through the light source. The conducting transistor becomes saturated and unfortunately, at low input levels (0 V), the transistor takes a long time to exit this saturation zone. Thus, the switch-off procedure is slow in order to enable the current flow through the light source. Therefore, this type of driver is a relatively slow one.

Regarding the laser, high-frequency modulation should be obtained (Figure 7.2). Therefore the transistors of the driver should not be allowed to become saturated. Moreover, including the aspect of a digital modulation capability, the use of differential amplifier circuits is favorable [6.19]. Such a circuit is shown in Figure 7.3. The transistor T3 represents a current source,

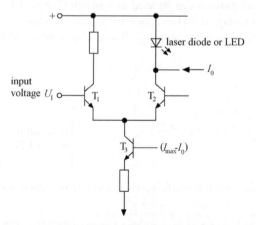

Figure 7.3   Improved driver circuit with a differential amplifier for modulation of semiconductor light sources

which will not be actuated. Relating to T1 and T2, the voltages are chosen in such a way that U_{CE} is accordingly high (e.g. U_{CE} higher than 0.6 V), thus saturation can be avoided. As a result, a fast driver is now realized. The amount of the difference between maximum and minimum current ($I_{max} - I_{min}$) is directly adjusted at transistor T3. By choosing an adequate voltage, U_{B3}, a corresponding base-emitter current I_{BE3} flows, adjusting the collector current $I_{CE3} = I_{max} - I_{min}$. This current is then switched to and fro between transistors T1 and T2. Relating to a logical zero, the following can be written as $U_{CE} \approx 0.6$ V, that is transistor T1 is turned on ($U_{CE2} \approx U_{12} \gg U_{CE}$) and transistor T2 blocks.
where:

$I_0 = I_{min}$ Bias current
$I_{max} - I_0$ Modulation amplitude

For a logical zero, the function of transistor T1 is replaced by transistor T2. The current now flows through the light source. The adjustment of the bias current $I_0 = I_{min}$ is directly carried out at the light source (e.g. driven by a monitor diode control, Chapter 6.1.2). The coupling of transistors T1 and T2 occurs at the emitter. Thus, this procedure is referred to as the emitter coupled logic (ECL) [6.19]. This is a very fast logic, with modulation frequencies of up to several tens of GHz being possible. The development of a driver for high-speed transmission systems has to be considered, not solely in relation to the laser diode [7.1,7.2]. The whole system, particularly the dispersion of the glass fiber, has to be taken into account (Chapter 3.2.1). Therefore, a specific driver is realized. Due to an optimized electrical pulse shaped by the semiconductor laser, a reduction of the bit error rate in high-speed optical systems can be achieved, called the dispersion supporting transmission (DST) [7.3].

7.2   Optical Receiver Circuit Design

An optical receiver consists of the photodiode and a subsequent preamplifier. Due to the fact that this part is placed in front of the subsequent electronic circuits for signal processing, it is called the *front-end amplifier*. A high bandwidth, a high receiver sensitivity and a high dynamic range represent the most important requirements of an optical receiver [7.4–7.6]. The receiver sensitivity is defined as the lowest optical signal level P_{min} at which the electric output signal of the receiver still achieves the required signal-to-noise-ratio. The specification is defined in dBm:

$$P_{min} = 10\text{dBm lg}(P/P_0), \qquad P_0 = 1\text{mW} \tag{7.1}$$

where:

P_{min} minimal optical input power in mW, e.g. 0 dBm = 1 mW

The difference (in dB) between the minimum detectable signal level and the maximum allowed input level is named the dynamic range of the receiver. The maximum level is defined when saturation has not yet occurred. Regarding optical transmitters, the bandwidth is defined as the total frequency range. These ranges reach from direct-current up to the cut-off frequency,

f_c, at which the absolute value of the transfer function is diminished by 3 dB. Figure 7.4a depicts the basic circuit diagram of an optical receiver. At the input of the preamplifier, the voltage drop is given by:

$$U_E = R_{eff} \cdot I_{Ph} \tag{7.2}$$

The resistance R_{eff} consists of the parallel connection of the load resistance R_L and the input resistance R_E of the preamplifier. The equivalent circuit diagram of the photodiode with the input resistance of the preamplifier is shown in Figure 7.4b. The photocurrent I_P generated by the incident photons represents a current source. The junction capacitance of the photodiode dominates as capacity C_D. In particular, it contributes to the cut-off frequency of the front-end amplifier (see below). The series resistance (bulk resistance) R_S of the photodiode lies at a few Ohm. The shunt conductance σ_D of the photodiode can be neglected due to its minor value.

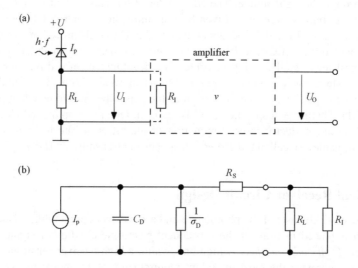

Figure 7.4   Basic circuit diagram (a) and equivalent circuit diagram (b) of an optical receiver

where:

hf	incident photon energy	U_O	output voltage
I_P	photocurrent	v	amplifier gain
R_L	load resistance	C_D	junction capacitance
R_I	input resistance	R_S	series resistance
U_I	input voltage	σ_D	shunt conductance

A PIN diode shows a linear interrelationship between the incident optical power and the photocurrent. This holds over decades and consequently for the voltage drop U_E too [7.7]. Impairments of linearity only occur at very high optical power densities [7.8]. This can lead to problems concerning analog transmission systems. However, by applying an APD, the receiver suffers from non-linearity disturbances. A high photocurrent I_M especially flows at a high amplification:

$$I_M = M \cdot I_{Ph}, \quad \text{with } I_{Ph} = I_M \text{ for } M = 1 \tag{7.3}$$

Thus, a negligible voltage drop is no longer generated at the series resistance R_S. The applied high voltage U_H is then reduced by the voltage drop $I_M \cdot R_S$. The electrical field strength is correspondingly reduced in the gain zone. Thus, the gain M depends on the photocurrent. At high incident optical power levels, the gain is less when compared to low levels. As a result, a non-linearity appears [7.9].

Therefore, the application of a PIN diode is suitable for analog systems as well as for digital ones. In contrast, the APD is inappropriate in terms of its use for high linearity requirements in analog systems. The following equation applies to the cut-off frequency f_c of the receiver:

$$f_c = \frac{1}{2\pi R_{eff} C_{eff}} \tag{7.4}$$

Here, the series resistance can be neglected, as it is much lower than the effective resistance R_{eff} given by the parallel connection of the load resistance and the input resistance of the amplifier. The total capacity C_{eff} arises from stray capacitances of the electric leads and the junction capacitance C_D of the diode. The latter is proportional to the area of the diode. In order to achieve high cut-off frequencies, the area of a photodiode has to be as small as possible. Typical diameters of fast diodes amount to about 30 μm to 50 μm. Figure 7.5 illustrates a fast optical contact probe. It is a matter of a PIN diode with a 50 Ω strip line and SMA socket without further successive electronics.

7.2.1 Receiver Circuit Concepts

Figure 7.6 shows the simplified diagram of a *low-impedance amplifier* [7.10]. For this type of amplifier, it can be written as:

$$f_c \geq f \tag{7.5}$$

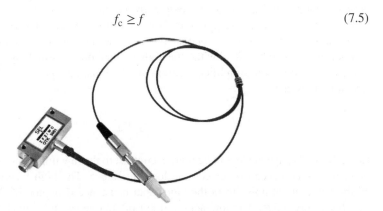

Figure 7.5 0–19 GHz optical contact probe with an InGaAs-PIN diode (50 μm diameter) coupled to a glass fiber and an optical plug/connector

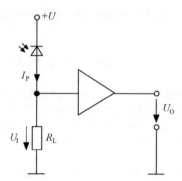

Figure 7.6   Simplified diagram of a low-impedance amplifier

where:

f Frequency of the signal to be transmitted
f_c 3 dB-cut-off frequency

Thus, regarding the effective resistance R_{eff}, it is obtained from Eq. (7.4):

$$R_{eff} \approx R_L \leq \frac{1}{2\pi f C_{eff}} \tag{7.6}$$

In this case, the load resistance is significantly lower than the input resistance of the preamplifier. Therefore, the value of the parallel resistance of the circuit results from the load resistance. The low-impedance load resistance (which also gives its name to the amplifier type) causes a high cut-off frequency and consequently a large bandwidth. However, this advantage appears at the expense of an unfavorable thermal noise from the load resistance (Chapter 7.2.2), the effect of which is to lower receiver sensitivity. Therefore, the minimum optical input power P_{min} has to be comparatively high.

Thus, the low impedance amplifier is suitable for the detection of *high bandwidth* optical signals and *high* available optical power at the photodiode, for example in labs when components have to be tested without a high attenuation, causing long optical fiber.

Figure 7.7 shows the simplified diagram of a *high-impedance amplifier* [7.10–7.12]. In this case it follows that:

$$f_c \ll f \tag{7.7}$$

Here, the load resistance has to be high in contrast to the low impedance type. Hence, the high input resistance R_E can supersede the load resistance. The high receiver sensitivity caused by the low thermal noise from the high input resistance (Chapter 7.2.2) can be treated as a significant advantage. The low achieved cut-off frequency unfortunately affects the signals inversely. Without further measures, this frequency would allow signals to be transmitted with exclusively small bandwidths only. The high time constant $\tau = R_E C_{eff}$ causes an integrating

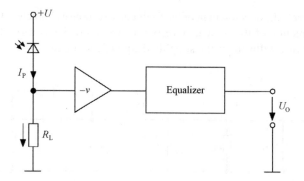

Figure 7.7 Simplified diagram of a high-impedance amplifier with a succeeding equalizer

behavior. Low frequency parts yield a high modulation amplitude, leading to saturation of the amplifier. Therefore, its dynamic behavior diminishes.

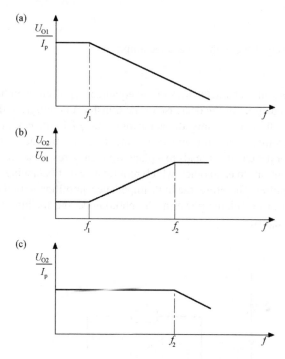

Figure 7.8 Frequency response correction by an equalizer for a high-impedance amplifier

where:

U_{O1} Output voltage of the high-impedance amplifier
U_{O2} Output voltage of the equalizer
f_1 Edge frequency of the high-impedance amplifier and the equalizer
f_2 Frequency of the overall circuit

Figure 7.8a depicts the development of the frequency-response characteristics. In order solve the problem, correction of the frequency-response characteristics is carried out by means of an equalizer. Figure 7.9 illustrates the simplified equivalent circuit.

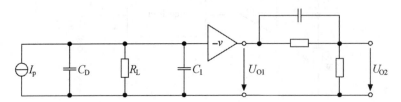

Figure 7.9 Equivalent circuit diagram of an optical receiver with high-impedance amplifier and following high-pass as an equalizer

where:

C_I Input capacitance of the high-impedance amplifier

The frequency-response characteristics of the equalizer represent a high-pass filter. It compensates to a certain degree for the decrease of the transfer function from the high-impedance amplifier (Figures 7.8b and c). Thus, the achievable cut-off frequency of the overall circuit is shifted to higher values. But only an exact coincidence of the cut-off frequency f_1 of the amplifier with that of the equalizer leads to a plane frequency-response characteristic. The two slopes have to be contrary to each other, which require an exact balancing of the equalizer for each individual amplifier. Therefore, the high-impedance amplifier is used for the detection of optical signals with low available power at the photodiode and medium to high bandwidths and bit rates, respectively.

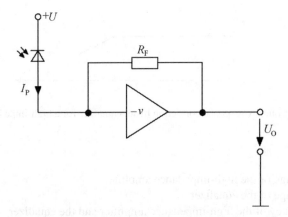

Figure 7.10   Simplified circuit diagram of a trans-impedance amplifier

The *trans-impedance amplifier* [7.2,7.10,7.13] describes a favorable compromise between both the types of amplifiers mentioned above (Figure 7.10). Concerning this type, it can be written as:

$$f_c \geq f, \quad \text{with } f_c = \frac{1+v}{2\pi R_F C_{eff}} \tag{7.8}$$

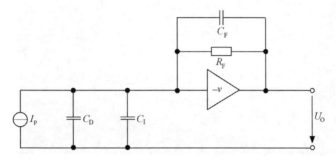

Figure 7.11   Equivalent circuit diagram of an optical receiver with trans-impedance amplifier

where:

R_F Feedback resistor
v Circuit gain

At the input of the trans-impedance amplifier, the feedback resistor seems to be reduced by about a factor of $1/(1 + v)$ compared to the physical value of resistance R_F. In comparison with the low-impedance-amplifier, this resistance can be chosen higher at about a factor of $1 + v$ to gain the same cut-off frequency. Figure 7.11 shows the equivalent circuit of the trans-impedance amplifier. Regarding the cut-off frequency, it follows for high amplifications v :

$$f_g = \frac{v}{2\pi R_F (C_D + C_E + v C_F)} \tag{7.9}$$

where:

C_I Input capacitance of the amplifier
C_F Stray capacitance of the feedback resistor R_F

Thermal noise is low due to the high physical resistance R_F. The useful bandwidth which is used by the trans-impedance amplifier without an equalizer is increased compared to the high-impedance amplifier. The bandwidth improvement is about the value of the amplification with a constant noise contribution of $R_F = R_E$. Hence, the trans-impedance amplifier is used for the detection of optical signals with low optical power at the photodiode and high bandwidth bit rate, respectively. Regarding the configuration of receivers, this type is particularly applied to optical communication engineering. Figure 7.12 illustrates a trans-impedance preamplifier for a data signal of 678 Mbit/s.

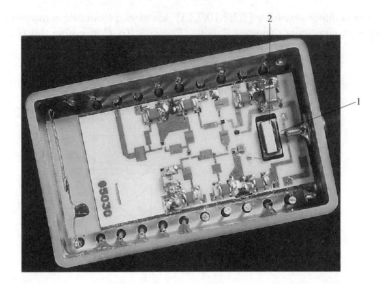

Figure 7.12   Trans-impedance amplifier with Ge-avalanche photodiode and coupled glass fiber

where:

1   Glass fiber
2   Ge-avalanche photodiode

7.2.2   Noise in Optical Receivers

Noise parameters are statistical parameters. Therefore, generally a noise current $<I_N>$ is defined as the square root of the statistically appearing noise currents squared, the root mean square value (RMS):

$$< I_N > = \sqrt{\overline{I_{\text{Noise}}^2}}, \overline{I_{\text{Noise}}^2} = \left(I_1^2 + I_2^2 + I_3^2 + \cdots + I_N^2\right)/N \tag{7.10}$$

The components also appear statistically. First, they are squared and then averaged. The thermal noise of a resistor R is one of the most important sources of noise in electronic circuits. It is defined as the RMS noise current, also named the Johnson–Nyquist noise [6.19]:

$$\overline{I_{\text{Nth}}^2} = 4kTB/R \tag{7.11}$$

where:

k Boltzmann constant $k = 1.38 \cdot 10^{-23}$ J/K
T Temperature
B Bandwidth

This noise current is the greater, the larger the bandwidth B is, and in contrast the smaller the bandwidth, the larger the resistance value that can be chosen. Due to the fact that optical radiation consists of a multitude of statistically consecutive photons, quantization occurs. This leads to a *quantum noise* in the photodiode, also described as *shot noise* [7.14]. This kind of noise already exists in the light itself, but becomes perceivable by light and photodiode interaction. Therefore, the current flow in a photodiode happens by a discrete, independent, statistically consecutive movement of electrons. For the shot noise squared, it holds that:

$$\overline{I_{\text{NSh}}^2} = 2eI_{\text{Ph}}B \tag{7.12}$$

Moreover, concerning the photodiode, further effects of noise have to be taken into account besides the shot noise [7.15,7.16]. Regarding the complete photodiode, the following result is achieved:

$$\overline{I_{\text{NPh}}^2} = 2e\left[(I_{\text{Ph}} + I_{\text{DV}})M^2F + I_{\text{DS}}\right]B \tag{7.13}$$

where:

I_{DV} Dark current of space charge
I_{DS} Surface leakage flow
F Excess noise factor

The dark current arises from thermally-generated charge carriers, even if no light exists. It consists of the *space charge dark current* and the *surface leakage current*. The latter can be mostly neglected. Regarding an APD, the dark current of space charge and the photocurrent are amplified, corresponding to the gain M, whereas the surface leakage flow remains unamplified.

Not every primary electron generates the same number of new charge carriers. Relating to the amplification, it is the matter of a statistical event, too. Thus, the root mean square noise current is amplified not only by M^2 but furthermore by an additional noise factor. Thereby the noise current increases much more than the useful photocurrent. In Eq. (7.13), this fact is taken into account by an *excess noise factor, F*:

$$F = M^x \tag{7.14}$$

where x can be considered as a material dependent parameter.

Regarding an Si-APD, the experimental values lie at $x = 0.2 \ldots 0.5$, $x = 0.9 \ldots 1$ when applied to a Ge-APD. Concerning a quaternary APD (InGaAs(P)), the value x is approximately 0.7. Thus, a long-wavelength APD owns more unfavorable properties than a short-wave one. A complete transmission system has to take into account all components. The lower attenuation of a fiber at long wavelengths is much more favorable, as the problems are raised by a long-wavelength APD. The useful photo current is amplified by the gain M, whereas the noise current increases with the power of M^{2+x}. Therefore, due to the aspects of noise, the amplification cannot be enhanced arbitrarily. Thus, an optimization of these opposite effects in conjunction with the thermal noise of the receiver has to take place.

8

Fiber-Optic Amplifiers

To improve optical communication systems, two principal ways have been studied, that is to achieve the most important goals to maximize the transmission capacity in terms of the product of achievable fiber bandwidth B and length L (Chapter 1). One technique is to work on the increase of receiver responsivity, for example by the development of coherent-optic transmission systems (Chapter 9.4). Another technique is to amplify the absolute optical power without leaving the fiber, so as to obtain an *optical amplification* [1.12,9.12] within the transmission link. For that purpose, non-linear effects in fibers or semiconductor optical amplifiers (SOA) can be applied (Chapter 3.2 and 6.1.2). Whereas SOAs still suffer from a lot of noise problems, optical fiber amplifiers have proved to be field-tested products.

Optical amplifiers can be introduced after the transmitter laser as booster-amplifiers along the fiber link as inline-amplifier and as pre-amplifier in front of the receiver. Moreover, they are used in MAN- and LAN-networks as cascaded systems for signal distribution (Figure 8.1).

After achieving sustained success concerning the use of fiber-optical amplifiers in practical systems, one of two fundamental problems of optical signal transmission, the attenuation problem is solved as far as possible. Furthermore, scientists have also been successful in solving the second fundamental problem, the dispersion, with the aid of a revolutionary technique, dispersion suppression by soliton transmission (Chapter 3.2.5).

8.1 Erbium Doped Fiber Amplifiers

Glass fibers doped with laser active materials have been realized. The dope material is Erbium, a rare-earth metal element [8.1]. Figure 8.2 shows the most important components of an *erbium doped fiber amplifier, EDFA*.
where:

λ_S Wavelength of the signal light
λ_P Wavelength of the pumping light

The key element is a single-mode fiber with a length of about 10 m and an erbium-doped fiber core. Typical Er^{3+} concentrations amount to about 100 ppm. Figure 8.3 presents the energy level diagram of Er^{3+}.

Optical and Microwave Technologies for Telecommunication Networks, First Edition. Otto Strobel.
© 2016 John Wiley & Sons, Ltd. Published 2016 by John Wiley & Sons, Ltd.

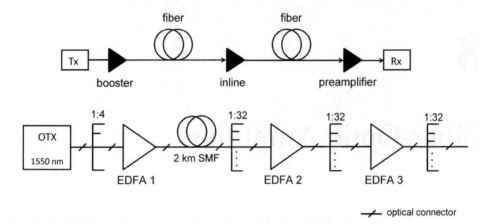

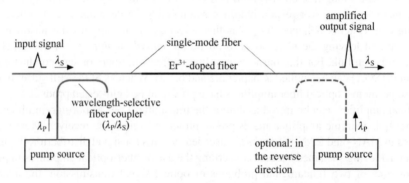

Figure 8.1 Amplifier arrangements

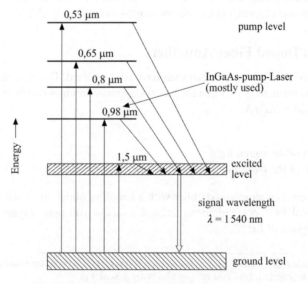

Figure 8.2 Set-up of an erbium doped fiber amplifier

Figure 8.3 Energy level diagram of erbium (Er^{3+})

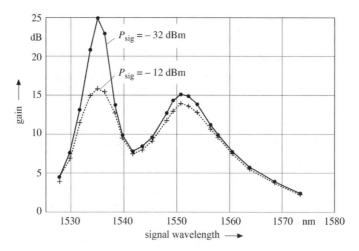

Figure 8.4 EDFA amplification (pump laser at 980 nm)

Erbium exhibits absorption bands with several wavelengths, in particular, favorable pumping wavelengths exist at 980 nm and 1475 nm. The pump light is low-loss injected into the erbium-doped fiber with the aid of wavelength selective couplers (Chapter 4.3.5, fiber-to-fiber coupling). The direction of the pump light can be fed to the fiber in the same direction as the signal light, as well as in the opposite direction; moreover, both directions are possible at the same time.

Due to the large fluorescence band in the range of about 1.5 μm, the erbium fiber amplifier is especially suitable for the wavelength region of most favorable fiber attenuation: the "third window" of optical transmission (Figure 9.1). The amplification region ranges from about 1530 nm to 1570 nm [9.12] (Figure 8.4) – meanwhile also wavelengths up to approximately 1630 nm (L-band) have been covered. A glass fiber in this region still features a low enough attenuation. It is especially interesting to realize wavelength division multiplex (WDM, DWDM, Chapter 9.1.1) over a larger wavelength region. Thus, the number of the channels can be enhanced. Maximum possible amplifications closed to 46 dB and high saturation powers of about 32 mW were achieved [8.2]. Figure 8.5 shows an overview of attained link lengths and bit rates in optical communication systems before and after the introduction of the EDFA [8.2–8.10]. Also coherent and dispersion supporting transmission systems (DST) are depicted (Chapters 9.4 and 3.2.5). [3.63,8.11–8.16]. The dashed line separates the fields of transmission results with and without optical amplification (exception: NTT IOOC 89 [8.13]).

A direct transmission with 10 Gbit/s over a link of 1500 km could be realized. Thereby, 22 erbium fiber amplifiers were used [7.10].

Besides point-to-point connections, the EDFA is also applied in optical distribution systems to compensate for losses (Chapter 9). With the aid of two amplifiers, one for the compensation of the distribution losses and another as an optical preamplifier straight in front of the receiver, a 10-Gbits/s signal could be distributed to 260 000 participants [8.16].

8.2 Fiber Raman Amplifiers

Parallel to the erbium amplifier, further developments use the transmission fiber itself as an amplification medium [8.16,8.17]. To this, the fiber-optical Raman effect [4.22–4.25] is again

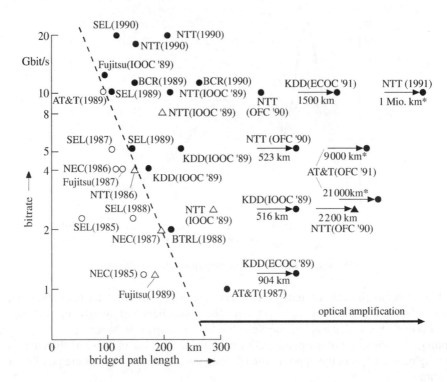

○ direct transmission,
● direct transmission with optical amplifier,
△ coherent transmission,
▲ coherent transmission with optical amplifier.

* multiple pass in a fiber coil
  with optical amplification

Figure 8.5 Bridged path length of optical high bit rate transmission experiments with and without optical amplification. The arrowed results outside this figure represent path length

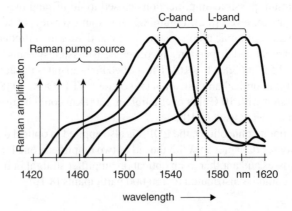

Figure 8.6 Raman pump and amplification spectra

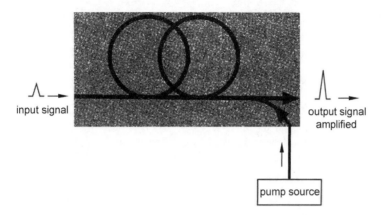

Figure 8.7 Raman fiber amplifier in reverse set-up arrangement

revived. A non-linear scattering due to molecule oscillations is being developed (phonons, see Figure 6.3). The scattered light is frequency shifted with the magnitude of the difference between the pumping frequency and those of the SiO_2 molecule oscillations. This is called the Stokes' field and Stokes' lines, respectively. The interaction of the Stokes' field and that of the optical data signal yields a stimulated Raman scattering and thus leads to an optical amplification. The pumping wavelengths lie approximately at 1450 nm upwards. Thereby amplifications are generated with wavelengths close to 100 nm shifted upwards, which are suitable for transmissions in the C-band (about 1530 to 1560) and in the L-band (Figure 8.6).

Regarding the coupling direction, the *Raman amplifier* works in reverse and thus amplifier noise problems at the receiver are avoided (Figure 8.7). In addition to the enhancement by means of the erbium amplifier, an increase of the transmission link of up to 50 km can be attained, provided that the fiber best value attenuation of about 0.2 dB/km is assumed.

9

Fiber- and Wireless-Optic Data Transmission

Figure 9.1 represents the three main components of an optical transmission by means of glass fibers. GaAs light sources and Si light drains as well as the glass fiber can be applied in a range of about 0.85 μm. Former fibers have shown a significant increasing attenuation at approximately 0.9 μm. A second favorable range at about 1.3 μm results from the use of InGaAs light sources and Ge and InGaAs(P) light drains, respectively. A third region exists at 1.55 μm. Here, the glass fiber features its absolute minimum attenuation below 0.2 dB/km. Optoelectronic devices consisting of the same semiconductor materials as in the second region are used. These three regions are also described as the *three windows* at which a reasonable optical transmission can be carried out (see the hatched areas in Figure 9.1).

Moreover, it is obvious that germanium diodes would also be applicable for the 0.85 μm range. However, based on responsivity and noise (Chapter 6.2), silicon is preferred in this wavelength range. In fiber measurement an exception is made, because the whole interesting area from 800 nm to 1600 nm can be investigated without a photodiode change.

A fiber-optic transmission link is primarily set up with the components described in Chapters 3, 6 and 7, composed of the glass fiber and the optical transmitter and receiver at the beginning and the end of the link, respectively (Figure 9.2a). In the case of a non-sufficient realizable fiber link for bridging the desired straight length, a second transmitter with an adjacent fiber optic link is located behind the receiver. The process is repeated until the required transmitting link is gained. The combination of receiver and transmitter with an intercalated *regenerator* is called a *repeater*. Afterwards, the signal distorted by fiber dispersion is amplified, regenerated and then fed to a subsequent transmitter (Figure 9.2b). In the narrow sense, these components belong to an optical transmission system. However, further electronic modules in front of the driver and behind the preamplifier (named front-end amplifier) are necessary to use in such a system regarding digital transmission with high data rates and high and long link lengths (Figure 9.7).

Optical and Microwave Technologies for Telecommunication Networks, First Edition. Otto Strobel.
© 2016 John Wiley & Sons, Ltd. Published 2016 by John Wiley & Sons, Ltd.

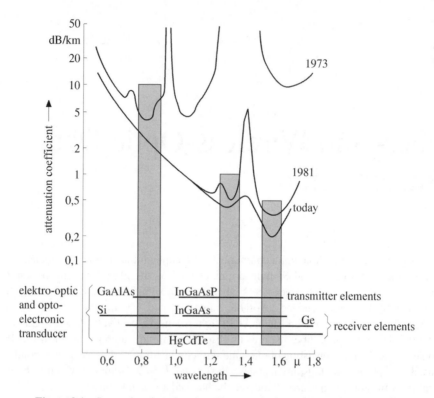

Figure 9.1   Spectral regions for glass fibers, optical transmitters and receivers

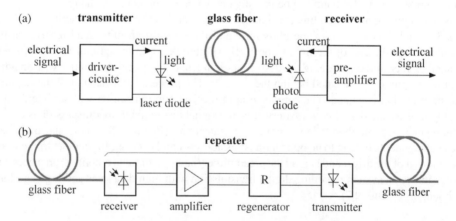

Figure 9.2   Fiber-optic transmission link (a) with repeater (b)

9.1   Direct Transmission Systems as Point-to-Point Connections

The *signal-to-noise ratio S/N* [9.1,9.2] is important for the quality of an optical transmission link. The electrical signal power S at the input of the respective amplifier arises from the effective input resistance and the photocurrent (Eqs (6.26) and (8.39)) in the field of an APD:

$$S = R_{\text{eff}} M^2 I_P^2 = R_{\text{eff}} M^2 S^{*2} P^2 \tag{9.1}$$

For the noise power N_P of the photodiode, it follows with Eq. (8.13) that:

$$N_P = R_{\text{eff}} 2e \left(I_P + I_{DV} \right) M^2 FB \tag{9.2}$$

With Eq. (8.11) concerning thermal noise of the load resistance R_L, a noise power is given by:

$$N_{Rth} = R_{\text{eff}} 4kTB/R_L \tag{9.3}$$

In addition, thermal noise of the preamplifier N'_V (including the input resistance R_E) has to be taken into account:

$$N'_V = kTBF_t \tag{9.4}$$

where:

F_t Noise factor of the preamplifier

To calculate the absolute minimum of the signal power, the noise power of the load resistance and the amplifier is often summarized ($N_V = N'_V + N_{Rth}$). The pure noise of the diodes is considered separately. Thus, the S/N can be written as:

$$\frac{S}{N} = \frac{S}{N_P + N_V} \tag{9.5}$$

The spectral noise current density I is achieved by conversion of the noise power N_V into a current with respect to the demanded receiver bandwidth:

$$I^* = \sqrt{\frac{N_V}{R_{\text{eff}} B}} \tag{9.6}$$

The declaration on data sheets is given in $pA/\sqrt{Hz}$.

The noise of a preamplifier is characterized that way and is a significant quality feature of amplifiers. By using a PIN diode the amplifier noise can be appreciated in an easy way [9.3]. At first, the spectral noise power density is measured without light incidence by means of a spectrum analyzer. The spectrum range of interest is from zero up to the cut-off frequency. After having memorized this frequency response, a second line is generated whose level at the analyzer is electronically attenuated about 3 dB. Then light from a bulb is applied to the receiver. The noise of this light can be considered as white (uniformly distributed in frequency). The power of the bulb is increased until the new line at the measured frequency coincides with the stored level. The according direct photocurrent of the PIN diode is measured.

Regarding excellent PIN diodes, the dark current noise is so low that it can be neglected compared to the amplifier noise. In this case, the latter has to be equated with the shot noise generated by the bulb. The spectral noise current density of broadband preamplifiers is so high that this measurement method is acceptable. Thus, according to Eqs (9.2) and (9.6), the spectral noise current density of the preamplifier can be determined. Thereby, it follows that:

$$I_{DV} \rightarrow 0, M = 1 \text{ and } N_P = N_V. \tag{9.7}$$

Concerning a trans-impedance amplifier with a bandwidth of 1 GHz, typical values lie at about 5 up to 10 pA/$\sqrt{Hz}$.

From Eqs (9.1) to (9.6), the following applies to the signal-to-noise ratio:

$$\frac{S}{N} = \frac{M^2 S^{*2} P^2}{\left[2e \left(I_P + I_{DV}\right) M^2 F + I^{*2}\right] B} \tag{9.8}$$

Besides the suppression of non-linear distortion products [9.4–9.6], the signal-to-noise ratio directly represents a dimension for the quality of the transmission link regarding analog systems. Concerning digital systems, the *bit error rate, BER,* can be regarded as a given parameter which must be adhered to. It characterizes the quality of a digitally transmitted signal finally applied by the user, for example the quality of an audio or a video signal. The bit error rate concerns incorrect decisions due to disturbances during the scanning of the received signal. A logic-one is interpreted as logic-zero and vice versa. The transmitted bit string with a given statistic is standardized and well known at the receiver. The latter compares it with the arriving signal. Dividing the number of incorrect decisions by the total bit number, the bit error rate is gained. Thus, such a *bit error test site* represents important measurement equipment for the evaluation of digital optical transmission systems (Figure 9.3).

It consists of a bit pattern emitter, the bit pattern generator and a bit pattern receiver, the bit error detector. In the development stage of an optical transmission system, the bit *clock* is often transmitted electrically from the bit pattern generator to the bit error detector. The

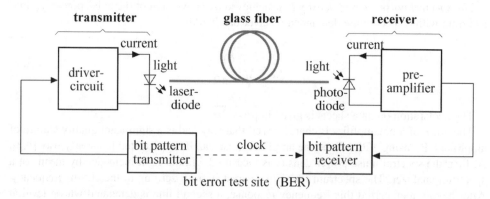

Figure 9.3   Fiber-optic transmission link with bit error test site

clock arises from the fundamental frequency of the data rate which has to be transmitted. The receiver must be able to find the correct sampling instant for the decision. In practice, the system itself has to be capable of carrying out a *clock-extraction* conductance from the transmitted signal. This is called *clock regeneration*. For high data rates and large fiber lengths, as well as for BER measurements in the lab clock, regeneration is necessary. This is due to variable pulse delays caused by temperature changes in the fiber. The optical transmission systems mentioned above are also described as direct transmission systems. Hence, they can be distinguished from coherent transmission systems (Chapter 9.4).

Regarding these direct transmission systems, the following correlation is given between BER and S/N [9.7,9.8]:

$$BER = \frac{1}{2}erfc\sqrt{\frac{1}{8}\left(\frac{S}{N}\right)} \tag{9.9}$$

Relating to very low bit error rates ($BER \ll 1$), the following approximation can be applied:

$$erfc(x) \approx \frac{e^{-x^2}}{x\sqrt{\pi}} \tag{9.10}$$

Thus, regarding the bit error rate, it can be written as:

$$BER = \sqrt{\frac{2}{\pi}} \cdot \frac{1}{\sqrt{\frac{S}{N}}} \cdot e^{-\frac{1}{8}\left(\frac{S}{N}\right)} \tag{9.11}$$

Typical bit error rate requirements of an optical transmission link amount to about $BER = 10^{-9}$. At this error rate, for example, it is no longer possible for the human eye to recognize errors on a screen. Noise powers according to the definition of noise currents (Chapter 7.2.2) are concerned as average powers. In contrast, the peak power is often chosen for the signal power. Therefore, the optical maximum power has to be correspondingly used in the signal-to-noise ratio involved equations (Figure 9.4). For the signal-to-noise ratio in Eq. (9.11), it follows from the above condition that:

$$S/N \approx 144, \text{i.e.} \approx 21 \text{ dB} \tag{9.12}$$

This value has to be adhered to all optical transmission systems. It represents an important parameter in optical communications engineering!

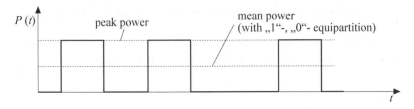

Figure 9.4 Peak and average power of a digital-optical signal

In the literature, the indication $S/N = 36$ according to 15.6 dB is sometimes found. In this case, not the peak power but the average output power, the half value of peak power, has been used for definition. Thereby, an equal distribution of logic-zeros and logic-ones is assumed over a longer period. Presupposing a bit error rate of $<10^{-9}$ governed by the Eqs (9.8) and (9.12) for the minimal optical power, it is given by the *receiver sensitivity* $P_{\min}$ at the receiver diode (peak power):

$$P_{\min} = \frac{12}{MS^*} \cdot \sqrt{\left(2e\left(I_P + I_{DV}\right) M^{2+x} + I^{*2}\right) B} \tag{9.13}$$

With typical values for receiver sensitivity, $P_{\min}$, shot noise, dark current noise as well as amplifier noise, a rough estimation for practical application is obtained [9.9]:

$$P_{\min} \text{ in dBm} = -34 + 10 \lg BR \text{ in Gbit/s} \tag{9.14}$$

where:

BR   Bit rate

For practical use, the processing of a digital signal with the bit rate *BR* depends on a receiver with following bandwidth B [9.10]:

$$B \approx (0.7 \text{ till } 0.8) \text{ times } BR \quad \text{or} \quad BR \approx 4/3 \text{ times } B \tag{9.15}$$

Theoretically, the Nyquist Theorem says *BR* could be twice B depending on the quality of the rectangular nearing pulses, again $BR \approx 4/3$ times B is a good approximation for practical application [6.26). That has to be considered as an estimation too. The exact correlation is basically determined by the time-dependent behavior of the pulse [9.11].

Dark current noise and amplifier noise represent limits. But it is possible to go below these limits up to a certain degree by means of improvements to these devices. However, the shot noise describes the absolute limit of physical noise in an optical transmission system. The generation of charge carriers exhibits a statistical process. The latter obeys the *Poisson distribution $w(q)$*:

$$w(q) = N_L^q \frac{e^{-N_L}}{q!} \tag{9.16}$$

where:

$w(q)$ Probability for the generation of charge carrier pairs to a certain time q
N_L Average number of charge carrier pairs generated over a long time period

In case of no generation of charge carriers, in spite of incident photons on the diode of the receiver, it follows that $q = 0$. Thus, with Eq. (9.16), it can be written as:

$$w(0) = e^{-N_L} \tag{9.17}$$

For $N_L = 20$ it follows $w(0) \approx 2.1 \cdot 10^{-9}$
For $N_L = 21$ it follows $w(0) \approx 7.6 \cdot 10^{-10}$

By presence of 21 charge carrier pairs, the probability of misinformation is $<10^{-9}$, which is the typical value of the required bit error rate of a transmission link. From Eqs (6.23) to (6.25), the following for the minimal power of an optical pulse P during Δt is given by:

$$P = \frac{N_\mathrm{L} hf}{\eta \Delta t} \qquad (9.18)$$

Thus, with the bit rate BR ($\Delta t = 1/BR$), it follows that:

$$P = 21hf\, BR/\eta \qquad (9.19)$$

As described above, instead of the maximum power, the average output power is sometimes indicated. Therefore, the half value from Eq. (9.19) is also found in the literature.

Experimentally gained results of the minimal receiver sensitivity as well as theoretically achievable receiver sensitivities of Eq. (9.19) are illustrated in Figure 9.30 [9.14] as a function of the bit rate. Thereby, the average output is presupposed and a quantum efficiency of $\eta = 1$ is chosen. At a given transmission power at the fiber end of the light source, a certain attenuation of the transmitting fiber can be bridged until the limit of the receiver sensitivity is reached by the power at the end of the link. This yields the achievable maximal length of the transmission link and therefore the system range (Figure 9.5a). Provided that this is the sole reason for the limitation of the system range, the system is then called *limited by attenuation*. In contrast, if the limit is exclusively due to dispersions, this is described as *limitation by dispersion* (Figure 9.5b). Due to dispersion after Eqs (3.48), (3.50) and (9.15) $BR \approx 4/3$ times B, it follows for the transmittable bit rate:

$$BR^2 \approx \frac{4}{9} \cdot \frac{1}{(M_\mathrm{mod}\, L^\gamma)^2 + (M_\mathrm{chr} L \Delta \lambda)^2} \qquad (9.20)$$

Besides the calculation of the system range due to limitation by dispersion, it has to be examined if the dispersion of the fiber and the spectral width of the light source is low enough at the calculated link. Otherwise, a bit rate corresponding to Eq. (9.20) is not ensured. Moreover, the transmitter has to be able to analyze the necessary modulation bandwidth. This also correspondingly applies to the demodulation bandwidth of the receiver.

Our main interest typically is high-bitrate long-distance systems, but there are also applications in much shorter distances and much smaller data rates to be transmitted. Figure 9.6

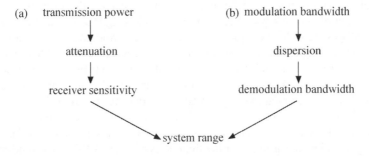

Figure 9.5 Limitation by attenuation and dispersion of optical transmission systems

Figure 9.6 Optical transmission system for the local area with V.24 interface, optionally with plastic fiber (left) or glass fiber (right)

shows a photograph of components of an optical transmission system for industrial application in the local area. It can be alternatively activated with a plastic fiber or a graded index glass fiber of 50/125 μm. A distance of up to 125 m is bridged by means of a plastic fiber at a data rate until 20 kbit/s. Taking glass fibers, the achievable distance amounts to 4.4 km.

In laboratories, the frequency response of the total transmission link is detected with the aid of a network analyzer. Thus, it can be determined if the demanded bandwidth is achieved. This also applies to the frequency response of the individual components. Moreover, quality fall-offs or transient overshoots in the frequency response can be analyzed.

In Figure 9.20 [9.12], experimentally gained link length and bit rates of intensity modulated direct transmission systems for the long-distance traffic are illustrated.
where:

a) Transmitter side
b) Receiver side
c) Repeater or booster amplifier

Figure 9.7 depicts the electronic parts of a digital optical long-distance transmission system in detail. These transmission systems work with line encoding, so that the signal can be optimally adapted to the link [9.15]. During data communication, this enables a simultaneous performance control and an error correction. In the following it is assumed that the signal is already digitally existent, for example as CMI code (see partial picture in Figure 9.7(a)). By means of a code converter, this code is transformed to a binary code and led to a *scrambler*. Scrambling is necessary to prohibit a transmission of a long cycle of logic-zeros or logic-ones (Chapter 7.22). The 5B/6B code often is suitable for optical communications engineering [9.15], in particular concerning the transmitted power-density spectrum.

Afterwards, a light source driver follows. Here, a control amplifier is included for working point adjustment (Chapter 7.22). The light source, normally a semiconductor laser, becomes modulated by the injection current. For low-bit rate systems, an LED can also be applied. The

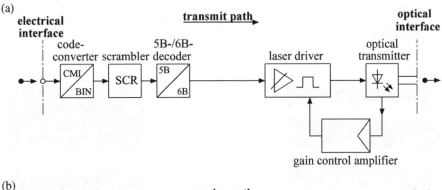

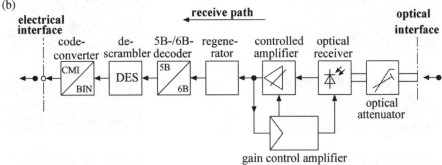

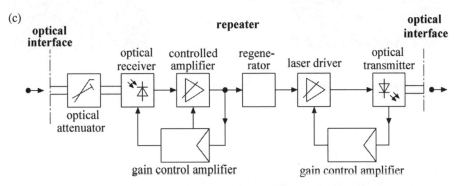

Figure 9.7 Digital-optical long-distance transmission system

transmission systems are laid out in such a way that a certain system margin – normally a 3 dB margin –finally exists. Moreover, for initial operation of a system, an *optical attenuator* is applied in front of the photodiode (Figure 9.7b). Hence, an override of the optical receiver and too high demands on its dynamic range are eliminated, respectively. Relating to a potential decrease of the optical power at the end of the fiber link with increasing service life, the set value of attenuation can be degraded to the necessary amount. Then in the receiver, the optical signals are again transformed into electrical ones. Owing to a following *Automatic Gain Control (AGC)* amplifier (automatic-gain-control), a constant electric signal level is achieved for electronic subsequent processing.

Occurrence of dispersions causing signal distortions are tolerated to some extend. However, they are not permitted to exceed a certain value so that the subsequent regenerator is able to bring the distorted signal back to its original form, that is, to regenerate it.

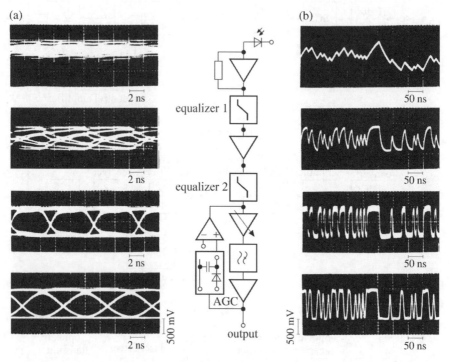

Figure 9.8   Eye pattern (a) and 168 Mbit/s data signal (b) of an optical receiver

where:

AGC:	Automatic Gain Control
top trace:	Signal without equalization
second trace:	Signal after a first equalization
third trace:	Signal after a second equalization
bottom trace:	Signal after low-pass [9.14]

How apparently indefinable signal types still can be regenerated is illustrated in Figure 9.8b (top trace). A well distinguished 168 Mbit/s NRZ (non-return to zero) signal is already available after two stations of the equalizer. A subsequent low pass suppresses higher frequent noise components (bottom trace). The *eye pattern* [9.11] of the corresponding signal is depicted with a higher time resolution in the partial picture (a). Therefore, several arriving digital bit strings of certain lengths are recorded one upon the other. Thus, in many places, a logic-one coincides with a logic-zero and vice versa. Regarding an ideal signal, the result would be a rectangle with the signal amplitude as height and the bit period as width. With deteriorating signal the vertical and horizontal limitations show an increasing noise behavior. An eye is developed, which becomes more and more shut with increasing signal distortion (see supreme line in Figure 9.8a).

Therefore, a visual estimation of the signal quality can be carried out by means of the eye pattern. However, a quantitative statement is gained with the aid of a bit error test site (Figure 9.3). Sampling instant and the threshold of the signal level are exactly set in the middle of the eye. After the regenerator, the processes at the input of the following transmitter are again reversed in the decoder, the *descrambler* and code converter. Thus, the original signal is gained. It can be committed to an electric interface, which complies with the input one (Figure 9.7a).

If the system link length is not large enough to bridge the required straight length, the driver circuit of a light source again follows the regenerator instead of the decoder. The light from that source is then coupled into a further fiber. Thus, a repeater is realized. This procedure is repeated for as long as the required straight length is achieved.

9.1.1   Unidirectional, Bidirectional and Multichannel Systems

Point-to-point connections as yet described are also called *unidirectional system* (Figure 9.9a). But in practice, for example regarding telephone or data communication, it is often necessary to transmit the information in forward and reverse directions. This can be realized by use of a second fiber.

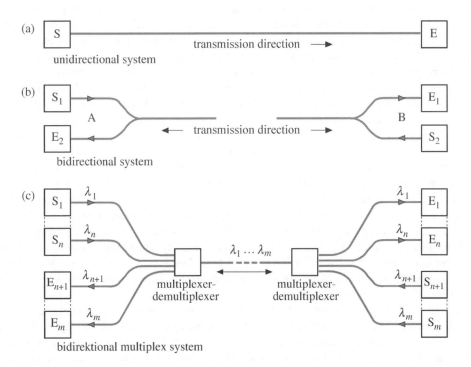

Figure 9.9   Concepts of fiber-optical transmission systems

Concerning long fiber lengths, it is worthwhile passing just one fiber and applying optical couplers on both ends (Chapter 5.2.3) [9.16]. Figure 9.9b represents such an arrangement, a bidirectional system, with a transmitter and a receiver at both fiber ends. Therefore, it has to be taken into account that each coupler represents a 3 dB splitter, so a minimum loss of 3 dB is generated inherently. Moreover, the fact that light from the transmitter A reaches the receiver A can pose a problem in the case of an override of the latter [9.17]. This can be caused by cross-talk in the coupler due to scattering effects. Fortunately, in general, this effect is negligible. The feedback effects caused by Fresnel reflections at impact locations, for example by connectors or bad splices, are much more important. Reflection decreasing measures are here indispensable (Chapter 5.2).

Instead of two transmitters and receivers at each side of the link, a multiplicity can be located. The assignment of the A-sided sender X to the B-sided receiver Y can be executed by the choice of appropriate electric frequency bands (time division multiplex and frequency multiplex systems [1.12]). However, the high losses of division and the undesired cross-talk of several frequency bands, particularly based on the non-linearity of the optical transmitters, pose a problem.

In contrast, the arrangement of so-called *Wavelength Division Multiplex* systems (WDM) represents an advantageous method [1.12,9.18–9.20]. Thereby, the lasers emit at different wavelengths. Choosing high differences of wavelengths according to the characteristics of the attenuation of a glass fiber and the varied electro-optical devices, *window multiplex* is described. Therefore, the transmission occurs at a wavelength λ of 800 nm and 1300 nm, with corresponding light sources and photodiodes being insensitive to the undesired wavelength (e.g. GaAs laser with silicon diodes and quaternary laser diodes with quaternary photodiodes). However, not only two channels but more channels are required for transmission systems. Therefore, semiconductor lasers with only a few nm neighbored wavelengths are chosen ($\Delta\lambda$ from about 2 nm to 10 nm). The insertion of the transmitter and the separation of the receivers are carried out by means of wavelength selective couplers, so-called multiplexers and de-multiplexers (Chapter 5.2.3). In this case, it is favorable that the 3 dB losses are omitted in comparison with non-wavelength selective couplers. Regarding the de-multiplexer, high demands on cross-talk are necessary. For this reason, grating multiplexers with higher attenuation concerning cross-talk are often used. Figure 9.9(c) shows a schematic diagram of a bidirectional wavelength division multiplex system.

Considering this subject in the frequency range, the distances of wavelengths from 2 nm to 10 nm correspond to 100 GHz up to 1000 GHz. These values are much greater than typical modulation bandwidths of a few 10 GHz. A narrower channel spacing is referred to as a *Dense Wavelength Division Multiplex* (DWDM). Channel spacing below 100 GHz and 50 GHz are gained. In the wavelength range, this corresponds to a distance of less than a half nanometer (Eq. (2.25)). Thus, the transmission capacity can be raised drastically by the number of channels, N, $C_t = N \cdot B \cdot L$ (Eq. 1.1). In addition, data streams with a low bit rate are combined in high bit rated ones in all these methods; this is described as the *Time Division Multiplex* (TDM). Meanwhile, we are talking of PBit/s or even TBit/s.

To utilize the available bandwidth of the fiber to a major degree, the channels have to be much narrower, adjacently positioned. This leads to the *optical frequency multiplex*, feasible by means of coherent techniques (Chapter 9.4). Such multiplex systems are applied for glass fiber networks, public communication networks or data processing systems, for example office communication or technically scientific fields in research, development, construction and production [1.24].

9.2 Orthogonal Frequency Division Multiplex (OFDM) Systems

Dr. Ronald Freund and Dr. Nicolas Perlot
Fraunhofer Heinrich Hertz Institute, Berlin, Germany

In recent years, Internet traffic has been growing exponentially. According to the latest traffic forecast from Cisco for the period from 2013 to 2018 [9.21], capacity demand is still growing with annual growth rates of 24 to 40 and 5 to 20% for the metro and long-haul region, respectively, resulting in an estimated average capacity growth rate of 21% per year. As a result, larger-capacity optical fibre transmission systems are required to build up future optical core and metro networks.

9.2.1 Approaches to Increase Channel Capacity

So far, *wavelength division multiplexed* (WDM) systems with up to 100 Gb/s per channel have been deployed in the backbone networks. Transmission systems with an optical bandwidth of 63 nm and 150 channels at 100 Gb/s each, which results in a total capacity of 15 Tb/s, are commercially available [9.22]. Those systems use polarisation multiplexed *quadrature phase-shift keying* (QPSK) modulation, coherent detection and complex digital signal processing units for compensation of transmission impairments (e.g. chromatic and polarization mode dispersion). However, it is predicted that optical transport systems have to support 100 Tb/s per fibre in future backbone networks [9.23]. Scaling current systems to the expected capacity demands in a cost-efficient way is challenging, as it suffers especially from electrical and optical bandwidth limitations of the sub-systems and components, as well as the processing speed of the digital signal processing units.

To cover future capacity demands, research activities are undertaken in three main directions:

1. increasing the spectral efficiency by using higher-order modulation formats and smaller channel spacing (within the bandwidth of the Erbium-doped amplifier)
2. extension of the optical bandwidth by using Raman amplification and wideband optical amplifiers for additional wavelength regions
3. using multi-core and few-mode fibres to increase the number of available modes within the transmission fibre

The latter requires the installation of new fibres and is therefore especially of interest for new builds. Consider, for example, 4 modes per fibre core and 8 cores per fibre, this would increase transmission capacity by a factor of 32. However, the realization of cost-efficient amplifier units for those systems is still a challenge. Figure 9.10 gives an overview of research directions toward higher transmission capacity.

Due to the limited bandwidth of the optical amplifiers, it is especially necessary to increase the spectral efficiency within their bandwidth in order to further increase available system capacity. Technologies on the way to higher spectral efficiency are higher-order optical modulation formats, which lead to a reduction in symbol rate and spectral width and therefore allow a smaller channel spacing in WDM systems. Thereby, the choice of a particular modulation format and symbol rate is a trade-off between spectral efficiency and robustness towards transmission impairments, such as chromatic and polarisation mode dispersion as well as fibre non-linearities.

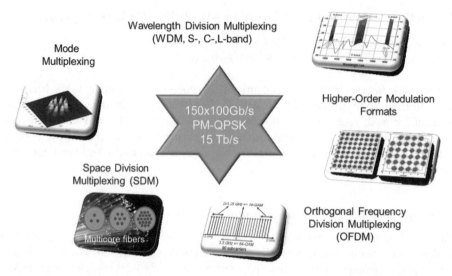

Figure 9.10   Techniques to increase transmission capacity for future demands

Moreover, increase in symbol rate is not necessarily related to a higher overall transmission capacity, unless spectral efficiency is increased. In order to increase the overall capacity per channel, as well as its spectral efficiency, while having to increase the bandwidth of the electrical and opto-electrical components, it is discussed to form so-called optical "*superchannels*" [9.24]. That means to group several lower bandwidth channels into one high capacity "superchannel". This, however, requires spectrally efficient multiplexing of the individual lower bandwidth channels, in order not to waste bandwidth by using spectral guard bands. Figure 9.11 gives an overview of several investigated modulation formats to reach higher spectral

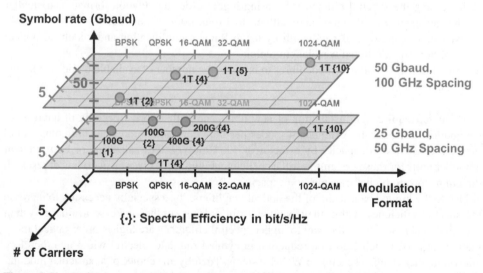

Figure 9.11   Options to generate high-speed transmission systems, depending on the number of subcarriers, symbol rate for selected modulation formats and polarization multiplexing

efficiency and aggregated capacity per channel (100G: 100 Gb/s to 1T: 1 Tbit/s) considering polarization division multiplexing (PM). For an aggregated capacity of 400 Gb/s, a two-carrier approach using 16 *quadrature amplitude modulation* (QAM) and PM is currently an attractive candidate for commercial products.

9.2.2   Fundamentals of OFDM

Orthogonal frequency division multiplexing is a multi-carrier modulation technique that uses orthogonal subcarriers to multiplex a number of low-rate data signals into a single channel for transmission. The first patent concerning this multiplexing method was published in 1966 [9.25]. The difference between OFDM and regular frequency division multiplexing (FDM) is that the different subcarriers in an OFDM system are spectrally overlapping and thus cannot be separated by simple filtering. The spectral spacing between the individual subcarriers is given by $\Delta f = 1/T_S$, where T_S is the OFDM symbol duration. One OFDM symbol consists of several Δf spaced unmodulated oscillations, each having different amplitude and phase according to the encoded data symbol. Consider a multiple of these OFDM symbols, which lead to a rectangular modulation of the individual oscillations in the time domain with a symbol rate of Δf. To separate the individual subcarriers, each subcarrier has to be shifted to the baseband and an integration over the OFDM symbol period (Figure 9.12) has to be performed. This ensures ideal separation of the subcarriers on the one hand and matched filtering and thus maximization of the signal-to-noise ratio (SNR) per subcarrier on the other. The discrete Fourier transform (DFT) exactly performs this operation on all subcarriers simultaneously. Since the DFT process can be simplified and the complexity can be significantly reduced by using the fast Fourier transform (FFT) instead [9.26], OFDM has become very popular over the last 20 years.

Today, OFDM is one of the most used multiplexing techniques in radio systems. Figure 9.13 shows a block diagram for OFDM processing at the transmitter and the receiver. After serial to parallel conversion, the modulation of the subcarriers is performed by using an inverse FFT (IFFT). Insertion of a cyclic prefix extension is used to mitigate for inter-symbol interference, thereafter the digital (and usually complex) time domain signal is converted into an analogous signal by using DACs. This baseband signal is usually up-converted to an appropriated RF or optical passband, using an electrical or optical *IQ modulator*, respectively. At the receiver side, the OFDM signal is down-converted, digitized by analog-to-digital converters and then demodulated using FFT and baseband signal processing to recover the data. The advantage of OFDM in radio systems is that it can be explained by its adaptability: the individual low bandwidth subcarriers can carry different modulation formats and/or power levels and can

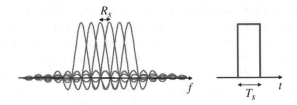

Figure 9.12   ODM spectrum (left) and time-domain pulse (right)

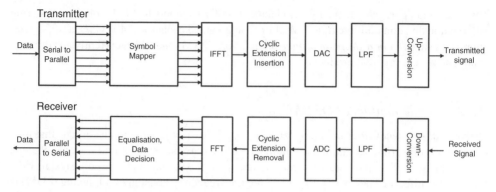

Figure 9.13 Block diagram for OFDM signal processing: (I)FFT: (inverse) fast Fourier transform, DAC: digital-to-analog converter, ADC: analog-to-digital converter, LPF: low-pass filter

therefore be adjusted to the frequency response of the channel. This makes OFDM transmission very robust, especially in a fading channel scenario, as in radio systems.

The first publications concerning OFDM for optical transmission systems were almost simultaneously published in 2006 by Lowery et al. [9.27] and Shieh et al. [9.28]. Both proposed OFDM for adaptive dispersion compensation in long-haul WDM systems. However, Lowery considered an optical direct detection receiver in combination with an electrical IQ demodulation, while Shieh proposed a coherent optical receiver set-up. While the direct detection scheme of optical OFDM offers simpler implementation and therefore lower cost, it is less spectrally efficient due to the need of spectral guard bands to suppress unwanted intermodulation products [9.27,9.29,9.30]. For this reason, almost all long-haul and high spectral efficient optical OFDM systems are designed as coherent systems. Many experimental realizations have been shown so far [9.3,9.32,9.33]. For a more detailed discussion about the adaption of OFDM to the optical channel, see [9.34].

9.2.3   Implementation Options for Coherent Optical OFDM

Coherent optical OFDM systems can generally be subdivided into either electrically multiplexed or optically multiplexed OFDM systems. In the first case, the individual sub-channels are generated and multiplexed digitally by the use of a digital IFFT operation at the transmitter. This OFDM signal is then modulated onto the optical carrier with an optical IQ modulator. This optical carrier is transmitted, coherently detected and the individual sub-channels are separated again with a digital FFT operation in the receivers' digital signal processing (DSP).

Optically multiplexed OFDM signals in contrast are generated out of several orthogonal optical carriers, which are individually modulated as rectangular as possible non-return-to-zero (NRZ) electrical driving signals using optical IQ modulators. Very stable optical carriers (in frequency and phase) have to be used, in order to ensure orthogonality between them and thus guarantee cross-talk-free de-multiplexing of the individual channels. Therefore, in most of the experimental demonstrations several coherent laser lines, generated by so-called "comb sources", were used. Several methods are being proposed to generate such sources [9.24,9.35]. The separation of the optically multiplexed *sub-channels* can then either be performed electrically or optically. In the first case, all sub-channels must be detected with one optical coherent front-end and a further separation done by using DSP [9.36]. In the latter case, the FFT operation is performed optically. An efficient method to implement an optical FFT was proposed in [9.37]. Since then, several integrated optical FFT devices have been demonstrated, for example in [9.38–9.41].

Recently coherent optical OFDM is not only considered as a multiplexing technique of lower rate subcarriers within one optical channel, but also as an option to multiplex several optical channels in order to form one "*superchannel*", as discussed in the Section 9.2.1. These "superchannels" can be formed out of electrically- as well as optically-multiplexed OFDM signals. However, when considering the multiplexing of OFDM signals into an optical "superchannel", the stability of the optical carriers is of great importance for both electrically- and optically-multiplexed OFDM signals. Several experimental demonstrations of either electrically- [9.44,9.45] or optically-multiplexed [9.36] Tb/s "superchannels" have been shown.

One impressive experimental demonstration of an all-optical OFDM "superchannel" experiment is reported in [9.46]. It uses a mode locked laser with a highly non-linear fibre as a comb source to generate the unmodulated subcarriers in the transmitter and a low-complexity all-optical FFT circuit to de-multiplex the tributary sub-channels in the receiver. In this way, an all-optical FFT has been successfully used to extract single subcarriers from a 336-subcarrier OFDM band for analysis. The subcarriers were subsequently demodulated with receivers corresponding to the particular modulation format used. This scheme has been demonstrated at OFDM channel bit rates exceeding 25 Tb/s and has thus been proven to be scalable far beyond the Gb/s per channel regime.

As an example of an electrically-multiplexed OFDM "superchannel", an experimental demonstration reported in [9.47] is described in the following. Five optical carriers are generated and individual lines are separated into odd and even channels and modulated with an optical IQ modulator driven by the in-phase and quadrature component of a pre-calculated OFDM signal consisting of 256 subcarriers. The bandwidth of this signal was 6 GHz and the modulation formats were 64QAM for the inner 90 subcarriers and 16QAM for the outer subcarriers. With this configuration, a net bit rate of 45 Gb/s per OFDM band could be realized, which results in a net bit rate of 225 Gb/s for the whole "superchannel". Since the individual OFDM bands were not temporally synchronized in the first measurement set-up, the resulting crosstalk forced the channel spacing to be at least 6.25 GHz, which resulted in a spectral gap of 250 MHz between the OFDM bands. In a second step, the individual bands were properly temporarily aligned, which drastically reduced the cross-talk between them and consequently allowed for a channel spacing of exactly 6 GHz, meaning no spectral gap between them.

The resulting spectrum after the transmitter is shown in Figure 9.14a. Additional to the higher spectral efficiency, also the modulation format of the outer subcarriers could be increased to a 32QAM format, which then results in an increased overall capacity. The results for both, the unsynchronized and synchronized case after 80 km transmission and also the resulting constellation diagrams, are shown in Figures 9.14b and 9.14c, respectively.

These demonstrations confirmed that OFDM, regardless of the actual implementation of the multiplexing technique, can be a viable option for spectrally efficient multi-Tb/s transmission. To overcome the need for coherent laser sources, recently also a method was proposed, where free running lasers instead of a coherent "comb source" could be used to generate and also to de-multiplex an optically multiplexed OFDM signal [9.42,9.43]. However, this method

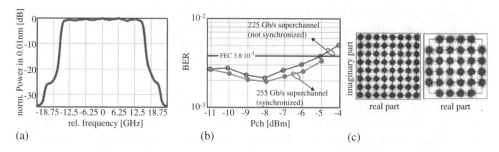

(a)   (b)   (c)

Figure 9.14 (a) shows the optical spectrum of the 255 Gb/s superchannel after the transmitter, (b) the measured BER as a function of the fiber input power per channel for both, synchronized and non-synchronized tributaries and (c) received constellation diagrams for 64QAM and 32QAM

requires a rather complex optical set-up and is therefore meant to be a proof-of-concept experiment.

9.2.4 Nyquist Pulse Shaping as an Alternative to OFDM Systems

Besides the previously discussed OFDM solutions for Tb/s "superchannels", recently also so-called *Nyquist pulse shaping*, or *Nyquist WDM*, is considered as an alternative approach to spectrally efficient multiplex several channels [9.48,9.49]. This method relies on the fact that a band limited data signal is sufficiently described by all frequencies smaller or equal to the Nyquist frequency f_N. It ideally uses a rectangular filter to remove all undesired frequencies higher than f_N, which is equal to half the symbol rate and thus allows for a channel spacing equal to the symbol rate. However, an ideal rectangular filter in the frequency domain results in infinitely wide sinc-shaped pulses in the time domain (Figure 9.15) and is therefore not realizable. Therefore, in real systems, raised-cosine filter shapes with a slightly larger bandwidth are used, which in turn lead to sub-channel spacing larger than the symbol rate. The steep filtering of the individual sub-channels within a "superchannel" can very accurately be performed in the electrical domain within the transmitter DSP [9.51,9.50]. Since there is no overlap of the individual sub-channels, in a Nyquist WDM system, they can be separated with a steep, but relatively simple filter at the receiver.

Recently, many investigations concerning the numerical [9.53,9.54,9.55] as well as experimental [9.52] comparison of Nyquist pulse shaping and OFDM were published. It turns out that the main advantage of OFDM, namely the adaptive choice of the modulation format and

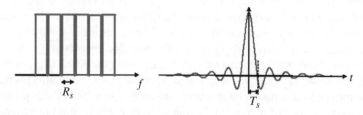

Figure 9.15 Nyquist-shaped WDM system (left) and time-domain pulse (right)

power level per subcarrier, is less important for the relatively flat optical channel compared to the radio channel. Furthermore, Nyquist WDM does not rely on coherent optical carriers, and is therefore, in contrast to OFDM, able to operate with free running optical laser sources. Several more points in favour and against the individual approaches are discussed, but it turns out that the Nyquist pulse shaping is more likely to be implemented in near-future optical "superchannel" systems.

9.3   Optical Satellite Communications

Dr. Ronald Freund and Dr. Markus Nölle
Fraunhofer Heinrich Hertz Institute, Berlin, Germany

In recent years, *free-space optical communication* has become attractive for data transmission between Earth orbiting satellites, optical ground stations (OGS), aircraft, high-altitude platforms (HAP) and unmanned aerial vehicle (UAV). In comparison to radio frequency (RF), optical communication benefits especially from high data rates, low power consumption and low payload size. Moreover, there are no frequency and bandwidth regulatory restrictions for free-space optical communication. Several satellites are already equipped with laser communication terminals and more will follow in the coming years [9.56]. The following sections give an overview of the channel characteristics and fields of application with the focus on *optical satellite communication*.

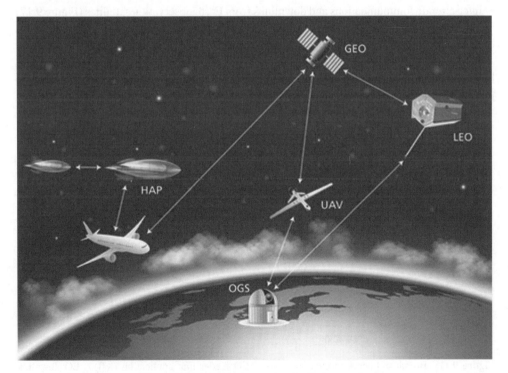

Figure 9.16   Laser communication link scenarios

9.3.1   Applications of Optical Satellite Communications

Satellite networks can provide various services, such as communication, broadcasting, navigation and Earth exploration. A major change is occurring to include the end consumer in this customer mix, with direct to the home, direct broadcast satellite, mobile telephony, and Internet access as the primary services. This change is accelerating as a result of the insertion of new technology onto satellites; namely increased power, phased array antennas, large diameter antennas, multi-spot beam antennas, and on-board processing and switching. These changes allow satellites to be a cost-effective competitor to cable TV, cellular telephony, and high-bandwidth Internet service providers in many parts of the world. Today, RF systems are widely used in satellite communication. However, optical satellite communication systems are rapidly evolving for satellite and terrestrial applications, mainly driven by the demand for high bandwidth and the progress in components and sub-systems for fibre-based telecommunication. Optical satellite communications is seen as a way to increase speed, transparency and security, especially for the following applications [9.57]:

- temporary communication in the case of cyber attacks or an emergency after natural disasters when terrestrial communication links have broken down
- secure high-speed links for special operations forces, encampment and temporary control rooms
- connection of temporary communication (e.g. high density TV transmission) capacity during mass events
- inter-aircraft communications and downlinks from Earth observation satellites (Figure 9.17)

Moreover, high-speed optical communication between *geostationary satellites* (GEO) and *optical ground stations* (OGS) is an attractive way of increasing the data traffic of high-bandwidth multimedia services to the customer. The OGS represents the gateway operated by the service providers. The transmission path for channels between the gateway(s) and satellite

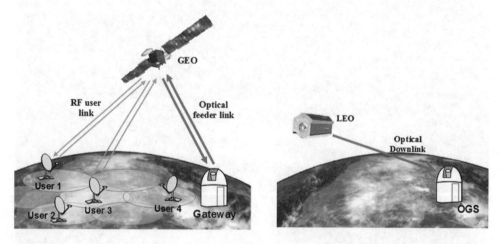

Figure 9.17   Broadband services over satellites: GEO feeder link application (left), LEO downlink from Earth observation satellite (right)

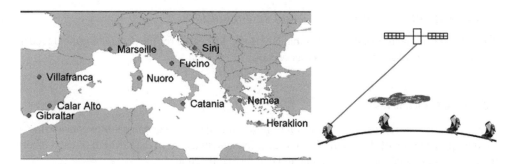

Figure 9.18 Distribution of ground stations to achieve an availability of 99.89% (left), principle of cloud diversity: the satellite selects a cloud-free ground station (right)

carrying trunk or network traffic is called a feeder link. In general, the feeder link is a critical path of the network, since it contains aggregated traffic and crosses the atmosphere. Today RF systems for the *Ku- and Ka-band* are used to feed the satellites but the lower Ku-band frequency capacity has filled up in most of the geosynchronous orbital slots. As a result, the growing demand for satellite bandwidth is driving usage of the Ka-band as a choice for near-future Gbit/s capacity. In this context, optical feeder links are also of interest [9.58], but on a longer term, because they allow for Tbit/s capacity and a reuse of the feeder link frequencies for additional RF user links.

The main challenges for optical feeder links are beam degradation due to weather conditions. Under clear sky conditions, the atmospheric transmission spectrum exhibits windows for telecom wavelengths. However, these wavelengths are attenuated or basically blocked in the presence of clouds (Section 9.2.2). The main concept to overcome this problem is spatial diversity: for optical feeder link applications, a network of ground stations at locations with good weather conditions is required. Figure 9.18 shows a network example consisting of 10 sites in South Europe. Joint cloud statistics could be calculated from observation data provided by European Cloud Climatology and acquired on-board the NOAA satellites [9.59]. The annual availability of the network, which is given by the probability that at least one station is not covered by cloud, is estimated to 99.89%.

The measured clear-sky transmission data show a wavelength window at 1550 nm with low absorption, which is very attractive for the application of optical feeder links (Section 9.2.2). Moreover, the 1550 nm wavelengths region is the predominant wavelength for terrestrial fibre communication technology. For this wavelength region, there exists a large and expanding global market for optical components and subsystems. Some of the components, such as high speed lasers and photodiode, are already space qualified and therefore are best suited for applications using free space optical laser links. For other components, like the Erbium-doped fibre amplifier (EDFA), large effort is undertaken to qualify them for space [9.601].

For the optical feeder links, the propagation path is primarily characterized by free-space diffraction. Atmospheric turbulence impacts the downlink, satellite-to-ground, in a different way than the uplink. However, for both directions, turbulence perturbations can only be mitigated by appropriate techniques at the ground terminal.

9.3.2 Channel Characteristics and Technical Issues

There are serious technical challenges that have to be addressed for the design of optical feeder links, which are described briefly in this section.

9.3.2.1 Geometric loss

The geometric loss represents the portion of power that the receiver can capture considering a vacuum propagation of the beam. When the optical antennas are in the far-field of each other, the geometric loss becomes significant and can be expressed as:

$$L_{geom} \approx \left(\frac{D_{Tx} D_{Rx}}{\lambda Z} \right)^2 \tag{9.21}$$

where:

D_{Tx} Transmitter diameter
D_{Rx} Receiver diameter
λ Wavelength
Z Link distance

The far-field condition, which can be written as $Z \gg (D_{Tx}^2 + D_{Rx}^2)/\lambda$, is generally fulfilled for satellite links. For 0.1 m telescope diameters and 1 μm wavelengths, this corresponds to distances significantly larger than 20 km.

9.3.2.2 Atmospheric attenuation

Figure 9.19 depicts the atmospheric clear-sky transmission windows vs. wavelength. Highly attenuated spectral regions are caused by molecular absorption, mainly from water and carbon dioxide. Lasers around 800 nm, 1064 nm and 1550 nm are of common interest for free-space optical applications.

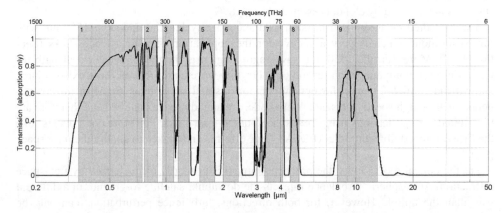

Figure 9.19 Atmospheric (clear sky) transmission windows in the visible and near-infrared regions

Additional attenuation occurs as the visibility through the air decreases. From the attenuation values given in Table 9.1, we deduce that it is highly weather dependent. Consequently, the availability of optical satellite communications links through the atmosphere depends on weather statistics:

Table 9.1 Typical attenuation of light under various weather conditions

Weather	Typical attenuation in dB/km
Thick fog/cloud	30 … 300
Thin fog/cloud	4.0 … 10
Haze	0.7 … 2.0
Clear sky	0.1 … 0.2

9.3.2.3 Atmospheric turbulence

Random *atmospheric turbulence* leads to wavefront distortions and, in turn, to scintillation. Polarization is maintained. The strength of wavefront distortions is given by the Fried parameter [And05]:

$$r_0 = \left[0.423 k^2 \int_{Path} C_n^2(z) \, dz \right]^{-3/5} \tag{9.22}$$

where:

k Wavenumber
z Optical path variable
$C_n^2(z)$ Turbulence strength parameter over the optical path

For up- and downlinks, r_0 is typically in the order of 10 cm. Figure 9.20 shows the intensity profile of a Gaussian beam after propagation through vacuum and turbulence. The profile can be viewed as an angular spectrum that applies to both up- and downlinks. The perturbed

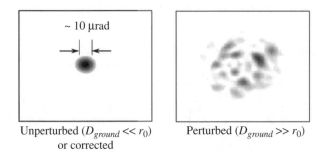

Unperturbed ($D_{ground} \ll r_0$) Perturbed ($D_{ground} \gg r_0$)
or corrected

Figure 9.20 Beam profile after vacuum propagation (left) and after turbulence perturbation (right)

profile consists of several random spatial modes. Strong perturbations occur when the Fried parameter r_0 is much shorter that the diameter D_{ground} of the ground terminal. For the uplink, the beam diameter is much larger than the satellite receiver and the turbulence-induced spatial dissemination of the beam energy leads to fading at the satellite receiver. For the downlink, the angular spectrum of Figure 9.20 refers to the focal-plane image at the receiver and the turbulence-induced random modes, in a similar way to the uplink, leads to deep fading when this focused beam is coupled into a single-mode fiber. Techniques of turbulence mitigation can be implemented on the ground terminal for both up- and downlinks:

- On the uplink, multiple transmitters are deployed with a separation distance of about 0.5 m. The multiple beams carry the same information but undergo different distortions through different turbulent volumes. In this multiple-input single-output (MISO) configuration, the spatially diverse beams are averaged at the receiver and the signal fade probability is reduced [9.62]. The deployment of several laser sources additionally increases the average Tx power. The multiple transmitters must be mutually incoherent to prevent inter-beam interferences at the satellite receiver. This mutual incoherence requires additional optical spectrum bandwidth.
- On the downlink, adaptive optics corrects the distorted wavefront so that the received optical beam can then be focused into a single-mode fiber with a good coupling efficiency [9.63]. Furthermore, the large Rx aperture provides spatial averaging of scintillation [9.64].

9.3.2.4   Launch power

The *launch power* is critical because of eye safety regulations. As depicted in Figure 9.21, lasers between 400 nm and 1400 nm cause damage mostly to the retina, whereas above 1400 nm the beam is significantly absorbed by the cornea. Unless the lasers are strongly pulsed (e.g. femtosecond lasers), the mean power of modulated beams can be considered for safety assessment. At 1550 nm, the maximum permitted exposure to the human eye is 1000 W/m^2, which is 20 times higher than at 1064 nm and 50 times higher than at 800 nm [9.64].

9.3.2.5   Background noise

The amount of background light reaching the photodetector can be derived from the background radiance, generally expressed in W/(cm^2·sr·nm), and the receiver's filtering parameters: aperture area (cm^2), field of view (sr) and spectral filter width (nm). The spectral filter

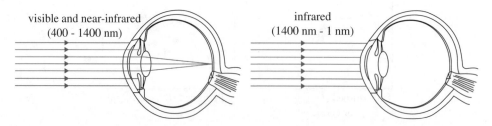

Figure 9.21   Different wavelengths produce different types of eye injury

is typically a few nm wide. Ideally, the field of view is diffraction-limited and thus inversely proportional to the aperture area. In that case, the collected background light is proportional to the square of the filter's central wavelength.

Links with higher data rates tend to be more robust to *background noise* for two reasons. First, the bit duration is shorter and thus fewer background photons per bit are detected. Second, high-bandwidth detectors are smaller in diameter, resulting in a smaller field of view. Most modern optical satellite systems can operate under bright daylight. However, the radiance increases by several orders of magnitude as a terminal turns its field of view towards the Sun. Except for a coherent-detection receiver, where enough background light can be spectrally removed [9.66], most systems must avoid the presence of the Sun in their fields of view.

A particular technology that is very sensitive to background noise is photon counting. This type of receiver requires very few signal photons per bit and is appropriate for photon-starved channels, such as links from deep space to the Earth [9.67].

9.3.2.6 Power efficiency vs. spectral efficiency

The *power efficiency* is evaluated in bit per photons and the *spectral efficiency* in bit per hertz. Energy and bandwidth management through free space and fiber differs in several aspects. The air channel has weak non-linearity effects, allowing higher peak powers. It also has negligible dispersion, allowing shorter pulses.

Space links are generally power-limited as a result of limited transmit power onboard satellites and high geometric losses. Such links require highly sensitive receivers that for example feature coherent homodyne detection [9.66] or photon counting [9.68]. Photon counting with PPM can in principle decode more than 1 bit per photon. The spectrum use is mostly limited by the detector bandwidth.

A widely-used modulation for free-space laser links both in terrestrial and space domains is the *On-Off Keying* (OOK) with direct detection. The advantage of OOK lies in the compromise between spectral and power efficiency as well as cost efficiency.

When the detector bandwidth is fully utilized and we cannot afford to spend more photons per bit, then the maximum channel capacity is reached and a solution to further increase the data rate is to use more channels by means of multiplexing.

In addition to the conventional fiber multiplexing techniques (WDM, Pol-Mux), channels can be divided over space or over orbital angular momentum. The latter uses different wavefront geometries and is predestinate for near-field applications. Space multiplexing follows the principle of MIMO systems.

9.3.2.7 Optical tracking

The drawback of high directivity is the requirement of precise pointing. To establish and maintain a link, each optical satellite communication terminal must first acquire the direction of the counter terminal, then point towards the acquired direction and repeat the process (tracking) at a rate faster than the terminals' attitude variations.

A precise knowledge of the counter terminal's direction is obtained by measuring the direction of its propagating optical beam. This is generally done by focusing the received beam on a detector array.

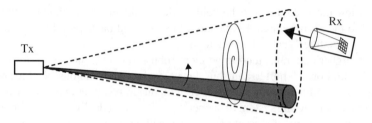

Figure 9.22 Acquisition of the transmitter's position by the receiver. The transmitter (Tx) scans its narrow beam over the uncertainty cone according to a spiral pattern, the receiver (Rx) waits to receive the beam and measures its direction on a focal-plane detector array

However, before using mutual optical recognition, terminals generally use other sources of information (e.g. finderscope, GPS, inertial measurement unit) to point at each other with a given error. This error then provides the uncertainty cone over which optical acquisition is to be performed. A beacon beam is used by one terminal to cover its uncertainty cone and inform the counter terminal about its position. Instead of a large divergence beam, it is technically often more convenient to use a narrow beam, which can also be the communication beam, and to cover the uncertainty cone through fast scanning as depicted in Figure 9.22.

The illuminated terminal "sees" the emitting terminal and sends a beam back in its direction for mutual recognition. *Optical tracking* can start with the important task of minimizing pointing losses during communication.

As a result of the high satellite speed, the transmitted beam direction should be ahead of the received beam direction (Figure 9.23, left). The point ahead angle (PAA) is a few tens of micro-radians for Earth orbiting satellites. For up- and downlinks, atmospheric turbulence perturbs the wavefront tilt. The isoplanatic angle θ_{iso} depends on the turbulence strength and

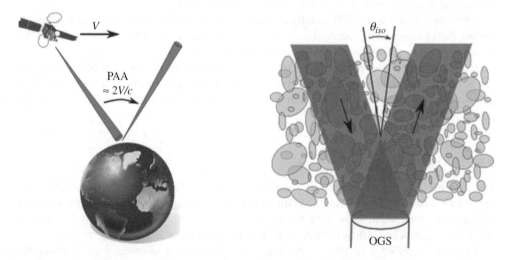

Figure 9.23 Optical tracking for satellite uplink: point ahead angle PPA (left), isoplanatic angle θ_{iso} (right)

can be higher or lower than the PAA. When $\theta_{iso} < $ PAA, the down- and uplink beams do not propagate through the same turbulence and their fluctuations are decorrelated from each other (Figure 9.23, right). The impossibility of correcting turbulence-induced beam deviations is a critical problem for stable uplink beam tracking. The alternative mitigation technique generally applied is transmitter diversity, where different uplink beams incoherently combine on the satellite [9.69,9.70].

9.4 Coherent Transmission Systems

The coherent transmission, also referred to as *optical heterodyne reception*, allows contrary to the direct detection a basically new type of optical communication [6.39,9.71–9.74]. By utilization of the applied light sources coherence properties, it is theoretically possible to enhance the receiver sensitivity of an optical transmission link significantly. Possible improvements of a factor of about 1000 can be gained. Therefore, a considerable increase in the fiber link length without an amplifier will be achieved. Analogous to heterodyne reception in radio engineering, there can be additional multiplexing by means of a multitude of adjacent channels. This provides the opportunity to achieve a substantial increase of the transmission capacity in future optical transmission systems (Eq. 1.1).

In context to the direct transmission systems describe above, first of all the intensity of the light was just relevant. The data transfer – digital as well as analog – generally results from the modulation of the laser output. Regarding the receiver, an identification of the data stream will be carried out with the aid of an intense sensitive detector. In contrast to that, coherent systems make use of the fact that except for the amplitude also the frequency and the phase of an electromagnetic wave contain useful information.

9.4.1 Main Principle of Coherent Transmission

The optical heterodyne reception represents the analogon to the electric one known from TV and Radio Broadcasting. There is a transmitter oscillator on the TV-tower and a local oscillator laser in the receiver. Figure 9.24 shows the fundamental set-up of an optical heterodyne receiver [6.39]. The light emitted by a semiconductor laser (transmitter oscillator) will be transferred by a fiber to the receiver location. At this place another semiconductor laser is situated, the local oscillator.

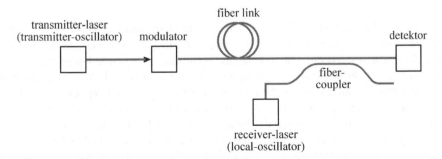

Figure 9.24   Fundamental set-up of an optical heterodyne reception

A superposition of the light waves emitted by both lasers takes place with the aid of an optical coupler (Chapter 2.2). According to Eq. (2.15), the time-dependent power $P(t)$ is given by the absolute square of the absolute incident field strength $E(t)$:

$$P(t) \sim E(t)E^*(t) \qquad (9.23)$$

where:

E^* Conjugate complex of E

Regarding the optical heterodyne reception, at the detector a superposition of the field strength $E_S(t)$ from the transmitted signal at the fiber end and $E_L(t)$ of the local oscillator is achieved. It includes the information which has to be transmitted; from this it follows with Eq. (2.16) that:

$$I \sim A_S^2 + A_L^2 + 2A_SA_L \cos[(\omega_S - \omega_L)t + \Phi_S - \Phi_S] \qquad (9.24)$$

where:

A_i Amplitude of the wave i
ω Angular frequency
Φ_i Phase of the wave i

Rewritten in optical powers, it follows that:

$$P(t) = P_S + P_L + 2\sqrt{P_SP_L} \cos[(\omega_S - \omega_L)t + \Phi_S - \Phi_L] \qquad (9.25)$$

where:

P_S Power of the transmitter light at the end of the fiber link
P_L Power of the local laser light
ω_S Angular frequency of the transmitter light
ω_L Angular frequency of the local laser light
Φ_S Phase of the transmitter light
Φ_L Phase of the local laser light

It is obvious that the power of the transmitted light attenuated by the transmission link is multiplied – and therefore amplified – by the power of the local oscillator ($P_L \gg P_S$). In contrast to optical direct transmissions, this implies the possibility of bridging longer fiber links. In other words, it is possible to transmit a higher data rate with equal link length. Concerning all optical communication systems, the realization of the information happens by modulation of a certain characteristic parameter of the transmitter laser light. If there is no local laser regarding direct detection systems, it can be written as $P_L = 0$. Thus, it follows $P(t) = P_S$ and there is only a power modulation, also named the intensity modulation (IM), left. In contrast, relating to coherent detection, power, frequency, as well as phase of the transmitter laser light, can be modulated (Eq. 9.25. In connection with digital transmission links, three types of modulation have to be distinguished: ASK (*Amplitude Shift Keying*: modulation of P_S), FSK (*Frequency Shift Keying*: modulation of ω_S) and PSK (*Phase Shift Keying*: modulation of Φ_S). Furthermore, a distinction is drawn between *Homodyne Systems* ($\omega_S = \omega_L$) and *Heterodyne Systems* ($\omega_S \neq \omega_L$). In the latter case, an *Intermediate Frequency* is given by $IF = \omega_S - \omega_L \neq 0$.

The quality of a transmitting link (characterized by its transmission capacity: bit rate times length) is evaluated by the use of the attained bit error rate. The signal-to-noise ratio is a parameter which decisively determines the bit error rate similar to point-to-point connections (Chapter 9.1). The signal-to-noise ratio can be written as:

$$\frac{S}{N} = \frac{square\ of\ the\ signal\ current}{sum\ of\ the\ noise\ current\ squares} \tag{9.26}$$

The signal current is obtained by Eqs (6.26) and (9.25) as:

$$S^* = I_P/P = \eta \frac{\lambda}{1.24\ \mu m} \frac{A}{W} \tag{9.27}$$

where:

S^* Responsitivity
I_P Photo current
P Light power
η Quantum efficiency
λ Wavelength

A multitude of noise processes is developed [7.6] (Chapter 7.2.2), such that amplifier noise, including noise of the load resistor, dark current noise of the photodiode and excess noise of the avalanche photodiodes, are dominant contributions. The amplifier noise is named for all these processes as a representative. From Eq. (9.6), it follows the mean square amplifier noise current is:

$$\overline{I_V^2} = I^{*2}B \tag{9.28}$$

In addition, there is in either case the absolute physical limit of the smallest possible noise, Eq. (7.12), the shot noise (Chapter 7.2.2)).

This way, all parameters can be determined which are necessary for the investigation of the signal-noise power ratio. Table 9.2 shows the signal-noise power comparison of different detection methods. In the line "noise power", attention should be paid to the shot noise term from the heterodyne and homodyne systems. Due to the high local laser power, the shot noise term is so high that both the shot noise term induced by the transmitter oscillator and the amplifier noise may be neglected [6.39].

By comparing the signal-noise ratio of the different systems, it becomes obvious that a quality improvement has already been observed if direct systems are replaced by heterodyne systems. Considering a homodyne system, a further factor of 2 is gained. Another upgrade is achieved in the way that in case of coherent detection in contrast to direct detection, not only

Table 9.2 Signal-to-noise power comparison of different detection methods

Method	Direct	Heterodyn	Homodyn
characterized by	$P_L = 0$	$P_L \gg P_S$ $\omega_S - \omega_L \neq 0$	$P_L \gg P_S$ $\omega_S - \omega_L = 0$
signal current I_P	$S^* P_S$	$2S^* \sqrt{P_S P_L} \cos(\omega_S - \omega_L)t$	$2S^{*2} P_S P_L$
signal power $\sim I_P^2$	$S^{*2} P_S^2$	$2S^{*2} P_S P_L$	$4S^{*2} P_S P_L$
noise power $\sim \overline{I_{RS}^2} + \overline{I_V^2}$	$2eS^* P_S B + I^{*2} B$	$2eS^* P_L B$	$2eS^* P_L B$
signal to noise-ratio $\dfrac{I_P^2}{\overline{I_{RS}^2} + \overline{I_V^2}}$	$\dfrac{S^* P_S}{\left[2e + \dfrac{I^{*2}}{S^* P_S}\right] B}$	$\dfrac{S^* P_S}{eB}$	$2\dfrac{S^* P_S}{eB}$

the intensity can be modulated but also frequency and phase. The whole sensitivity hierarchy is presented in Table 9.3 [9.73,9.74].

The following simplified perspective should illustrate the advantages of the different systems. The gain from the transition of ASK to FSK results from the fact that regarding ASK modulation, exclusively in the case of a logic-one there is incident light on the detector. Concerning FSK modulation, this occurs if there is a logic-one as well as a logic-zero. This yields a more favorable statistical error rate [9.75] at the receiver measured by the following bit-error detector and affects an improvement of approximately 3 dB. FSK-odulation signifies that the *IF* is shift keyed between two values, depending on whether a logic-one or a logic-zero is existent or not. In contrast, for PSK modulation, there exists a single *IF* or none, *IF* = 0, regarding homodyne detection. Therefore, in this case only the half detection bandwidth is required. This causes an enhancement in *S/N* of 3 dB (note: relating to heterodyne ASK-and-FSK systems, respectively, the IF is analyzed by ignoring the phases).

Table 9.3 Theoretical receiver sensitivity gain of different coherent transmission systems

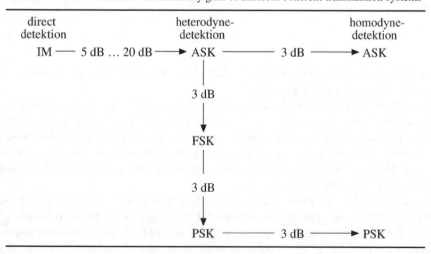

Therefore, theoretically, a gain in receiver sensitivity close to 30 dB can be obtained by adding all improvement opportunities together. Establishing such a transmission system by light sources with a wavelength of 1.55 μm, the minimum of the glass fiber attenuation ($\alpha \approx 0.2$ dB/km) is arranged. This causes a theoretically possible extension of the glass fiber link of about 150 km, taking into account the transition from direct transmission systems to coherent ones with equal data rate. This fact clarifies that optical heterodyne reception provides the opportunity to gain a substantial improvement in optical communication.

Another enhancement is reached if not only a single laser is applied on the transmitter side but also a multitude of lasers (Figure 9.11). These lasers respectively emit at different frequencies. Because of the tuning of the local oscillator lasers frequency, it is possible to set an IF – consistent to each transmitter laser – which only conforms to the center frequency of a downstream band-pass filter to the detector. For this reason, the completed optical analogon to heterodyne reception in the broadcasting techniques, such as for radio and TV, is gained. The high achievable optical channel density yields a considerable increase of the transmission capacity.

As an example, the first transmitter laser could oscillate in the long wavelength range corresponding to about 200 THz. The other transmitter lasers are spaced by 2 GHz, 3 GHz ... The local oscillator laser then emits at 200 THz + 1 GHz. Thus the IF with the first laser amounts precisely to the difference of 1 GHz. The band-pass filter also is set to 1 GHz center frequency fitting with the IF. To receive the second transmitter laser information, the tunable oscillator laser must change its frequency to 200 THz + 2 GHz. The IF now fits to the second laser and so on. The channel spacing of such *Frequency Division Multiplex Systems (FDM)* is much narrower as compared to Wavelength Division Mulltiplex Systems (WDM), also compared to Dense WDM Systems (DWDM) (Chapter 9.1.1). Thus, many more channels can be transmitted over a single fiber.

9.4.2   System Components

To utilize completely the improvement gained by application of coherent systems, it is necessary to request high component demands compared to direct detection systems. In particular, this applies to lasers and fibers.

In this context, the semiconductor laser represents the most important key element. It is an absolute requirement that the laser only emits one single longitudinal mode. But unfortunately, conventional Fabry-Perot semiconductor lasers with a wavelength range of 1.3 μm and 1.55 μm normally show a mode spectrum which consists of about 2 to 5 modes. Fortunately, the so-called DFB laser (Distributed Feedback) has succeeded in becoming a dynamic single-mode laser [6.49] (Chapter 6.1.2).

Except for the single-mode characteristic, in order to achieve a certain essential bit error rate, it is also required that the single-mode spectrum is sufficiently narrow. Theoretical investigations have shown that the line width has to be less than 10^{-2} of the transmitted bandwidth [9.76]. Line width reduction in a few KHz can be realized by means of a long grating resonator [9.77] or coherent irradiation of a HeNe laser (injection locking) [9.78]. For line widths of such a dimension, additional steps have to be taken regarding all types of semiconductor lasers. Developments based on erbium-doped fibers (Chapter 9.4) permit the realization of a tunable fiber-ring laser with line widths below 10 KHz [9.79]. According to

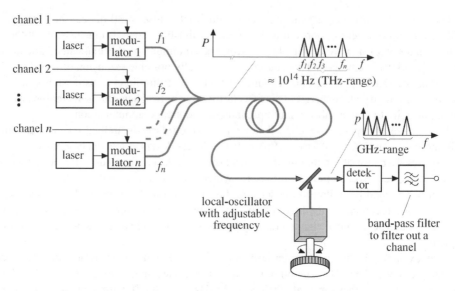

Figure 9.25   Tuneable detector with frequency division multiplexing

Eq. (9.25), a deviation causes a modification of the interference term. But due to the fact that only this term contains the complete information which should be transmitted, a stabilization of the center frequency is inevitable. The demands on the wavelength stability of both lasers amount to ca. 10^{-9} of the nominal value [9.80].

It is also possible to stabilize only the transmitter oscillator and track the local one to the frequency of the transmitter laser with the aid of an IF regulation [9.81]. Center frequency stabilization can be gained by means of reference elements. Typical values are in the range of 1 MHz [9.82,6.18]. In particular, regarding frequency division multiplex (Figure 9.25), the absolute stabilization of *one* transmitter laser is necessary. Then the others can be stabilized relative to this one. Concerning the local oscillator, tuning is very important [9.83]. Injection current and temperature fluctuations in particular influence the center frequency drastically (Chapter 7.1.2).

Power and intensity fluctuations of the local laser falsify the transmission signal corresponding to the second term from Eq. (9.25). Compared to the interference term (third term from Eq. (9.25)), the part of the local laser is dominant. This is due to the high power attenuation by the long fiber link. Therefore, consequences of the relative intensity noise caused by the laser are no longer negligible. Thus, the most usable local laser power is limited. By means of push-pull-receiver concepts [9.84], it is possible to minimize this problem.

Another problem results from back reflected and back scattered light, respectively, which is injected into the laser again. Semiconductor lasers are influenced by already low feedback amounts in the range of 10^{-6} [5.1]. Well-directed back reflections can favorably affect the spectral laser behavior (i.e. reduction of the line width (Figures 6.33 and 6.34)). However, undesired ones, such as reflections at glass–air interfaces, must be suppressed apart from insignificant influences.

Inappropriate back reflection and back scattering can result in frequency jumps, involving line broadening [9.85]. Back-reflection and back-scattering suppression may be achieved by application of optical isolators with a suppression value below 10^{-6}, based on the Faraday effect [9.86] (Chapter 2.4).

Relating to the optical heterodyne reception, attention should be paid to an important parameter, the polarization (Chapters 2.4 and 3.2). The application of polarization preserving fibers shows a possible solution to the polarization problem [3.19] (Chapter 3.2). Nevertheless, such fibers are much more expensive than the standard single-mode fibers. Moreover, polarization preserving fibers have higher attenuation values.

Another opportunity also provides the realization of polarization-diversity reception [9.87]. In this case, both orthogonal polarization components are separated with the aid of well-aligned polarization filters. Both polarization paths are then doubly processed by reception with two photodiodes. Thus, the reception is guaranteed, also during variable polarization states caused by the use of standard fibers. Moreover, a polarization control could be installed. By means of active techniques, time variable polarization behavior can be brought back again to the desired polarization status. This is attained by the adjustability of the relative time delay of both orthogonal polarization conditions with the aid of polarization optical retarders [9.88].

9.4.3 Modulation Methods for Coherent Transmission Systems

The signal modulation can be carried out in various ways. In principle, it is essential to distinguish between direct modulation using the laser current and the application of external modulators.

The frequency modulation (FSK) by means of the transmitter laser injection current [9.72] represents the most popular method. A few current amplitudes already suffice (some mA) to gain frequency changes in the range of GHz. The power modifications appear in the percent scale and can be neglected. A compact FM transmitter may be realized using this technique. The advantage is that there are no additional losses and undesired reflections in contrast to the application of external modulators. A disadvantage is the fact that the laser characteristics are influenced using direct modulation (noise problems). In addition, regarding the modulation, with the aid of the injection current, a simultaneous change of amplitude, frequency and phase always occur.

In addition, different frequency responses are gained. However, by means of a pre-distortion module network, it is possible to compensate such frequency response for the injection current modulation. Amplitude modulation (ASK) resulting from direct modulation in coherent systems appears to be inapplicable. Hereby the necessary current amplitudes would inhibit the maintaining of a stable intermediate frequency. Coherent irradiation according to the Master–Slave principle allows a phase modulation (PSK) by the injection current of the Slave-laser [9.89]. In this connection, it is disadvantageous that on the transmitter side two lasers have to be already applied.

Avoiding the modulation of the injection current, external modulators have the advantage that the laser characteristics remain unaffected. The additional insertion attenuation (>2dB), as well as the occurrence of reflections at the interface fiber modulator, are disadvantageous. Integrated optical components are applied as external modulators (Chapter 5.2) [9.90], which based on its geometry are well-adapted for fibers. Lithiumniobat ($LiNbO_3$) is the most frequently used substrate material for electro-optical modulators. It offers high electro-optical coefficients. External modulators are used in ASK and PSK systems [9.77,9.91].

Technologically, it would be easiest to realize phase modulators. They feature the lowest insertion loss. The most frequently used intensity modulators are Mach–Zehnder interferometers and directional couplers. Cut-off frequencies above 50 GHz are gained by means of electro-optical modulators [9.92]. Also frequency modulation is in principle realized with external modulators, which rely on the acousto-optical effect. But the application of such modulators in systems is doomed to failure due to the bandwidth, which is too narrow [9.72].

9.4.4   Detection and Demodulation Methods for Coherent Transmission Systems

Figure 9.26 shows the fundamental set-up of a heterodyne system. The IF has to be stabilized with the aid of a control circuit (*AFC: automatic frequency control*) [9.77,9.81]. Both lasers change their frequencies as a result of temperature and current fluctuations. The stabilization is achieved by a *frequency discriminator* (Figure 9.27a) inducing the control variable by which the local laser is adjusted. Stabilization will be done, referred to the zero-crossing (S) of the discriminator characteristic line (Figure 9.13b).

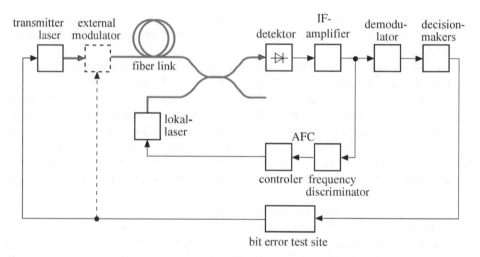

Figure 9.26   Fundamental set-up of a heterodyne transmission system

where:

  IF   Intermediate frequency
AFC   Automatic frequency control

The IF signal is the carrier of the whole information. Concerning ASK modulation, the IF is on-line and off-line, respectively, according to a logic-one or logic-zero which has to be transmitted. Regarding PSK transmission, it is a question of a phase shift keying of the IF. For FSK-modulation, the IF-spectrum essentially occurs for two frequencies being consistent with the logic-one or logic-zero.

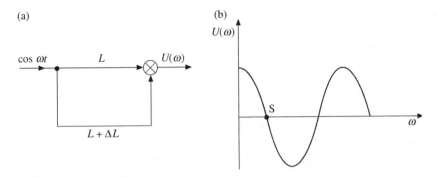

Figure 9.27 Electrical frequency discriminator (a) and its characteristic (b)

where:

U Output voltage
L Length
S Stabilization point

Figure 9.28 illustrates the schematic power density spectrum of the different modulation types. From a simplified point of view, relating to ASK and PSK modulation, the double base bandwidth is required for the detector compared to the intensity modulation (IM) at direct transmission. In contrast, regarding FSK transmission, the quadruple bandwidth will be necessary [9.89].

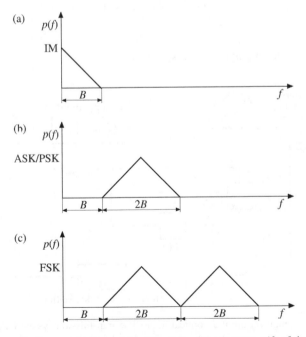

Figure 9.28 Schematically presentation of the power density spectrum $p(f)$ of the different coherent transmission systems

where:

B Base bandwidth
f Frequency

In order to obtain the transmitted information, the IF in the receiver is demodulated, by means of the frequency discriminator or by band-pass and low-pass filter, respectively [6.39]. The demodulation can be used for ASK, FSK and PSK modulation [6.39,9.93,9.94].

In homodyne systems, the information can be directly gained from the base band, because there is no intermediate frequency. Figure 9.29 presents the fundamental set-up of an ASK and a PSK system, whereby an external modulator is used. Homodyne systems require highly sophisticated detection concepts. An optical phase control circuit (*OPLL: Optical Phase Locked Loop*) is necessary, which couples the transmitter laser and the local laser phase is locked [9.95]. Regarding ASK modulation, the phase difference $(\Phi_S - \Phi_L)$ of both light waves is controlled to zero. Neglecting the strong attenuated first term from Eq. (9.25) and with Eq. (6.27), the following photo current at an AC-coupled detector is given by:

$$I_P(t) = 2S^* \sqrt{P_S(t)P_L} \qquad (9.29)$$

Thereby, it is essential that $P_S(t) \neq 0$ and $P_S(t) = 0$, respectively, for a logic-one or logic-zero.

Concerning PSK systems, the required phase difference is $\Phi_S - \Phi_L = \pi/2$. A phase shift keying about $\pm\pi/2$ is carried out. Thus, the following current in the AC-coupled detector is gained:

$$I_P(t) = \pm 2S^* \sqrt{P_S P_L} \qquad (9.30)$$

where:

$\pm$ applies to a logic-one and logic-zero, respectively.

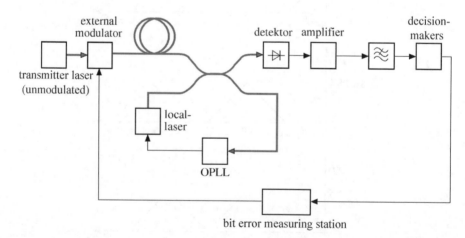

Figure 9.29   Fundamental set-up of an optical homodyne transmission system OPLL (Optical Phase Locked Loop)

9.5 Top Results on Fiber-Optic Transmission Capacity for High-Speed Long Distance

Depending on modulation/demodulation methods, coherent transmission systems theoretically allow a considerable gain of receiver sensitivity. Also an increase of the transmission capacity of optical fiber communication is achieved compared to direct transmission techniques. But in practice, a series of drawbacks avoids the complete success in obtaining results of an ideal coherent reception (Chapter 9.3.2). Figure 9.30 compares practically obtained data of receiver sensitivities of coherent and direct system experiments with theoretically found results [9.12].

The values which are experimentally achieved obviously lie about 10 dB above the quantum noise limit. They also show an improvement of only approximately 10 dB compared to the best possible results of direct transmission systems. Furthermore, the receiver sensitivity improvement degrades with increasing bit rate. This is particularly related to the higher bandwidth needs for coherent systems (Figure 9.14). Hence, a more favorable field of coherent systems could not be point-to-point connections but that of frequency division multiplex systems (Figure 9.11) [9.97,9.98].

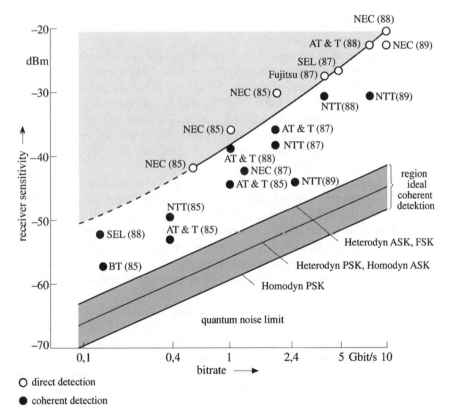

Figure 9.30 Receiver sensitivities of direct and coherent transmission experiments in high bit rate transmission systems by comparison

In order to improve optical communication systems, a further technique does not work on the increase of the receiver sensitivity, but aims at absolute *optical amplification* [9.12] of the transmission link. For that purpose non-linear effects in fibers or semiconductor amplifiers can be applied (Chapter 8).

In the following, top results in optical transmission over the last years are presented. The main focus will be on the state-of-the-art achievable data stream and their technique.

In 2009, Alcatel-Lucent announced a new optical transmission record. Their scientist broke the 100 Petabit per second kilometer barrier. Wavelength Division Multiplexing (WDM, Chapter 9.1.1) was applied with 155 channels, each carrying 100 Gbit data per second. The achieved distance was 7000 km. The transmission network was set up with repeaters separated every 90 km. Therefore, the achieved direct distance with 15.5 Tbit/s would be 90 km, which is 2% more than in the commonly-used tracks back then. The capacity was also increased by advanced digital signal processors with coherent detection at the receiver side [9.99].

In 2010, Nippon Telegraph and Telephone Corporation successfully demonstrated an optical transmission of 69.1 Tbit/s over a single 240 km-long optical fiber. The transmission capacity is also based on WDM of 432 channels, with a capacity of 171 Gbit/s per channel generated by the combination of the 16QAM (Quadrature Amplitude Modulation) and PDM (Polarization Division Multiplexing) in the transmitter (Chapter 9.2). This technique allows ultra-dense WDM with wavelength spacing of 25 GHz. At the receiver coherent detection and digital signal processing is applied [9.100].

In 2011, Hillerkuss et al. [9.37] reported a 26 Tbit/s single channel/source transmission using all-optical fast Fourier transform processing. The all-optic OFDM transmitter, including an all-optic comb generator, generates a stable comb spectrum with 336 subcarriers. Two 16QAM modulate the subcarriers with a spacing of 12.5 Ghz. The OFDM signal is fed through a 50 km SMF. The all-optic OFDM receiver, including a serial-to-parallel converter and an all-optic FFT (Fast Fourier Transformation) circuit, receives the signal by the aid of coherent detection. Therefore, energy consumption decreases by the aid of the all-optic OFDM transmitter and receiver. More Information can be found in [9.101].

In 2011, J. Sakaguchi et al. [9.102] introduced a 109 Tbit/s transmission over 16.8 km using Space Division Multiplexing (SDM) and a 7-core Multi Core Fiber (MCF). The MCF feature an ultra-low cross-talk of less than –90 dB/km. All the single-mode cores were made of pure silica and the core pitch of 45 μm. Ninety-seven WDM channels with 100 GHz spacing and 2×86 Gb/s PDM-QPSK signals were used. SDM MUX and DEMUX for the 7-core MCF was set up based on a free-space optical configuration (Figure 9.31). The MUX and DEMUX consist of a set of SMF collimators and an aggregating lens in front of the MCF [9.102].

NTT, Fujikura Ltd., Hokkaido University and Technical University of Denmark demonstrated ultra-large capacity transmission of 1 Pbit per second over a 52.4 km length of 12-core MCF. A new 12-core MCF structure with the cores arranged in a nearly concentric pattern has been developed. The MCF feature less adjacent cores and therefore cross-talk losses are decreased in comparison with conventional structures of MCF. Moreover, a novel fan-in fan-out device, and a digital coherent optical transmission scheme for transmitting dense wavelength division multiplexed signals in each core was reported. The fan-in fan-out device was developed for efficient and stable coupling (low cross-talk and low insertion loss) of each

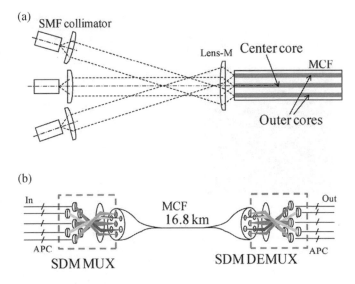

Figure 9.31 Space Division Multiplexer: schematic drawing of the optical configuration (a); illustration of the SDM MUX and DEMUX including MCF (b)

of the 12 cores of the MCF to a conventional single-mode fiber. A transmission capacity of 84.5 Tbit per second per core could be achieved with the aid of 32QAM. Each core includes 222 WDM channels with a capacity of 380 Gbit per second yielding to a total capacity, when summarizing all the cores, of 1.01 Pbit per second through a 52.4 km length of MCF [9.103].

Many scientist groups work on new transmission techniques to increase the transmission link or to enhance the transmission capacity. But also research on new fibers has been made, such as presented in [9.102,9.103]. Therefore, an increasing data stream is not only an object of the transmission technique but the transmission medium itself. More data can be transmitted by increasing the amount of cores in the fiber, but also by increasing the speed of data as presented in [9.104].

A hollow-core photonic-bandgap fiber (HC PBGF) in which the data transmission is close (99.7 %) to the speed of light has been reported. The fiber has a core diameter of 26 μm surrounded by 6.5 rings of cladding holes with average hole-to-hole distance and relative hole size of 4.4 μm and 0.97, respectively. It features low loss of 3.5 dB/km at 1500 nm and a high bandwidth of 160 nm (Figure 9.32). ToF (Time of Flight) measurements yield to a group velocity of 0.290 m/ns for the HC-PBGF and 0.205 m/ns for a Non-Zero Dispersion-Shifted Fiber (NZ-DSF) for comparison. Therefore, the group index is 1.003 for HC-PBGF and 1.464 for NZ-DSF, respectively. Indeed, HC-PBGFs are not suited for long-distance transmission links, because of their relatively high attenuation compared to conventional fibers. Nevertheless, HC-PBGFs have the potential to become the ultimate high-capacity, low-latency transmission medium for future optical communication systems [9.104].

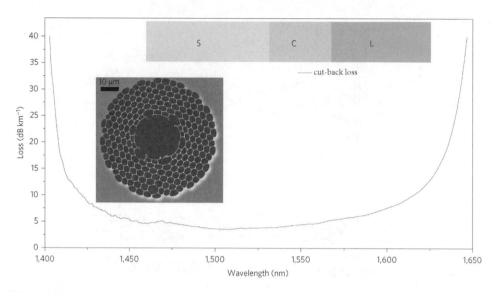

Figure 9.32 Attenuation spectrum and SEM image of the fabricated HC-PBGF (S, C and L bands as indication)

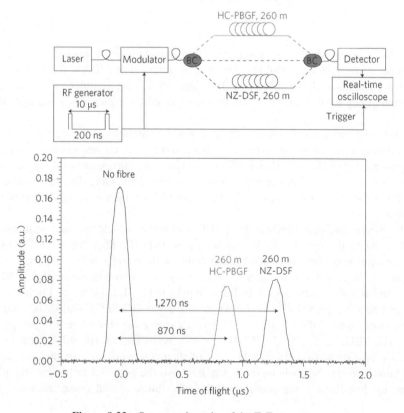

Figure 9.33 Set-up and results of the ToF measurement

9.6   Optical Fibers in Automation Technology

An *industry-standard fiber optic technology* opens up new perspectives for data communication in automation engineering. The automation technology today – depending on the requirements of the application – use different protocols for data transmission, including mainly Ethernet and Profibus, but also customized solutions based on Modbus or RS-232. The infrastructure, more specifically the cabling, is still mainly based on copper and is therefore very inflexible. Taking this approach, especially given the increasing demands of industrial data communication of a transmission medium, at the same time provides a high degree of flexibility and security for the future, for example in terms of bandwidth, transmission distance and noise immunity. Therefore, optical fibers are particularly suitable. Corresponding optical network components, with which industry-oriented infrastructures can be realized, are now available. This includes both active components such as Ethernet switches and fieldbus converters, as well as connectors and splice boxes, in other words passive components (Figure 9.34)

9.6.1   Optical Fiber Cables

Today, efforts are still being made to get along with existing copper cable installations. This may be useful when it comes to operate existing systems. For new projects this outdated technology is not suitable, because it does not allow future-oriented flexibility. In contrast, with optical fibers, slow RS-232 data can be transmitted today, Profibus data tomorrow and Fast or Gigabit Ethernet the day after tomorrow.

Figure 9.34   Compact and industrial splice boxes for DIN rail mounting

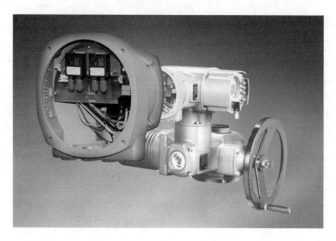

Figure 9.35   Actuator with fiber optic: there is no risk of destruction of the fiber in the case of lightning strikes by electrical insulation

However, the high bandwidth is only one advantage over copper cables. In addition, with *fiber optic*, even extremely long distances can be bridged easily. Moreover, it is not affected by electrical or magnetic disturbances. Therefore, optical fibers can also be installed in the immediate vicinity of power lines or other electromagnetic sources, which simplifies cable management.

Since all types of fibers are of electrically non-conductive (dielectric) material, the data is always transmitted via an electrical insulator. Thus, potential equalization currents do not occur, which are feared especially in extended systems. Even lightning strikes cannot generate any danger of destruction of the connected device (Figure 9.35). Moreover, in optical fibers – unlike copper cables – no grounding or additional shielding is required.

Therefore, a consistent fiber optic cabling is preferred in automation technology. Nevertheless, depending on the requirements of the application, even mixed infrastructures can be appropriate. However, the cost differences between fiber and high-quality copper cables are minimal. In addition, the commodity balance speaks for itself, because copper is actually a rare commodity and therefore too good to install as cables. On the other hand, fibers are made of silica or plastic, so are starting materials which are available in almost unlimited quantities.

For use in the industrial sector, different optic cables are supplied: *Single-mode fibers* (SM), which provide data rates up to 40 Gbit/s and distances of several 100 km. For distances up to 5 km, with transfer data rates of up to 1 Gbit/s, *multimode fibers* (MM) offer a low-cost alternative. For shorter distances and lower data rates, POF (*Polymer Optical Fiber*) or HCS fibers (*Hard Clad Silica*) can be used.

SM and MM fibers of silicate have core diameter of 9, 50 and/or 62.5 μm, while the cladding diameter is in all cases 125 μm. The POF is a pure synthetic fiber with a diameter of 980 μm, which can be easily connected without special tools. HCS is a hybrid fiber consisting of a glass core of 200 μm and a cladding glass having a diameter of 230 μm, unlike the SM and MM fiber having an optical cladding consisting of a few μm thick very hard plastic. When assembling, only the protruding fiber at the connector end has to be cut, which works well with a so-called Cleave Tool. Because of the combination of the easy fabrication and coverage of 200 m to 300 m, the HCS offers a perfect alternative to the above-mentioned fibers in automation engineering.

Indeed, cables are available. including plugs and attenuation protocol. but it is only recommended if the cable trays are easily accessible and not longer than 300 m because the cable still has to be rolled off. Otherwise, a specialized company can connect the plug on the spot and then measure the attenuation values of the cable. Thus, the risk that the cables are damaged during installation is eliminated – unlike pre-assembled cables – and therefore the infrastructure does not work properly.

9.6.2 Connectors

For the connection of the different optical fibers, different connectors are also available, which can be screwed, have a push-pull mechanism or are self-locking (FC, LC, SC, SC-RJ, SMA, ST, E2000).

In particular, the *SC-RJ connector* is becoming increasingly important in automation engineering. It is defined in the Profinet standard IEC 61784-5-3/DIN EN 61784-5-3 (VDE 0800-500-3). Preferably, this connector is used in combination with POF or HCS fibers. This connector is a combination of the optical SC connector in the design of the classical RJ45. Even IP67 versions of the SC-RJ connector are already available and a part of the Profinet standard.

In general, all of these connectors are stuck to the fiber or sometimes crimped. Detailed installation guidelines for wiring in automation technology have now been defined in the standards DIN EN 50173 and ISO/IEC 24702. In terms of industrial network technology, the reader is referred to the following standards: ISO/IEC 61918 and DIN EN 61918 (VDE 0800-500) (general guideline) IEC 61784-5-x/DIN EN 61784-5-x (VDE 0800 500-x) (profiles per fieldbus family), IEC 61784-5-3/DIN EN 61784-5-3 (VDE 0800 500-3) (Profibus-DP) and the standard IEC 62439/DIN EN 62439 (available by ring redundancies).

To connect the terminal devices with the active optical network, industry-standard splice boxes play an important role. Proven technologies from the IT world can be used here. An example of this is the compact splice box FIMP XL. This box is designed for up to 24 optical fibers and has an optional MPO Set (multipath push-on), which is used because of its high density, especially in data centers (Figure 9.36).

Due to the high flexibility of the cabling, this *splice box* can be used as a feed-through distributor. This example shows that the IT and automation world coalesce. The aim is, as already mentioned, a continuous interconnection. The key parameters here are generic infrastructures, homogeneous networks as well as standardized cables and connectors. However, please note that the IT and automation sector are fundamentally different (Table 9.4).

This applies especially with regard to the prices of products, cabling concepts (static and/or dynamic), cabling standards and the installation of network components (19 inches and/or DIN rail).

9.6.3 Network and Network Components

Furthermore, the solutions of the network design are different. The IT develops their networks independent of any application. In this respect, the design is based on standards and guidelines.

Figure 9.36 MPO connector with 24 fiber slots

Such guidelines also exist for the automation engineering. However, there the applications are in focus with their specific mechanical requirements.

Aggravated by the fact, that as with the change of interfaces – away from fieldbuses to ethernet – staff with other qualifications are required. Therefore, the foreman is increasingly replaced by the network administrator.

Due to the lower price, network components from the IT are still used in automation engineering, although they are not designed for the harsh environments in the industry. Indeed, the plugs fit and thus the network connection works, but at least when the non-air-conditioned

Table 9.4 Comparison between the field and the control level

	Field level	Control level
wiring	fixed basic installation in buildings; variable device connection at standard workstations; mainly star-shaped wiring	system-specific cabling and cable layout; field assembly connectors up to IP67; redundant cabling, often ring structures
data	large data packets; average network availability; mainly acyclic data transmission; no real-time behavior	small data packets; very high network availability; mainly cyclic data transmission; real-time behavior necessary
environment	normal temperature range; no external influences by dust, moisture or radiation; moderate temperature requirements (air-conditioned)	extended temperature range; resistance to oil, solvents, mechanical stress, such as vibration or shock; moisture possible; high EMC stress; UV exposure outdoors
network view/ solutions	the focus is on the application	the focus is on the products

Figure 9.37   Managed Ethernet Switch

control cabinet is closed and the temperature rises or falls, it becomes problematic. Because the operating temperature of IT products range from 0°C to 60°C, in contrast to this, industry-standard components must withstand –40°C up to 75°C. Figures 9.37 and 9.38 show typical IT products.

However, the use of IT products in automation engineering is only one problem, another being heterogeneous networks. These island solutions are historically grown. Often it deals with proprietary systems that have arisen from the application as well as from existing infrastructure and factory standards. This also explains why a variety of fiber-optic connectors are used in automation engineering, as outlined above. The problem is not in the optical transmission properties, but in the mechanical differences. How much should a push-pull connector with the requirements of a dynamic cabling withstand, in which the application is constantly in motion? Frequent changes in direction, torsion, dirt or moisture take their toll. In this respect, the failure of the system is predestined more or less.

However, a high availability is essential in automation engineering. Therefore, not only simple linear and star topologies can be build up with fiber optics, but also redundant ring structures in which each participant is available on two physical paths – ring redundancy according to DIN EN 62439 is today a synonym for high availability in automation engineering (Figure 9.39).

Figure 9.38   Mini-Media converter

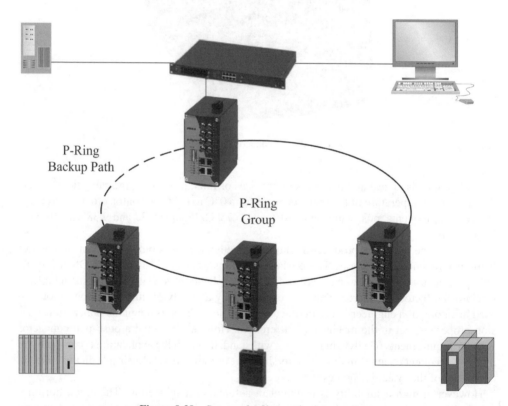

Figure 9.39   Set-up of a fiber optic ring structure

Especially when upgrading existing plants, free usable fibers are often not available. Therefore, some active network components support the *BiDi-technology* (bidirectional communication): where one fiber can be communicated in two directions. In most cases, no additional fiber must be installed.

But the infrastructure is only one aspect. Active network components provide a budget (difference between transmission power and receiving sensitivity) with which the required attenuation of the fiber-optic path, depending on the type of fiber, can be bridged. This increases imperceptibly over time, for example by loose connectors, dust and dirt, light, mechanical stress or changes in the network topology. That was figured out by expensive measurements – for example by means of optical reflectometry (OTDR). Modern network management and SCADA systems (*Supervisory Control and Data Acquisition*) are able to register the status of active components, but not the single fiber-optic links in detail.

"*Fiber View*" is a monitoring system that was developed especially for this task. It consists of a hardware/software combination, which is integrated into the active network components and permanently monitors the budget of the respective optical fiber link per port. Three LEDs or – at Ethernet switches – an additional user interface, which can be accessed via the web interface, indicate the status by a traffic light, whether the budget is in the green, yellow or red range. The yellow range defines the budget just above the systems margin of 3 dB (Figure 9.40). This early warning level is also signaled via a potential-free contact and can therefore be analyzed centrally in SCADA systems.

In contrast to status messages, which often need to be interpreted, the traffic light system is clear and understandable. In addition, the yellow status enables a predictive behavior because the attenuation is not too high, or in other words, the fiber optic link still works. However, then the focus is on maintenance or repair to prevent an outage. As a result, downtime of the whole system can be avoided. Even expensive service calls at night or on the weekend are therefore no longer necessary. This also applies to short-term and expensive hotel or flight bookings. Thus, the monitoring system helps to increase productivity while reducing costs.

Figure 9.40   Arrangement of the fiber view LEDs on a converter

High bandwidth, covering large distances and maximum security are just some of the reasons that speak for data transmission via fiber optic cables. Therefore, also the automation engineering benefit from it in terms of a continuous network. But that was in the past often easier said than done. Meanwhile, appropriate standards and guidelines have been created, as can be read in this chapter. There are now industrially-capable active and passive optical components that enable maintenance-free infrastructures, which offer enough flexibility and performance for future applications. Therefore, the wiring should be realized consistently with fiber optic in new automation projects.

10

Last Mile Systems, In-House-Networks, LAN- and MAN-Applications

For more than 20 years, fiber-to-the-home (FTTH) has generated much interest worldwide. This also includes plans to bring optical communication into the family home. This did not happen in the past for reasons of economy. However, because of the soaring use of the Internet, the higher data rates needs increasingly rises in the family home too. In order to achieve a corresponding quantum jump, it is absolutely essential to reduce costs to the participant. The keyword is "opening of the last mile". Plastic fibers have been developed (*Polymer Optical Fiber*, POF), which feature nowadays a considerably low attenuation in the wavelength range of interest (Chapter 10.2). Developments already started in the 1990s.

Figure 10.1 [10.1] shows the spectral attenuation of different plastic fibers and a glass fiber.

Attenuation values are indeed considerably higher than those of the glass fiber. This can be tolerated because the application of plastic fibers is only designed for the 100 m range. However, the compatibility to the wavelength fields is decisive in which glass fibers are used. At short fiber links it is possible to solve the problem of the mode dispersion. Synthetics with gradient profile are already produced: Perfluorinated-Graded-Index Plastic Optical Fibers (PFGI-POF).

As an economic alternative to the application of semiconductor lasers, another important step is the development of LEDs which can be speedily modulated. The gained cut-off frequencies of GaAs LED shave already reached the GHz range [10.2].

Combinations of fiber optical waveguide with mm waves or coaxial cable systems are developed for local area networks as an alternative (Chapter 9.3). For this purpose, a superposition of two light waves emitted from narrow-band semiconductor lasers occurs at the receiver (Figure 10.2) [10.3,10.4].

Both lasers differ in their frequency only by a few 10 GHz. An intermediate frequency (Chapter 9.4.1), which only lies in the region of mm waves, arises from that. One of the two lasers is modulated with the data signal. The superimposed light waves are transmitted up to the

Optical and Microwave Technologies for Telecommunication Networks, First Edition. Otto Strobel.
© 2016 John Wiley & Sons, Ltd. Published 2016 by John Wiley & Sons, Ltd.

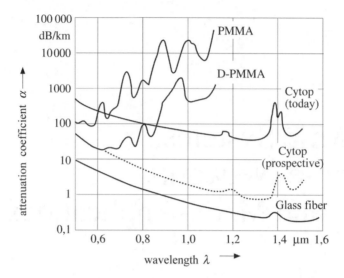

Figure 10.1 Spectral attenuation of different plastic fibers compared to a glass fiber

last mile by means of fiber communication techniques. The receiver transforms the modulated IF, for example of 30 GHz, from the optical region into the electrical one. This signal is fed to an mm-wave transmitter, in order to bridge the last mile to the participant in their home.

Due to the high demands on narrow line widths of the lasers, the IF is provided by only one laser (Figure 10.2). An intensity modulator (Mach–Zehnder Modulator) follows, operating by modulation near the operating point π (Figure 5.5). Thus, the carrier frequency disappears and two sidebands occur. A following optical filter (Mach–Zehnder-Interferometer Filter) separates both sidebands. One of the two sidebands is modulated with the data signal. By further superposition, an electrical mm wave signal of 30 GHz, see the example in Figure 10.3, is again attained at the receiver. Figure 10.4a shows a photograph of the antenna device.

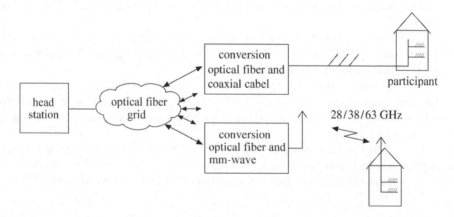

Figure 10.2 System architecture of future broadband networks with inclusion optical mm wavelength technique

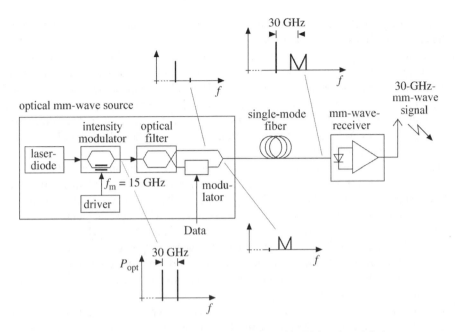

Figure 10.3 Optical mm wave transmission with sideband modulation

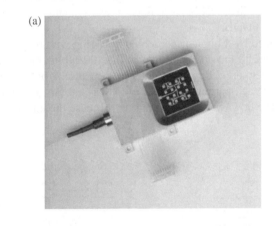

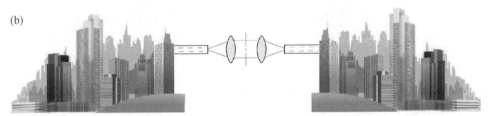

Figure 10.4 (a) Optical antenna device of an mm wave transmitter; (b), Free-space transmission in skyscraper areas

As another alternative to cable systems, the declared dead free space transmission could also be revived with distances in the 100 m range. For that purpose, the emergent light from a fiber is fed to a lens and forms an expanded beam. At the reception site it is again coupled into a fiber or directly to a detector. This conjunction of fiber-optic and free-space transmission may be particularly favorable in skyscraper areas. Here a large expenditure is necessary to lay cables, but on the other hand freeboard transmission is just as easy to implement, for example on the roofs of any urban district (Figure 10.4). Problems like rain, snow, fog, cloud, smoke and dust emissions can be tolerated to a certain extent over short distances. However, this devastates the application area of helicopters.

Wireless data transmission is also an interesting option for distances in the 10 m range. Connections between, for example, PC, printer, scanner or adjacent participants in intranets (LANs, see below) should not be bridged by interfering cables. Thus, typical infrared interfaces are used working with 850 nm LEDs and Si-photodiodes. Also, developments deal with the transmission in the microwave range (*Bluetooth*) [10.6] at about 2.5 GHz (ISM band). Regarding communication between eight participants, frequency hopping techniques are applied. Thereby, the transmission frequency is statically varied to minimize collisions. Compared to optical absorbent obstacles, walls, furniture and so on, only play a secondary role. In extreme cases, up to 100 m can be bridged.

Moreover, network devices are also discussed, besides typical point to point connections [10.7,10.8]. In the near field, for example inside a business building, the commonly used term is LANs (*Local Area Networks*). In the local net or metro region, it is MANs (*Metropolitan Area Networks*). The network structures can be designed as a star as well as a ring or a bus (Ethernet) (Figure 10.5). Electrical but also fiber optical systems are thereby used (*Fiber Distributed Data Interface*, FDDI).

The FDDI is characterized by a high-speed double fiber ring with more than 1000 participants on a length of about 200 km and data rates in the 100 Mbit/s range. Optical systems are used for high-bit rate networks. Regarding the ring structure, a high-bit rate transmitter and receiver is arranged at each participant. Related to star and bus, it is possible to locate only one receiver taking the downward data stream. In this case the upward data stream has

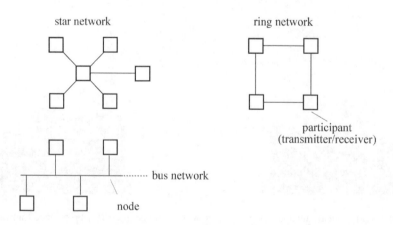

Figure 10.5   Network topologies for LAN and MAN

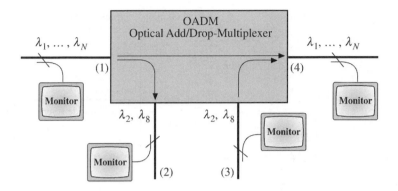

Figure 10.6   Function principle of an optical nodes in OADM network: Optical Add/Drop Multiplexer

been omitted. Furthermore, a narrowband feedback channel is used, for example to choose the channels in TV distribution systems.

For networks, in-service optical nodes are also necessary [10.8]. WDM- or DWDM-systems are often applied in that process. The access equipment for separation and connection of optical channels from several participants is realized by an optical multiplexer and demultiplexer, respectively: OADM (*Optical ADD/Drop Multiplexer*, Figure 10.6). Thereby it concerns optical four gate equipment with N-wavelength selective channels in the network (input and output gate, (1) and (4) and a variable quantity of less than N in the breakout and access gate, respectively, (3) and (4). By means of optical couplers, an ever smaller power part is decoupled for monitor purpose (mon).

Nowadays, MANs have channel numbers greater than 100 with a bit rate of up to the Gbit/s range for each channel. Thus, the DWDM technique (Chapter 9.4) no longer suffices. Limits are reached with increasing channel quantity; on the one hand, by the smallest possible channel distance and, on the other hand, by the defined spectral bandwidth of the optical amplifier. Therefore, CDM techniques are applied (*Code Division Multiplex*) [10.9].

Broadband intensity modulated LEDs, transforming the data of the channels from the electrical signal into the optical, can be used as light sources. Concerning channel coding and decoding, respectively, passive periodic filters based on Mach-Zehnder- (MZI) or Fabry-Perot interferometers (FPI) are therefore applied (Figure 10.7, Chapters 4.3.5 and 5.4).

The periodicity of the transmission function is given by the free spectral range, $FSR = \frac{1}{T}$:

$$T = \frac{\Delta g}{c} \quad \text{for MZIs} \tag{10.1}$$

$$T = \frac{2g}{c} \quad \text{for FPIs} \tag{10.2}$$

Thereby, g and Δg represent the optical path and the optical path difference, respectively, between both interferometer arms (Eqs. (2.23) and (2.24)).

In the network, different *FSR* are allotted to diverse optical transmitters and thus a channel coding is carried out (Figure 10.8a). The easiest way to realize these filters is the use of an MZI with two fiber couplers (Figure 10.7). Therefore, both interferometer arms show an optical

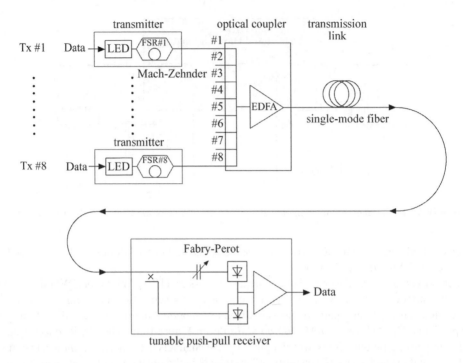

Figure 10.7   Fundamental set-up of a CDM System with periodic spectral coding

path difference, Δg (Eq. (10.1)), due to the different fiber lengths. Distinct *FSR* are gained by varying these length differences of the diverse transmitters. With a fiber optical 1/n coupler, several channels are combined. A fiber-optical Erbium amplifier improves their power and transmits it by a single-mode fiber.

At the receiver side a filter is used, whose *FSR* can be controlled the following way. By means of interferometer tuning, the optical path difference is adjusted to fit with the desired channel. For that purpose, piezo actuator elements micro-optical superstructures are applied, which realize displacements in the nanometer range. This applies to MZI- as well as to FPI-filters. The filter types of transmitter and receiver do not need to be identical, but the *FSR* must be adjusted correctly. Varying the *FSR* and the time of circulation T_{Rx} at the receive filter, respectively, a signal is gained, corresponding to Figure 10.8b. The optimally adjusted *FSR* arises at the maximum $T_{Rx} = T_{Tx}$. By increasing the circulation time drastically the result recurs and the next channel appears. A push-pull receiver is used for suppression of the offsets in Figure 10.8a. The offset is generated in this case by the variation of the *FSR*. Furthermore, it remains in systems with multiple channels by incomplete channel disconnection. By means of the CDM technique described in this context – embedded in a complete system – more than 400 channels can be transmitted inside the spectral bandwidth of a typical optical amplifier. TDM techniques (Chapter 9.4) are also applied. Regarding DWDM systems, this applies to comparable channel distances of up to 10 GHz.

Also, in the MAN region, free space transmission is discussed. Point-to-point components can also be used in the distribution field [10.5]. Transmitters with divergent radiation are

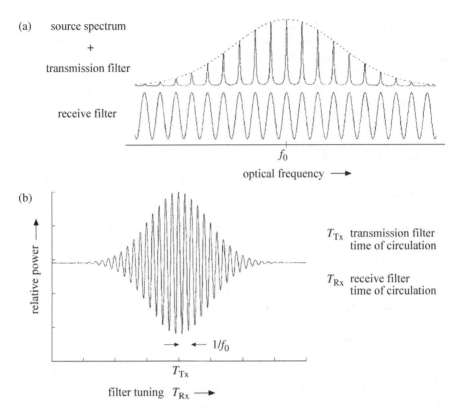

(a)   source spectrum

\+

transmission filter

receive filter

f_0

optical frequency ⟶

(b)

relative power ⟶

T_{Tx} transmission filter time of circulation

T_{Rx} receive filter time of circulation

⟶   ← $1/f_0$

T_{Tx}

filter tuning T_{Rx} ⟶

Figure 10.8   Basic principle of a periodical spectral coding and decoding (a) and optical power at the receiver as a function of the time of circulation in the receive filter (b)

applied as central units operating many receivers. With increasing distance, the divergence causes a smaller existing optical power. A network circuit is alternatively contemplated by means of a multitude of transmitters and receivers. The keyword is called: WOMAN, Wireless Optical MAN, nowadays named OWC, optical wireless communication (Chapter 10.3).

Moreover, free space transmission is gaining a particular renaissance in space. Outside of the Earth's atmosphere, typical problems like natural disturbances by precipitation or artificially caused impurities basically do not exist. Therefore, laser free space connections between satellites have already been tested successfully in communication (Chapter 9.3.1).

10.1 Last Mile Systems

Krzysztof Borzycki
National Institute of Telecommunications, Warsaw, Poland

Since the introduction of optical fibers around 1980, steady proliferation of this medium in most segments of telecom networks has been observed. The so-called "access", "last mile" or "first mile" segment, constituting fixed links to residential customers and small businesses, is also being converted from copper to fiber cables. However, progress toward the New Generation Access (NGA) network, designed for handling broadband IP traffic and largely fiber-based, is relatively slow and uneven, particularly in Europe. In addition, multiple variants of fiber access, collectively referred to as the "Fiber To The X" (FTTx), exist. They are competing against each other for acceptance by network operators and more recently also against LTE broadband wireless access.

10.1.1   Special Case of Access Network

Before analyzing technologies of fiber access, unique characteristics of fixed access networks must be considered:

- large investment, constituting some 70% of total investment in fixed telecom infrastructure
- costly and slow construction of cable plant: permits, digging, customer visits, etc.
- limited feasibility of separate, overlaying networks, making infrastructure-based competition problematic
- low profitability and long depreciation period, ca. 25 years
- heavy regulation protecting competition and consumers, not investors. In particular, regulation of rates and mandated "unbundling" do not help in ensuring return on investment and attracting financing for network projects

Traditional twisted pair copper loops with lengths of up to 5–8 km in most countries, and up to ≈60 km in remote rural areas of some others (US, Australia, India), have narrow bandwidths, considerable cross-talk and unstable transmission characteristics. This legacy infrastructure was originally designed for 4 kHz bandwidth and voice services. Today, a downstream bit rate of approximately 20 Mbit/s is required for high-quality Full-HD video, and 100 Mbit/s is a threshold for top-quality broadband. Only short copper loops (0.3–1.5 km) in good condition support such services (Figure 10.9), otherwise substantial conversion from copper to fiber is required.

If a complete switch to fiber infrastructure is chosen, the default approach is to build a long-lasting passive cable plant, designed for 25–40 years of service, used by multiple generations of active devices. Tight control of capital expenditure (CAPEX) is critical. The practice of replacing a mobile phone, tablet, PC or LAN wiring every three years does not belong here. Demand for services is uncertain because of "cord cutting" by customers migrating to mobile networks, first with voice, now with broadband traffic. Exceptions exist, in particular in newly-built, affluent neighborhoods, where a new, "greenfield" broadband infrastructure is considered essential. This, however, is a minor part of the market for FTTx solutions. The bulk are "brownfield" areas, where the existing copper network has to be replaced and customers migrate, while prospects for increased revenues are limited. Several solutions can be adopted to reduce CAPEX:

a) re-using as much as possible of existing infrastructure in "brownfield" areas: cables, ducts, street cabinets, poles, indoor wiring, etc.
b) removal of old copper infrastructure to avoid duplicated maintenance costs and regulatory obligations, except for parts re-used in the new network

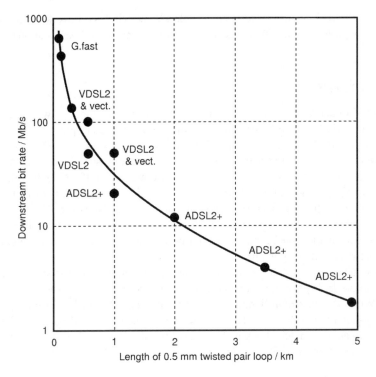

Figure 10.9 Typical bit rate of DSL link (multiple technologies) vs. distance

c) routing of single fiber to customer premises instead of fiber pair, employing wavelength division multiplexing (WDM) duplex

d) splitting of single feeder line from central office (CO) into multiple customer loops, employing passive fiber splitter or fiber-fed remote *Digital Subscriber Line Access Multiplexer* (DSLAM), feeding 10 s or 100 s relatively short copper loops to customer premises

e) elimination of locally-powered active devices, together with associated buildings or cabinets from the network

f) use of low-cost active devices at customer locations (high count), while higher-cost but fewer devices at central office or other operator site is acceptable.

The network should provide adequate bandwidth for all customers, although definition of "adequate" download bit rate lies anywhere between 30 Mbit/s and 2 Gbit/s today. This threshold, and characteristics of existence of copper infrastructure (or lack of it in a given area), dictates the choice of a particular network architecture by the operators.

10.1.2 Fiber Access Networks

This section presents basic types of FTTx networks, definitions and comparisons of their relative technical advantages and limitations. Fiber technologies developed for FTTH (*Fiber*

To The Home) networks were also adopted in cable TV and wireless networks, where passively-split fiber network links to equipment feeding coaxial drops or LTE base stations.

10.1.2.1   FTTx networks

In the 1990s, Copper access networks consisted of core and regional networks, linking COs and remote units (RU – switches or access multiplexers) with fiber optic cables. Single-mode fibers were introduced around 1984, but all connections between CO or RU and subscriber's premises were built of twisted paired copper cables, and occasionally aerial open wires as well. The first step toward broadband was the introduction of ISDN (Integrated Services Digital Network) and later DSL (*Digital Subscriber Line*) devices at both CO/RT and the subscriber's home, while still using the existing copper loops. However, with the exponential rise of the desired bit rates, copper cables became a bottleneck. Despite advances in DSL signal processing, achievable bit rate falls rapidly with distance, due to signal attenuation and cross-talk between pairs (Figure 10.10).

With each generation of DSL equipment offering higher speeds over shorter distances, it was necessary to deploy the last active network device, known as DSLAM, ever closer to the customer. While the maximum length of the 0.5 km copper loop adopted for digital remote units in telephone networks was about 5 km, deployment of 20 Mb/s VDSL2 equipment reduced this to approximately 1.5 km. The limit for 100 Mb/s services with VDSL2 devices working in phantom mode is 300–400 m.

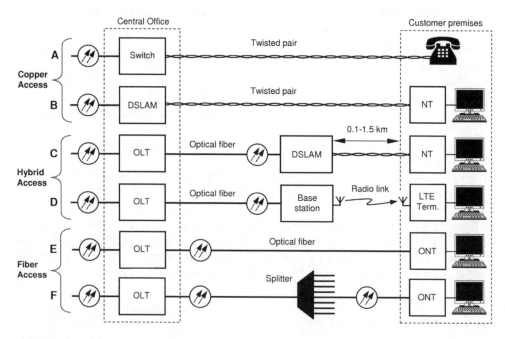

Figure 10.10 Telephone and broadband access loops. Variants C-F are collectively referred to as "FTTx" (Fiber To The x).

G.fast devices now under development may provide 500–1000 Mb/s bit rates – but only over 70–100 m. In all cases, the rest of the subscriber loop is made of (shared) *optical fiber*, and the proportion of copper left in the network accordingly reduced, while numerous DSLAMs are installed in new locations, predominantly in street cabinets. Coaxial cables to subscriber premises can be used as well, offering superior bandwidth and noise performance to twisted pair cables, which explains the success of digitized cable TV networks. It is based predominantly on the DOCSIS 3.0 standard, in provision of broadband services. Logically, the best and future-proof, but most expensive and disruptive option, is total replacement of copper loop with single-mode optical fiber extending all the way from the CO to the subscriber's premises, avoiding any intermediate active devices (Figure 10.10).

LTE networks, offering access speeds of over 100 Mbit/s, deployed since 2012, have no copper loops, but the backhaul links to base stations are fiber optic. The compromise between capacity and reach exists here too. To handle rising traffic, the network operator must reduce cell size, deploy more base stations and lay more fiber cables.

where:

A)   Traditional voice network for Plain Old Telephone Service (POTS)
B)   DSL network without remote units
C)   DSL network with remote units (FTTN/FTTC/FTTB)
D)   LTE, or other wireless network with fiber backhaul
E)   Fiber network with point-to-point loops (FTTH-P2P)
F)   Fiber network with passive fiber split (FTTH-PON)

Excluding solutions including radio and coaxial cables, the following fiber broadband technologies are most common [10.10]:

- *FTTN (Fiber to the Node)*: Remote units with DSLAMs, usually deployed in street cabinets or containers, are connected to customer premises with copper loops no longer than 1– 1.5 km. Downstream bit rates are usually between 20 and 50 Mb/s.
- *FTTC (Fiber to the Curb)*: Remote units with DSLAMs are deployed closer to customers than in the FTTN scenario, mostly in street cabinets near multi-dwelling units (MDUs); length of copper loops is up to 300 m. With the latest VDSL2+ equipment, downstream bit rates are 40–100 Mb/s, with prospects of some increase.
- *FTTB (Fiber to the Building)*: An active device, usually Ethernet switch, is installed inside a building. The network covering a single multi-dwelling building, or sometimes a small cluster of buildings, operates as Ethernet LAN at 100 Mb/s (Fast Ethernet) or sometimes 1000 Mb/s (Gigabit Ethernet) over unshielded twisted pair (UTP) cables. An equivalent network with a DSLAM device, utilizing legacy telephone wiring, is also designated as FTTB, provided the DSLAM is located inside the building.
- *FTTD (Fiber to the Door)*: Ultra-fast variant of DSL, temporarily designated as G.fast, is currently in development at ITU-T [10.11] and offering bit rates of 250–1000 Mb/s. This figure is a sum of downstream and upstream rates, as time division duplex (TDD) is employed. The reach over available copper loops – often old and of poor quality, may be too short (25–70 m) to cover a whole multi-dwelling building from a single DSLAM. In such cases, mini-DSLAM is deployed on each floor or even next to the doors of each apartment. With better cables, a fast (≥200 Mb/s) FTTB network can be built.

- *FTTH (Fiber to the Home)*: No copper loop exists between CO and customer premises. One single-mode optical fiber (SMF) runs from OLT (*Optical Line Termination*) to ONT (*Optical Network Termination*) deployed at customer premises – either indoors or in an enclosure on the wall of the building or apartment [10.10]. This fiber may be split (Figure 10.10F), in this case a single OLT is connected to multiple ONT units. The outside plant is fully passive. Several standards of FTTH active equipment exist (Chapter 10.1.9.2). Bit rates range from 10–2 Gb/s and, unlike FTTN/C networks, are almost independent of loop length, which is up to 20–40 km. Extended-range FTTH equipment exists with loop length increased to 60–100 km. All FTTH networks in operation today use exclusively single-mode fibers of dispersion-unshifted type, either SMF [10.12] or BIF [10.13] (Chapter 4.2.1). This is the most common and cheapest optical fiber available.

Additional and sometimes confusing names and abbreviations exist. For example, FTTP also means "Fiber To The Pedestal", and FTTN – "Fiber To The Neighborhood".

The networks indicated above are not limited to the provision of any particular service and usually provide a mix of several services. In particular, combination of Internet access, video and telephony – either POTS or VoIP, is common and known as "triple play".

Popularity of specific FTTx technologies varies substantially with country. Within the European Union, or EU28, share of homes passed with broadband networks using all-fiber outside plant in 2013 was FTTB – 43%, FTTH-P2P – 31%, and FTTH-PON – 26% [10.14].

Extensive descriptions of FTTH-PON technologies are to be found in [10.15], including both complete equipment/systems and most important components. Being published in 2007, it does not, however, cover XG-PON and NG-PON variants.

FTTH Council Europe organization, devoted to promotion of fiber access networks, publishes regularly and freely updates a distributed technical handbook of FTTH [10.16]. This handbook is especially focused on passive cable plants and covers issues such as standards, testing, troubleshooting, work planning or documentation. It is useful for network designers and investors.

10.1.2.2 ITU-T reference architecture

Architecture of broadband fiber access networks, standardized in ITU-T Recommendations G.983.1 [10.17] and G.984.1 [10.18], reflects the variable depth of fiber penetration toward the subscriber (Section 10.1.2.1). The FTTH network is regarded as all-optical (Figure 10.11). While this is true for the core, regional/metro and access networks of the telecom operator, the home network is usually not fiber optic as yet. Devices at the subscriber's home: router, PC, TV set, telephone set, network-enabled appliances, etc., are either wired to the ONT by means of twisted-pair or coaxial cables, or linked via wireless link using Wi-Fi or other similar technologies. Because of short distances, usually 2–25 m, optical fibers have no particular advantage here. Multimode plastic optical fibers (POFs) were proposed for home wiring, but failed to find acceptance for cost reasons. Almost all IP-enabled devices have built-in Ethernet interfaces with RJ45 connectors for twisted-pair cables.

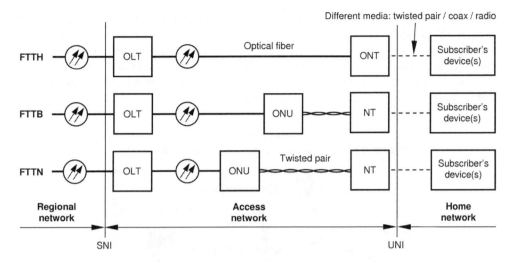

Figure 10.11   Architecture of FTTx networks set in ITU-T recommendations

where:

OLT   Optical Line Termination
ONT   Optical Network Termination
 NT   Network Termination
UNI   User Network Interface
SNI   Service – Network Interface

In publications dealing with FTTH networks, the active device at a subscriber's premises is alternatively designated as ONT or ONU. In this chapter, the ONT abbreviation will be used.

10.1.3   FTTB Networks

A network covering a building, or sometimes a small cluster of buildings, most often operates as Ethernet 100BASE-TX Local Area Network (LAN) at 100 Mb/s [10.19]. Internal wiring of a building is made of Cat. 5, 5e or 6 unshielded twisted pair (UTP) copper cables. The cable contains four twisted pairs, normally used together by a single device. Technology, cables, accessories and active equipment are directly adopted from office LANs, being inexpensive, tried and tested, except for Ethernet switches optimized for a small number of users in one apartment block (Figure 10.34). The structure of such a network is shown in Figure 10.12.

The key advantage of Ethernet-based FTTB is almost universal inclusion of Ethernet 100BASE-TX (Fast Ethernet) interfaces in computer devices and network-enabled consumer electronics, thus no separate NT is needed. The 100 m reach is adequate for most apartment buildings. Large buildings require sub-division into several zones served by separate switches, with fiber optic cable running to each switch.

PCs and other devices are connected to wall outlets using a jumper terminated with 8P8C (RJ45) modular connectors. To serve multiple devices in a single apartment or provide interactive video services (IPTV), additional customer premise equipment (CPE), hub or router/bridge, is necessary. Examples of active devices are presented in this section.

A faster variant of Ethernet, 1000BASE-T (1000 Mb/s), may be adopted; in this case, a higher grade twisted pair cabling is required, at least Cat. 5e.

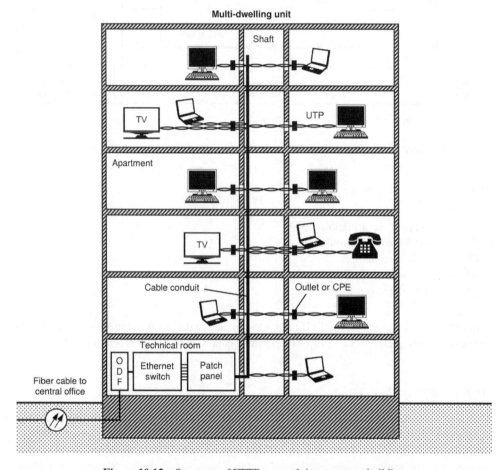

Figure 10.12 Structure of FTTB network in apartment building

where:

ODF   Optical Distribution Frame

The DSL-based variant of FTTB is less popular, mostly because separate NT is needed at each served apartment. However, re-wiring of the building is avoided.

Cost of backhaul fiber links to the central office can be reduced by replacing multiple point-to-point links with an FTTH-PON network linking all served buildings (Chapter 10.1.5), but including Ethernet switches instead of ONTs.

With the newest 10G EPON system operating at 10 Gb/s, a fiber split of 1:16 and 24 ports per switch enables connection to 384 subscribers. At low 2.5:1 over-subscription of OLT bandwidth, each subscriber can be provided with a 50 Mb/s service. Ethernet switches capable of operation in 1 Gb/s EPON [10.20] and 2.5 Gb/s GPON [10.21] networks are also available. However, bit rates offered to most subscribers are lower, for example 20 Mb/s.

10.1.4 Point-to-Point FTTH Networks

This type of access network is based on Ethernet standard IEEE 802.3 – Section 5 [10.19] and ITU-T Recommendation G.686 [10.22]. While the exhaustive IEEE 802.3 standard covers all aspects and applications of Ethernet, ITU-T G.686 specifically covers P2P access networks based on 1 Gb/s Ethernet technology and single-mode fiber cabling.

10.1.4.1 Wavelength division duplex

Duplex transmission over a single fiber helps reduce costs of outside plant, the dominant component of CAPEX. The technology of choice is wavelength division multiplexing (WDM). A separate optical band is assigned to each direction: 1260–1360 nm downstream (to subscriber) and 1480–1500 nm upstream (from subscriber). A WDM coupler is fitted at each end of the fiber loop (Figure 10.13); typical insertion loss of this device is 0.5–0.8 dB [10.22].

Optical transmitter and receiver constitute an interchangeable SFP module, available in several versions with different power budgets and operating wavelengths. Multiple OLT modules are parts of the Ethernet switch, often installed in a remote location. While such a solution requires local power with battery back-up and temperature-hardened equipment, it brings savings on optical fiber cables. For example, the remote switch needs only 2 fibers, or 4 fibers with 1·1 protection, for connection toward the metro or regional network.

The time division duplex (TDD) technique is poorly suited to this application, due to the large round-trip transmission delay introduced by the subscriber loop. This delay is approximately 600 μs for the 30 km length considered as maximum in ITU-T G.686 [10.22].

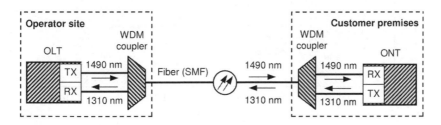

Figure 10.13 Signal paths and wavelengths in FTTH–P2P network without video overlay

10.1.4.2   Selection of wavelengths

Transmission wavelengths in FTTH networks were chosen for both economic and legal reasons, with common arrangements made for P2P and PON variants to allow volume manufacturing of active and passive components to common specifications.

A single OLT serving the PON may be relatively expensive, therefore its cost is shared between 8, 16, 32 or even more users. The costs of ONT are allocated to a single user and should be low. To reduce cost and insertion loss of the WDM coupler and relax limits of wavelength stability for transmitters in OLT and ONT, wavelength separation must be large. The best option is to use two low-loss windows of single-mode fiber around 1310 nm (O-Band) and 1550 nm (C-Band) (Figure 10.14).

However, chromatic dispersion of SMF is high in the 1550 nm band [10.3] (Figure 4.13). A transmitter operating at a 1–10 Gbit/s bit rate over 20 km or longer distance must include a costly distributed feedback (DFB) semiconductor laser, combined with an external modulator of the Mach–Zehnder or electro-absorption types. Conversely, low fiber dispersion in the 1310 nm band allows the use of a cheaper transmitter with a directly modulated semiconductor laser. This band is traditionally assigned to an upstream link. Another technical factor is asymmetry of bit rates in broadband access, the downstream rate usually being higher. Lower loss of single-mode fiber at 1550 nm: 0.18–0.22 dB/km compared to 0.32–0.35 dB/km at

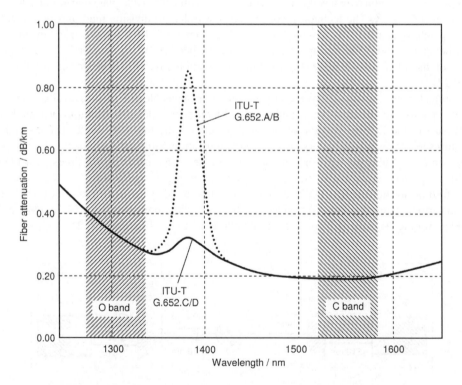

Figure 10.14   Spectral characteristics of attenuation in single-mode fiber

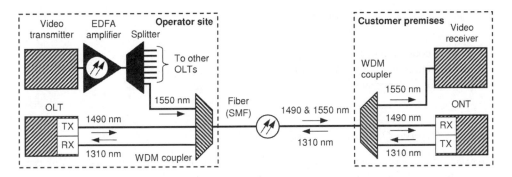

Figure 10.15 Signal paths and wavelengths in FTTH–P2P network with RF video overlay

1310 nm [10.24,10.25], allows choosing a higher bit rate, associated with increased receiver input power and reduced power budget for this direction, if necessary.

Allocation of 1490 nm nominal wavelength instead of 1550 nm to downstream transmission is due to legal issues with distribution of TV content. When early work on FTTH systems was made ca. 2000, copyright laws in several countries, such as US and Japan, allowed re-distribution of analog TV signals over coaxial or fiber cable, but not their digitization and transmission over the IP network. Operators avoided this restriction by sending analog video signals over a 1550 nm optical carrier, using equipment originally developed for feeder sections of cable TV networks. Beginning from publication of ITU-T Recommendation G.983.3 in 2001 [10.26], the 1540–1560 nm band is reserved for one-way, optional downstream video delivery in FTTH networks. With steady increase of IP traffic in access networks, there is also some interest in delivery of digital video at this wavelength, to relieve OLT capacity occupied by video streams [10.27].

A corresponding arrangement for point-to-point network is shown in Figure 10.15. Erbium Doped Fiber Amplifier (EDFA) boosts the level of signal produced by a costly, highly linear RF video transmitter to allow splitting and routing to multiple customers. This dramatically reduces equipment cost per subscriber. The EDFA device, presented in Chapter 8.1, works in a narrow band of approximately 1528–1563 nm. This results in a choice of a 1550 nm wavelength for video transmission.

10.1.4.3   Merits and disadvantages

Advantages of point-to-point fiber FTTH networks include:

- *high performance*: user has access to full OLT bandwidth
- *reliability*: malfunctioning ("rogue") ONT does not interfere with other ONTs
- *security*: physically separate link to each customer, no eavesdropping or jamming
- *easy troubleshooting*: of fiber plant with OTDR (no splitters)
- *easy unbundling*: as each subscriber has a separate fiber loop
- *simple upgrade*: to new equipment, without changing splitters

Disadvantages of P2P solution are of economic nature:

- high fiber count and duct/pole occupancy
- high cost and time-consuming construction of passive fiber plant
- high cost of OLT per subscriber – the device is not shared

10.1.5   Passive Optical Networks (PON)

This section contains general descriptions of issues specific to FTTH networks of this type. Standardized systems are presented in Sections 10.1.9 and 10.1.10.

10.1.5.1   Fiber splitting

The PON network is optimized to reduce the costs of passive fiber infrastructure and occupancy of duct space. Work on development of passive optical access network systems was initiated in 1995 by the Full Service Access Network (FSAN) consortium, formed by seven large operators including British Telecom, NTT and Bell South [10.15]. Later, ITU-T continued; the first result was Recommendation G.983 [10.17].

In a PON (Figure 10.16), both OLT and a considerable length of fiber loop are shared by a group of users. With a passive optical *splitter*, a single "feeder" fiber leaving the OLT is usually divided into 8–64 "drop" fibers running toward multiple ONTs. The passive fiber plant between OLT and ONTs is known as the Optical Distribution Network (ODN).

Fiber splitter is compact, inexpensive and requires no power supply, but introduces considerable insertion loss dependent on split ratios (Table 10.2). More data on passive fiber devices are described in Chapter 4.3.

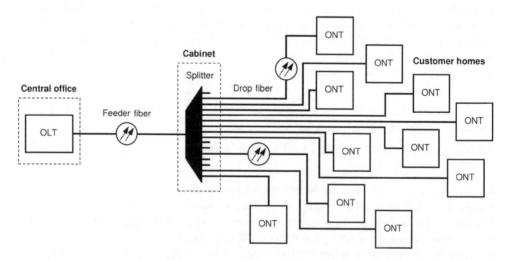

Figure 10.16   Basic structure of PON. In most cases, a large proportion of splitter ports remains unused due to unpredictable demand for services

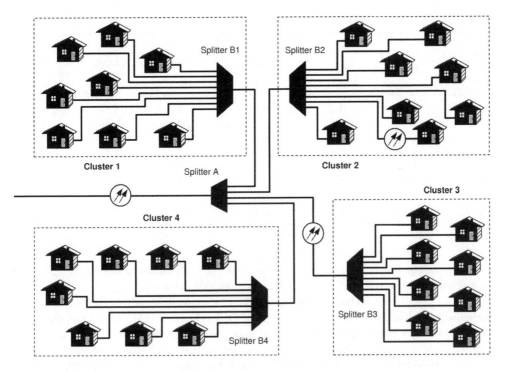

Figure 10.17   PON with two-stage fiber split, serving 32 subscribers in 4 clusters

Fiber splitting may be arranged in multiple stages, for example to optimize cable plant when customers are clustered in several separate areas (Figure 10.17) or to increase the split ratio of the existing network. However, adding an extra split stage increases a loss of the ONT-OLT path by approximately 1–1.5 dB with respect to an equivalent single splitter, due to loss of additional connectors and uneven division of power in each splitter.

10.1.5.2   Loss budget

Loss of PON fiber plants is dominated by contribution from splitters, devices absent in most other fiber networks. Splitter can consume up to 75% of the OLT-ONT link power budget (Table 10.1). For our calculations, the following data were adopted:

- *Fiber attenuation*: 0.34 dB/km @ 1310 nm for SMF [10.24] and BIF [10.25]
- *Loss of fusion splice*: 0.10 dB (1 splice per km + 1 at splitter site)
- *Loss of complete indoor wiring*: 1.2 dB (national standard in Poland)
- *Splitter loss*: value for 1:128 device is estimated (Tables 10.2, [10.23])

The loss budget of standard, low-cost optical modules (transponders) in OLT and ONT is usually in the 24–33 dB range. With 2.5 dB of them being set aside as a reserve for aging of

Table 10.1   Loss budget of PON at 1310 nm – examples

Fiber distance Km	Split ratio –	Loss of fibers dB	Loss of splices dB	Loss of splitter dB	Indoor wiring dB	Total dB
5	1:16	1.7	0.6	13.8	1.2	17.3
20	1:16	6.8	2.1	13.8	1.2	23.9
5	1:32	1.7	0.6	17.0	1.2	20.5
20	1:32	6.8	2.1	17.0	1.2	27.1
5	1:64	1.7	0.6	20.5	1.2	24.0
20	1:64	6.8	2.1	20.5	1.2	30.6
5	1:128	1.7	0.6	24.0	1.2	27.5
20	1:128	6.8	2.1	24.0	1.2	34.1

the cable and equipment, repair splices, contaminated or worn-out connectors, etc., "Extended range" equipment is available, but considerably more expensive.

With a 21.5–30.5 dB net budget, split ratios like 1:64 are realistic only when the PON covers a relatively small and densely populated area and OLT is located within a few kilometers. The latter requirement is in conflict with plans of most operators to concentrate OLTs in a single facility. They are serving a large area to reduce costs of real estate, power, security, etc. Distances of 20–60 km are permitted by all ITU-T and IEEE standards [10.17–10.20,10.29,10.30]. This potentially allows to serve a city with a population of 1 000 000 from a single site. Therefore, it reduces the split ratio and OLT cost per subscriber, making such an approach less attractive, unless most customers order very fast Internet access like 200 Mb/s. In addition, total centralization brings serious reliability and security issues, as a fire or other problems at the central node can disrupt services to a large city.

The maximum distance between OLT and ONT, known as the physical reach, may be increased by installing active devices, for example mid-span optical amplifiers or repeaters. This solution, while standardized for some systems like XG-PON [10.31], violates the principle of fully passive ODN infrastructure, decreases network reliability and is expensive.

Table 10.2   Designs and insertion loss of fiber splitters – theory and typical specifications [10.23]

Split ratio	Number of stages	Loss (theoretical) dB	Loss (actual) max. dB
1:2	1	3.01	4.0
1:4	2	6.02	7.3
1:8	3	9.03	10.7
1:16	4	12.04	13.8
1:32	5	15.05	17.0
1:64	6	18.06	20.5
1:128	7	21.07	(no data)

10.1.5.3   Bandwidth sharing, reliability and security

OLT capacity is shared between all active users in PON by means of time division multiplexing (TDM). Equipment must handle differential lengths of fibers between OLT and each ONT, up to 20–40 km, and resulting differential transmission delay: 1 km of SMF with refractive index n ≈ 1.47 [10.24] introduces approximately 98 µs of two-way delay. PON equipment has a "ranging" function: relative transmission delay between OLT and each newly added ONT is measured to apply proper delay compensation and avoid collision of frames received from several ONTs. For GPON equipment, ranging procedure is standardized in ITU-T Rec. G.984.3 [10.32].

Failure of ranging and multiplexing mechanisms, for example continuous transmission from ONT, disrupts the whole PON. PON is also sensitive to jamming by injection of a powerful signal in the upstream band into any ONT port, for example from a 1310 nm OTDR. Another security problem is created by sending all downstream data to all users; unauthorized access with specially modified ONT or a dedicated spy device is prevented only by ONT authentication and data encryption. Suitable mechanisms are standardized for GPON [10.32] and XG-PON [10.33] networks; but currently not required for EPON.

Another disadvantage of the PON vs. P2P network is that a single cut of feeder fiber or OLT failure affects service to all users in a given PON, say 32, instead of one. To prevent this, systems like GPON standardized in ITU-T Rec. G.984.1 in 2008 [10.18] and later variants include 1:1 protection, for example two OLTs are connected through separate, differently routed feeder fibers to a second common port of the splitter. Suitable 2:N splitters are available commercially [10.34]. Several protection options are described in the literature [10.26] and standards [10.17], allowing the network operator to minimize costs by covering only network segment(s) with the highest failure rate, for example feeder cables being subject to frequent cuts during construction work. For cost reasons, protection is offered mostly as a premium service to business users.

10.1.5.4   Multi-operator access

Copper networks of incumbent operators in most countries are subject to "unbundling" – mandated rental of copper loops, and to various extents, other network plant like ducts and space at facilities to alternative operators at rates set by the regulator. This idea works in a P2P fiber network, but not in PON. The attempt to connect another OLT would result in interference and outage. A similar problem also appears in copper networks, if a VDSL2 or similar DSLAM with "vectoring" cross-talk cancellation mechanism is used.

Forcing incumbent operators to "open" their expensive FTTH networks to competitors usually results in cancelled construction projects, being partly responsible for slow progress of NGA rollout in most of the EU. The matter is improved by public support covering a fraction of network cost, but technical issues remain.

A solution mandated in France by regulator ARCEP [10.35] for areas with medium population density is to build "fiber distribution points", where at least 300, and preferably about 1000 fiber loops (drops) from a certain area are terminated at the Optical Distribution Frame (ODF). Several operators can then install their splitters and connect their customers (Figure 10.18). Because of the long reach of the FTTH equipment, an increased loop length is not critical. A fiber distribution point has multiple fiber links to facilities of different operators,

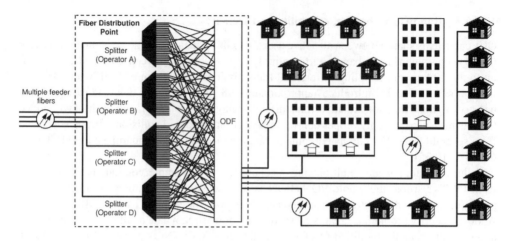

Figure 10.18   Fiber distribution point for multi-operator access

where their OLT and other active equipment is located. This solution is applicable also to P2P networks.

10.1.5.5   Multiple subscriber devices

A large proportion of broadband customers want to connect and simultaneously use several network-enabled digital devices: PC, laptop, game console, TV set, DVD player, etc. For this purpose, the ONT includes a small Ethernet switch, for example with four Fast Ethernet (100 Mb/s) or Gigabit Ethernet (1000 Mb/s) ports, to which customers' devices are connected with unshielded twisted pair (UTP) cables. Another option is WiFi: wireless transponder operating in the 2.4 GHz band, and integrated into the ONT (Figure 10.19). Both solutions are also common in NT devices for FTTN/FTTC networks.

While lack of rather bulky UTP wiring appears to favor the wireless option, it is often problematic due to propagation problems, especially in apartment blocks constructed with walls made of steel reinforced concrete. Moreover, security issues exist, because without a

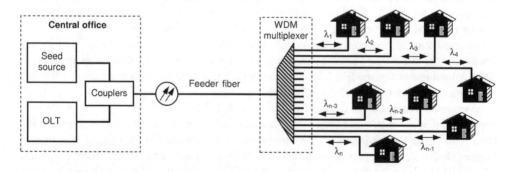

Figure 10.19   WDM-PON with duplex transmission to each user at the same wavelength

proper Wi-Fi set-up and encryption, the service may be illegally accessed by strangers. In several jurisdictions, the subscriber can be made responsible for their actions, like hacking or illegal distribution of copyrighted materials.

10.1.6 WDM-PON Networks

10.1.6.1 Advantages

Each user has assigned separate wavelength(s), and the wideband optical splitter is replaced with a spectrally-selective WDM multiplexer (Figure 10.19). In theory, this arrangement eliminates most drawbacks of ordinary PON, because:

- WDM multiplexer has lower loss (3–6 dB) than splitter (Table 10.2)
- each user has access to full capacity of OLT working at his wavelength
- no TDM multiplexing and ranging are necessary, and equipment is simpler
- security is improved: no access to data streams sent at other wavelengths
- PON capacity rises in proportion to number of wavelengths
- multiple types of equipment can be used in a single WDM-PON
- individual fiber loops ("lambdas") can be rented to several operators

10.1.6.2 Implementations and trials

Practical and cost-effective implementation of WDM-PON, likely as an upgrade of an existing network PON, must meet the following requirements:

- re-use of existing cable network: single fiber per user, similar reach
- similar or higher split ratio: 1:32 or higher
- low-cost and reliable ONT units
- single model of ONT operating at any wavelength

Splitters and other passive components can be replaced, if necessary, but new devices must be passive as well.

Implementation of the WDM-PON idea turned out to be difficult. Direct adoption of designs in successful DWDM metro and core networks, where line terminals have banks of separate transponders with DFB lasers and precise wavelength monitoring, is too expensive. Besides, operation on single fiber requires assignment of separate downstream and upstream wavelengths to each customer. Thus, fiber split ratio in a network using 40 wavelengths with 100 GHz ($\approx$0.8 nm) spacing is only 20 – lower than in typical PON.

CWDM equipment, working with 20 nm wavelength spacing, is cheaper, as distributed feedback (DFB) lasers with temperature control and wavelength monitoring. They are replaced with directly modulated, uncooled Fabry-Perot (FP) lasers. CWDM-PON equipment was tried in Malaysia in 2008, but with a set of 8 wavelengths in accordance with ITU-T Recommendation G.694.2 [10.36], fiber split was only 1:4 [10.37]. This variant of PON failed to win acceptance of the operators.

The most efficient and prospective embodiment of WDM-PON is the one using the same wavelength for both directions of transmission to a given ONT. Instead of DFB transponders generating a signal at each wavelength at OLT and ONTs, light re-modulation by a reflective semiconductor optical amplifier (RSOA) at ONT was employed. The RSOA is a low-cost FP laser diode modulated by applying drive current below the threshold of laser emission. Operating in this regime, it modulates the amplitude of incoming light from an external broadband source, known as the "seed source" (Figure 10.19); usually composed of an amplified spontaneous emission (ASE) device and EDFA amplifier. Wavelength control is provided by a WDM multiplexer, which passes only a proper "slice" of seed source spectrum to each ONT. Wavelength tuning of the ONT and OLT equipment is not necessary, and one model of ONT works at any wavelength.

Unfortunately, duplex transmission at single wavelengths is easily degraded by reflections from passive components and Rayleigh backscattering in fibers. This problem can be mitigated by using low-reflection APC connectors and allocation of separate wavelength bands and feed fibers for each transmission direction, but with added cost and complications.

A WDM-PON test network was built by KT in South Korea in 2005 [10.38,10.39]. It had a 1:32 split ratio, provided symmetrical 100 Mb/s services and used separate bands: 1426–1449 nm and 1534–1559 nm, for downstream and upstream directions, respectively.

The seed light source is costly and, like the video transmitter shown in Figure 10.14, a single unit provides light to all OLTs at a given facility through splitters and EDFA amplifiers. Another issue is temperature-induced wavelength drift of the WDM multiplexer, which must work in uncontrolled climatic conditions outdoors, replacing ordinary splitter. While multiplexers for older DWDM equipment traditionally included thermostats, devices for WDM-PON must be "athermal" – insensitive to temperature variations by design [10.15], as most of their locations are without an electrical supply.

Despite the promise of ultimate capacity, large commercial WDM-PON networks have not yet been built, and no international standards have appeared. Research work, however, continues.

10.1.7   Upgrade and Migration Issues in FTTH Networks

10.1.7.1   Service life

Most designs of FTTH networks assume a long service life of the passive cable infrastructure, except for some components such as jumpers or splitters, of 25 to 40 years. During this period, the active equipment: OLT, ONT, routers, line systems, video equipment, amplifiers, etc. is expected to be replaced several times, every 5–10 years, predominantly to cope with rising traffic and demand for higher performance services. These assumptions have been supported by good experience with durability of optical fiber cables networks since their introduction around 1985, and the frequency at which several generations of PON equipment have been introduced in the past (Table 10.3). In addition, no prospective transmission medium expected to replace SMF in access and regional networks is on the horizon (Chapter 4.2).

The same cannot be said of the FTTN/FTTC infrastructure, as subsequent generations of faster DSL equipment have shorter reach (Figure 10.9). If this trend, and move to access bit rates significantly over 100 Mb/s continues, most likely copper loops will ultimately be eliminated. This does not apply to some old buildings, particularly of historical value, whose

Table 10.3   First commercial deployments of FTTH-PON systems

Year	Operator	State	Standard	OLT bandwidth, Mb/s
1995	Deutsche Telekom	Germany	OPAL	29
1997	NTT	Japan	Pi PON	49
2001	NTT	Japan	ATM PON	155
2003	NTT West	Japan	BPON	622
2004	NTT East	Japan	EPON	1000
2007	Verizon	US	GPON	2488
2012	China Telecom	China	10G EPON	10 000

owners or inhabitants strongly object to drilling and other disruptive work. While a large proportion of copper cables in FTTN networks can serve for another 25 years or longer, evolution of technology will most likely make them obsolete sooner.

Investment in an FTTB network is safer, because the amount and cost of cables to be replaced are much lower, and periodic replacement of wiring is an established practice in office LANs, from which the FTTB technology is largely borrowed. If necessary, the next upgrade may be with fiber cables, creating an FTTH network.

10.1.7.2   Overlay operation

Upgrade of the existing FTTH-P2P network is essentially limited to replacement of active equipment, and maybe connectors. Importantly, it can be made selectively, for new customers or those upgrading to higher grade services, and paying higher rates.

Conversion of FTTH-PON to a new standard would require replacement of all ONTs, also for customers who do not order new services and will not pay more, thus being rather uneconomic. To relieve this problem, standards allow "overlay" operation of equipment belonging to two different generations, for example GPON and XG-PON, by specifying separate wavelength bands for each system [10.40,10.41]. Two networks can co-exist interference-free on the same ODN, if suitable WDM multiplexers and blocking filters are installed. It allows gradual migration of customers to a new standard, or increase of PON capacity beyond original limits However, this method has drawbacks:

- WDM couplers and filters must be installed, causing service disruption
- these components introduce a loss of 0.5–1 dB each
- old equipment may be unsuitable: first editions of GPON and EPON standards set wider wavelength limits

Each standard has new wavelength bands reserved. NG-PON [10.30] is the last ITU-T one complying with this rule, as most of the fiber bandwidth is already allocated [10.41]. As can be seen in Figure 10.20, transmission bands set in IEEE and ITU-T standards are synchronized for the same generations: EPON and GPON, 10G EPON and XG-PON.

Wavelengths above 1610 nm are set aside for monitoring purposes [10.42], because change of fiber loss due to bending, aging, etc. rises rapidly with wavelength.

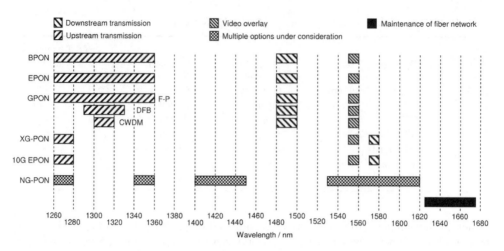

Figure 10.20   Current and planned allocations of fiber bandwidth to FTTH-PON systems

GPON data show three laser options for ONT transmitters, with DFB and CWDM variants introduced for compatibility with XG-PON. Overlay distribution of RF video is optional. To allow the overlay option, network designers must specify all passive plant to operate across the full 1260–1650 nm wavelength range. This applies in particular to splitters, as many devices on the market are designed for bands of first-generation PON systems.

10.1.8   Passive Fiber Plant

This section is devoted to the peculiarities of the design and construction of passive fiber infrastructure for access networks, especially of the FTTH type. More extensive descriptions of optical fibers and cables are presented in Chapter 4.2, including bending-insensitive single-mode fibers (BIF), originally developed for FTTH applications.

10.1.8.1   Common issues

Reduction of total costs is a universal requirement, but cost structure depends on labor rates: high in Western Europe, US, Japan or Canada, much lower in Russia, Ukraine, Central Europe, China or Indonesia, while other costs are comparable worldwide. With high labor rates, installation must be quick, and highly-skilled workers employed sparingly, even if necessary materials are more expensive.

Where labor is cheaper, costs of cables and associated hardware such as joint closures, distribution frames, jumpers, etc. are more important, and preference is given to less advanced but cheaper materials and cable laying equipment. Also, hiring technicians with decent experience in fiber optics is affordable.

Conversion of the copper network to an FTTH one is associated with vacation of most central office buildings. They are often located in expensive urban areas, which can be sold, generating significant revenue for the network operator. However, regulators may not allow scrapping of copper network, citing needs of competitive operators and emergency communications.

During Hurricane Sandy in the USA in October 2012, only the copper network with traditional telephone sets amplified spontaneous emission remotely powered from central offices worked after massive power failures and exhaustion of back-up batteries in cellular systems [10.43].

Important are local conditions, codes and regulations regarding telecom infrastructure;

- whether aerial cables are permitted (US, Japan, Korea) or not (Germany, Italy)
- requirements for crossings between power and telecom systems which, in countries like Poland, force operators to use dielectric fiber cables only
- natural disasters: earthquakes, hurricanes, etc. dictating type of cable plant
- terrain and climatic conditions
- dominant designs of residential buildings; in particular, large apartment blocks common in Eastern Europe, South Korea and China allow rapid, cost-effective construction of FTTB networks
- coverage, characteristics and condition of existing copper infrastructure, if FTTN/FTTC solutions are considered

10.1.8.2   Bending-tolerant fibers

Cheaply executed FTTH projects can involve workers lacking experience with fiber optics, who assume fiber cables tolerate handling and fixing methods common for telephone wiring, like stapling, bending on wall corners, tight coiling of spare lengths, etc. In addition, drilling holes for cables earned a bad a reputation as being dirty and noisy, so being widely rejected by potential customers. A European operator indicated that 30% of his potential new FTTH subscribers cancel orders after seeing technician preparing to drill holes [10.44]. Residential customers want compact, less visible cables and hardware. Around 2000, those factors instigated the introduction of "bending-tolerant" and "bending-insensitive" single-mode fibers, and cables tolerating harsh mechanical treatment. More details are presented in Chapter 4.2.

On the other hand, operators in several countries have no problems with finding qualified installers and using regular indoor cables with ordinary, low-cost single-mode fibers compliant to ITU-T Rec. G.652.D [10.12].

10.1.8.3   Optical fiber cables

- *Feeder cables:* In point-to-point FTTH networks, without fiber splitting, extreme fiber counts are often required. Feeder cables used in Japan in the early 2000s had up to 4000 fibers [10.45], although current catalogs of Japanese manufacturers include designs with up to 1000 fibers only [10.46–10.48]. Duct space is generally at a premium, so cables need to be as compact as possible, and sub-ducts are used extensively to lay multiple cables in a single duct. Besides, rapid splicing is essential, with a strong preference to "dry" cable designs, completely without gel filling.

  Bending-tolerant fibers (ITU-T G.657.A1/A2) retain required attenuation stability, despite reduced degree of mechanical protection in "microcable", which is thinner: 3–9.5 mm depending on fiber count. They are lighter than conventional designs, due to reduced thickness of sheath and cross-section of strength member, as well as smaller tubes with only marginal space for fiber movement.

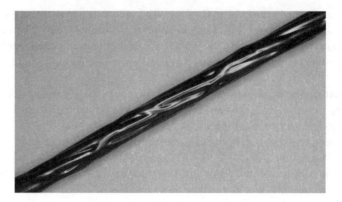

Figure 10.21 Microcable with 6 loose tubes and undulated sheath. Source: Reproduced with permission of P.H. Elmat Sp. Z o.o.

Pulling of cables into thin ducts is made easier by lower coefficient of friction. HDPE ducts with grooved inner surfaces are widely used. This idea is also applied by some manufacturers to cables, by making a thin HDPE sheath to follow the shape of the tubes inside. Moreover, undulations reduce the area of contact and friction between cable and duct, plus increasing air drag during pulling with pneumatic machinery (Figure 10.21).

Microcables have a very small margin for error in design and manufacturing, requiring a high level of consistency in process parameters and properties of materials used.

- *Indoor cables:* Rapid installation in a wooden house, especially when drilling and use of matched cable fasteners is to be avoided, involves:
  a) periodic, tight stapling of cable to walls or panels, with compression of cable
  b) routing of cable around corners of walls, doors, windows, etc., with multiple bends at radius below 5 mm
  c) placement of cable under carpet, with the prospect of people walking and furniture standing on it, exposing the cable to considerable crush forces
  There are two distinct schools of cable design to deal with such threats.
  Most European and US suppliers offer dielectric, round indoor cables with fiber encased in a rigid 0.9 mm tight buffer, which have a larger diameter (4.5–5.0 mm) and a sturdier jacket than the conventional simplex cable for other applications, which is normally 1.2–3.0 mm in diameter (Chapter 4.2). The large diameter makes a critical difference in dealing with threats (a) and (b), as there is enough loose aramide yarn inside to allow fiber movement. This reduces the bending diameter to approximately 5 mm, even when the cable is bent around a 90-degree corner or stapled (Figure 10.22). Therefore, a cable with "bend-insensitive" (G.657.B3), or even a "bend-tolerant" (G.657.A2) fiber in many cases works satisfactorily. Resistance to long-term crush (c) is limited, and should be avoided.

Typical Japanese products, shown in Figure 10.23 [10.46–10.48], are designed for maximum resistance to crush and insect bites, employing:

- large flat surfaces for handling crush forces with limited deformation
- multiple rigid strength members
- fibers in thin primary coating located centrally in hard jacket
- longitudinal notches for rapid access to fiber(s)

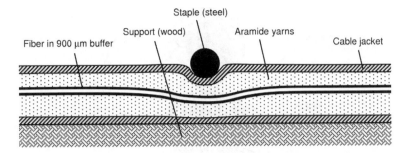

Figure 10.22   Self-limiting of fiber bend curvature in aramide-reinforced indoor cable

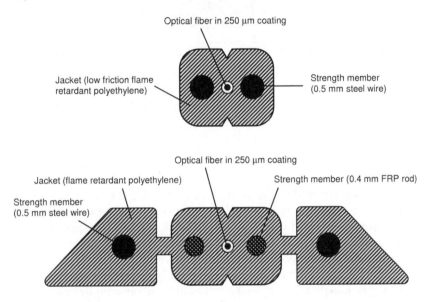

Figure 10.23   Single-fiber indoor cables for apartment wiring in Japan

where:

Top: 1.6 × 2.0 mm low-friction cable optimized for pulling through conduits
Bottom: Crush-resistant flat cable for routing through the door, permanently bendable

Aerial "figure 8" drop cables of related design exist. Aerial cable has an additional strength member, a steel wire or composite (FRP) rod (Figure 10.24). Multi-fiber versions can include 4-fiber ribbons [10.48].

The price for mechanical robustness, including excellent resistance to long-term crush, is lack of flexibility and more complicated handling. However, in some situations, like entry through a window, rigid and malleable metallic members allow the cable to be manually formed to exactly and permanently follow the shape of the supporting surface.

Installation in an apartment with plastered or brick walls, where drilling is allowed, which is typical for apartment blocks in Russia, South Korea or China, does not include cable stapling or passing through extremely narrow, multi-step clearances in doors or windows. In this

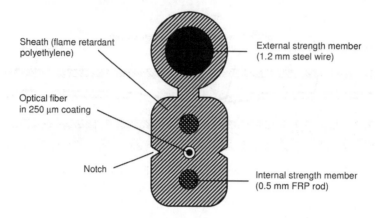

Figure 10.24   Single-fiber aerial drop cable for FTTH networks (Fujikura DC-1). Dimensions: 2.0 × 5.3 mm, short-term tensile load: 600 N

situation, threats (a) and (b) are avoided, but (c) remains possible, and requires clear warnings of customers.

10.1.8.4   Splitters

Single-mode fiber splitters for PON networks should operate and retain a constant split ratio in a wide band of wavelengths, taking into account that all systems currently standardized or considered may be extended from 1260 nm to approximately 1650 nm. Similar devices have been used earlier in multi-channel loss measurement systems and for point-to-multipoint distribution of video signals in fiber optic cable TV networks.

The basic device is a symmetrical 2 × 2 coupler made of tapered, intersecting optical fibers or waveguides, providing a 1:2 split. Higher split ratios are obtained by cascading multiple units in n stages and blocking unused ports, if necessary (Figure 10.25).

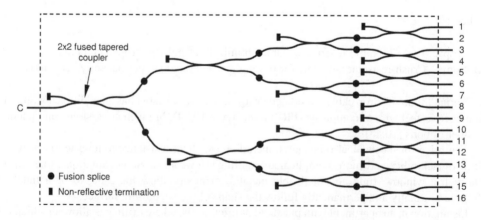

Figure 10.25   Structure of 4-stage FBT fiber splitter with a 1:16 split ratio

The PLC structure is somewhat similar, but lacks fusion splices and includes butt couplings to optical fibers extending outside.

Number of ports N of fully assembled multi-stage splitter, providing 1:N split is:

$$N = 2^n \qquad (10.3)$$

and theoretical insertion loss IL between common fiber ("C" in Figure 10.25) and any other port, resulting from a perfectly even division of optical power, expressed in dB, is:

$$N = 10 \log 2^n \qquad (10.4)$$

or

$$IL \approx 3.01\, n \qquad (10.5)$$

In real devices, extra loss is introduced by connections between stages and fiber terminations, plus uneven splitting between ports (Table 10.2). Insertion loss of a single-mode splitter does not depend on direction, but can change with wavelength.

Two technologies are used to make splitters:

- *Fused Biconic Taper (FBT)*: bare fibers are fused, twisted and pulled to make a 2 × 2 coupler, than several couplers are fusion spliced to obtain the given split ratio (Figure 10.25)
- *Planar Lightwave Circuit (PLC)*: 2 × 2 couplers and connections between them are made as optical waveguides by ion diffusion into glass substrate. This structure is fitted with fiber pigtails, butt-coupled to waveguides on PLC edges

PLC technology is better suited for making large splitters, as loss and labor costs related to fusion splicing are avoided. 2 × 2 and 2 × 4 devices are often of the FBT type, offering lower loss and superior stability in harsh conditions, like vibrations and extreme temperatures.

High split ratios create a problem with storage of fiber pigtails and connectors, occupying much more space than the splitter itself. One solution is to use thin cables, like 1.2 mm or tight-buffered fibers (0.9 mm) and compact connectors with 1.25 mm ferrule, most often LC-APC (Figure 10.26). Splitters are often installed outdoors in extreme climatic conditions, and must meet severe environmental specifications.

10.1.8.5   Fiber connectors

- *Requirements:* Operation and properties of optical fiber connectors are presented in Chapter 4.3. Requirements for single-mode fiber connectors for FTTH networks include:
  - *low-cost*: high numbers used, price-sensitive market
  - *small size*: limited space in enclosures, distribution frames, etc.
  - *low loss*: typical loss budget of fiber drop with 2 connectors is 1.2–1.4 dB
  - *wide range*: of operating temperatures, because of outdoor installation
  - *high return loss*: to prevent interference from unused ports
  On the other hand, durability requirements are relaxed due to infrequent re-connections.

- *Connector designs: PC v. APC:* Most projects include single-fiber, plastic body, push-pull connectors of rectangular shape: SC and LC (Figure 10.26), predominantly with ferrules and alignment springs made of Zirconia (ZrO_2) ceramic and APC (Angled Physical Contact) ferrule finish to minimize reflections, especially from open ports in splitters and outlets. APC connectors are mandatory in networks including analog video transmission and optical amplifiers. Connector finish is indicated by body color:
  - *PC (Physical Contact*: dome shaped with 8–25 mm radius: blue
  - *APC (Angled Physical Contact*: dome shaped, with 8-degree tilt: green

  PC and APC connectors can be mated, but fiber tips fail to contact, and scratching of ferrule edges produces highly abrasive ceramic dust. The insertion loss of such a connection has high values, and 10–20 dB are typical. In case of accidental "hybrid" mating, both connectors and adapter should be carefully cleaned before further use.

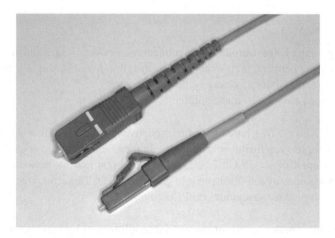

Figure 10.26 Optical fiber connectors

where:

    Top: SC-APC
Bottom: LC-APC

For installations in apartment blocks, complete lengths of drop cables (10–100 m) can be factory terminated with connectors to minimize labor-intensive fiber splicing on site.

- *Protection against contamination*: Most locations for cable hardware in FTTx networks, such as apartment and office buildings, street cabinets, cable enclosures on poles, etc. are "dusty" environments. Careful protection of connectors and adaptors with caps or plugs is strictly required. The material of the protective cap or plug may be important, black polymer with carbon black filling and without a high amount of plasticizer being the best. A cap of bright color reflects radiation back into the fiber, considerably reducing insertion loss of an open APC connector. Plasticized PVC can slowly exude an oily liquid, which covers the connector endface. While the liquid itself does not degrade connector performance, any dust will stick to it and form an abrasive paste.

10.1.8.6   Diagnostics of fibers in a PON

Fiber breaks and other damage in fiber networks are normally investigated with OTDR (*Optical Time Domain Reflectometer*). This instrument can detect extra loss or reflection and measure distance to such an event. In FTTx networks, fibers in feeder cables extending from the central office to DSLAM, Ethernet switch or splitter (Figures 10.10 and 10.12), can be tested without problems; this applies also to complete fiber loops between OLT and ONT in FTTH-P2P networks.

With respect to monitoring of PON fiber plant, the OLT normally monitors signal levels and detects loss signal received from the ONT, but cannot locate fiber breaks or bends. OLT equipment with a built-in OTDR module for locating fiber breaks was developed by Alcatel-Lucent in 2012 [10.49]. In this solution, OTDR functionality is built into the SFP optical transceiver module at GPON OLT, using additional, shallow amplitude modulation superimposed on the 1490 nm downstream signal. However, most operators still rely either on separate in-service OTDR monitoring systems, presented below, or on manual testing of a faulty PON with OTDR from the splitter location. The latter method requires sending technicians to remote locations and testing disrupts service to all subscribers in an affected PON.

- *Overlapping of fiber traces*: OTDR connected to a common port of splitter (Figure 10.16) shows a sum of traces of all fibers extending from the splitter towards subscriber premises. Consequently, non-reflective break of one of *N* drops is seen by OTDR as an event with apparent loss *S*:

$$S = \log \left[\frac{(N-1)}{N} \right] \qquad (10.6)$$

Fiber specifications [10.24,10.25] allow trace non-uniformity of up 0.05 dB, and additional allowance must be made for OTDR noise. Figure 10.27 shows that non-reflective

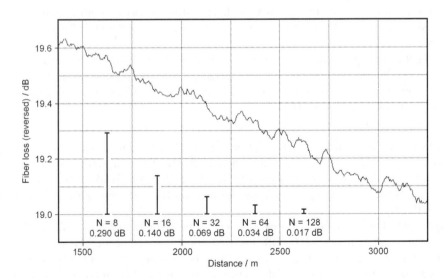

Figure 10.27   OTDR trace of medium-quality SMF acquired at 1310 nm and 100 ns pulse width

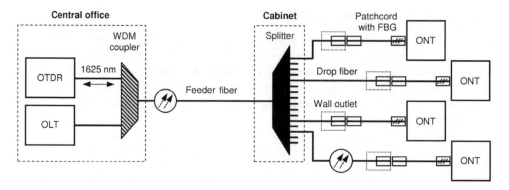

Figure 10.28   Arrangement for in-service testing of PON fibers with OTDR

break of one fiber in a PON is hard to detect at the 1:32 split ratio, and almost impossible at higher.

The vertical bars show apparent loss displayed when one of the N fibers in PON is cut.

Reflective break is detected, albeit at reduced amplitude and distance is measured correctly, but OTDR does not provide information as to which of the drop fibers is damaged.

- *Stepped-length fibers and reflectors*: Faulty fiber can be distinguished using OTDR, when:

  - all fibers after splitter have unique, documented lengths
  - fiber termination at ONT has strong, stable reflectivity, close to 100%

  In such a case, the OTDR trace shows a series of reflective events, each corresponding to different fiber drop. Fiber break or increase of loss, due to excessive crush or bending, will manifest as a missing or weakened reflection, allowing identification of damaged fiber(s) and estimation of introduced loss.

  Testing should preferably be made without cutting services to other customers connected to the same PON and interfering with the operation of the OLT. This can be done using:

  - OTDR operating at separate wavelength, preferably 1625–1650 nm (Figure 10.28)
  - wavelength-selective optical reflectors matched to OTDR wavelength, transparent and non-reflective at wavelengths used by OLT and ONT. Specifications for such devices are under development within IEC [10.50]

Reflectors are manufactured by inscription of Fiber Bragg Grating (FBG) in a short length of fiber placed inside a connector of a patchcord. It links the wall outlet and ONT, on the ONT side. The FBG also serves as a filter blocking entry of OTDR pulses to the ONT port. The OTDR is connected to a feeder fiber at the OLT site through a WDM coupler (Figure 10.28).

Tight control and inventory of lengths of cables between splitter and ONTs is required. As resolution of commercial OTDR intended for PON diagnostics is 0.3–0.5 m, the minimum difference between fiber lengths must be 2–4 m. PON networks in apartment blocks are often wired using cables of identical lengths to simplify ordering of materials and installation work. Thus, this network cannot be tested in this way. Another drawback is loss of

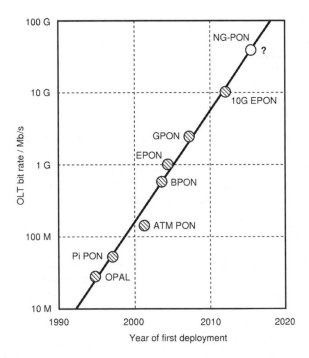

Figure 10.29   Evolution of OLT downstream bandwidth in FTTH-PON networks

components used: 0.6–1.0 dB for coupler and FBG, which must be included in the PON loss budget.

10.1.9   Development and standardization of FTTH technologies

For almost two decades, the dominant direction of development was increase of bandwidth, especially with respect to PON systems, where the OLT bandwidth is shared (Figure 10.29). However, market failure of 10 Gbit/s equipment: 10G EPON and XG-PON, despite approval of relevant standards in 2009 and 2010, respectively [10.19,10.28,10.29,10.54], indicates a rift between manufacturers developing new products and operators finding them too expensive. WDM-PON technology, promising ultimate capacity, is still to be adopted on any scale, despite trials of DWDM-PON in South Korea in 2005 [10.38,10.39] and CWDM-PON in Malaysia in 2008.

Standards for 40 Gb/s PON (NG-PON) systems started to appear in 2013 [10.30], but there is no prospect of deployment. On the other hand, large interest of operators lead to standardization of upgraded versions of 1 Gb/s EPON by the IEEE [10.52,10.53,10.56].

10.1.9.1   Race towards higher capacity

Deployment of successive PON technologies worldwide (Table 10.3) has long followed Moore's law (Figure 10.29), with OLT bandwidth almost doubling every two years.

Of 10 Gb/s PON systems, 10G EPON and XG-PON, only 10G EPON has seen limited use by China Telecom since 2012 as a feeder system for FTTB networks in large apartment blocks (see descriptions in Section 10.1.3), rather than true FTTH technology with fibers running all the way to subscribers [10.53].

10.1.9.2   Standards

Worldwide standardization of FTTH active equipment is traditionally split between:

- US-based Institution of Electrical and Electronic Engineers (IEEE)
- International Telecommunications Union – Telecommunications Standardization sector (ITU-T), a specialized agency of the United Nations

There is a substantial difference between priorities and directions of activity of both standard bodies.

The IEEE (*www.ieee.org*) includes FTTH technologies as part of an extensive Ethernet portfolio, including systems for several other uses: data transmission, local area networks, core networks, and suited also for transmission media other than optical fibers. Standards are prepared relatively quickly, with focus on speed and relatively low cost of implementation, primarily in data transmission environment.

ITU-T (*www.itu.int*) has traditionally been dominated by representatives of governments, and focused mostly on technologies for large telecom operators. ITU-T, despite improvements observed since 2000, works slower than IEEE, but with much greater attention given to issues such as network management and maintenance. However, more extensive functionality included in equipment standardized by ITU-T, in particular GPON and XG-PON, results in higher cost and somewhat delayed availability.

A good example is the detailed ITU-T Recommendation G.988 [10.55] devoted to management, supervision and configuration of ONT devices in GPON and XG-PON networks. This recommendation includes equipment operating modes in emergency conditions (power failure) and power management, which was first published in 2010. Its equivalent for EPON networks – IEEE P1904.1 [10.53], has appeared three years later, following pressure from network operators experiencing interoperability problems and emergence of several incompatible variants of EPON targeted at different markets and customers [10.56]. ITU-T recommendations also cover single-mode fibers used in telecom networks, as described in Chapter 4.2, but no cables or passive devices.

Data presented in Table 10.4 and wavelength bands shown in Figure 10.20 indicate a considerable degree of synchronization of work at ITU-T and IEEE, driven mostly by the issues of equipment compatibility with passive infrastructure, and the use of common optical components. Most standards include multiple options for physical interfaces to the ODN, with variable power budget, reach and capacity.

Initial work on XG-PON technology also included a symmetrical 10/10 Gb/s variant known as XG-PON2, but ultimately only the cheaper 10/2.5 Gb/s XG-PON1 system was standardized. In the case of 10G EPON, the IEEE 802.3av standard [10.57] includes a symmetrical 10/10 Gb/s variant, and equipment is available, although expensive. Standardization of the new-generation PON (NG-PON2) system with an OLT capacity of 40 b/s or higher, is currently in progress within ITU-T. Only general system characteristics were agreed [10.30], while detailed specifications are yet to be issued.

Table 10.4 Comparison of FTTH-PON systems

Name	BPON	EPON	GPON	10G EPON	XG-PON
Standard	ITU-T G.983.1	IEEE 802.3ah	ITU-T G.984	IEEE 802.3av	ITU-T G.987
Year of approval	1998	2004	2003	2009	2010
Multiplexing	TDM	TDM	TDM	TDM	TDM
Downstream rate(s), Mb/s	155 622 1244	1000	2488[1]	10312	9953
Upstream rate(s), Mb/s	155 622	1000	1244[1] 2488	2488 10312	2488
Users per PON	16–32	32/64[2]	64	32/128[3]	64/256[4]
Physical reach, km	20	20	20	20	40
Differential distance, km	20	20	20	20	40

(1) Lower bit rates included in older version of standard are no longer used
(2) Line rates of 10 312.5 Mb/s and 1250 Mb/s after 64B66B and 8B10B line coding
(3) Up to 64 per new standard IEEE 802.3bk. Previously up to 32
(4) 32 mandatory, 128 possible with PR30 transponders, see IEEE 802.3av
(5) 64 mandatory; TDM controller must support 256 per ITU-T G.987.1

where:

BPON	Broadband Passive Optical Network
EPON	Ethernet Passive Optical Network
GPON	Gigabit Passive Optical Network
XG-PON	X Gigabit Passive Optical Network
10G EPON	10 Gb/s Ethernet Passive Optical Network
NG-PON	Next Generation Passive Optical Network ($\geq$40 Gb/s)

Of other standard bodies, the two most relevant to fiber access networking are:

1. International Electrotechnical Commission (IEC, *www.iec.ch*), a non-profit international standards organization active in electrical, electronic and related sectors. Of technologies important to fiber access networks, IEC documents cover fibers, cables (also of twisted-pair type) and passive components, including splitters, connectors, OTDR reflectors and WDM couplers, but no transmission functions of active equipment. Several other subjects, such as electrical safety or mechanical and environmental testing, are also covered by IEC documents.
2. Cable Labs (*www.cablelabs.com*), a non-profit research and development consortium, are charged with the development of standards for equipment used in cable TV networks. They started initially with analog networks utilizing coaxial cables, and later increasingly digital and fiber-based systems are added, including the adoption of PON technologies. The consortium is US-based, but has worldwide membership consisting exclusively of cable TV operators.

10.1.10   Active Equipment

The descriptions below cover general features of active equipment for FTTH and FTTB networks, and a few specific issues requiring attention. Detailed presentation of the device inner workings such as multiplexing, ranging or management is not possible here due to space limitations. The reader looking for such information should first consult standards applicable to a given type of device, which at the time of writing are available free of charge from websites of ITU-T (*www.itu.int*) and IEEE (*www.ieee.org*), and check for the most recent edition of a given document. Structure and operation of FTTH-PON equipment is also presented in the literature [10.15].

10.1.10.1   OLT

This equipment is normally installed in a central office. Only some models are designed for a wide range of temperatures and humidity encountered in street cabinets or similar locations. This type of installation brings with it issues of power supply and construction permits.

Optimized location of the central office is important. While FTTH-PON networks have a long reach, 20 km and more, reduction of loop length to 5–10 km enables re-allocation of 3–6 dB of OLT-ONT loss budget from cables to splitter, doubling fiber split ratio and number of customers served by a single OLT. Examples are presented in Table 10.1.

Large OLTs are of modular design, holding multiple plug-in modules, each of them serving one or several PONs. This feature enables flexible expansion of the OLT capacity with rising demand, deferring a large part of equipment costs. Such equipment is designed for installation in a standard telecom rack and runs on 48 V or 60 V DC power, albeit AC mains power option is often included.

A single OLT of this kind [10.28] (Figure 10.30), occupying 310 mm of rack height, serves 1280 subscribers with a typical 1:32 fiber split. At 20% take-up rate, this means coverage of an area with approximately 5000 apartments and a population of 10 000–15 000. The number of customers served may be increased, if the OLT is used to connect Ethernet switches at apartment blocks instead of ONTs used by individual subscribers. At the same time, this OLT requires allocation of 40 fibers in the feeder cable. Because single OLT failure or cable cut can result in outage affecting thousands of customers, protection arrangements should be carefully considered.

OLTs have extensive sets of management features, covered by ITU-T Recommendation G.988 [10.55] for GPON and XG-PON systems, and IEEE standard P1904.1 [10.53] for EPON. OLT serves an intermediate unit for setting up all ONTs and service options for each customer, as well as monitoring of ONT operation.

However, with limited initial take-up rates encountered by several FTTH operators, down to 2%, and plans to cover areas with medium population density such as small towns and sub-urban areas, there is also a demand for small, compact and low-cost OLTs. A typical low-profile (43 mm height) GPON OLT [10.59] is shown in Figure 10.31. With 4 ports (fixed) and a 1:32 split, it can provide a service to 128 subscribers, or coverage for an area with 1000–1200 apartments and 2000–3000 inhabitants at a 10% take-up rate.

The numbers indicated above create an entry barrier for small alternative service providers, who often cannot attract enough customers. For comparison, the smallest DSLAMs and Ethernet switches, like the device shown in Figure 10.35, have about 20 subscriber ports.

Figure 10.30 Large GPON OLT with up to 10 PON modules and 40 ports (Dasan Networks V8240). Source: Reproduced with permission of P.H. Elmat Sp. Z o.o.

10.1.10.2 ONT

Exceptionally for network equipment, ONTs are:

a) installed in customer's apartments in very large numbers
b) handled by non-professional users
c) locally powered from 100–240 V, 50–60 Hz AC mains, depending on the country

Figure 10.31 Small GPON OLT with 4 PON ports (Dasan Networks V5812G). Source: Reproduced with permission of P.H. Elmat Sp. Z o.o.

Figure 10.32 Indoor GPON ONT with 4 data ports (10/100/1000 Mb/s Ethernet) and 2 telephone ports – Dasan Networks H640GV. Source: Reproduced with permission of P.H. Elmat Sp. Z o.o.

ONT installed indoors must be compact and aesthetic, while being sufficiently robust to withstand occasional drop or other mechanical abuse. However, it should not be exposed to extreme temperatures or high humidity, the range of operating temperatures being quite narrow, for example 0–40°C. ONT appearance, easy to use controls, compactness and convenience of placement are important for creating image of operator and service. In this respect, the device firmly belongs to the category of consumer electronics. Examples are shown in Figures 10.32 and 10.33.

ONT is mostly installed outdoors, often on the wall of the subscriber's home, must work reliably in adverse environmental conditions (depending on location), including rain, snow, dust, temperature extremes, solar radiation, mold and insects. Therefore, it is typically encased in a strong, watertight protective enclosure with entries for multiple cables. The latter is important due to the variable number of subscriber devices served. The aesthetics is much less important.

If the FTTH network includes an analog video distribution at 1550 nm wavelength, the demodulated RF signal, usually in the 47–870 MHz band, is fed to an antenna input of the TV set through a coaxial cable. Analog telephone sets are connected to ports with traditional RJ11 connectors, but voice services in FTTH networks are usually of the VoIP type.

To support provision of different service mix for each customer and traffic management, ONT functionality includes remote:

- selection of active ports (data, video, phone, etc.)
- setting bit rate limit for data ports
- selective deactivation of circuits and ports following power failure
- device deactivation performed through the OLT serving particular PON

10.1.10.3 Power supply

Traditional telephone sets are remotely powered from a central office over the copper loop. Hence, it provides service regardless of disturbances to electric power supply, as the CO or

Figure 10.33   Indoor GPON ONT with 4 data ports (10/100/1000 Mb/s Ethernet, RJ45), 2 telephone ports (RJ11) and WiFi wireless interface – Dasan Networks H640GW. Source: Reproduced with permission of P.H. Elmat Sp. Z o.o.

remote switch or access multiplexer feature power back-up provided by rechargeable batteries and sometimes also a generator set.

The same requirement applies to remote active devices such as DSLAM or Ethernet switches in FTTN, FTTC and FTTB networks. Sealed lead-acid rechargeable batteries provide back-up power for 2 h or more, depending on local regulations and operator standards. However, batteries installed in street cabinets can be damaged if a power cut coincides with an outside temperature of −25°C or lower. If this power blackout is long enough to exhaust the battery, a shutdown of equipment and loss of the heating is produced in this way.

Optical fiber is dielectric and carries low optical power to ONT locations, typically in the 0.001–0.1 mW (−30 to −10 dBm) range. The typical power consumption of a fully activated ONT is 10–25 W. This precludes a remote power supply over the subscriber loop. ONT is locally powered, typically from an external mains adaptor. Consequently, ONT instantly stops working during a power failure.

While TV sets and most desktop PCs or game consoles require mains power to work as well, battery-powered portable devices such as laptops or tablets and wired telephone sets do not. Replacement of the traditional wireline phone service with fiber-based "triple-play" service raises the issue of guaranteed emergency, or "lifeline" communications, required to report accident, fire, power failure, etc. In several countries, including the USA, operators of

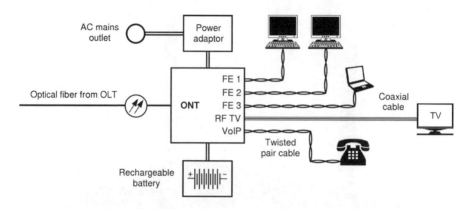

Figure 10.34 Typical connections between ONT and other devices in apartments, with power back-up provided by rechargeable battery. Interfaces: FE – Fast Ethernet; RF TV – analog TV in the RF band, demodulated from 1550 nm optical carrier; VoIP – Voice-over IP.

FTTH networks are legally required to ensure the continuity of voice service for emergency purposes.

In this case, ONT is equipped with an external, sealed rechargeable lead-acid battery providing power back-up (Figure 10.34). At typical current draw of 1–1.5 A, a low-cost 12 V, 5 Ah battery can power the ONT for 3–5 hours. This duration can be considerably extended by selective de-activation of broadband and video circuits in the ONT under power fault conditions, leaving only the voice service. ITU-T Recommendation G.988 [10.55] and IEEE standard P1904.1 [10.53] include detailed specifications for programmable power-saving features of GPON, XG-PON, EPON and 10G EPON devices. ONT includes circuits for battery charging and monitoring, with reporting of degraded or missing batteries to the network management system. The main problem is the relatively short service life of the batteries, approximately five years, which must be replaced at all customer locations. The battery must always be located indoors.

However, many models of ONTs, especially those sold in Europe and Asia, do not include power back-up functions. Example of connections between ONT, power supply devices and customer devices are shown in Figure 10.34. Actual configurations of equipment vary considerably.

10.1.10.4 Equipment for FTTB networks

Ethernet switches for apartment buildings are in many respects similar to the equipment for office LANs, usually being rack-mounted. The power supply is either from a 48/60 V DC source or AC mains, as the location in an apartment building often lacks DC power.

In an FTTB network, service to the customer is provided through a Fast Ethernet or Gigabit Ethernet port, using twisted-pair cable up to 90 m long. If multiple or portable devices are to be connected, a Customer Premise Equipment (CPE) is required. It includes a bridge or router to handle the traffic through multiple ports and separates data and IPTV traffic [10.62]. WiFi wireless interface can be included, like in the device shown in Figure 10.36.

Figure 10.35 Fast Ethernet Switch working in GPON network, with 24 ports (Dasan Networks V2824). Source: Reproduced with permission of P.H. Elmat Sp. Z o.o.

Figure 10.36 CPE with 4-port Fast Ethernet router, support for IPTV and WiFi wireless interface (Fibrain IPTV CPE FSR-R2). Source: Reproduced with permission of P.H. Elmat Sp. Z o.o.

10.1.11 Conclusions

A single-mode optical fiber, with its low attenuation, wide bandwidth and lack of cross-talk and EMC issues, is the best transmission medium for broadband access networks. It is almost free of capacity and distance limitations imposed by twisted pair or coaxial cables and radio systems. Alas, full conversion of an existing copper network to an optical one is also expensive and disruptive. With intense competition between service providers on price, and pervasive regulation of incumbent operators, FTTH often loses to cheaper "hybrid" options such as FTTN or FTTB, and overall progress is very uneven. Operators are also reluctant to adopt the latest generations of FTTH technology for cost reasons.

10.2   Polymer Optical Fibers, POF

Dr. Alicia López, Dr. M. Ángeles Losada and Dr. Javier Mateo
GTF, Aragón Institute of Engineering Research (i3A), University of Zaragoza, Spain

This section reviews the most important issues related to communication systems that use *Polymer Optical Fibers* (POF) as transmission media. Topics covered include physical characteristics of POF together with transmission basics and also application areas where advantages over other transmission media are identified. Moreover, several techniques usually found in classical optical communication systems are evaluated as for their use in communications over POF. Finally, two of the most relevant application scenarios in the field of communications are addressed, with emphasis on currently existing systems and future demands and involved technologies. In-home networks and networks within transportation systems are envisioned as two areas where POF can exploit its strengths and thus the associated technology can experiment fast development in the next years.

10.2.1   Basics of POF

Polymer (or Plastic) Optical Fibers (POF) have been extensively used for many years in the industrial area under standards like PROFIBUS, INTERBUS or SERCOS, and in automotive environments under the widely deployed MOST specifications. Their physical characteristics make them convenient in applications sensitive to Electromagnetic Interference (EMI) or those where harsh environments are involved. Data rates in these scenarios have been traditionally limited to tens of Mb/s. However, latest developments in the POF market and research results reported over the last few years have demonstrated that POF can also play an important role in higher-rate communication systems, particularly when short distances (several tens of meters) have to be covered.

10.2.1.1   Physical characteristics

The term Plastic Optical Fiber is related to a type of fibers that are made of plastic (polymer) materials for core and cladding and usually exhibit a large core diameter. The evolution of the POF market from the first manufactured fiber in the late 1960s up to now, has led to the variety of fibers we know today. This evolution has been influenced by the increasing need for data transmission at high speed in all network scales, thus coming closer to the final user, where aspects such as ease of installation, eye safety and cost are crucial. POFs can be roughly classified according to their material (which determines the attenuation and the temperature of operation) and geometric characteristics, mainly their index profile (which determines the bandwidth).

Polymethylmethacrylate (PMMA), which can bear temperatures of up to 80°C and exhibits a refractive index of 1.492, is the most common material used for the fabrication of POF. As for index profiles, the first manufactured fibers back in the 1960s had step-index profile (SI-POFs) and this type is still the most used today. Standard-sized POFs present a core diameter of approximately 1 mm, although smaller cores are also commercially available. All these characteristics lead to the most common and used POFs: 1-mm PMMA SI-POFs, which typically present a numerical aperture (NA) of 0.5 and a attenuation factor (@ 650 nm) in the range of 0.11–0.25 dB/m [10.63], being the transmission bandwidth limited by modal dispersion to bandwidth-distance products of about 40 MHz × 100 m [10.64]. This rather poor bandwidth can be greatly improved by the use of graded-index profile fibers (GI-POFs), which have bandwidth-distance products that are nearly 100 times higher than SI-POFs.

Different materials with improved characteristics have been used together with more complex profiles and smaller core diameters, in order to obtain better transmission performance in terms of bandwidth and attenuation. However, the benefits encountered have been reached at the expense of increasing complexity and cost, as well as and also reducing the amount of optical power launched into the fiber. In this context, and as a compromise between the cost-effective and widespread SI-POF and the high-bandwidth GI-POF, double-step-index POF (DSI-POF) and multi-step-index (MSI-POF) were developed to lower the NA by attenuating the power propagating at high angles. Even 120 μm core diameter perfluorinated graded-index fibers (PF GI-POF) with an important improvement in transmission bandwidth have been manufactured [10.65,10.66]. In addition, these fibers are manufactured using *perfluorinated* polymers (i.e. CYTOP from Asahi Glass Co. [10.67]), which has led to the lowest attenuation found in POF, less than 0.02 dB/m together with a wider transmission window, thus increasing the usable wavelength range. Bandwidth-distant product in these fibers has been reported to be as high as 2 GHz × 100 m [10.68]. There are also PMMA 1 mm GI-POFs that aim at compromise between ease of light injection in large-core fibers and high- bandwidth in graded-index fibers, but they are limited by attenuation [10.69].

The multi-core fibers (MC-POF) represent another approach to obtaining better transmission performance and still maintain the advantages of large size. Several variants exist, depending on the number of cores and the index profiles of each core. The division of the single core into many individual light guiding areas allows performance improvement, particularly regarding their lower curvature losses, while the numerical aperture and the transmission bandwidth are nearly the same as in SI-POFs. Although promising, these MC-POFs are still far from being widely used. One of the presumable applications of MC-POFs in the future lies in the Space Division Multiplexing (SDM) technique. This would allow the transmission of separate signals through each one of the individual cores, thus increasing the available bandwidth that has already been proved in glass fibers [10.70].

Figure 10.37 shows schematically some of the index profiles found in POF. For each fiber type, the refraction index as a function of radial distance and fiber cross-section are represented. Core diameter has been assumed to be similar for all fibers considered, which holds for commercially available large-core PMMA POFs.

It is generally assumed that plastic optical fibers have a lower bandwidth than glass fibers. For *PMMA POF*, the bandwidth is challenged by large modal dispersion due to their strong mode coupling. Mode coupling is caused by scattering in PMMA that is also responsible for differential mode attenuation [10.71,10.72] and has been modeled using ray-tracing [10.72] and power flow differential equations [10.73,10.74]. Modal dispersion is reduced in graded-index perfluorinated fibers, where chromatic dispersion has a significant contribution to pulse broadening [10.75]. However, due to their reduced core diameter, PF GI-POFs partially lose the advantages of the traditional 1 mm core diameter PMMA POFs. Despite the transmission impairments present in POF links, they are gradually gaining ground in short-reach scenarios, as will be discussed in the following.

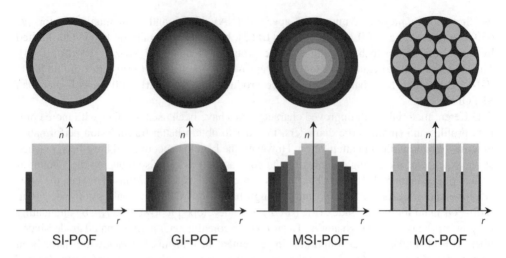

| SI-POF | GI-POF | MSI-POF | MC-POF |

Figure 10.37 Classification of POFs according to their index profile. (a) Step-index, (b) Graded-index, (c) Multi-step-index and (d) Multi-core POFs

10.2.1.2 Comparison with other optical transmission media

POF represents a good compromise between EMI- and bandwidth-limited copper cables and expensive hard-to-install optical fibers. The most relevant differences between POF and other optical transmission media are related to the physical characteristics already outlined above. POF has a much larger diameter than any other fiber, 1 mm for the most common fibers. This large diameter allows easy handling and power launching and thus enables transmission, even if the fiber ends are slightly damaged or the light axis is slightly off-center. Therefore, POF-related components, such as optical connectors, can be made at low cost and the installation work of POF links is simplified. A comparison is made of POF with other optical fibers. In particular, main physical/transmission parameters and the suitability in short-reach scenarios, can be read in the following:

- *Polymer-Clad Silica fibers (PCS)*, which are silica glass fibers with polymer cladding, usually have a core diameter of 200 μm and present numerical apertures of around 0.3–0.4. They represent a hybrid solution between POF and standard glass optical fibers and are considered a good candidate for short-range data transmission
- *Glass Optical Fibers (GOF)* are the most commonly used fiber in communications applications. They can be further classified according to the number of light-propagation modes that they are able to transmit
- *Multimode GOF (MMF)* supports hundreds of modes. These fibers usually have core diameters of 50–62.5 μm and present numerical apertures of around 0.2–0.3. They are the most commonly used optical transmission media for LAN backbones, requiring Gigabit Ethernet rate
- *Single-Mode GOF (SMF)* only supports the propagation of the fundamental mode. These fibers usually have core diameters in the range 5–10 μm and present numerical apertures in the range 0.12–0.14. They are extensively used in long-haul backbones

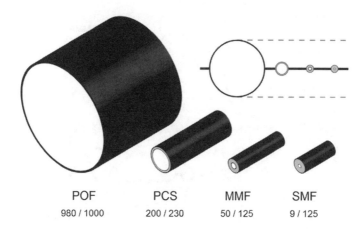

POF	PCS	MMF	SMF
980 / 1000	200 / 230	50 / 125	9 / 125

Figure 10.38 Size comparison of different optical fiber types, including typical values for core/cladding diameter dimensions in μm

Figure 10.38 compares the size of the types of optical fibers considered in the discussion below: POF, PCS, MMF and SMF. Core diameters for these fibers exhibit great differences, ranging from 10 μm in a single-mode GOF to about 1 mm in a large-core POF.

The polymer cladding in PCS fibers confers them with roughness and capacity to tolerate stress, whereas it greatly affects fiber characteristics such as attenuation and thermal range. PCS fibers present attenuation values of about 1–2 dB/km over their transmission windows, which are located at the usual wavelengths for standard silica fibers. Reported bandwidth-distance products are in the range of 5–10 MHz km, which allows lengths of up to 200 m to be covered. When comparing POF and PCS, the latter shows slightly higher bandwidth and more obvious improved attenuation characteristics. In addition, a wider temperature range is possible when using PCS, which made them a good alternative in the automotive environment. On the other hand, POF is a better solution regarding aspects such as handling, ease of installation and cost, both of fiber and of passive/active components. In addition, transmission over PCS is usually carried out in the infrared range and at higher emitted powers. Therefore, PCS lose the advantages of visible-light POF systems regarding eye safety and link test by simply visual inspection.

Regarding glass fibers, they have been extensively used as the backbone in high-performance long-distance communications networks. For many years, research efforts within the fiber optics community were taken up by the steady search for higher data rates and longer reach over GOF, whereas other aspects such as complexity and cost were usually accepted. Figure 10.39 shows a typical attenuation spectra for the types of fibers we consider according to their material [10.76,10.77]. It should be noted that attenuation also depends on the index profile [10.78]. As can be seen, PMMA-POFs present much higher attenuation values than the other fibers, three orders of magnitude higher than GOF when their respective transmission windows are considered. In addition, the wavelength range that is usable for transmission applications is much broader in PF-POF, PCS and GOF, compared to the case of PMMA-POF.

Glass fibers present much lower attenuation than POF, but there are some issues that arise when using GOF, such as brittleness and complicated fiber termination using expensive equipment that make these fibers unsuitable for short-reach user-made networks.

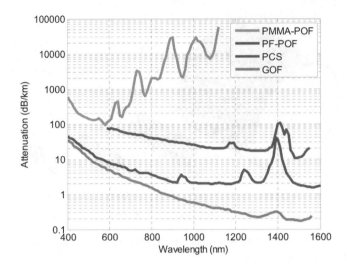

Figure 10.39 Spectral attenuation of different optical fiber types: PMMA-POF, PF-POF, PCS and GOF

If we compare POF to silica MMFs, the latter exhibit bandwidth-distance products of about 1 GHz km, which are significantly better than that of POF or PCS fibers. Considering short-reach scenarios, such as automotive or in-home networks, MMFs can even transmit 10 Gb/s and can thus be considered future-proof in terms of data rates. They represent a very interesting solution, because they enable the use of relatively cheap optical sources such as Vertical-Cavity Surface-Emitting Lasers (VCSELs). Moreover, they allow easier and lower-cost connection compared to glass single-mode fibers. However, they are far from the low precision requirements for the inexpensive POF connectors. In summary, the short fiber lengths involved in POF application scenarios is not enough to contest the position of MMFs as the best candidate, at least for short- and medium-term bandwidth demands.

Comparing POF to silica Single-Mode Fibers (SMFs), and present bandwidth-distance products of tens of GHz over many kilometers, meaning they pose no limit to communications in short-reach scenarios, as considered here. These fibers are widely used in backbone and access networks, as they present very interesting transmission properties (low attenuation and no practical bandwidth limit). However, they have some drawbacks, such as high installation costs due to the high precision required in the optical connectors and splices, as well as sophisticated high-cost sources and detectors.

Typical values for the bandwidth-distance product exhibited by the set of optical fibers considered in the above comparison [10.63] are represented schematically in Figure 10.40.

Graphical representation of bandwidth-product values in ascending order allows for ranking of the fiber types according to their capacity. Great differences can be found ranging from about 4 MHz km for PMMA SI-POF to around 1 GHz km for MMF GOF or even to an unknown limit for SMF GOF. The latter is able to transmit at such long distances, that its performance is usually power-limited rather than bandwidth-limited. If we consider complexity and cost as the comparison parameters, the inverse ranking (with some slightly differences) can be built.

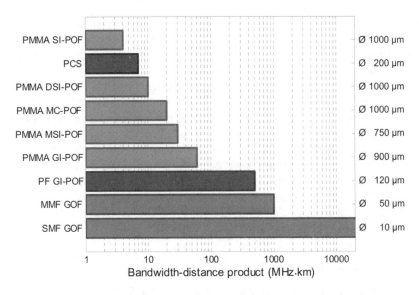

Figure 10.40   Bandwidth comparison for the optical fibers considered

10.2.1.3   Components for POF communication systems

Apart from the physical properties, it is important to consider the available optical components for each fiber type, which eventually will define whether the practical deployment of the system is possible/affordable or not. A wide range of components has been developed in the last years [10.63]. These include passive components such as connectors, couplers, filters and attenuators and also active components such as optical sources and photodiodes. Great effort has been made recently in the development of optical transceivers for Gigabit transmission, gathering together the most important communication functions in a single part: light emission, line coding, modulation, demodulation and reception [10.79,10.80]. Development of POF components for communications has always been conditioned by the constant search for higher bandwidth, better power budget and reasonable complexity/cost.

Under these criteria, optical sources such as Light Emitting Diodes (LED), *Resonant Cavity LEDs* (RC-LED), Laser Diodes (LD) and *Vertical-Cavity Surface-Emitting Laser* diodes (VCSEL) have been successfully introduced in POF communication systems. Among these sources, LEDs are the most inexpensive but the commercially available components are not fast enough to be modulated at high data rates. On the other hand, LDs can be modulated at Gb/s data rates, but their cost and power consumption can prevent their use in cost-sensitive environments. RC-LEDs and VCSELs show a good tradeoff between bandwidth and cost. Therefore, they are considered the potential future choice in POF broadband links [10.63]. At the receiver end, large-area photodetectors are required in order to attain high coupling efficiency, but they suffer from high capacitance values and hence limited operation bandwidth. Metal-Semiconductor-Metal (MSM) photodetectors have been proposed and experimentally tested showing great bandwidth characteristics of up to 11 GHz, although the sensitivity of these devices still needs to be improved [10.81].

Another approach for bridging the bandwidth limitation in POF systems is to use advanced transmission techniques, as will be addressed in the next section. Several components for such transmission techniques have been proposed and developed in the literature, such as POF WDM multiplexers [10.82,10.83], demultiplexers [10.84], and transmitters [10.85], or transceivers for multilevel signals using electronic equalization [10.86,10.87].

Apart from the bandwidth requirements, the big issue in PMMA POF systems is related to the limited power budget. One way for improving the performance of such systems is the use of green wavelength optical sources, which would allow transmission using the minimum attenuation window of PMMA (for 50 m of fiber, loss is about 7 dB below that for the 650 nm window [10.88]). In addition, it has been demonstrated that the bandwidth limitation of LEDs can be overcome with the new developments on GaN-LED with reduced active area [10.89,10.90]. Green LEDs exhibit, as an additional advantage, great output power stability over a wide temperature range, which would further improve the power budget in POF links.

10.2.2   Techniques for Data Transmission over POF

Most POF communication systems use the Intensity Modulation with Direct Detection (IM-DD) scheme, which is the simplest way to establish an optical data link. Such systems encompass automotive infotainment, industrial automation and in-home networks, for which usually a 100 Mb/s data rate can be easily achieved and at a low cost. However, and in order to overcome the limited bandwidth found in these systems, advanced modulation schemes are being considered for next-generation Gigabit- or even multi-Gigabit capable POF communication networks. Particularly, several works in the last few years [10.98,10.104–10.108,10.110–10.115,10.117] demonstrated that increased data rates are feasible over POF when using a combination of multiplexing and/or advanced modulation formats and/or adaptive electronic equalization. It must be noted that, although promising, the reported works strongly rely on offline digital signal processing performed by a computer using specific software. Figure 10.41 shows the most common experimental set-up used for the evaluation of advanced techniques over POF.

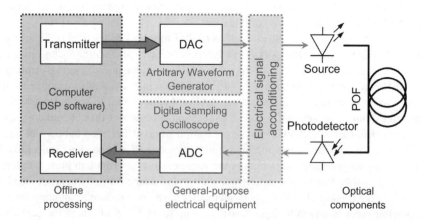

Figure 10.41   Experimental set-up for the assessment of advanced transmission techniques over POF

The optical part of the experimental set-up is just a POF link of a certain length and characteristics such as type of source/detector and fiber. Usually direct modulation is used once the appropriate electrical signal is obtained, which can be made in several steps, the most important one being the Digital to Analog Conversion (DAC) and Analog to Digital Conversions (ADC). When dealing with Gigabit signals, DAC and ADC tasks require high resolution in terms of GSa/s and are thus carried out by high-performance Arbitrary Waveform Generators (AWGs) and Digital Sampling Oscilloscopes (DSOs), respectively. Offline processing at the transmitter and receiver ends is used to perform basic and advanced functions in digital communications systems, such as binary sequence generation, line coding, modulation, synchronization, equalization and error correction.

10.2.2.1 WDM technology over POF

Wavelength Division Multiplexing (WDM) has been for many years considered as the definite and ultimate approach to increase the capacity of an optical link. The application of WDM in silica fibers dates back to the 1970s and it is nowadays widespread. The application of such techniques in POF seemed reasonable, as some experimental test beds showed [10.91–10.98]. However, the lack of commercial components, mainly fast optical sources operating in the visible range, has slowed down the progress in this area. It was found [10.91] that using LEDs as optical sources in WDM-POF systems does not allow for high-speed data transmission.

Gigabit data transmission can be accomplished using laser diodes (LD) as optical sources. Several works have focused on transmission over PF-POF [10.92,10.93],and, more recently, high-speed WDM transmission over large-core SI-POF has been possible due to the current progress in the manufacturing of LDs operating in the visible range [10.98]. In this work, 10 Gb/s data transmission has been successfully carried out over 50 m SI-POF employing three wavelength channels and spectrally efficient DMT per wavelength channel, together with offline processing including FEC at the receiver. A bulk demultiplexer is built based on commercial components such as beamsplitters and lenses. Figure 10.42 shows the schematic

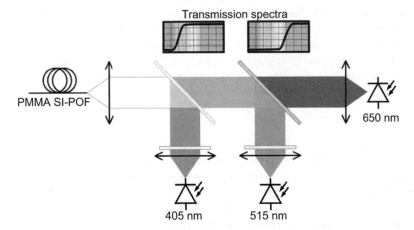

Figure 10.42   WDM demultiplexer for POF using bulk components

set-up for performing three-channel demultiplexing as proposed in this work, including transmission spectra for both longpass interference filters.

As has been presented, WDM-POF systems are not popular because of the limited availability of low-cost and fast optical sources; as an example, the green LD in [36] was a sample in the research and development phase provided by the manufacturer. In addition, there is no commercial solution for effectively demultiplexing signals in such systems, although several proposals have been made in this area [20-22,31]. In general, there is a need to optimize performance of components in the POF link, so that low insertion losses are introduced and transmitted power is kept within the eye-safety limit.

Another multiplexing technique to increase SI-POF capacity is the so-called Mode Group Diversity Multiplexing (MGDM). Some experiments have shown that it is possible to improve SI-POF performance by multiplexing different channels into different mode groups (high-order, middle-order and lower-order modes) [37]. But this method implies sophisticated processing and is too vulnerable to the strong mode coupling caused by the various disturbances that are present in POF networks (curvatures, connectors, etc.) to be ready for practical application.

10.2.2.2 Electronic equalization techniques

Electronic equalization by means of well-known techniques usually applied in non-optical transmission systems can be successfully applied to POF systems, allowing significant mitigation of impairments induced by the transmission properties of plastic optical fibers, mainly modal dispersion. Equalization is usually performed at the receiver end, but can also be applied at both transmitter and receiver ends in such a way that the signal is predistorted, so that the combined effect of pre- and post-equalization improves transmission quality [1.100]. The tradeoff between complexity and performance has always to be taken into account when using these techniques. This holds especially in POF-based systems, because of their high cost-sensitivity.

In practice, equalization is done by the introduction of Finite Impulse Response (FIR) digital filters at either link end, whose implementation complexity is proportional to the number of taps of the FIR filter architecture. Apart from implementation complexity, one major problem of equalization is that the response of the channel is usually unknown and moreover it varies over time so that blind characterization is usually needed and adaptive techniques give better results than static ones [10.101]. The different equalization techniques differ from each other in the filter structure and the algorithms for tap weights calculation and adaptability. Thus, equalization filters can be classified according to the following categories:

- *Linear equalizers (FFE)*, which only contains feed-forward elements in its structure. Transversal filters are the most popular structures in this group. Tap coefficients in these equalizers are optimized using criteria such as peak distortion or MSE.
- *Decision-Feedback equalizers (DFE)*, which has feed-forward and feed-back elements, so that information of the previously detected symbols can be used to improve equalization performance. As in FFE, tap coefficients are usually calculated according to the MSE criterion.

- *Adaptive equalizers (FFE or DFE)*, which are able to automatically adjust their coefficients and thus can compensate for time variations in the link characteristics. The update of the tap weights in these filters is usually based on the Least Mean Square (LMS) algorithm.

Performance gains of a relatively simple linear FFE and an optimal full-state Maximum Likelihood Sequence Estimator (MLSE) have been obtained and compared by means of simulation as a function of fiber length [10.102]. Results show that MLSE exhibits the best achievable performance, while FFE can obtain performance gains of the order of 1 to 3.4 dB for data rates ranging from 100 to 500 Mb/s at 10^{-5} BER. Experimental comparison of some equalization schemes has also been performed in the literature for 50 m of nearly 1 mm core PMMA SI-POF [10.103]. Results show that an FFE+DFE equalization structure applied only at the receiver side is very effective in Gigabit POF transmission systems and represents a very good balance between transmission performance and complexity. It has been demonstrated that the introduction of equalization techniques in traditional On-Off Keying (OOK) POF systems allows for data rates of up to 1.25 Gb/s over 75 m of 1 mm core SI-POF [10.104], or even 10.7 Gb/s over 35 m of 1 mm GI-POF [10.105].

10.2.2.3   Advanced modulation techniques over POF

Advanced modulation techniques are already widely applied in data-transmission electric systems as, for example, Ethernet over unshielded twisted pair uses multilevel Pulse Amplitude Modulation (PAM) together with digital equalization and Forward Error Correction (FEC) techniques, or Digital Subscriber Line (DSL) communications make use of Discrete Multitone (DMT). When dealing with short-range optical communication systems, particularly when POF is used as the transmission medium, the question is whether the achieved performance enhancement compensates for the incurred system complexity and cost. This question has been analyzed in-depth from a theoretical point of view [10.106], showing that the best-suited modulation scheme depends on the application and that transmitted bit rates can be increased to about a ten-fold in the system's 3 dB bandwidth. Moreover, focusing on multilevel PAM and DMT formats, the European Research Project POF-PLUS has achieved successful 10 Gb/s transmission over a 1 mm core diameter POF [10.107].

- *Transmission of multilevel signals*: Pulse Amplitude Modulation (PAM) is the most common signaling technique in digital communications. In particular, optical systems traditionally employ NRZ 2-PAM or OOK formats, which in this context is known as Intensity Modulation and consists of using two levels of optical power for transmitting a binary data sequence. It is well known that by increasing the number of amplitude levels of the signal to be transmitted, spectral efficiency can be improved and thus greater data rates can be achieved over the same bandwidth-limited system. Figure 10.43 compares schematically in time and frequency domains 2-PAM and 8-PAM signals corresponding to a data rate of 300 Mb/s. Transmission of this signal would require three times more link bandwidth in the former case than in the latter. The insets contain the computed eye diagrams at the receiver for both signaling techniques, revealing that power limitation is one of the main drawbacks when improving spectral efficiency by means of *multilevel signaling*.

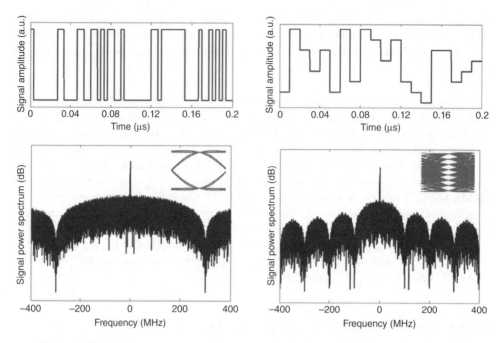

Figure 10.43   Time and frequency comparison of binary NRZ and multilevel 8-PAM signals

As explained in Figure 10.43, the advantages of multilevel signaling come at the expense of transmission quality degradation reflected in greater bit error rates due to reduction of the eye opening, which results in overall power penalties of several dBs [10.106]. Therefore, in order to compensate for this degradation, multilevel PAM is usually combined with digital equalization and forward error correction techniques. The latter are mainly responsible for the increased complexity of such systems.

Several studies, both theoretical/simulation and experimental, have been carried out to test the performance limits and other issues encountered when transmitting multilevel signals over POF. Some of the literature results have shown transmission at a data rate of 10 Gb/s over 300 m of PF-POF using 4-PAM signaling [10.108]. Moreover, 25 Gb/s have been transmitted over 50 m of this fiber [10.109]. Also, by using the same multilevel signals, 1.25 Gb/s transmission over 75 m of 1 mm core diameter SI-POF was achieved [10.110]. In most cases, electrical adaptive equalization was performed at the receiver in order to achieve such high data rates. This task was performed off-line, after recording the received signal with a real-time oscilloscope and performing computer post-processing (Figure 10.41).

• *Discrete Multitone modulation (DMT)*: Discrete Multitone Modulation is a multicarrier modulation technique where a high-speed serial data stream is divided into multiple lower-speed parallel streams. They are modulated onto subcarriers of different frequencies for simultaneous transmission [10.111]. Each subcarrier uses spectral-efficient modulation formats, usually high-level Quadrature Amplitude Modulation (QAM), with the ability to optimally adapt transmission parameters to the channel characteristics. As numerous subcarriers are distributed over the transmission bandwidth, modulations with different orders can be allocated on each subcarrier, which is an important advantage of DMT. The possibility to allocate an arbitrary number of bits per subcarrier according to the corresponding signal-to-noise ratio (SNR) profile of the channel is typically known as bit loading. Thus, DMT modulation format allows for an efficient use of the available bandwidth, but it requires considerable signal processing at the transmitter and receiver ends. It also adds latency to the transmission, due to its block-wise nature. There are several issues related to DMT transmission over POF: on the one hand, non-linearities of the optical source causing signal distortions [10.112]; on the other hand, high bit resolution is required in the DAC and ADC processes, so that reported experiments using this technique use offline digital signal processing together with high-performance AWGs and DSOs [10.113-10.115] (Figure 10.41).

DMT has been widely used in experimental test beds to reach some of the most impressive data rate transmission records using POF. Some of the best results achieved include more than 10 Gb/s over 50 m of 1 mm core diameter GI-POF [10.105], more than 10 Gb/s over 25 m of 1 mm core diameter SI-POF and bend-insensitive MC-POF [10.113], or even more than 40 Gb/s over 100 m of PF-GI-POF [10.114]. These results were obtained using commercial components, but were not particularly chosen according to the low-cost philosophy inherent to POF links, as they used either high-performance and high-cost transceivers or high-power lasers. Despite this, the DMT technique has also allowed data transmission over 50 m of 1 mm core diameter GI-POF at a data rate of more than 5 Gb/s, using eye-safe transceivers and off-the-shelf optoelectronic components [10.115].

- *Carrier-less Amplitude and Phase (CAP) modulation*: Recently, CAP modulation has received attention in the research field of optical communications due to the potential high spectral efficiency and the possibility of implementation using well-known technology. Originally proposed in the 1970s, as a modulation technique for high-speed communications over copper wires, it received most of its attention during the 1990s at the initial phases of DSL and ATM communications. This technique essentially consists of a QAM modulation using a low frequency carrier, so that it is not necessary to modulate two orthogonal carriers (quadrature carriers) with the baseband signal. Instead, CAP generates the modulated signal by combining two PAM signals filtered through two specifically designed filters. These pulse-shaping filters are usually square-root raised cosine filters that must have impulse responses forming a Hilbert pair [10.116]. Thus, CAP represents a very attractive solution for short-range cost-effective links, as it employs low-complexity electronics, and achieves high spectral efficiency.

Several works found in the literature demonstrate the feasibility and capacity potential of applying this modulation technique in POF links. In particular, 300 Mb/s over 50 m of 1 mm core diameter GI-POF were achieved using low-cost optical sources such as RC-LEDs and 64-level CAP modulation [10.117]. The experimental set-up was similar to that shown in Figure 10.41, because the CAP transmission system (modulation and demodulation) was modeled by a computer to ensure the flexibility needed for a detailed analysis of data rates and CAP levels.

Theoretical comparison of multilevel NRZ, CAP and optical Orthogonal Frequency Division Multiplexing (optical OFDM) over a POF link was presented in [10.118], where by means of computer simulation, system capacity and power dissipation were obtained for transmission over a 1 mm core diameter SI-POF link with LEDs as optical sources. CAP-64 was found to outperform the performance of NRZ and 64-QAM-OFDM systems supporting more than 2 Gb/s data rate over a 50 m link. Analogous comparison studies by means of experimental measurements were carried out recently [10.119,10.120], yielding similar results. In these works, system performance of several multilevel-PAM and multilevel-CAP formats was tested for various digital equalization schemes, link lengths and fiber-coupled power values. When the system is bandwidth-limited rather than power-limited, CAP modulation has advantages over PAM. Thus, a maximum data rate of 5 Gb/s was found over a 50 m link length when transmitting 16-CAP signals with +6 dBm fiber-coupled power. However, 4-PAM modulation format showed the best performance when transmission over 100 m was considered, with a maximum data rate of 3.4 Gb/s. These recent and impressive results, together with the simpler implementation at the transmitter and receiver ends, when compared with the DMT technique, has positioned CAP modulation as a competitive technique for transmission in future-proof POF links.

In summary, Figure 10.44 collects the state-of-the-art advanced transmission techniques applied to POF systems. In this plot, successfully transmitted data rates are represented as a function of fiber length. The most representative types of plastic optical fibers have been considered, that is PMMA SI-POF, PMMA GI-POF and perfluorinated GI-POF. Those results reported together in a single paper using the same experimental set-up have been joined by lines in the representation, whereas colors are related to the optical power at the input of the fiber link. As expected, the collected data demonstrate that the highest transmission rates are reached in experiments based on PF GI-POF [10.108,10.109,10.114]. The best bandwidth results are obtained using infrared DFB laser sources [10.114], but transmission over PF-POF, using more cost-effective optical sources such as VCSELs, also lead to very good results [10.108,10.109].

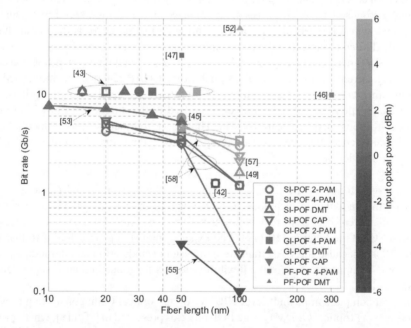

Figure 10.44   Bit rates of different POF systems (status 2013). Different marker symbols are used for each transmission technique and fiber type. Colors represent the fiber input optical power

On the other hand, the lower maximum data rates have been found for CAP transmission over 1 mm PMMA GI-POF [10.117], since in this work visible low-cost optical sources (RC-LEDs) with strong bandwidth limitations are used. In this case, the overall system is bandwidth-limited due to the characteristics of the optical source rather than the plastic optical fiber considered. The application of CAP modulation in such a low-cost bandwidth-scarce system allows for spectral efficiency improvement. In general, when the components and optical power remain fixed, a decreasing behavior of bit rate as a function of POF length is observed [10.115,10.117,10.120]. The reason for this tendency is a combination of increasing attenuation and modal dispersion. When PAM and CAP transmission over POF are compared [10.120], it can be concluded that the latter only presents a slightly better performance when the received Signal to Noise Ratio (SNR) is high enough, that is for the shortest link lengths and the highest launched optical power values considered. The most complex signal modulation considered in this review is DMT, which showed no improvement in data rate compared to PAM transmission [10.105].

However, it can be a good choice for strongly bandwidth-limited links, as the transmitted bit rate can be adjusted to the link characteristics. In general, it can be observed that classic PAM signaling together with equalization techniques achieves the best performance at a reasonable complexity. In particular, multilevel PAM has only advantages over binary PAM if the sufficient SNR is received, because in this case the system can take advantage of the better spectral efficiency [10.105,10.120]. Comparing the two large-core PMMA POFs [10.105] leads to the conclusion that GI-POF allows for improved performance when the main transmission issue is bandwidth and not received SNR. Despite this, as can be seen in Figure 10.44, the performance of the more mature large-core PMMA SI-POF allows for data rates of 10.7 Gb/s for up to 20 m using eye-safe visible VCSEL [10.105], or even more than 1 Gb/s for up to 100 m using low-power visible laser diodes [10.120].

The number of applications where the use of POF has been successfully demonstrated is very diverse; these include fashion and illumination, sensors and data communications. The following sections will focus on data transmission applications, where POF is especially suitable for short-range links with moderate bandwidth requirements. At the beginning of the century, in-house communications and communications in transportation systems were put forward as two promising scenarios, where the strengths of POF could bring some advantages over its electrical and optical counterparts.

10.2.3   In-House Communications

The rapidly increasing interest in Fiber-to-the-Home (FTTH) services and digital consumer electronics is leading to a significant demand for high-speed home networks able to connect several devices, such as PCs, high-definition TVs and other home electronic appliances. Plastic optical fiber is a cost-effective solution in many ways to provide Gigabit connectivity within in-home networks, also known as Home Area Networks (HAN). In these networks, the need for high capacity is not the only issue to be taken into account; another major challenge is the great heterogeneity of the signals that must be supported. When compared with other transmission media usually found in HANs, SI-POF exhibits similar achievable data rates with the added advantages of EMI-insensitivity, very easy installation procedures, bending tolerance and small size [10.121].

10.2.3.1   Existing home networks

Today, the distribution of data signals within HANs is done by transmitting them separately, according to their specific nature and format: IP data for triple-play services, RF signals for broadcast TV, or other formats such as HDMI for connecting specific devices. Thus, separate networks implemented over separate physical media are dedicated for each application, for example, Ethernet cables for IP data, coaxial cables for RF signals or AOC copper cables for HDMI links. In addition to the wired infrastructure, wireless connectivity enables the final link to many devices in the currently deployed HANs. This kind of connectivity is usually found more convenient and flexible by users.

Despite this, full coverage of the entire home through wireless links cannot be achieved, because the requirements of high capacity and Quality of Service (QoS) cannot be met. Recently it was demonstrated through extensive lab environment tests that currently deployed home networks cannot support short- and long-term services [10.121]. In this context, the introduction of optical fiber in the home area addresses both the bit rate increase and the great heterogeneity of the signals that have to be delivered to the various home devices. Among the variety of optical fibers, POFs are gaining momentum in HAN scenarios, as they represent a good tradeoff between performance and ease of implementation.

In summary, although nowadays wireless is assumed to be the best-positioned technology for in-home networks, it does not fulfill the bandwidth and reach requirements due to wall-material problems and its reliability remains in question. Among the wired media, optical fiber is well suited to carry different services within a single converged network. Particularly, POF offers considerable advantages due to the DIY capabilities it encompasses.

10.2.3.2   Future demands and technologies

Basically, there are two issues to tackle in the home networks field: convergence of all services on one single infrastructure, and high data rate in order to meet the increasing bandwidth demand. In addition, simplicity and low cost are desired and will eventually be key factors for success. Fulfillment of these conditions poses strong requirements for the transmission medium in the backbone of HANs [10.122].

Power Line Communications (PLC) would be a good candidate for a do-it-yourself home network, since power outlets are everywhere inside the home. However, this solution is not future-proof in terms of bandwidth and suffers from EMI-related signal degradations. On the other hand, the new wireless standard 802.11n provides full connectivity for home applications and enough data rate to support real-time applications. However, it exhibits some reliability and reaches issues with frequent signal interruptions experienced in practice [10.123], which are unacceptable for real-time applications. Finally, copper cables, which are widely used in LANs, have a number of disadvantages when compared to POF, the most significant ones being reduced malleability and EMI issues. Thus, POF combines the ease of installation of copper networks with the transmission advantages of fiber optics.

As the residential home is a very cost-sensitive environment, expenses incurred in home networks are of paramount importance in the design of such infrastructures and the choice of suitable technology. Analysis of the economics of home networks, breaking them down in CapEx and OpEx terms, shows that costs are largely dominated by network topology, transmission medium and power consumption [10.124]. Several topologies and transmission media have been evaluated and the results obtained are based on several assumptions regarding installation and network elements, which were taken from commercial equipment and recent market surveys [10.125]. These results reveal that POF cabling is already cost-competitive with the mature CAT-5e solutions, despite not yet experiencing the advantages of economy-of-scale. From a total cost perspective over the network lifetime, POF solutions are attractive not only from the technical point of view but also when economics are taken into account.

Several approaches are being proposed for the introduction of POF into the home networking environment. Gigabit requirements in these scenarios have promoted the use of graded-index POF with less modal dispersion than step-index POF and thus better bandwidth characteristics. On the other hand, perfluorinated POFs, which exhibit the best bandwidth behavior, have the added advantage of being compatible with multimode GOF and thus can be used with commercial infrared transceivers. However, the advantage of using visible light is lost with infrared sources, and in addition, this fiber is thinner and not so easy to connect. On the other hand, large-core PMMA GI-POFs retain the advantages of SI-POFs, but they show higher attenuation compromising the system power budget. Until today, it is not clear which of these types of POF will be used in the extensive deployment of POF-based solutions in HANs.

Figure 10.45 represents a possible scheme of a domestic network, which corresponds to an experimentally implemented prototype [10.123,1.126]. The system is conceived as a hierarchical structure based on a server and several clients interconnected in a Fast Ethernet network. The topology is a daisy chain in order to reduce the installation cost and the aesthetical impact over built houses. The backbone of the network is made by SI-POF operated at a visible wavelength, while the connection to the electronic devices is made by standard RJ45 copper cables. The media conversion is performed by low-cost transceivers or electro-optical switches with several electrical ports to connect the different devices. This prototype demonstrates a viable uncluttered application of the POF, where the up-to-date technologies converge in a homogeneous domestic system network for distributing video, audio and data information.

POF application in domestic networks has been slow and there is still an uncertain future for POF penetration into the European home, although it is already widely used in other countries, mainly in Asia [10.127,10.128]. In Europe, from the beginning of 2006, the POF-ALL project and its continuation, POF-PLUS, have gathered together some European research centers and companies with the purpose of developing a low-cost solution based on POF. Therefore, it is possible to enable the delivery of broadband access to everyone [10.107,10.129], promoting a considerable advance in the production of standard POF and commercially available devices by European companies. As a conclusion, POF penetration in the home area has been slowed down by the still immature market status, the shortage of commercial off-the-shelf components for broadband communications and the lack of standardization. Over the last few years, however, considerable advances in the production of standard POF and commercially available devices promise a brighter future for POF home networks.

10.2.4 Communications in Transportation Systems: From Automotive to Spatial

POF application in the automotive field has been particularly successful. Since the vehicle bus specifications called *Media Oriented Systems Transport* (MOST) were first introduced by BMW in 2001, POF has displaced copper in the passenger compartment for multimedia

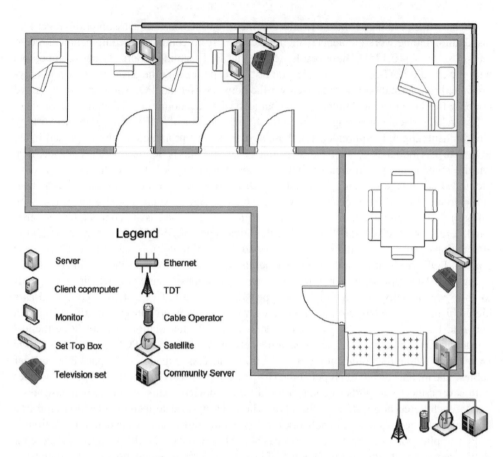

Figure 10.45　Schematic of a domestic network [10.126]

applications, and today MOST technology is used by almost all major car manufacturers world-wide [10.13]. In addition, POF is also used in protocols designed to support the rapidly growing number of sensors, actuators and electronic control units within cars. Up to date, a number of different in-car networks for multimedia and security applications have been developed.

Aircraft networks can also be considered as an example of short distance communication, as its size is limited to a maximum length of about 100 m. The requirement of faster communication systems in the range of 100 Mb/s–1 Gb/s, insensitive to electromagnetic radiation, initiated in the 1990s the migration of avionics data buses from copper- to fiber-based networks [10.131].

10.2.4.1　Existing aircraft networks

The first aircraft optical networks were based on silica fibers and demonstrated a high reliability which has steadily increased the use of optical data links in commercial aircrafts over the last decade. This evolution culminated with the recent 787 Dreamliner airplane in December

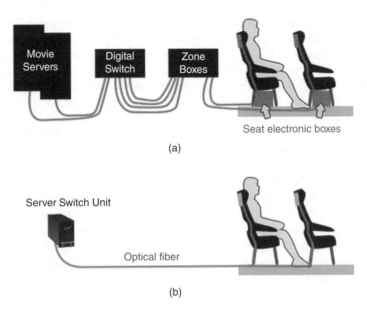

(a)

(b)

Figure 10.46   In-Flight entertainment solutions: (a) traditional approach over copper wires, (b) emerging approach using fiber optics

2009, by The Boeing Company introducing the third generation of airborne fiber-optics communication networks [10.132], and with the introduction in the Flydubai B737 of Lumexis Fiber-To-The-Screen (FTTS) In-Flight-Entertainment (IFE) systems [10.133]. Figure 10.46 shows two different IFE implementations: the traditional copper-based approach (a) and the emerging fiber-optics approach (b). The latter solution allows for direct connection of the passenger units to the fiber link, eliminating the need for distribution or seat boxes. In addition to this, optical fiber is extremely lightweight when compared to metal/copper wires and thus great weight reduction can be achieved [10.133].

Over recent years, the Aeronautical Radio INC. (ARINC), an organization composed of major airlines and airplane manufacturers, has invested heavily in standardization, developing a series of ARINC standards for silica fiber [10.134]. Despite these efforts, precision installation and maintenance of single-mode silica fiber present significant challenges that make the cost structure of airborne optical data networks based on small-core silica fiber still too high to replace the vast majority of aircraft copper data networks [10.135]. However, only a small percentage of aircraft networks need the high-speed links that require the SMF high-cost components. The remaining data wiring are for low-speed control/sensing access networks spreading to all parts of the aircraft, where the fiber cable needs to be even more rugged. The ease of installation and lower costs of multimode fibers excited the interest of the avionic industry in polymer fibers, as it is possible to realize high-speed, short-range optical links for data communication at relatively low cost using commercial off-the-shelf components.

Only POF can meet the techno-economical requirements in order to compete in the replacement by fiber of the aircraft data links that are now based on low-speed, low-cost copper. As already presented, compared to silica fibers, POF has a large core diameter and a high numerical aperture that enables affordable light sourcing and diminishes the chance of power loss due to fiber misalignment at the connector end. POF is also ductile and easier to handle in installation and maintenance. The use of visible light has also obvious advantages for failure detection and maintenance. However, there are important issues that prevent the direct implementation inside the airplane, of technology developed for in-house or other transportation networks. Thus, the integration of POF links in commercial aircraft remains the scope of an important line of research, until their reliability for everyday use in the aircraft industry is firmly demonstrated.

The EU has funded many POF projects in the last decade, including MOTIFES, but has not made good progress on POF for avionic environments. The major goal of MOTIFES (or Multimedia optical-plastic technologies for in-flight entertainment) was to research and develop the enabling technologies for the future implementation of high-speed POF for passenger entertainment systems using the consumer electronic IEEE 1394 serial data bus communication technology [10.136]. In the USA, the Defense Advanced Research Projects Agency (DARPA) funded the High Speed Plastic Network (HSPN) project in 1997 led by Packard Hughes Interconnect and including Boeing, Boston Optical Fiber and Honeywell [10.137]. The HSPN team was established to specify, design and produce advanced POF-based components and systems for three commercial applications: office networks, commercial aircraft and automotive, but failed to meet avionic environment requirements.

10.2.4.2   Future demands and technologies

A typical aircraft data link is susceptible to extreme temperature, vibrations, contaminants, etc. Networks inside an aircraft have to satisfy firm environmental requirements to assure its safety and reliability following procedures described in the MIL STD 1678/TIA 455 series [10.138] and compliance with RTCA/DO 160E [10.139]. For use in the airplane, it is necessary to design fire and heat resistant POF cable that must also be waterproof. In addition, high-temperature POF must be implemented to force stable operation at temperatures up to $+130°C$. Also, the hostile operating environment inside the aircraft imposes constraints on the link as temperature sensitivity and thermally induced stresses such as micro-bending, deformation, aging, etc. add power penalty. The power budget for passive optical networks in large-sized commercial airplanes has to be ample enough to provide power for dozens of nodes on a passive optical star and to cope with fiber lengths of up to 100 m, requiring the transmission of very high powers.

On the other hand, high optical power faces thermal, reliability and eye safety issues, while high sensitivity photodetectors with optical burst mode operation introduce bandwidth constraints. Moreover, many in-line connectors are needed in aircrafts compared to other networks based on multimode fibers. Although the use of large-core POF enables low-cost termination, the soft end face of plastic optical fiber poses significant problems for achieving low-loss interconnects in the aircraft environment. These difficulties can be solved with interconnect ferrules using air gaps, which add vibration stability, but these introduce higher attenuation and the possibility of moisture condensation. Development of a low-cost POF infrastructure includes rugged cable and devices such as connectors, splices, couplers, passive and active stars, transceivers, and media converters that meet harsh environmental and operational requirements [10.135]. According to this analysis, an integrated system architecture, with special focus on minimizing wiring that can be achieved by WDM techniques, is essential to advance POF-based airplane links.

At present, different research and development groups are actively considering the use of POF as the transmission media for aircrafts both in Europe, such as the German Aerospace Centre, and in the USA, where they are led by The Boeing Company working in collaboration with other aerospace industries and universities. The Institut für Flugsystemtechnik (Institute of Flying Systems) of the Deutsches Zentrum für Luft- und Raumfahrt (German Aerospace Center, DLR), jointly with the Institut für Hochfrequenztechnik (Institute for High Frequency Technology), of the Technische Universität Braunschweig, have performed initial experimental evaluations and tests to identify the performance of POF-based systems for civil aircraft. They concluded that POF fibers are a promising media for future aircraft applications, due to their excellent price-to-performance ratio in an environment where the cost of the physical network depends mainly on the cost of its passive and active components as well as on installation and maintenance costs. They have defined a way forward for further research and development in this field, to overcome the limitations of POF and its passive components [10.140–10.142]. In particular, data networks in aircraft are characterized by requiring multiple in-line connectors, which represents a challenge in the deployment of POF-based systems in these scenarios, due to the limited optical power budget [10.143].

Also, The Boeing Company is developing special measurement set-ups to investigate and analyze POFs for application under the conditions of daily use in aircraft. An analysis of the evolution of fiber optics for commercial aircraft is carried out. It assesses the aspects needed to make optical fiber wiring in aircrafts. This leads to the conclusion that there is a good potential market for the suppliers of POF components and tools on commercial aircraft, as well as other harsh environment transportation and military applications [10.135].

10.2.5   Standardization Activities

Standards in the POF industry are required to ensure that such aspects as quality, size, measurement methods or production procedures remain constant. These standards have been developed within the POF field since the early 1990s, when they were first put forward in the Japanese Industrial Standard (JIS) [10.144] for characterization methods. Since then, a significant amount of POF-related standards have been made by several groups. The increased research activity in the last few years has led to a plethora of results which have enabled POF experts to provide recommendations to international standardization organizations. Such organizations are the International Electrotechnical Commission (IEC), the Telecommunications Industry Association, which is part of the Electronic Industries Alliance (TIA/EIA), the Japanese Industrial Standards Committee (JIS), the European Telecommunications Standards Institute (ETSI), or the Deutsche Kommission Elektrotechnik, Elektronik und Informationstechnik (DKE, German Commission for Electrical, Electronic and Information Technologies). Next, POF-related standards will be addressed according to the categories specified in Figure 10.47.

10.2.5.1   Standards for POF passive components

Regarding passive components, plastic optical fiber and POF cable specifications can be found in standard IEC 60793-2-40. In this standard, several types of POF are defined, including four families of PMMA-based SI-POF (A4a to A4d) and four families of perfluorinated GI-POF (A4e to A4h). On the other hand, several standards for POF connectors have been made. These include SMA connectors (TIA/EIA-475C000 and IEC-QC210100), ST connectors

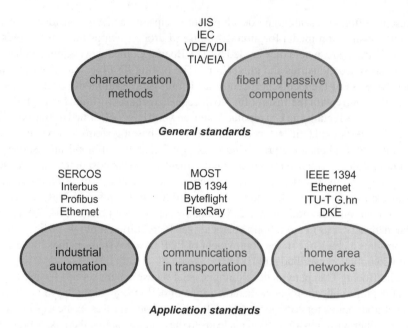

Figure 10.47   Standards for POF, related components and typical applications

(TIA/EIA-475EA to TIA/EIA-475EC, TIA/EIA-604-2 for intermateability FOCIS-2 and IEC 874-10 for BFOC connector type) and Toshlink connectors (JIS C5974-F05, EIAJ RC-5720 and JIS C5976-F07). The SMI fiber connector family (Small Multimedia Interface), which was specifically designed for consumer electronic devices using IEEE 1394b (also known as FireWire or i-LINK) is specified in standard IEC 61754-21. Several standards dealing with POF networking applications have been made. In the industrial area, these include Industrial Ethernet, SERCOS, Profibus/Net, ControlNet and DeviceNet.

10.2.5.2   Standards for transportation systems using POF

In the automotive field, there are several standards that specify POF as the main transmission medium, both for infotainment and safety-critical applications. MOST (*Media Oriented Systems Transport*) was developed in Europe and represents the de-facto standard for multimedia and infotainment networking. MOST specifies the technology able to provide an efficient and cost-effective solution to transmit at up to 150 Mb/s audio, video, data and control information between any devices within an automobile, despite the harsh environment involved. The main rival of MOST in the automotive infotainment area is *IDB-1394*, which was created as a result of adapting the FireWire protocol to this environment. IDB-1394 supports data rates of up to 400 Mb/s at the expense of higher cost, by using RC-LEDs as optical sources.

   This MOST-compatible protocol was developed with the approval of US and Japanese car manufacturers by the 1394 Trade Association working together with the Intelligent-transportation-systems Data Bus Forum (IDB Forum). The MOST Cooperation and the IDB Forum are the organizations through which these specifications are defined. They consist of a large number of international car makers and many key component suppliers that contribute to their innovation and continuous update in order to be aware of the latest technology requirements. Although MOST and IDB-1394 were originally conceived for the automotive industry, they can be used for applications in other areas, such as other transportation scenarios, audio and video networking, security and industrial applications.

Also in the automotive field, but focused on the area of safety and information, the ByteFlight protocol was introduced by BMW and partners at the beginning of this century. ByteFlight was intended for safety-critical applications such as information for the airbag inflation system and was designed to transmit low data rates of about 10 Mb/s and thus it is based on very cost-effective components such as SI-POF and red LEDs. Since 2008, this protocol was replaced in production cars by the *FlexRay* protocol, whose specifications are gathered in the set of ISO standards ISO 17458-1 to ISO 17458-5. This protocol is defined independently of the topology and the physical layer, and thus enables the use of both electrical and optical transceivers. FlexRay was conceived as the communications protocol for the new X-by-wire applications that replace mechanic and hydraulic systems by Engine Control Units (ECUs) and communication links.

In aircraft environments, there is still much work to be done for the standardization-driven commercial introduction of POF to become a reality. The HSPN consortium and the succeeding organization called PAVNET (Plastic Fiber and VCSEL Network) aimed at the development of POF systems for use in avionics. The goal was to use perfluorinated GI-POF and VCSELs operating at wavelengths of 850 and 1300 nm, but this group has been rather inactive over the last few years [10.62]. It must be noted that concerning standards for aviation, there are differences between the USA and Europe, ruled by Part 25 of the Federal Aviation Regulations (FAR-25) and Part 25 of the Joint Aviation Requirements (JAR-25), respectively. Moreover, there also exist company-specific standards, which slow down the availability of commercial components. Thus, the research and development of POF-based technologies for aircraft networking is a multi-dimensional task, whereby non-specific standards are investigated for their use in POF links within airplanes. In this way, flammability issues have been evaluated in POFs having different jacket material and thickness, by means of the method described in standard IEC 60332-1-2 for testing flammability in electrical and GOF cables [10.145]. Fire resistance is one of the major challenges POF components have to face in order to be flight-qualified and thus this kind of research work lays the foundations for future standardization.

10.2.5.3   Standards for home networks over POF

Finally, in the home area, many efforts have been made in recent years. The *HomeGrid Forum* drives the development and evolution of the ITU-T G.hn standards family with the purpose to create a single standard for wired home networks. POF was recently added as a fourth physical medium together with coaxial, phone and power line cables. The recommendation G.9960 describing system architecture and physical layers was published by the ITU-T in January 2012. An experimental trial of the standard was conducted in August 2012 at the premises of the Eindhoven University of Technology (TU/e), the Netherlands. A full-scale triple-play network scenario based on commercial products and SI-POF as a transmission medium was successfully accomplished, including connection to the campus network, bidirectional high definition (HD) 1080i video stream transmission, and voice-over-IP telephony service [10.146]. The trial confirms the potential of G.hn over POF technology for the short- and medium-term in the home or building area network.

In addition, the DKE, which constitutes a joint organization of DIN and VDE (Verband Deutscher Elektrotechniker, Association of German Electrical Engineers), has made important insights toward the standardization of a common POF interface for Gigabit transmission in home networks. Group AK421.7.1 within this organization, with more than 40 European partners, is responsible for defining the passive and active parts of the interface, such as type of connectors, wavelength range, transmitted optical power or modulation format [10.88].

10.3   Radio over Fiber (RoF) Systems

Dr. Joaquín Beas, Dr. Gerardo Castañón, Dr. Ivan Aldaya and Dr. Alejandro Aragón-Zavala
Tecnológico de Monterrey, Mexico

In recent years, considerable attention has been devoted to research into the relationship between communication systems and management infrastructure of future smart cities [10.194]. A smart city is one where investments in human and social capital and traditional and modern communication infrastructure fuel sustainable economic development and a high quality of life [10.150]. When there is always a dependably nearby access point to a secure network, this will lead to even further increases in new terminals and communication services that are secure and safe, and even more enjoyable and convenient. The final aim of a smart city is to make better use of the public resources, increasing the quality of the services offered to the citizens, while reducing the operational costs of public administration [10.162]. This objective can be achieved by a deployment in which the objects of everyday life communicate with each nother and with the users, becoming an integral part of the Internet, thus realizing the so-called Internet of Things (IoT) concept. Therefore, application of the IoT paradigm to the smart city is particularly attractive to local and regional city administrations that may become the early adopters of such technologies.

Figure 10.48 shows the relationship of the management and communication infrastructure of a smart city. The smart city management infrastructure handles information and controls across the different types of infrastructure needed by a city. Examples include smart grids in the energy sector, navigation systems and green mobility involving the use of electric vehicles in the transportation sector; and advanced water management systems using water from rain and recycling in the water system. Its roles include information management, operational management and equipment operation within the city [10.238]. This infrastructure uses information technology to provide information platforms for linking within and between different types of infrastructure. Information systems can collect operational data from various areas of life, and then transform this data into information and knowledge that applications can use to provide smart services. For example, data can enhance the operation of factories, electric power systems, railways and other services. Through this integration of information, it is possible to develop infrastructure systems that are optimized across the whole of society.

The process of developing a smart city requires the identification of the right level of "smartness" for that city, and necessitates undertaking long-term projects aimed at achieving this goal [10.199].

While futurologists predict smart cities in 2020 [10.176], networking engineers inquire whether state-of-the-art wireless technologies are able to provide the required connectivity. On the one hand, to support IoT requirements of low throughput, low power consumption, and thousands of indoor and outdoor devices working at the same area, the IEEE 802 LAN/MAN Standards Committee (LMSC) has formed the IEEE 802.11ah Task Group (TGah) [10.180]. Furthermore, the wireless access in vehicular environments (WAVE), a vehicular communication system, has been presented in IEEE 802.11p and studied recently in [10.149].

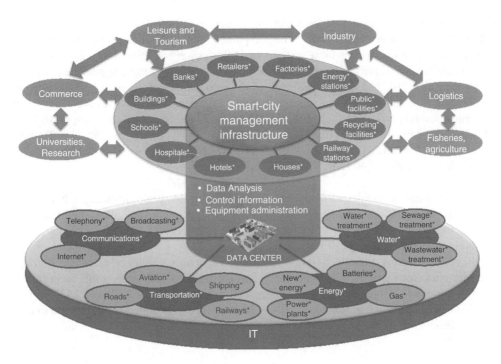

Figure 10.48   Smart city relationship between communication system and management infrastructure

On the other hand, one of the candidates to be part of the wireless network broadband infrastructure of a smart city is the fiber-wireless integration, referred as *Radio over Fiber* (RoF) [10.258]. The objective of mobile broadband wireless access has been addressed by the IEEE standards 802.15.3c [10.147] and 802.16 [10.148], as well as by the ITU evolving standard International Mobile Telecommunication-Advanced M.1635 [10.220]. The need is for efficient communication systems that could support such bandwidth requirements, are encouraged to exploit the advantages of both optical fibers and millimeter-wave (mm-wave) frequencies. While optical fiber technologies provide high bandwidths and support long transmission links, the use of mm-wave frequencies offers large bandwidths in the wireless domain and also overcomes the problem of spectral congestion at lower frequency ranges [10.260].

Various fiber-wireless access network architecture designs have been prototyped: wireless-optical broadband access network (WOBAN) [10.241], metro and access ring integrated network (MARIN) [10.256], fiber-optic networks for distributed extendible heterogeneous radio architectures and service provisioning (FUTON) [10.239], *fiber wireless access network* (FiWi) [10.222], grid reconfigurable optical and wireless network (GROWNet) [10.242], and converged copper-optical-radio OFDMA-based access network (ACCORDANCE) [10.211]. They operate at wireless frequencies of up to 5 GHz and can provide many hundreds of megabits per second bandwidth (up to 1 Gbps). Thus, it fulfills the objectives of the so-called 4th Generation (4th-generation wireless systems) through an innovative Distributed Antenna System (DAS). However, the trend for high-capacity wireless systems is likely to continue to even greater carrier frequencies (>30 GHz) and bandwidths (>1 Gbps) [10.208,10.245].

mm-wave communication systems that can achieve multi-gigabit data rates at a distance of up to a few kilometers already exist for point-to-point communication. However, the component electronics used in these systems, including power amplifiers, low noise amplifiers, mixers and antennas, are too large and consume too much power to be applicable in mobile communication [10.260]. Therefore, much of the engineering efforts have been invested in developing more power-efficient mm-wave band RoF systems for mobile communications. Recently, several proposals have been presented for the 24–30 GHz band (600 Mbps) [10.218], 75–110 GHz band (40 Gbps) [10.231], at 120 GHz (10 Gbps) [10.151], at 250 GHz (8 Gbps) [10.197], and for 220 GHz (20 Gbps) [10.237]. However, the frequency band that has attracted major interest is around 60 GHz, mainly for two reasons:

1. This frequency coincides with one of the oxygen absorption peaks, resulting in high atmospheric attenuation exceeding 15 dB/km [10.182]. High attenuation allows cell reduction and increases the frequency re-use factor (K) in cellular systems, increasing the aggregated wireless system capacity [10.184].
2. In North America, there is 7 GHz of unlicensed spectrum around 60 GHz that overlaps with unlicensed spectra in Europe, Japan and Australia, which opens up the opportunity for worldwide standardization and commercial products [10.230,10.236,10.243]. RoF at 60 GHz has been demonstrated in [10.174,10.153,10.159], to support transmission of data throughputs of up to 27, 32 and 84 Gbps, respectively.

Figure 10.49   General RoF communication system architecture

where:

CS   Central Station
ODN   Optical Distribution Network
BS   Base Stations

A typical *mm-wave RoF network* is shown in Figure 10.49. Three main subsystems are defined for the RoF land network: *Central Station* (CS), *Optical Distribution Network* (ODN) and *Base Station* (BS). In the downlink (DL) transmission, the CS up-converts the electrical signal to the optical domain and uses the ODN to communicate with the BSs, using *Remote Nodes* (RNs). The optical signal is amplified in the case of an active network, and split or demultiplexed towards the corresponding BS that converts it back to the electrical domain. In the wireless network, the BS radiates the signal to the Mobile Terminal (MT) in the mm-wave bands [10.170]. In the uplink (UL), the BS receives the mm-wave wireless signal from the MT, and depending on the BS configuration, the mm-wave UL signal received can be down-converted before feeding to an electro-optical device to transmit it via the ODN back to the CS. On each land network subsystem (CS, ODN and BS), specific network functions were identified depending on whether they operate for signal transmission, distribution or reception for DL and UL.

RoF distribution systems have several advantages over conventional coaxial cable or point-to-point wireless backhaul:

- Low attenuation in the distribution network by the use of optical fibers
- Simplicity and cost-effectiveness, since it centralizes resources at the CS where they can be shared; remote simple BSs consisting only of an optical-to-electrical converter, Radio Frequency (RF) amplifiers, and antennas are readily available.
- Low-cost expandability, as they are virtually modulation format agnostic
- High capacity, because higher frequencies can be transported through RoF systems allowing data rates to accommodate future service demands
- Flexibility, because it allows independent infrastructure providers and multi-service operation. The same RoF network can be used to distribute traffic from many operators and services
- Dynamic resource allocation, since functions such as switching, modulation and others are performed at the CS, being able to allocate capacity dynamically

Table 10.5 presents how each network function is further broken down into network elements (NEs). At the same time, a list of enabling technologies required to support the design of each NE is presented. Furthermore, Figure 10.50 presents the RoF land network diagram which specifies the location of the NE.

10.3.1   Key Enabling Technologies

The major challenges of an mm-wave RoF land network are:

1. as the RF frequency increases, so does the requirement of sophisticated DL optical transmitters at the CS [10.163,10.173,10.209,10.210,10.223,10.233]
2. the design of efficient topologies for the ODN [10.167,10.189,10.224,10.252,10.257]
3. the implementation of low-cost and "green" UL transmitter configurations at the BS [10.164,10.178,10.185,10.216]. In each subsystem, the technologies addressing these challenges are identified as key enabling technologies and require special consideration

Table 10.5   Network elements and enabling technologies for the mm-wave RoF land network

Subsystems	Functions	Network Elements (NEs)	Enabling technologies
Central Station (CS)	DL Transmission	NE1. Electrical multiplexer NE2. Optical transmitter NE3. Optical multiplexer NE4. Optical transmission band	• Electrical multiplexing schemes • mm-wave generation and transmission data techniques • UL transmission technologies • Optical multiplexing schemes
	UL Reception	NE5. Optical demultiplexer NE6. Optical receiver	• Optical transmission bands • Optical receiver technologies • Network topologies
Optical Distribution network (ODN)	Distribution	NE7. Network Topology NE8. Fiber Type	• Fiber types • Optical passives technologies
	Remote Node (RN)	NE9. DL optical DROP configuration NE10. DL Optical Amplifier NE11. UL optical ADD configuration NE12. UL Optical Amplifier	• Optical amplification technologies
Base Station (BS)	UL Transmission	NE13. Electrical multiplexer NE14. Optical transmitter NE15. Optical multiplexer NE16. Optical transmission band	
	DL Reception	NE17. Optical demultiplexer NE18. Optical receiver	

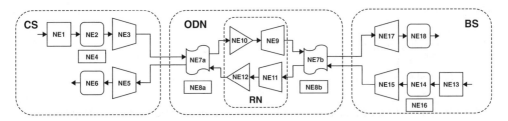

Figure 10.50   Network elements at the RoF land network

10.3.1.1   DL Optical Transmitter at the CS (NE2)

The most challenging stage in the *DL transmission* is the generation of the optical signal at the CS that, after being detected by a square-law device such as a photodetector (PD) in a remote BS, results in the desired mm-wave signal. An increasing number of techniques has been proposed in recent years to perform this task based on different approaches, such as directly modulated lasers [10.177,10.225], externally modulated lasers [10.152,10.171,10.210], multi-wavelength sources [10.159,10.172,10.195,10.253] and heterodyne of independent lasers [10.158,10.187,10.199,10.201]. Figure 10.51 shows block diagrams of the most important configurations for the DL optical transmitter of an mm-wave RoF land network.

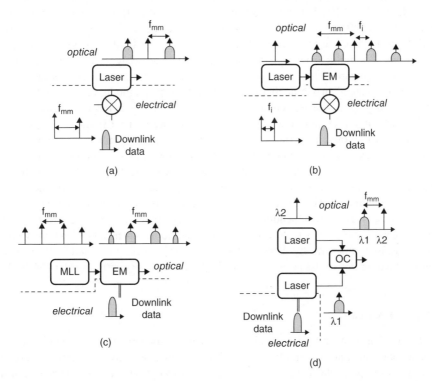

Figure 10.51   Directly modulated lasers (a), externally modulated lasers (b), multi-wavelength sources (c) and heterodyne of independent lasers (d)

where:

  EM   External Modulator
  MLL Mode-Locked Laser
  OC   Optical Coupler

A laser is directly modulated with the DL signal at the desired mm-wave frequency (Figure 10.51a). The transmitter configuration is extremely simple and cost-effective, but its performance is severely limited by the laser modulation impairments. On the one hand, the frequency chirp, the significant non-linearities and high RIN cause poor frequency stability and transmission integrity [10.191]. On the other hand, the maximum modulation frequency is generally limited by the laser resonance peak [10.205].

In order to overcome the impairments of the direct modulation, external modulation is the straightforward solution. The simplest implementation consists of a Continuous Wave (CW) laser followed by an External Modulator (EM) that modulates the laser light with an *intermediate frequency* (IF − fi) or an mm-wave tone (*fmm*) (Figure 10.51b). The EM used in this technique can be an intensity modulator (a Mach-Zehnder Modulator (MZM), an Electro Absorption Modulator (EAM)) or a Phase Modulator (PM), whose output is optically filtered [10.210].

Multimode light sources, such as Fabry-Perot Lasers (FPL) [10.207,10.253], Mode Locked Lasers (MLL) [10.159,10.250], dual-mode lasers [10.179,10.190] or Supercontinuum Sources (SCS) [10.195,10.249], have gained attention as a low-cost alternative to expensive broad bandwidth EMs and high numbers of independent lasers. Multimode light sources can be used in two different ways, as a multicarrier generator and as a way to generate the different beating tones for mm-wave frequency generation. For the latter, many variants have been reported in the literature, but the mechanism behind them is the same, two optical modes are used to beat them at the BS's PD and generate the desired mm-wave signal. Therefore, in most cases, the multimode source is designed to present modes separated by the desired carrier frequency. Figure 10.51c shows a MLL whose optical carrier separation equals the desired mm-wave frequency. Thus, its output can be externally modulated with an EM for downlink signal transmission.

Multi-mode sources are not the only way to generate two optical tones whose beating results in mm waves. Two lasers with emission frequencies separated by the desired frequency can be mixed to generate mm-wave signals. These lasers can be phase-correlated lasers [10.187] or uncorrelated lasers [10.229], depending on the technique implemented. Figure 10.51d presents two independent lasers with wavelengths $\lambda 1$ and $\lambda 2$, combined using an OC, whose separation equals the desired mm-wave frequency. This technique has been proposed to transmit information modulating one of the lasers directly [10.186,10.229] or externally [10.164,10.201].

10.3.1.2   ODN (NE7)

The *ODN* design of an mm-wave RoF system can be explained by the definition of two important concepts, topology and land clustering. A complete analysis of network topologies for RoF systems is presented in [10.189]. Several topology options, such as star, ring, multilevel rings, multilevel stars, hybrid multilevel star-rings, and ring-stars, are compared in terms of link distance, reliability and availability of the network. The analysis is performed using the

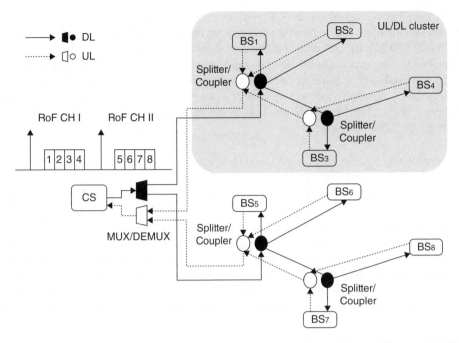

Figure 10.52 Multi-level star-stars network topology composed of 2 DL/UL land clusters with 4 BSs on each ($K = 4$)

software tool presented in [10.188], NetExpert, which supports the network planning, design and optimization processes and saves valuable time in preparing the network topologies.

On the other hand, the ODN land clustering can be understood by the definition of DL and UL land clusters as follows:

- *DL land cluster*: corresponds to the group of BSs supported by a common mm-wave RoF channel, which may differ from the RF cluster
- *UL land cluster*: corresponds to the group of BSs sharing the same UL optical channel

Figure 10.52 shows an example of a multi-level star-stars network topology designed with MUX/DEMUXES and optical couplers and splitters to support 8 BSs distributed in 2 DL/UL land clusters with 4 BSs on each [10.234]. Observe that in this specific example, the number of BSs within a DL/UL land cluster is exactly the same as the K factor for the wireless domain. However, there will be cases when more than one optical channel in the land network is required to support the K number of BSs [10.184].

10.3.1.3   UL Transmitter at the BS (NE14)

BS transmitter configurations can be divided into two main categories, with a laser installed and laser-free

When the BS is equipped with a laser, the *UL transmission* can be performed either by directly modulating the laser [10.181, 10.248] or by using an external modulation scheme.

For the laser-free configuration, the CS provides the optical tone for UL transmission. Under this scheme, the optical tone from the CS can arrive at the BS as a non-modulated optical tone [10.217,10.227,10.239] or as a modulated signal [10.213,10.259]. The latter requires a modulation erasing process before the UL signal can be transmitted back to the CS. In order to relax the bandwidth requirement of the electro-optical conversion in the UL transmission, typically the UL mm-wave wireless signal is first down-converted to baseband or Intermediate Frequency (IF) [10.164,10.178]. In this case, an electrical Down-Conversion Circuit (DCC) is required, which can be based on:

- a heterodyning scheme, requiring a Local Oscillator (LO) [10.169]
- an envelope detector [10.255]
- RF self homodyning [10.201], enabling in this way a spectrally narrow UL optical channel

However, there are cases when the mm-wave wireless UL signal received from the MT directly modulates a mm-wave device for UL transmission [10.156,10.185,10.215,10.216]. For this scheme, in terms of spectral width, the UL channel is exactly the same as the mm-wave RoF channel used for DL transmission. Figure 10.53 shows several BS configurations.

In a conventional BS configuration (Figure 10.53a), the DL signal coming from the CS is demultiplexed and received in an mm-wave bandwidth PD at the BS when using WDM

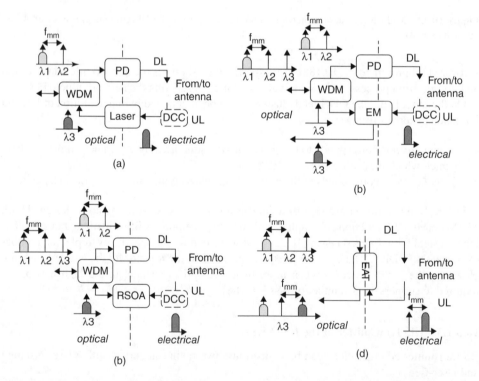

Figure 10.53 BSs with laser (fix wavelength) at the BS (a), externally modulated laser (b), RSOA using a non-modulated optical tone from CS (c) and Electro Absorption Transceiver (EAT) at the BS (d)

technology. On its side, the UL optical signal is generated by directly modulating a laser. The WDM component multiplexes the signal toward the CS. The laser installed at the BS can be a *Vertical-Cavity Surface-Emitting Laser* (VCSEL) [10.181] or a DFB laser [10.248], and can be modulated by baseband or intermediate down-converted signals, depending upon the scheme selected. For mm-wave frequencies higher than 30 GHz, direct modulation can be a difficult and expensive choice, because of the modulation bandwidth of the laser [10.219], thus the implementation of a down-conversion process from mm-wave frequencies to baseband or IF signals is necessary (DCC). Therefore, this BS configuration requires two independent optical components (laser and PD), and needs a DCC for the mm-wave signals from MT.

So far, considerable efforts have been made to simplify the BS design combining the use of external modulation and custom-designed optical components. The major advantage of removing the laser from the BS is that the wavelength assignment and resource monitoring can be done in a centralized way at the CS. Several laser-free BS schemes have been proposed. For an external modulation scheme, MZM, PM or an EAM is required at the BS. A WDM component at the input of the BS is used to demultiplex downlink a signal to the BS mm-wave bandwidth PD and the uplink optical signal from the CS to EM input for its corresponding modulation before being transmitted back to the CS (Figure 10.53b).

In Figure 10.53c, the RSOA replaces the WDM light source at the BS. Its multi-functionalities, such as colorless operation, re-modulation, amplification and envelope detection, make the BSs implementation more compact, less complex and cost-effective, improving the utilization of wavelength resources in the RoF network with centralized light sources. RSOA can be used in two different ways: reusing the downlink signal [10.213,10.259], or using an extra non-modulated optical tone generated at the CS [10.154,10.239]. When an RSOA is used at the BS, the mm-wave signal from the MT needs to be down-converted to meet the electrical bandwidth limits of the RSOA.

A single optical component approach at the BS is the EAT (Figure 10.53d [10.157]). This scheme utilizes the EAT that works simultaneously as an mm-wave bandwidth PD for downlink data and as an optical modulator for the uplink at different wavelengths. Since the cost of the BS is most significant in an RoF system, EAT should be cheap, reliable, small and lightweight, with low power consumption.

The transceiver receives two optical signals from the CS, the downlink which is absorbed, and a non-modulated signal from the CS, that is modulated at the BS with uplink data. The essential characteristic of this device is that it not only acts as a modulator but also as a PD [10.215,10.216]. The EAT has an optical input to collect both downlink modulated and remote non-modulated signals.

10.3.2   RoF Land Network Design

The design of an mm-wave *RoF land access network* is a very knowledge-rich domain, not only due to the particularities and complexity of each network subsystem and their interconnection, but also as the emerging nature of the art. A very large number of activities are involved in the design process, and among the most important are:

 (i)   determining the type of communication system [10.184,10.232]
 (ii)   geographic layout [10.155,10.183]
 (iii)   choosing transmission equipment and link power budget [10.156,10.240]

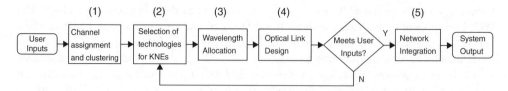

Figure 10.54   Design process of the proposed framework

(iv) selecting components for outside plant and premises installations [10.178]
(v) documentation [10.161,10.254]

This section focuses on (iii), splitting it further in five specific activities (Figure 10.54).

10.3.2.1   User inputs

The inputs to the proposed design process can be divided into two categories, functional network requirements specify what the system must do, while non-functional requirements are related to what the system should be in terms of prioritized figures of merit. Functional requirements define what the network is capable of in terms of bandwidth, coverage and signal integrity. These requirements are user inputs obtained prior to the design assistance process. They depend upon the specific application and scenario of the network under design. Table 10.6 summarizes the description of each of these requirements.

For non-functional network requirements, four Figures of Merit (FoM) were defined for the mm-wave RoF land network [10.204], which are used to assess the performance of the available key enabling technologies. These are as follows:

- *Cost Effectiveness*: Measured in terms of capital and operational expenses of the network, CAPEX and OPEX respectively [10.214]

Table 10.6   Functional network requirements

ID	Requirement	Description
NC	Network Coverage	The coverage area of the network design
K	Frequency Reuse Factor	The rate at which the same wireless frequency can be used in the network
DLBR	Downlink Bit Rate	The DL bit rate in Gbps that a single BS shall support
DLMS	Downlink Modulation Scheme	Modulation scheme necessary to calculate the DL occupied bandwidth
ULBR	Uplink Bit Rate	The UL bit rate in Gbps that a single BS shall support
ULMS	Uplink Modulation Scheme	Modulation scheme necessary to calculate the UL occupied bandwidth
DL SNR	Downlink Signal-to-Noise Ratio	The SNR desired at the output of the optical detector at the BS expressed in dBs
UL SNR	Uplink Signal-to-Noise Ratio	The SNR desired at the output of the optical detector at the CS expressed in dBs

- *Energy Efficiency*: Expressed either as power consumption per average access rate (W/Mbps) or as energy consumed per transmitted bit (J/bit) [10.202,10.203]
- *Reliability*: Defines the ability of the system to function for a specified period of time. In general, the reliability penalty imposed by a system, which can be derived from its Failure Rate (FR) in Failures-In-Time (FIT) (1 FIT = one failure in 10^9 hours) [10.189]
- *Scalability*: Defined as the capability of the RoF system to increase network coverage and capacity, in order to meet a higher usage demand without a significant cost increase. Measured in terms of the OPEX, which is necessary to provide flexibility in network coverage and capacity

The design engineer is allowed to prioritize one of several of these FoMs for a specific network design, in order to select the key enabling technologies that best meet the requirements.

10.3.2.2   Channel assignment and clustering

First the calculation of the total number of BSs (TBS) is required. At this point, the separation distance between BSs should be defined, which will directly depend on the wireless transmission technology. In the proposed design framework, it is assumed that the BSs are distributed uniformly over the *Network Coverage* (NC) with a separation distance of 100 m between them. This is the average size of urban blocks in typical US cities. Based on this, the TBS is calculated to meet the NC requirements.

For 60 GHz RoF networks, a maximum bandwidth per RoF channel of 7 GHz is considered (unlicensed spectrum around 60 GHz) [10.230]. As shown in Figure 10.55, the DL Modulated Sideband Bandwidth (MSBW) and UL MSBW are determined considering the channel bit rate (DLBR and ULBR) and the modulation scheme (DLMS and ULMS). Note that the Forward Error Correction (FEC) and the bandwidth Roll-Off must be considered in this calculation. The re-use factor K is used as a baseline as the maximum number of BSs that a single optical channel should support. In case the channel bandwidth, required to satisfy K BSs per optical channel, exceeds the maximum of the 7 GHz available, the number of BSs per land cluster is distributed accordingly considering the available bandwidth. Thus, the number of optical

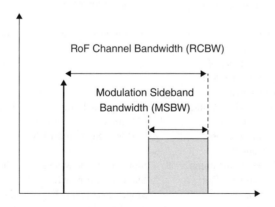

Figure 10.55   mm-wave RoF channel at optical domain

channels to support K BSs is higher than 1. After the definition of DL MSBW and UL MSBW, and the number of BSs per DL/UL land cluster, it is necessary to calculate the total number of DL/UL channels (land clusters) required in the specific network design. Finally, the number of BSs per DL/UL land clusters is optimized.

10.3.2.3 Selection of key enabling technologies

The selection process for the appropriate techniques/topologies for DL transmitter, the ODN topology, and the UL transmitter configurations is performed by assessing each enabling technology in terms of the previously presented FoM. The parameters used for such assessment are shown in Table 10.7.

DL and UL optical transmitters are assessed in terms of FoM as a function of the number of mm-wave RoF channels required and the TBS, respectively. The component cost and power consumption data are obtained from component manufacturer's datasheets [10.214], and reliability values are calculated from literature references [10.175,10.189]. On the other hand, for the network topology, the assessment is performed considering TBS and NC. The total fiber length for the network design and the fiber link connections between CS and the different BSs, which depend on the network topology, are calculated using NetExpert [10.188]. In order to determine the fiber installation cost, underground installation of fiber-optic cables is assumed.

Table 10.7 List of parameters used to perform the attribute assessment of the key enabling technologies

	Evaluation Parameters for the Attribute Array		
Figures of Merit	DL Optical Transmitter at CS (NE2)	ODN Network Topology (NE7)	UL Optical Transmitter at BS (NE14)
Cost effectiveness	Total cost as a function of number of mm-wave RoF channels	Fiber installation cost as a function of total optical fiber length	Total cost as a function of number of BSs
Energy Efficiency	Total power consumption in terms of number of mm-wave RoF channels	Insertion loss to the farthest BS	Total power consumption in terms of number of BSs
Reliability	FR in FIT as a function of number of mm-wave RoF channels	Healing capabilities Number of nodes and BSs to the farthest BS	FR in FIT as a function of number of BSs
Scalability	Slope of cost/number of mm-wave RoF channels	Slope of cost/new node Slope of cost/new BS Slope of cost /new DL cluster	Slope of cost/number of BSs Cost/wavelength change

Once the functional requirements are introduced, the design engineer calculates the parametric values for each FoM (Table 10.7). The calculation is used to assign an attribute value from 1–10 (1-Worst, 10-Best) on the FoM for each enabling technology under evaluation. This attribute array provides a reference baseline of performance for the network design. The top three enabling technologies that best meet the prioritized FoM are selected for evaluation in the link design stage.

It is worth noting that once the *DL* and *UL transmitter* technologies are selected, the optical receiver technology is directly included. That is to say, the complete link information, in terms of optical received power versus *SNR* for a specific optical receiver technology, is available in the transmitter datasheet, which is obtained from the corresponding reference literature.

For each pair of transmitter/receiver for DL/UL under analysis, it is tested to see whether the required SNR is met. In order to estimate the resulting SNR, it initiates from SNR values reported in the literature and then extrapolates them to the desired fiber length and signal capacity, considering that the limiting impairments are the noise, induced by the transmitter and receiver, and the attenuation of the fiber

1. *Obtaining the SNR from literature*: Results in the literature are often given in terms of the error vector magnitude (EVM) or bit error rate (BER) for a particular link configuration (modulation format and bandwidth, as well as fiber length). Typically, the resulting noise induced by the combination of the transmitter and receiver can be assumed to have a Gaussian distribution. Under this assumption, the effective SNR can be obtained from the BER (P_b) value according to [10.228]:

$$SNR = \frac{(L^2 - 1)\log_2 M}{6 \log_2 L} \left\{ Q^{-1} \left[\frac{P_b \log_2 L}{2\left(1 - \frac{1}{L}\right)} \right] \right\}^2 \tag{10.7}$$

where M is the number of the points in the constellation, L is the number of levels in each dimension of the M-ary constellation, and Q^{-1} is the inverse Gaussian co-error function. However, the SNR can be expressed in terms of EVM as:

$$SNR \approx \frac{1}{EVM^2} \tag{10.8}$$

In most cases, the system performance is limited by the optical receiver noise. These assumptions are realistic for most configurations operating near the FEC limit. The power penalties due to other impairments are neglected in a first approach.

2. *Extrapolation to an arbitrary fiber length and signal bandwidth*: The proposed design framework establishes a maximum link distance of 50 km from the CS to the farthest BS, as well as a maximum RoF channel bandwidth of 10 GHz. However, typically, in the literature, information for particular link/bandwidth combination is not available and in order to obtain the SNR of a signal with bandwidth BW at a determined distance, it is required to extrapolate from the reference values (SNR_{Ref}, BW_{Ref} and P_{Ref}) found in the literature:

$$SNR = SNR_{Ref} \frac{P_{RX}/P_{Ref}}{BW/BW_{Ref}} \tag{10.9}$$

The previous expression is written as a function of received optical power that further depends on the launched optical power and the losses of both fiber and passive devices, which will analyzed in detail in the next subsection.

10.3.2.4   Wavelength allocation

Wavelength allocation is performed separately for DL and UL signals; the proposed design framework assumes that DL and UL signals are traveling over independent optical fibers. The process considers the available *Dense Wavelength Division Multiplexing* (DWDM) and *Coarse Wavelength Division Multiplexing* (CWDM) Standards G.694.1 [10.192] and G.694.2 [10.193], in order to design multiplexing schemes employing standard and commercial available devices. The channel grids defined by these recommendations support either fixed channel spacing ranging from 12.5–100 GHz, and even 20 nm for CWDM, or flexible grid (DWDM). In addition, uneven channel spacing using fixed grids is also allowed. For DL channels, the DL MSBW is used to calculate the RoF channel bandwidth (RCBW) (Figure 10.52). For the UL channel, depending on the UL technology selected (with or without DCC), the spectral width may vary from around 60 GHz to the UL MSBW, such as a few GHz-s.

The wavelength allocation algorithm should find an optimum channel separation to meet the network design requirements. On the one hand, broader channel separation, spreading them over the whole transmission band mitigates the impact of Optical Beat Interference (OBI) on the received signal SNR [10.212,10.244], and reduces the transmitter cost, since lasers with relaxed emission frequency requirement can be used. CWDM lasers without temperature control present a lower cost than their counterpart DWDM lasers with integrated temperature controls. However, an excessive channel separation wastes a significant part of the available spectrum, reducing the spectral efficiency.

When the number of optical channels required for a specific network design cannot be allocated in the optical transmission band, the design engineer has the alternative to select one of the following wavelength allocation options to satisfy the requirements:

- *(ATB)*: Considers the addition of adjacent optical transmission bands to increase the optical bandwidth and include more channels in an attempt to meet design requirements
- *Optical Frequency Interleaving (OFI)*: Increases spectral efficiency with optical channel spacing $\leq$25 GHz to support a higher number of BSs operating at 60 GHz mm-wave frequencies in an RoF system [10.168,10.196,10.246,10.247]
- *Space Division Multiplexing (SDM)*: Increases the number of optical fibers to support the rest of the channels using space multiplexing. SDM can be performed on strands of optical fiber ribbon cable [10.235] or by using multi-core fibers [10.165]. In the design framework, when a second fiber or core is added to the network design, the number of channels is allocated equitably

10.3.2.5   Optical link design

The proposed design framework is mainly concentrated on the effects of the RoF land network by considering signal integrity at the output of the optical receiver [10.210,10.218, 10.226,10.234]. In order to determine the suitability of DL/UL transmitters and receivers, an *optical link budget analysis* is required to ascertain acceptable SNR based on the functional network requirements. This process initiates considering the top one key enabling technology for each subsystem, which were previously selected.

The signal power P_{RX} at the input of the receiver, in dBm, is given by:

$$P_{RX} = P_T - L + G_{OA} \qquad (10.10)$$

where P_T is the transmitted optical power per channel in dBm, L is the path loss in dB for the transmitter/receiver separation distance d, and G_{OA} is the optical amplifier gain in dB which, for passive ODN, does not exist. During the link design, the transmission power P_T is obtained from the transmitter datasheet. The link path loss L is given by:

$$L = L_{Fiber} + L_{Insertion} + L_{Splice} + L_{Connector} \qquad (10.11)$$

where L_{Fiber} is the optical fiber loss in dB, $L_{Insertion}$ is the accumulative insertion loss in dB of optical passive components (couplers, MUX/DEMUX, etc.), L_{Splice} is the accumulative insertion loss in dB of the optical splices, and $L_{Connesctor}$ is the accumulative insertion loss in dB of the optical connectors. The insertion loss of optical connectors can be assumed to be 0.2 dB, while the splice insertion loss is 0.15 dB. The number of splices, connectors and the distance of CS to the first RN are determined by the design of the ODN.

In active ODNs, the optical amplifier gain G_{OA} can be varied within a specified range to achieve the appropriate signal power value P_{RX} to satisfy the SNR functional requirement. The link design for UL and DL is performed using the farthest BS on the network (worst case). Although novel amplifier placement methods for metropolitan WDM networks have been studied in [10.160,10.206], in this work it is assumed that a single optical amplifier is usually sufficient to compensate for the losses. The exception occurs when the network topology, either for DL or UL, requires more than one fiber to support all the optical channels.

For the DL design, when the SNR requirement is not achieved, even with the use of an optical amplifier at its maximum gain, the design engineer has the opportunity to try with the second and third DL transmitter/receiver technologies. In case the DL transmitter/receiver technologies do not meet the network requirements using the top one network topology, a second and third network topology options can be evaluated. The same process described above is performed for the UL design. However, since the DL design is typically the most exigent design stage, usually the ODN topology is fixed at the DL design stage and it is not recommended to modify it during the UL design. This design decision is established to ensure DL/UL network topology commonality.

10.3.2.6   Network integration

At this stage, the technologies selected to support each NE presented in Table 10.5 are consolidated. The design engineer should summarize the number of optical components required by the network design as the optical transmitters, receivers, passives and amplifiers. Furthermore, it is necessary to specify the optical device configurations such as number of ports, number of channels, amplifier gain and channel bit rates.

10.3.3   Case Study of the Proposed Design Framework

Over the past few decades, Monterrey City has emerged as a major industrial and engineering center in Mexico. The city of over 4 million habitants sits adjacent to the dynamic US–Mexico border region and has 57 industrial parks specializing in everything from chemicals and cement to telecommunications and industrial machinery. The area has long been business-friendly. It has also become a major education center, with over 82 institutions of higher learning and 125 000 students, led by the Instituto Technologico de Monterey.

Several proposals for Monterrey City have been presented as part of its development toward a smart city [10.221]. The case study presented here is focused on the development of an urban laboratory downtown, an extended area of 1.1 km^2 which encloses museums, government and administration buildings, shopping and touristic areas, and sports parks. This location allows researchers the potential to both test and demonstrate innovative products and services for a smart city such as, but not limited to, lighting, parking, surveillance, waste management and open connectivity of broadband Internet.

The case study is a 60 GHz RoF network designed to provide Internet connectivity to downtown Monterrey (Figure 10.56). The Internet allows connecting information to people, and people to people. Furthermore, it can connect objects, places, things and everything that could benefit from being connected. When everything is connected to and is aware of everything else, the Internet emerges as an intuitive, context-aware, intelligent platform enabling smart services [10.166].

The design inputs for the functional network requirements are presented in Table 10.8.

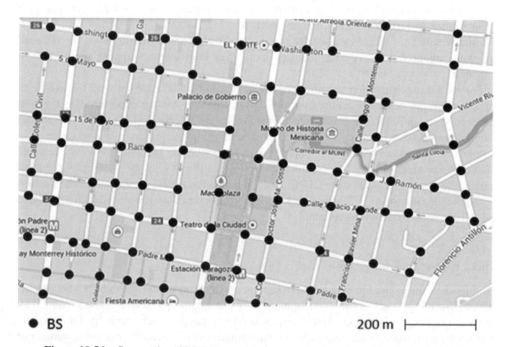

● **BS**   **200 m** ├────────┤

Figure 10.56   Case study: 108 BSs to provide coverage to downtown Monterrey (México)

Table 10.8 Functional network requirements

ID	Description	Requirement
NC	Network Coverage	1.1 km^2
K	Frequency Reuse Factor	9
DLBR	Downlink Bit Rate	1 Gbps
DLMS	Downlink Modulation Scheme	16-QAM
DL SNR	Downlink Signal-to-Noise Ratio	35 dB
ULBR	Uplink Bit Rate	300 Mbps
ULMS	Uplink Modulation Scheme	16-QAM
UL SNR	Uplink Signal-to-Noise Ratio	30 dB

The non-functional requirements for the case study presented here were prioritized as follows:

1. *Cost effectiveness*: The objective is to achieve the lowest CAPEX/OPEX necessary for the network design. In principle, the proposed network development can be supported by education institutions and private industries as a test bed for future deployments
2. *Energy efficiency*: In conjunction with cost-effectiveness, the design objective is to implement simple and green BSs over the network
3. *Scalability*: Due to the nature of the proposal, the network scalability is only in terms of network coverage not capacity
4. *Reliability*: Assumed with the lowest priority for the proposed urban laboratory. It is assumed that there will be easy access to CS equipment, ODN links and BSs for maintenance and reparations if necessary

10.3.3.1 Channel assignment and clustering

The baseline multiplexed total DL and UL bit rates are defined as:

$$Total_{\text{DL Bit Rate}} = DLBR \cdot K = 9 \text{ Gbps} \tag{10.12}$$

$$Total_{\text{UL Bit Rate}} = ULBR \cdot K = 2.7 \text{ Gbps} \tag{10.13}$$

Considering a 16-QAM modulation scheme, an FEC factor of $^3/_4$ and a Roll-Off bandwidth factor of 35%, the occupied bandwidth for the multiplexed DL and UL bit rates is equal to 4.4 GHz and 1.32 GHz, respectively. In both cases, the required bandwidth is below the 7 GHz limit established for the 60 GHz land network design.

Afterwards, the TBS is calculated based on the coverage area, 108 BSs. Then the number of RF channels in each RoF channel is calculated, that is, RoF land clustering. In the case under study, the numbers of RF channels in the UL and DL are the same, which is calculated as:

$$\frac{DL}{UL_{\text{channels}}} = \frac{TBS}{K} = 12 \tag{10.14}$$

Thus, 12 DL/UL land clusters will be required, with each optical wavelength supporting 9 BSs. In this case, the K number of BSs is satisfied by a single optical channel.

10.3.3.2 Selection of key enabling technologies

For this case study, a total of 19 different mm-wave DL generation techniques, 13 ODN topologies and 15 UL transmitter configurations [10.204] were evaluated. Based on the functional network requirements and the FoM, the enabling technologies selected to support the network design were Externally modulated Mode-Locked Laser (MLL) for the DL transmitter [10.200], Star-Bus topology for the ODN, and directly modulated laser with mm-wave PD for the BS configuration.

In the DL direction, the MLL generates multiple tones whose phase noise is highly correlated and, since the free-spectral range of the MLL is 50 GHz, the proposed approach is compatible with the ITU frequency grids. Some of the generated modes are segregated using a demultiplexer. Each mode is modulated with a downlink signal centered at 10 GHz, which will result in a 60 GHz signal after beating with the carrier of the contiguous channel at the PD (Figure 10.57a). The modulated optical tone and the carrier of the contiguous mode are multiplexed and sent through a fiber link. A star with 12 links is required to support the total number of BSs (Figure 10.57b). Each of those 12 links connects 9 BSs in a bus topology using optical couplers (OC). At the BSs, an mm-wave PD converts the received signal to the electrical domain. For UL, the mm-wave signal received at the antenna is first down-converted by a DCC, in order to avoid a high-frequency electro-optical conversion for UL transmission. The down-converted UL signal modulates directly a laser installed at the BS (Figure 10.57c).

10.3.3.3 Wavelength allocation

Because of the nature of the selected DL transmission technique and the number of DL channels, the *wavelength allocation* of DL optical signals was designed for 50 GHz *DWDM* channel spacing traveling in a single fiber without interleaving. It is assumed that a single MLL can support 6 RoF channels, therefore all signals are accommodated in the same optical transmission band using at least two MLLs. On the other hand, the cost-effectiveness was prioritized for the UL transmission. Thus, a *CWDM* scheme was selected for the UL wavelength allocation, allowing in this way the use of low-cost lasers.

10.3.3.4 Optical link design

DL design is presented in Figure 10.58.
   The DL budget was calculated as follows:

- $P_T = -3$ dBm
- $L_{Fiber} = 0.33$ dB from 1.6 km of standard single-mode fiber (the separation between the CS and the farthest BS)
- $L_{Insertion} = 16.4$ dB from 8 OCs of 80% with insertion loss of 1.1 dB and an additional OC of 20% with 7.6 dB insertion loss

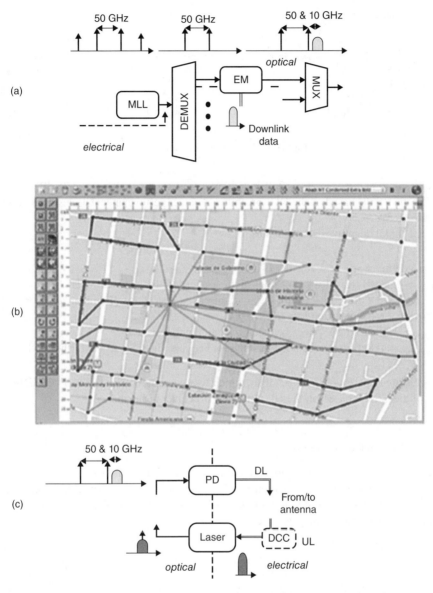

Figure 10.57 DL transmitter configuration: Externally modulated mode locked laser for the DL transmitter (a), NetExpert output for a Star-Bus topology of the 108 BSs case study network (b) and BS configuration (c)

- $L_{Splice} = 3$ dB from 20 splices between the CS and the farthest BS
- $L_{Connector} = 0.8$ dB from 4 connectors splices between the CS and the farthest BS
- $G_{OA} = 12$ dB configured during the optical link design to achieve a P_{RX} that meets the desired SNR of 35 dB for the technique under analysis

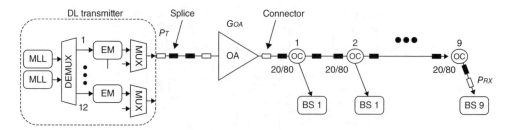

Figure 10.58 DL optical link design

This resulted in a $P_{RX} = -11.53$ dBm, which was enough to achieve the 35 dB SNR at the output of the BS PD based on the DL transmitter/receiver performance. UL design is presented in Figure 10.59.

The UL link budget was calculated as follows:

- $P_T = 3$ dBm
- $L_{Fiber} = 0.33$ dB from 1.6 km of fiber to the longer link between CS to BS
- $L_{Insertion} = 16.4$ dB from 8 OC of 8 % with insertion loss of 1.1 dB and an additional OC of 20% with 7.6 dB insertion loss
- $L_{Splice} = 3$ dB from 20 splices between CS and the farthest BS
- $L_{Connector} = 0.8$ dB from 4 connectors splices between the CS and the farthest BS

This resulted in a $P_{RX} = -17.53$ dBm, which was enough to achieve the 30 dB SNR at the output of the CS PD based on the UL transmitter-receiver performance.

10.3.3.5   Network integration

A summary of the *network integration* activities is presented in Table 10.9

The network integration information for the case study is summarized in Table 10.10. It specifies the optical device configurations, such as number of ports, number of channels, amplifier gain and channel bit rates, etc.

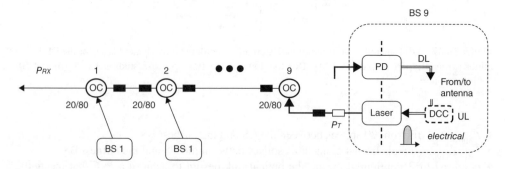

Figure 10.59   UL optical link design

Table 10.9 Summary of the network integration activities for each NE

Network Elements (NEs)	Network Integration Activities
• NE2: DL optical transmitters at the CS • NE7: Topology for the ODN • NE14: UL transmitter for BS	Enabling technologies collected from the optical link design stage. Those technologies that satisfied the functional network requirements.
• NE6: UL optical receiver • NE18: DL optical receiver • NE8: Fiber Type	Technologies obtained from the corresponding transmitter datasheet selected during the link design stage (NE2–NE18 and NE14–NE6).
• NE1: DL electrical multiplexing • NE13: UL electrical multiplexing	Defined depending whether single carrier or multiple carrier DL and UL channels were designed in the channel assignment and clustering stage.
• NE4: DL optical transmission band • NE16: UL optical transmission band	Defined primarily by the preferred transmission band specified in the datasheet of the transmitter-receiver technology selected. Secondarily, based on the wavelength allocation information, in case adjacent bands were included to support the required optical channels.
• NE3: DL optical multiplexer • NE15: UL optical multiplexer • NE5: UL optical demultiplexer • NE17: DL optical demultiplexer • NE9: DL optical DROP configuration • NE11: UL optical ADD configuration	Technologies obtained from the link design stage. Demultiplexing technologies are defined using design rules of compatibility.
• NE10. DL Optical Amplifier • NE12. UL Optical Amplifier	The amplification technology is recommended based on the execution of several design rules which use information such as: amplifier gain range, optical transmission band, and the number of optical channels supported.

10.3.4   Conclusions

This section presented a knowledge-based framework to support the design of 60 GHz RoF land networks for smart cities. The design framework is developed using design rule-based approach. The novelty and focus of the knowledge-based framework is on providing a general overview and tutorial for a network designer, and supporting the design of novel 60 GHz RoF land network architectures, by combining existing approaches and technologies.

10.4   Free Space Optical Communications

Prof. Dr. Zabih Ghassemlooy, Dr. Hoa Le Minh and Dr. Muhammad Ijaz
Northumbria University, Newcastle, UK

Free space optical communications (FSO) is an emerging technology that offers a large bandwidth for data, voice and video transmissions in mostly short- to medium-link distance

Table 10.10   Output of network integration for the example network design

Network Element	Enabling Technology	Network Design Outputs	
	Downlink (DL)		
NE1: Electrical Multiplexing	Subcarrier Multiplexing	9 Sub-Carriers	9 Gbps RoF channel
NE2: Optical Transmitter	Externally Modulated MLL	2 Transmitter	6 channels supported
NE3: Optical Multiplexer	50 GHz DWDM	12 Multiplexers	2×1 ports
NE4: Optical Transmission Band	C-Band	1530–1569 nm	
NE7: Network Topology	Star-Bus	12 links Star	9 BSs per bus
NE8: Fiber Type	SMF-28	1 fiber	
NE9: Optical DROP configuration	50 GHz DWDM	12 Ports to DROP	
NE10: Optical Amplifier	EDFA	1 amplifier	12 dB Gain
NE17: Optical Demultiplexer	80/20 optical coupler	108 Couplers	2 ports
NE18: Optical Receiver	mm-wave PD	108 Receivers	
	Uplink (UL)		
NE13: Electrical Multiplexer	OFDMA	9 Sub-Carriers	2.7 Gbps
NE14: Optical Transmitter	Directly modulated laser	108 UL Transmitters	
NE15: Optical Multiplexer	80/20 optical coupler	108 Couplers	2 ports
NE16: Optical Transmission Band	All optical fiber bands	1260–1611 nm	
NE7: Network Topology	Bus-Star	12 links Star	9 BSs per bus
NE8: Fiber Type	SMF-28	1 fiber	
NE11: Optical ADD configuration	CWDM multiplexer	12 CHs to ADD	
NE12: Optical Amplifier	Not Applicable	0	0
NE17: Optical Demultiplexer	CWDM demultiplexer	12 CHs	
NE18: Optical Receiver	Baseband optical PD	108 Receivers	

applications. FSO could be deployed as the primary, back-up and disaster recovery links offering a range of speed from 10 Mbits/s to beyond 10 Gbit/s [10.261]. Compared to the conventional radio frequency (RF) wireless and wired technologies, the line-of-sight (LOS) FSO links offer numerous advantages including:

- a wide unlicensed frequency spectrum capable of transmitting data in excess of hundreds of Gbit/s using the wavelength division multiplexing scheme
- a relatively lower power consumption
- security and immunity to the electromagnetic interference [10.262,10.263]

Nonetheless, outdoor LOS FSO links experience substantial optical signal losses due to adverse weather conditions, which severely degrade the link performance and the achievable range. The loss is mainly due to atmospheric absorption, scattering and temperature dependent scintillations [10.264,10.265]. The latter also results in pulse dispersion due to random varia-tion on the phase front, thus significantly affecting the FSO link performance. Scintillation can be thought of as changing intensities of light in both space and time at the receiver plane. The received optical signal level at the photodetector fluctuates due to changes in the temperature and the pressure of the propagation medium due to the random variation of the air refractive index along the beam propagation path. The timescale of intensity fluctuations is the same as the time it takes a volume of air, the size of the optical beam, to move across the beam path and therefore is related to wind speed. It is worth outlining that in indoor FSO links, only turbulence (scintillation) needs to be considered due to heating and air conditions.

Atmospheric turbulence has been investigated extensively and a number of theoretical models have been proposed to describe scintillation-induced fading [10.266–10.268]. In clear weather conditions, theoretical and experimental studies have shown that turbulence (i.e. scintillation) could severely affect FSO link reliability and disrupt connectivity [10.266,10.268,10.269]. Most research works reported are based on simulation and/or theoretical, with very little work on experimental measurement of turbulence. It is very challenging and time-consuming to carry out the turbulence measurement in an outdoor environment, particularly over a long transmission span. This is mainly due to the long waiting time to observe and experience reoccurrence of different atmospheric events, which sometimes can take weeks or months, as well as temperature measurement over a long FSO link.

In addition, in an outdoor environment, the weather effects could be due to a combination of atmospheric conditions (fog, rain, turbulence, smoke, dust, etc.). Therefore, carrying out a proper link assessment for specific weather conditions becomes a challenging task. Because of this, there is a need for measurement under controlled environmental conditions, where data can be collected at will. To address this problem, we have developed a dedicated indoor laboratory atmospheric chamber that can be used to assess the performance of the FSO link under a controlled environment [10.270,10.271]. The chamber offers the advantage of full system characterization and assessment, in much less time compared to the outdoor FSO link, where it could take a long time for the weather conditions to change, therefore prolonging the characterization and measurement.

The constitution of the real outdoor atmosphere (ROA), particularly aerosols (fog, smoke and dust) has similar particle size distributions to the wavelength of the propagating optical signal, which can potentially result in scattering and absorption of the visible and IR optical beams, thus degrading the FSO link performance and its availability [10.271,10.272]. As a result, the study of the relation between different aerosols and optical wavelength is important in order to characterize the link availability of FSO.

In [10.273], a laboratory-based set-up has been used to show that the terahertz (THz) signal displays significantly lower attenuation due to fog when compared to the FSO link operating at 1550 nm. In [10.274], a real-time measurement of fog-induced attenuation was reported, showing that the far infrared (FIR) at 10 μm offers higher transmission range in the presence of fog. Despite the advantages of FSO links, operating them at the THz and FIR wavelength bands will be costly due to components not readily available at the present time. However, this will change with time and therefore transmission at these bands may become common. Thus, almost all commercially available FSO systems operate in the wavelength range of 600–1550 nm. Consequently, the wavelengths from the visible to near infrared (NIR) bands are selected to experimentally verify the dependency of the wavelength on the fog and smoke attenuation in an indoor controlled laboratory environment.

Simultaneous investigation of the entire wavelength spectrum range (600 nm $< \lambda <$ 1600 nm) under real outdoor fog (ROF) conditions is challenging, and the research works reported explicitly lacks the verification of the wavelength dependent attenuation [10.275,10.276]. This is due to a number of reasons, mainly: *(i)* the unavailability of the experimental set-up for outdoor links due to the long observation time and reoccurrence of dense fog events for visibility $V < 0.5$ km, and *(ii)* the difficulty in controlling and characterizing aerosols in the atmosphere due to the inhomogeneous presence of aerosols along the FSO link path. Hence the use of an indoor atmospheric chamber, where aerosol and fogs can be used to test the FSO link performance by carrying out reparative measurements of the visibility and attenuation at any desired wavelengths. As for smoke, which is more common in urban areas, no experimental data is available for FSO links. Therefore, the atmospheric chamber could also readily be used to investigate the effect of smoke on FSO link performance.

This section will discuss the main challenges in the atmospheric channel, including turbulence and fog conditions.

10.4.1 FSO under Turbulence Conditions

10.4.1.1 Turbulence model and performance analysis

Outdoor FSO links experience substantial optical signal losses due to adverse weather conditions, which severely degrade link performance [10.264,10.265]. In a turbulence channel, the received optical signal level will fluctuate as a result of random variation of the air refractive index along the beam propagation path due to temperature and pressure. The refractive index fluctuation along the propagation path is given by:

$$n(r, t) = n_0 + n_t(r, t),$$ (10.15)

where:

r Radius of particle

The first and second terms are the average refractive index and the turbulence induced component due to spatial variation of pressure and temperature of the air, respectively. The most widely used parameter as a measure of the strength of scintillation is the refractive index (or atmospheric) structure parameter C_n^2. It is highly dependent on small-scale temperature fluctuations, the temperature structure constant, and the atmospheric pressure P and is given by [10.277]:

$$C_n^2 = \left(86 \times 10^{-6} \frac{P}{T^2}\right)^2 C_T^2, \quad \text{for } \lambda = 850 \text{ nm}$$ (10.16)

where:

p Atmospheric pressure in millibar
T Absolute temperature in Kelvin

C_n^2 can vary from 10^{-17} m$^{-2/3}$ up to 10^{-12} m$^{-2/3}$ for weak and strong turbulence regimes, respectively. The temperature structure constant C_T^2 is related to the universal 2/3 power law of temperature variation, as given in:

$$D_T = \langle (T_1 - T_2)^2 \rangle = \begin{cases} C_T^2 l_0^{-4/3} L_p^2 & \text{for } 0 \ll L_p \ll l_0 \\ C_T^2 L_p^{2/3} & \text{for } l_0 \ll L_p \ll L_0 \end{cases} \tag{10.17}$$

where T_1 and T_2 are the temperatures at two points separated by the propagation distance L_p. According to [10.277], the parameters l_0 and L_0 are inner and outer scales of small-temperature fluctuations.

Due to its nature, the turbulent media are extremely difficult to describe mathematically. The difficulty, according to [10.278], is primarily due to the presence of non-linear mixing of observable quantities, which are the fundamentals of the process. For the plane wave, generally applicable to laser beams propagating over long distances [10.278,10.279], the variation in C_n^2 can be readily used to predict the temporal log-amplitude variance σ_x^2, commonly referred to as the Rytov parameter, as given by [10.277]:

$$\sigma_x^2 = 0.56k^{7/6} \int_0^{L_p} C_n^2(x)(L_p - x)^{5/6} dx, \tag{10.18}$$

where:

$k = (2\pi/\lambda)$ Spatial wave number
λ Wavelength

Note that σ_x^2 is linearly proportional to C_n^2 and λ^{-1}.

For a field propagating horizontally through the turbulent medium, where C_n^2 is constant, the normalized variance (also known as the log irradiance variance or the scintillation index) for a plane wave is given by:

$$\sigma_I^2 = 1.23 C_n^2 k^{7/6} L_p^{11/6}. \tag{10.19}$$

The most commonly reported model for describing weak atmospheric turbulence is the log-normal distribution [10.279,10.280–10.282]. This is a well-known modeling approach and has been adopted in many calculations for the turbulence channel. As light propagates through a large number of elements within the atmosphere, each element induces independently random scattering and phase delay to the optical beam. The distribution of amplitude fluctuation is

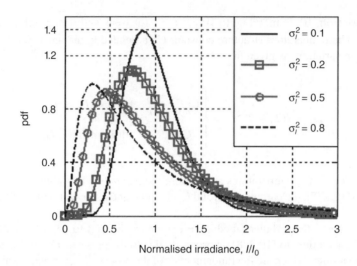

Figure 10.60 Intensity log-normal PDF against the normalized irradiance for a range of irradiance variance values

log-normal. Thus, the probability density function (PDF) of the received irradiance I due to the turbulence is derived by [10.278]:

$$p(I) = \frac{1}{\sqrt{2\pi\sigma_I^2}} \frac{1}{I} \exp\left\{-\frac{\left(ln\left(I/I_0\right) + \sigma_I^2/2\right)^2}{2\sigma_I^2}\right\} \qquad (10.20)$$

where:

I_0 Irradiance when there is no turbulence in the channel

Figure 10.60 shows plots of the log-normal PDF against the normalized irradiance and for a range of σ_I^2. For lower values of σ_I^2, the PDF distribution is nearly Gaussian centered about an I/I_0 value of 1. As σ_I^2 increases, the distribution is more skewed with a long tail toward the infinity, and reduced peak intensity as a result of signal fading.

For strong turbulence conditions, the Gamma-Gamma and negative exponential distribution models are used instead of the log-normal model [10.268]. In this section, the focus is on the weak turbulence, therefore the log-normal model is adopted. The normalized variance σ_I^2 of I detected at the receiver is defined as:

$$\sigma_I^2 = \frac{\langle I^2 \rangle}{\langle I \rangle^2} - 1 \qquad (10.21)$$

where:

$\langle . \rangle$ denotes the ensemble average

Note that Eq. (10.21) is based on the assumption that the photodetector (PD) is a single point. However, in real systems, the PD has a finite collection area. The coherence distance is another useful parameter that is adopted to describe turbulence-induced fading, which is defined as [10.277,10.283]:

$$\rho_0 = \left[1.45k^2 \int_0^R C_n^2(x)\,dx\right]^{-3/5} \tag{10.22}$$

The fading effect can be reduced significantly by means of aperture averaging, provided the receiver aperture $A_{rx} > \rho_0$. The aperture averaging factor is widely used to quantify the fading reduction and is given by:

$$A = \frac{\sigma_I^2(D)}{\sigma_I^2(0)} \tag{10.23}$$

where $\sigma_I^2(D)$ and $\sigma_I^2(0)$ denote the scintillation index for a receiver lens of diameter D and a "point receiver" ($D \approx 0$), respectively. For the plane wave propagation with a smaller inner scale $l_o \ll (L/k)^{0.5}$, the scintillation index is given by [10.277]:

$$\sigma_I^2(D) = \exp\left[\frac{0.49\sigma_I^2}{\left(1 + 0.65d^2 + 1.11\sigma_I^{\frac{12}{5}}\right)^{\frac{7}{6}}} + \frac{0.51\sigma_I^2\left(1 + 0.69\sigma_I^{\frac{12}{5}}\right)^{-\frac{5}{6}}}{1 + 0.90d^2 + 0.62d^2\sigma_I^{\frac{12}{5}}}\right] \tag{10.24}$$

where:

$d = \sqrt{kD^2/4L_p}$ is the scalar parameter

In the presence of turbulence, the fluctuated instantaneous irradiance leads to a variation in instantaneous signal-to-noise ratio (SNR). Here we use the average SNR to evaluate the FSO link performance. The ensemble mean of SNR can be expressed as [10.284]:

$$\langle SNR \rangle = \frac{SNR_0}{\sqrt{\sigma_I^2(D)(SNR_0)^2 + I_0/I}}, \tag{10.25}$$

where SNR_0 denotes the turbulence free link. With no turbulence, $\langle SNR \rangle = SNR_0$.

The bit error rate (BER) for the intensity modulation/direct detection (IM/DD) FSO link with on-off keying (OOK) under turbulent conditions is given by:

$$BER = Q(\sqrt{\text{SNR}}) \tag{10.26}$$

where $Q(.)$ is the Marcum's Q-function, which is the area under the Gaussian tail, given by:

$$Q(x) = \frac{1}{\sqrt{2\pi}} \int_x^\infty e^{-\alpha^2/2} d\alpha. \tag{10.27}$$

A number of methods can be used to combat the effect of turbulence, such as the multiple input multiple output (MIMO) system [10.285], the spatial diversity and the aperture averaging [10.286]. However, selecting a modulation format that is the most immune to the scintillation effect is also important [10.287–10.290].

10.4.2   System Set-up

The schematic diagram of the experimental LOS FSO link is shown in Figure 10.61. At the transmitter side, a narrow divergence beam laser source with a collimated lens is used. The input data, either Ethernet (10BASE-T) or Fast-Ethernet (100BASE-SX), is used to intensely modulate the light source. The channel is represented by an indoor laboratory atmospheric chamber with a dimension of $550 \times 30 \times 30$ cm^3. The chamber is composed of seven compartments, each having a vent for air circulation, and a hot-plate with built-in fans and thermometers. The temperature and the wind velocity within the chamber can be readily controlled to mimic outdoor atmospheric conditions as closely as possible. External fans can also be used to blow hot and cold air in the direction perpendicular to the optical beam propagating along the chamber to create variation in the temperature and the wind speed. The cold air temperature is kept at 20–25°C, whereas the hot air temperature could be in the range of 20–65°C. By using a series of air vents along the chamber, additional temperature control

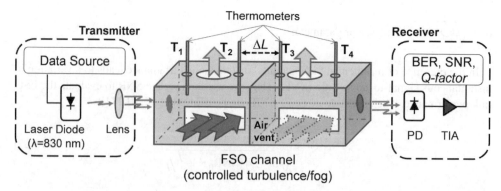

Figure 10.61   Block diagram of a turbulence chamber, which is excited by hot and cool air

Table 10.11   Parameters of the FSO link

Parameters		Values
Transmitter		
Data source	Ethernet line rate R_b	12.5 Mbit/s
	Fast-Ethernet line rate	125 Mbit/s
	Format	NRZ
	Line coder	4B5B
Laser diode	Peak wavelength	830 nm
	Maximum optical power	10 mW
	Class	IIIb
	Beam size at aperture	5 mm × 2 mm
	Beam divergence	5 mrad
	Laser beam propagation model	Plane
	Modulation bandwidth	75 MHz
Optical lens	Diameter	3.4 cm
	Focal length	20 cm
Channel	Dimension	$550 \times 30 \times 30$ cm^3
	Temperature range	20–65°C
	Temperature gradient $\Delta T/\Delta L$	7 K/m
	Wind speed	<10 m/s
	Link segment ΔR	1.5 m
	Rytov variance	<0.23
Receiver		
Photodetector	Spectral range of sensitivity	750–1100 nm
	Active area	1 mm^2
	Half angle field of view	±75 Deg
	Spectral sensitivity	0.59 A/W at 830 nm
	Rise and fall time	5 ns
Amplifier	Transimpedance amplifier	AD8015
	Bandwidth	240 MHz
	Transimpedance amplifier gain	15 kΩ
	Filter bandwidth	R_b

is achieved in order to maintain a constant temperature gradient ΔT/ΔL between both ends of the chamber, as required for creation of turbulence.

The receiver front-end consists of a PIN PD followed by a trans-impedance amplifier (TIA). The output of the TIA is captured using a high-frequency digital oscilloscope where full signal analysis (i.e. the eye-diagram, the Q-factor, and BER) are carried out. The main parameters for the proposed system are given in Table 10.11.

The strength of turbulence can also be controlled by placing a heating source near and/or away from the FSO transmitter, as illustrated by the ray tracing diagram in Figure 10.62. In Figure 10.62 it is observed that propagating optical beams will experience approximately the same degree of beam bending, since the same level of turbulence is generated within the chamber. However, due to the geometry of the chamber, a lesser amount of optical power will be collected at the receiver when the turbulence source is at the transmitting end (Figure 10.62b).

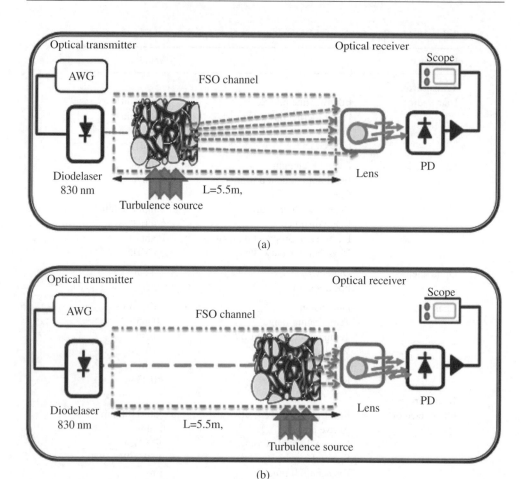

Figure 10.62 Sketch of diverted beams due to turbulence source positioned near: (a) the transmitter, and (b) the receiver.

10.4.3 System Performance under Weak Turbulence

In this section, the FSO link performance under the influence of weak atmospheric turbulence is outlined. For valid comparison of the FSO link performance compared with the outdoor FSO system, the atmospheric chamber is first characterized theoretically and experimentally prior to measurement of turbulence strength.

10.4.3.1 Channel characterization

In order to compensate for the short length of the chamber, $\Delta T/\Delta L \sim 7$ K/m, temperatures measured at each point along the chamber are maintained within the tolerance margin of ± 1 K. The average temperatures recorded to within $\pm 1^\circ$C at four different positions along the chamber are given in Table 10.12.

Table 10.12 Measured temperatures over five experiments at four positions within chamber

	T_1 (°C)	T_2 (°C)	T_3 (°C)	T_4 (°C)	C_n^2 (m$^{-2/3}$)
Set 1	28	25	23	23	1.40×10^{-12}
Set 2	33	28	24	23	4.61×10^{-12}
Set 3	37	30	24.2	23.4	9.05×10^{-12}
Set 4	45	34	25	24	2.20×10^{-11}
Set 5	54	40	27.3	25	3.97×10^{-11}

Using Eq. (10.15), the average values of C_n^2 against a range of temperature gradients within the chamber is plotted, as illustrated in Figure 10.63 and shown in Table 10.12. The average values of C_n^2 vary from 10^{-12} to 10^{-10} m$^{-2/3}$, which is in good agreement with the experimental data reported in [10.291].

In order to predict the scintillation index σ_I^2 using the indoor chamber, an assumption is made that C_n^2 is constant between two measured temperature points. As shown in Figure 10.63, the temperature is recorded at four points along the chamber, in order to predict σ_I^2 using Eqs (10.15), (10.16) and (10.18). Figure 10.64 depicts both the predicted and the measured σ_I^2. The measured values are obtained from the received signal using Eq. (10.20). Note that the difference between the plots is due to dependency of σ_I^2 on the temperature gradient. For example, ±1°C change in the temperate T1 can result in a ~2.5 time difference in σ_I^2.

Since the PD has the detection area of 1 mm^2, which is comparable to the transverse coherence distance ρ_0 of ~2.8 mm (calculated using Eq. (10.21)), then the aperture averaging factor A should be taken into consideration in the analysis. The factor σ_I^2, given in Eq. (10.18), is therefore multiplied by $A = 0.88$ (approximated using Eqs (10.22) and (10.23) for a plane wave propagation). In order to accurately predict the log-normal variance, the AWGN variance is subtracted from the total variance of the received signal and hence the measured variance is effectively the log-normal variance only.

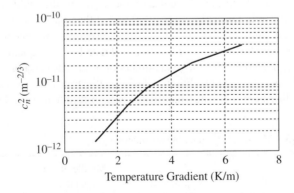

Figure 10.63 The average refractive index structure C_n^2 against the temperature gradient within a controlled atmospheric chamber

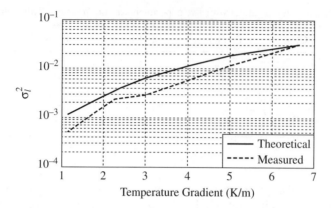

Figure 10.64 The theoretical and measured scintillation index σ_I^2 against the temperature gradient for the controlled atmospheric chamber

Note that for outdoor terrestrial FSO systems, C_n^2 is assumed to be constant and Eq. (10.18) is normally applied to approximate the log irradiance variance. However, in real systems, the temperate gradient $\Delta T/\Delta L$ can vary along the propagation length as atmospheric conditions greatly change over time and space. Therefore, the atmospheric link segmentation approach can be used as an alternative method to characterize the FSO link.

One advantage of using the indoor atmospheric chamber is that the σ_I^2 can be readily varied by changing the position of the turbulence source (Figure 10.62). The measured maximum values of σ_I^2 for the turbulence source located at different positions along the chamber are given in Table 10.13. As expected, σ_I^2 is the highest when the turbulence source is closer to the transmitting end.

10.4.3.2 Signal fluctuation histogram

In order to reduce the effect of ambient light, experiments should be carried out in total darkness. However, it is found that the noise variance due to the ambient light is $<3.35 \times 10^{-5}$, which is significantly lower (by up to 7000 times) than the log-normal variance, hence the effect of ambient light can be ignored in the presence of turbulence. Figures 10.65a and 10.65b–10.65d show histograms for the received data bit '1' sequence without and with turbulence, respectively. Notice that the total number of occurrences is normalized to the unity, which represents the occurrence PDF. Figure 10.65 also illustrates the curves fitting plots. With no turbulence the distribution is Gaussian, whereas with turbulence the PDF is a best fit to

Table 10.13 Measured σ_I^2 along the chamber length

Turbulence position	Near receiver	Middle of the chamber	Near transmitter
Scintillation index σ_I^2	0.009	0.067	0.240

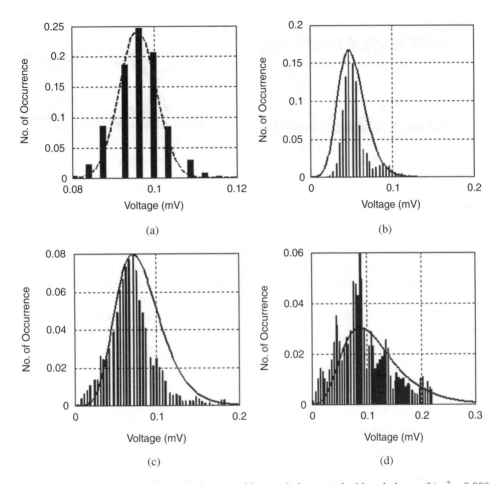

Figure 10.65   Received signal distribution: (a) without turbulence, and with turbulence, (b) $\sigma_I^2 = 0.009$, (c) $\sigma_I^2 = 0.067$, and (d) $\sigma_I^2 = 0.240$ (the curve fitting is shown by solid lines)

the log-normal distribution. As the measured values of σ_I^2, given in Table 10.13, falls within the range of [0, 1], the turbulence is classified as weak, therefore it is best modeled by the log-normal distribution model.

10.4.4   FSO Link Evaluation

In this section, the FSO Ethernet and Fast-Ethernet link availability and performance under different turbulence conditions will be evaluated. The measured *eye-diagrams* for the received Ethernet signal are depicted Figure 10.66a with no turbulence, and in Figure 10.66b with the weak turbulence. Note that the eye opening is smaller in the presence of turbulence, which results in a considerable level of signal intensity fluctuation. This is due to the substantial increase of the width at the top (bit 1) and base (bit 0) of the eye-diagram.

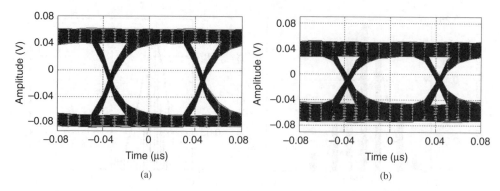

Figure 10.66 The measured eye-diagram of received Ethernet signal (NRZ 4B5B, line rate 12.5 Mbit/s) under conditions: (a) no turbulence, and (b) weak turbulence with the scintillation index $\sigma_I^2 = 0.0164$

Under the same turbulence conditions, the Fast-Ethernet signal (125 Mbit/s) is transmitted through the FSO channel and obtains the eye diagrams at the receiving end (Figure 10.67). Notice that with the increment in the turbulence level, the eye-opening reduces as in the case of the Ethernet link.

In order to quantify the FSO link performance, the Q-factor is measured from the received signal using:

$$Q = \frac{v_H - v_L}{\sigma_H - \sigma_L};$$ (10.28)

where:

v_H, v_L Mean received voltages
σ_H, σ_L Standard deviations for the 'high' and 'low' level signals, respectively

The measured Q-factor against a range of σ_I^2 values is shown in Figure 10.68 for both Ethernet transmission types. The results depict the corresponding predicted Q-factor values

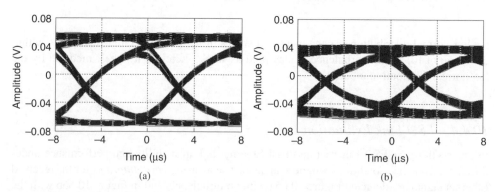

Figure 10.67 The measured eye-diagram of received Fast-Ethernet signal (NRZ 4B5B, line rate 125Mbit/s) under conditions: (a) no turbulence, and (b) weak turbulence with the scintillation index $\sigma_I^2 = 0.0164$

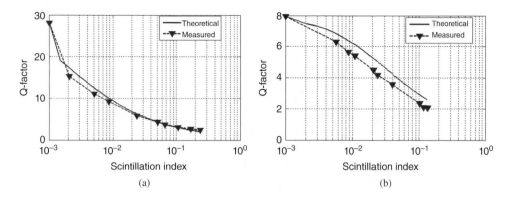

Figure 10.68 The predicted and measured Q-factor against different scintillation index values for (a) Ethernet, and (b) Fast-Ethernet

in the figure. For the clarity of the figure, the x-axis is in the logarithmic scale. Therefore, the initial value of σ_I^2 is selected to be 10^{-3} rather than zero for obtaining Q_0.

Figure 10.68 presents a good match between the predicted and experimental data of the Q-factor, as the difference between them is less than 0.5. The Q-factor decreases exponentially with the increase of scintillation index values in both Ethernet and Fast-Ethernet communications links. In fact, the drop in the Q-factor does not depend on the data rates but on Q_0 and values of the scintillation index.

Although Q_0 values are about 28 and 8 for Ethernet and Fast-Ethernet links, respectively, the obtained Q-factors show that they are almost identical as $\sigma_I^2 > 0.05$. This indicates that the increment in transmitted optical power will have little effect in mitigating the increased turbulence level. A number of techniques have been proposed in the literature to deal with the turbulence, including the aperture averaging, the spatial diversity and the cooperative diversity [10.267,10.292,10.293], which are beyond the scope of this chapter.

The evaluation of BER for both Ethernet/Fast-Ethernet FSO links under different turbulence conditions is depicted in Figure 10.69. In this experimental evaluation, Ethernet and Fast-Ethernet links show highly sensitivity to the strength of the turbulence. Observations have shown that for Ethernet links (using TCP/IP), link availability becomes a problem under a strong turbulence regime, that is, the connection is dropped.

10.4.5 Relation to Outdoor FSO Link

In Eq. (10.18), σ_I^2 is shown to be dependent on C_n^2 and the temperature gradient. Assuming a constant C_n^2 over a short propagation span of ΔL_{indoor} and $\Delta L_{outdoor}$ for indoor and outdoor FSO links, respectively, the relation $R_{in/out}$ between indoor experimental and outdoor links can be derived as:

$$R_{\text{in/out}} = \frac{\sigma_{I_{m_{outdoor}}}^2}{\sigma_{I_{m_{indoor}}}^2} = \frac{C_{n_{outdoor}}^2}{C_{n_{indoor}}^2} \times \left(\frac{\Delta L_{outdoor}}{\Delta L_{indoor}}\right)^{11/6} \tag{10.29}$$

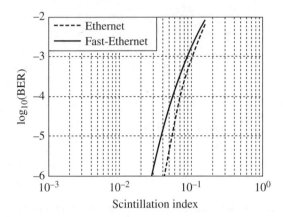

Figure 10.69 BER performance against a range of scintillation indexes for Ethernet and Fast-Ethernet FSO links (note that BER values below 10^{-6} are truncated to that level)

Note that the FSO link segmentation is used in both indoor and outdoor environments to keep the temperature gradient constant. For the outdoor link, $C^2_{n_{outdoor}}$ is much smaller than for the indoor link. The typical range for $C^2_{n_{outdoor}}$ is 10^{-16} to 10^{-14} m$^{-2/3}$ for weak turbulence [10.277]. For the outdoor FSO system operating at sea level (minimum turbulence level), the typical $C^2_{n_{outdoor}}$ is about 10^{-14} m$^{-2/3}$, whereas our measured $C^2_{n_{indoor}}$ is approximately 4×10^{-11} (Figure 10.63). As we would like to calibrate the FSO link performance to make it similar to outdoor cases, $R_{in/out}$ should be unity. From Eq. (10.28) we can therefore determine that the overall outdoor FSO link range is 550 m, which induces the same weak turbulence effect as our proposed system. By generating a higher temperature gradient in the indoor experimental FSO link, we can achieve the same performance as the longer outdoor FSO link.

10.4.6 FSO under Fog Conditions

The dependence of the FSO transmission wavelength on fog and smoke is considered an important parameter in order to achieve the maximum transmission span. However, selecting the best wavelength for FSO under dense fog (visibility $V < 0.5$ km) is still an open issue that needs addressing [10.294–10.296]. In [10.294,10.297], different models have been studied to predict the fog attenuation as a function of measured V(km) and the operating wavelengths. Kim et al. (2001) showed that fog attenuation is independent of the wavelength for dense fog conditions of $V < 500$ m [10.294]. However, the experimental data reported for the real atmospheric fog set-up has shown that there is a significant difference between fog attenuation using 830 and 1550 nm wavelengths [10.298,10.299].

Moreover, most reported experimental data lacks the prediction of fog attenuation at the lower visibility range ($V < 100$ m). Furthermore, the empirical approach to model real fog attenuation is based on the simultaneous measurement of V and fog. However, the FSO link length can be greater than 1 km in outdoor applications. In this case, due to the spatial heterogeneity, fog densities can vary from one position to a few hundred meters apart. Thus, the measured fog attenuation along the FSO link and the corresponding visibility data using a visibility device at a fixed position can significantly fluctuate from the actual value. This effect has been observed in an experimental system at 830 nm for two fog events, as in [10.299]. However, there is still a need to measure and characterize fog attenuation as a function of V along the propagation path rather than at fixed positions.

10.4.6.1 Modeling of fog attenuation

Fog is composed of very fine spherical water particles of various sizes suspended in the air, which results in light attenuation due to *Mie scattering* [10.300]. Fog particles reduce the visibility V near the ground and the meteorological definition of fog is when V drops to nearly 1 km [10.301]. Assuming fog particles have spherical shapes, we can apply the exact Mie theory to measure the scattering cross-section C_s of the particle by knowing the particle radius r. Thus, we can estimate the theoretical value of the normalized scattering efficiency Q_s as [10.302]:

$$Q_S = \frac{C_S}{\pi r^2} \tag{10.30}$$

The total attenuation induced by particles in the atmosphere is the sum of molecular absorption and scattering of light. However, wavelengths used in FSO links are almost selected in the atmospheric transmission windows where the molecular absorption cross-section C_a due to gases is negligible, such that $C_a \sim 0$ [10.302]. The atmospheric attenuation coefficient β_λ of the optical signal due to scattering of fog particles is given by [10.302,10.303]:

$$\beta_\lambda = \int_0^\infty \pi r^2 Q_S \left(\frac{2\pi r}{\lambda}, n' \right) N(r) dr \tag{10.31}$$

where:

n'	Real part of the refractive index
$2\pi r/\lambda$	Size parameter
$N(r)$	Particle size distribution function

Generally this distribution is represented by analytical functions such as the modified gamma distribution for aerosols and given by [10.303]:

$$N(r) = ar^\alpha \exp(-br) \tag{10.32}$$

where:

a, b and α Parameters that characterise the particle size distribution

Q_s is mainly dependent on the size parameter, which is dependent on λ and r. The size parameter is the ratio of the size of the fog particle to the incident wavelength. Therefore, the resultant fog attenuation will be remarkably dependent on the selected λ. Generally, due to the complexity involved in the physical properties of fog, like particle size and the non-availability of particle distribution, the fog-induced attenuation of the optical signal can be predicted using empirical models [10.304,10.305]. Empirical models use the visibility data in order to estimate the fog induced attenuation. The original empirical relationship, which relates V with the *fog attenuation*, has been given by the Kruse model [10.304]:

$$V \text{ (km)} = \frac{10 \log_{10} T_{th}}{\beta_\lambda} \left(\frac{\lambda}{\lambda_0} \right)^{-q} \tag{10.33}$$

where:

T_{th} Visual threshold taken as 2%
λ_0 Maximum spectrum of the solar band
q Coefficient related to the particle size distribution in the atmosphere

The Kruse model estimates the fog attenuation from visible to NIR wavelengths and the q value, which is a function of V, is defined as:

$$q = 0.585 V^{1/3} \quad \text{for } V < 6 \text{ km} \tag{10.34}$$

However, the estimation of the fog attenuation using the Kruse model is considered to be inaccurate for fog [10.295]. This is because the value of q has been defined from haze particles ($V > 1$ km) present in the atmosphere rather than from fog ($V = 1$ km). Kim modified the Kruse model using theoretical assumptions for fog by defining the q values as follows [10.294]:

$$q = \begin{cases} 1.6 & \text{for } V > 50 \text{ km} \\ 1.3 & \text{for } 6 < V < 50 \text{ km} \\ 0.16V + 0.34 & \text{for } 1 < V < 6 \text{ km} \\ V - 0.5 & \text{for } 0.5 < V < 1 \text{ km} \\ 0 & \text{for } V < 0.5 \text{ km} \end{cases} \tag{10.35}$$

Values of q indicate that β_λ is wavelength independent under fog conditions of $V < 0.5$ km. However, recently accrued experimental data at selective wavelengths of 0.83 and 1.55 μm has revealed a different behavior than the one predicted by the Kim model, especially for dense fog conditions of $V < 0.5$ km [10.305,10.308]. This demonstrates that in spite of a significant number of investigations, the model presented in Eq. (10.34) needs to be explicitly verified experimentally, not for the selective or specific wavelength, but for the entire spectrum of the visible to NIR range.

10.4.7   Characterization of Fog and Smoke Attenuation in a Laboratory Chamber

In [10.276], measurements show that the occurrence of fog starts when the relative humidity, H of the real outdoor atmosphere (ROA) approaches 80%. The density of the resulting fog reaches 0.5 mg/cm³ for $H > 95\%$. Thus, under high water vapor concentration conditions, the water condenses into tiny water droplets of radius 1–20 µm in the atmosphere. There are different types of real outdoor fog (ROF), which are categorized on the basis of their formation mechanism, such as convection fog, advection fog, precipitation fog, valley fog and steam fog [10.306]. Steam fog is localized and created by the cold air passing over much warmer water or moist land [10.306,10.307]. It is possible to simulate this form of fog in the lab by achieving H close to 95%. Hence, to mimic ROF, artificial fog can be generated by means of water-based steam.

In addition, to demonstrate the physical similarity of the lab-based fog to ROF, the mean ROF attenuation data from Prague for $0 < V < 1$ km published in [10.308] and Metrological Institute, Czech Republic for $V < 0.6$ km [10.305] were compared with the measured attenuation data for the lab-generated fog at a wavelength of 0.83 µm, and showed very good agreement. Thus, this confirms that laboratory-generated fog resembles the ROF. Smoke is generally formed in ROA from the combustion of different substances such as carbon, glycerol and household emissions [10.307]. Smoke is generated in the laboratory-based atmospheric chamber by imitating the natural process using a smoke machine, in which a glycerine-based liquid is used to produce dry smoke particles. Further details on the inverse relationship between refractive index n and wavelength from visible to NIR for water and glycerine is given in [10.309,10.310].

The link visibility (i.e. the meteorological visual range) is used to characterize fog and smoke attenuation. Visibility is defined in [10.311] as the distance to an object at which the visual contrast of the object drops to 2% of the original visual contrast (100%) along the propagation path commonly known as the Koschmieder law. This 2% drop value is known as the visible threshold T_{th} of the atmospheric propagation path. The 2% T_{th} value is adopted here in order to follow the Koschmieder law as opposed to the airport consideration of $T_{th} = 5\%$ [10.312]. The meteorological $V(km)$ can therefore be expressed in terms of β_λ and T_{th} at a λ of 0.55 µm and is given as:

$$V = -\frac{10\log_{10}(T_{th})}{\beta_\lambda} \qquad (10.36)$$

where β_λ is normally expressed in dB/km, and is mathematically defined by knowing the transmittance T of the optical signal and the propagation distance $L(km)$ using the Beer-Lambert law as [10.312,10.313]:

$$\beta_\lambda = -\frac{10\log_{10}(T)}{4.343L}. \qquad (10.37)$$

10.4.8   Fog and Smoke Channel – Experiment Set-up

A block diagram of a laboratory test bed for the FSO link, composed of an optical transmitter end T_x, an optical receiver R_x, an atmospheric chamber and optical and electrical modules, is shown in Figure 10.70a. The inset in Figure 10.70b shows the scattering of the optical signal at 0.55 µm in the presence of dense fog. Two experimental approaches have been adopted to characterize *fog and smoke attenuation*: (i) a continuous LS-1 tungsten halogen source with a broad spectrum (0.36–2.5 µm) and an Anritsu MS9001B1 optical spectrum analyzer (OSA) with a spectral response of 0.6–1.75 µm to capture the attenuation profile, and (ii) a number of laser sources at wavelengths of 0.55, 0.67, 0.83, 1.31 and 1.55 µm with the average transmitted

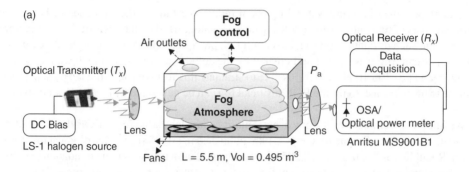

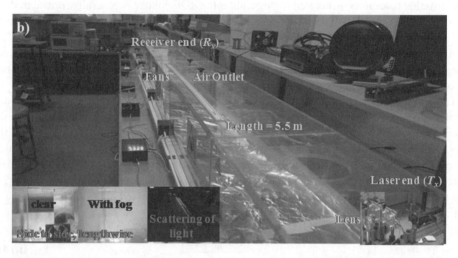

Figure 10.70 (a) The experimental set-up to measure fog attenuation and visibility and (b) the laboratory controlled atmospheric chamber and FSO link set-up. The inset shows the presence of fog and the scattering of light in the atmospheric chamber

optical powers P_T of −3.0 dBm, 0 dBm, 10 dBm, 6.0 dBm and 6.5 dBm, and an optical power meter, respectively.

The amount of aerosols in the atmospheric chamber is controlled by a number of fans and a ventilation system. The aerosols within the chamber are very fine and light, slowly moving particles suspended within the chamber. The time duration of 30 seconds was allowed for fog/smoke particles to settle down homogeneously within the chamber before data acquisition (DAQ). An automatic DAQ system is developed using the LabVIEW to control the MS9001B1 OSA. This process allows control of the wavelengths under the test, the sampling frequency and optical loss estimation. In order to measure the effect of fog on different wavelengths, the average received optical power P_R is measured at the R_x before and after the injection of fog into the atmospheric chamber. Fog density is varied by small outlets in the chamber so that dense to very light fog conditions can readily be created and the optical power measurements are taken at one second time intervals until the chamber is fog free. The normalized transmittance T was calculated from P_R with and without fog.

The value of β_λ is measured using Eq. (10.31), corresponding to the measured T from light to dense fog conditions for all wavelengths. The link visibility was measured simultaneously with β_λ along the length of the chamber using a laser diode at 0.55 µm to ensure the maximum transmission to the human eye.

Note that the goal of the experiment is to characterize the attenuation, therefore having identical powers at different wavelengths are not essential. The geometric and other losses were also not taken into account for T_x, as P_R was measured both before and after fog and smoke at R_x to attain the wavelength dependent losses.

10.4.9 Results and Discussion

In this section, the experimental results were reported with the proposed wavelength dependent model for fog and smoke attenuation. The atmospheric chamber allowed us to control and replicate the same atmospheric conditions; therefore the procedure was repeated 10 times using the same fog and smoke conditions.

10.4.9.1 Experimental attenuation measurements for fog and smoke attenuations

The logarithmic plot of the measured average *fog attenuation* (in dB/km) against the measured V for 670 nm and 1550 nm wavelengths are shown in Figure 10.71a. The measured attenuation is notably higher for 670 nm than at 1550 nm for the given visibility range (0.032 km < V < 1 km). There is an attenuation difference of 50 dB/km, 10dB/km and 7 dB/km at V of 0.048 km, 0.103 km and 0.5 km, respectively. This contradicts the wavelength independency of the Kim model for V < 0.5 km [10.294]. Furthermore, log-log plots of the measured attenuation of smoke (in dB/km) against the visibility for 830 nm and 1550 nm wavelengths are depicted in Figure 10.71b. The experimental result clearly demonstrates the dependency of λ on the resultant smoke attenuation, even if V is below 0.5 km. In general, the smoke attenuation difference is more than the fog attenuation. The difference for the smoke attenuation values are 108 dB/km, 23 dB/km and 8 dB/km for 830 nm and 1550 nm at V of 0.07 km, 0.25 km and 0.5 km, respectively, as illustrated in Figure 10.71b (inset). This depicts that the selection of 1500 nm is more favorable in the fog and smoke channels for the dense (V < 0.07 km), thick (V = 0.25 m) and the moderate fog and smoke (V = 0.5 km) conditions, respectively.

10.4.9.2 Empirical modeling of fog and smoke

The attenuation due to the scattering and absorption is highly dependent on the size parameter (Eq. (10.35)). However, Kim had also defined the value of q for fog in (10.20), which describes the wavelength dependency of the fog attenuation and the type of scattering. Values of q are −4, −1.6 and 0 for Rayleigh scattering ($r \ll \lambda$), Mie scattering ($r \sim \lambda$), and geometric scattering ($r \gg \lambda$), respectively [10.294]. In order to predict a suitable model for the attenuation of fog and smoke, based on measured data, we carried out the following. Values of q in Eq. (10.35) were obtained for individual wavelengths from 600–1600 nm, respectively, by using the empirical curve fitting method with a reference wavelength of 550 nm (Table 10.14). The values of the root mean square error (RMSE) and R^2 confirm that the curve fit is in good correlation

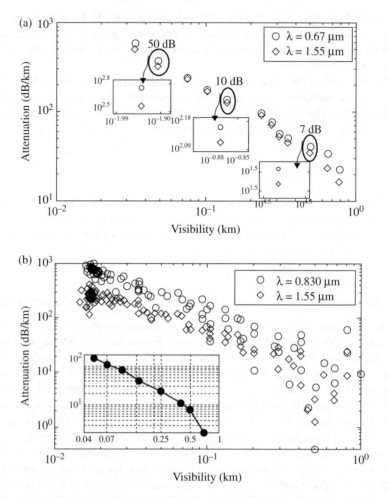

Figure 10.71   The measured attenuation (dB/km) and visibility (km): (a) fog and (b) smoke

with the measured data for fog and smoke. The value of q is found to be 0–0.14 for fog and 0–0.6 for smoke, indicating a predominance of Mie scattering ($r \sim \lambda$). This verifies that in the ($r \sim \lambda$) region for dense fog conditions, the q value is the function of λ, but not V.

The plot of predicted q values against the wavelength and the curve of best-fit against a wavelength range of 600–1600 nm for fog and smoke conditions are shown in the Figures 10.72 and 10.73, respectively. The best curve fit satisfies Eq. (10.23) with R^2 and RMSE values of 0.9732 and 0.0076 for the fog, and similarly for the smoke with R^2 and RMSE values of 0.9797 and 0.0497, respectively.

$$q(\lambda) = \begin{cases} 0.1428\lambda - 0.0947 & \text{Fog} \\ 0.8467\lambda - 0.5212 & \text{Smoke} \end{cases} \tag{10.38}$$

Table 10.14 Values of q obtained for different wavelength from measured fog and smoke attenuation data

Wavelength-µm	q	R^2-value	RMSE
		For Fog	
0.6	0.002	0.9670	0.1950
0.8	0.020	0.9850	0.1400
0.9	0.030	0.9846	0.1470
1	0.045	0.9827	0.1560
1.1	0.050	0.9813	0.1620
1.2	0.070	0.9803	0.1590
1.3	0.093	0.9802	0.1690
1.4	0.105	0.9800	0.1750
1.5	0.130	0.9760	0.1760
1.6	0.135	0.9751	0.1890
		For Smoke	
0.55	0	0.9985	0.0467
0.67	0.100	0.9680	0.2000
0.83	0.180	0.9220	0.3100
1.31	0.580	0.9121	0.2100

The new proposed model for the prediction of the attenuation due to fog and smoke can be presented as:

$$\beta_\lambda(dB/km) = \frac{17}{V\,(km)} \left(\frac{\lambda}{\lambda_o} \right)^{-q(\lambda)} \tag{10.39}$$

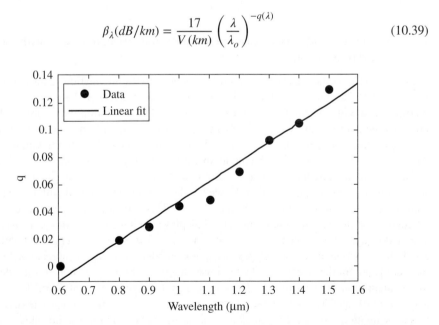

Figure 10.72 The predicted q value and linear curve of best-fit against wavelength for fog

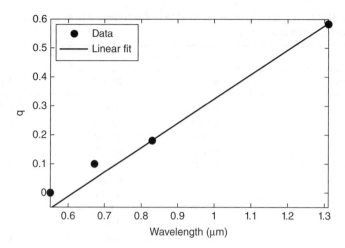

Figure 10.73   The predicted q value and linear curve of best-fit against wavelength for smoke

where, the wavelength with a range from 550 nm $< \lambda <$ 1600 nm is valid for the visibility range of 0.015 km $< V <$ 1 km. The function for $q(\lambda)$ for the fog and smoke is expressed in Eq. (10.39). Experimental data shows that as $V \rightarrow 0$ for very dense fog conditions, the received optical signal is significantly lower than the OSA minimum sensitivity at V of 0.0135 km. This validates that the measurements below 0.015 km are not very practical in showing the wavelength dependent fog attenuation, thus validating the model for the visibility range of 0.015 km $< V <$ 1 km.

10.4.9.3   Comparison of the measured visibility and attenuation data with the proposed model

Figure 10.74 shows the logarithmic plot for the measured fog attenuation against the concurrent visibility data for the selected wavelengths of 670, 830, 1100, 1310 and 1550 nm for the modified and Kim fog models. The log-log plot of the attenuation curve obtained from the proposed model defined by Eq. (10.38) and the comparison with the measured data shows a good agreement. However, comparison of the Kim model at λ of 1550 nm with the measured data for $V <$ 0.5 km shows that the model overestimates fog attenuation.

This is explained by the fact that the Kim model does not take into account the wavelength when estimating the fog attenuation. However, the Kim model fits well with the experimental data for $V >$ 0.5 km (see inset in Figure 10.74). This indicates the dependency of the fog attenuation on the wavelength. Figure 10.75 displays the log-log plot for the measured smoke attenuation against the concurrent visibility data for the selected wavelengths of 550, 670, 830 and 1310 nm. Comparison of the proposed smoke model shows a good agreement between the measured data for the selective individual wavelengths. This clearly indicates the dependency of attenuation on the wavelength. The plot shows that a difference of 50 dB/km is observed between 1310 and 830 nm for dense smoke condition ($V <$ 0.07 km), with progressive reduction in the attenuation difference for thick and moderate smoke conditions, thus clearly indicating suitability of the NIR wavelengths under dense smoke condition.

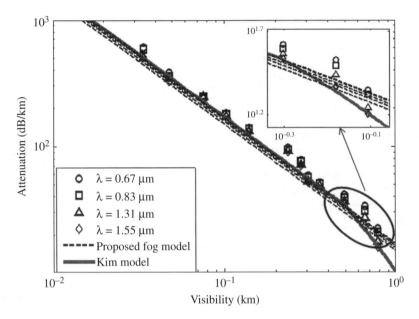

Figure 10.74 Real-time measured fog attenuation vs. visibility ($V = 1$ km) and curves of modified fog and Kim models for different wavelengths

The proposed model, where the available visibility data needs to be considered instead of the liquid water content (LWC) and the particle radius, is a simple approach using Eqs (10.38) and (10.39). It is most suitable for the FSO link budget analysis in an urban area, where fog and smoke are more likely to occur all year round.

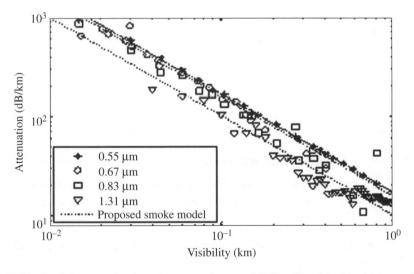

Figure 10.75 Real-time measured smoke attenuation vs. visibility ($V = 1$ km) and curves of smoke model for different wavelengths

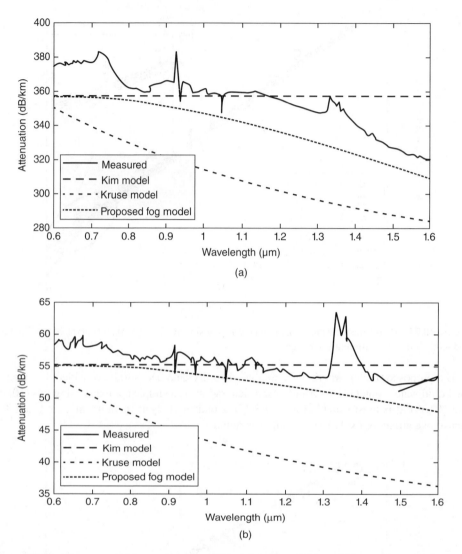

Figure 10.76 The measured fog attenuation (dB/km) from the visible to NIR spectrum and the comparison with the selected empirical models: a) for $V = 0.048$ km, and b) for $V = 0.3$ km

10.4.9.4 Comparison of the proposed model with the spectrum attenuation

The measured *fog attenuation* (in dB/km) for very dense fog ($V \sim 0.05$ km) in the visible to NIR spectrum under test (SUT) is shown in Figure 10.76a. The measured attenuation for the SUT at very dense fog condition ($V \sim 0.05$ km) shows three possible attenuation windows:

1. *600–850 nm*: which has an attenuation range of 375–361 dB/km, with the peak attenuation of 382.4 dB/km at 720 nm
2. *850–1000 nm*: showing a lower attenuation (360 dB/km) at 830 nm than at 925 nm, with a peak attenuation of 383.6 (dB/km). However, at 940 nm, the attenuation has a lower peak value of 354 dB/km

Table 10.15   The suitable possible wavelengths to operate in a foggy channel,
measured at V = 0.048 km

Wavelength windows (μm)	Window attenuation (dB/km)	Peak attenuation (dB/km)	Suitable wavelengths (μm)
0.6–0.85	375–361	382.4	0.830
0.85–1.0	361–360	383.3	0.940
1.0–1.55	360–323	357.0	1.55

3. *1000–1550 μm*: with an attenuation range of 360–323 dB/km. Results show that 1330 nm has a higher attenuation peak of 357 dB/km than the 1050 nm with an attenuation dip of 347.6 dB/km.

However, 1550 nm has the lowest attenuation of 324 dB/km in very dense fog at $V = 0.05$ km. Table 10.15 shows the possible wavelengths with the minimum fog attenuation suitable for use in FSO links. The behavior of the attenuation spectrum is almost the same for $V = 0.3$ km with the maximum attenuation of ~59.5–53.5 dB/km for 0.6 and 1.55 μm, respectively (Figure 10.76b).

The proposed model is verified experimentally for SUT using the controlled indoor chamber for fog and smoke. In order to relate the proposed model with the outdoor FSO channel, we have selected Kim and Kruse models, which are widely used in the literature. The Kim model at $V \sim 0.05$ km and 0.3 km shows that fog attenuation for SUT is wavelength-independent, contradicting the experimental data. The Kruse model underestimates fog attenuation at $V \sim 0.05$ km and 0.3 km for SUT (Figures 10.76a and b). However, the new proposed fog model shows a close correlation for SUT. This verifies that the proposed model estimates fog attenuation more accurately than both of the Kim and Kruse models for $V < 0.5$ km. This verifies that the proposed model follows the profile of measured fog attenuation more precisely than both of the Kim and Kruse models for $V < 0.5$ km. The RMSE values for Kim, Kruse and the new proposed fog models are 2.3473, 13.3434 and 3.8009 from the measured attenuation; and the standard deviation (SD) values are 2.3048, 3.7834 and 2.0378, respectively. The RMSE and SD values of the attenuation spectrum from the visible to NIR range shows a better agreement compared to RMSE and SD of published data for ROF at individual wavelengths [10.299, 10.314].

In the case of smoke, the measured *smoke attenuation* (in dB/km) at $V \sim 0.185$ km and 0.245 km for SUT is depicted in Figure 10.77. The resultant smoke attenuation at $V = 0.185$ km is almost 90 dB/km in the visible range and drops to 43 dB/km in the NIR range of SUT (Figure 10.77a). A similar behavior of smoke attenuation is observed at $V = 0.245$ km with the attenuation of 70 dB/km in the visible range decreasing to 33 dB/km in the NIR range of SUT (Figure 10.77b). The new proposed smoke model is also compared to the Kim and Kruse models for smoke attenuation. The Kim model overestimates the measured smoke attenuation and shows wavelength independent attenuation for $V < 0.5$ km. However, the Kruse model underestimates the smoke attenuation for 700 nm $< \lambda <$ 1000 nm and overestimates the smoke attenuation for 1100 nm $< \lambda <$ 1600 nm (Figures 10.77a and 10.77b). The new proposed model fits the experimental data, showing a close correlation with the measured smoke attenuation spectrum, thus verifying the validity of the proposed model for smoke conditions in the visible

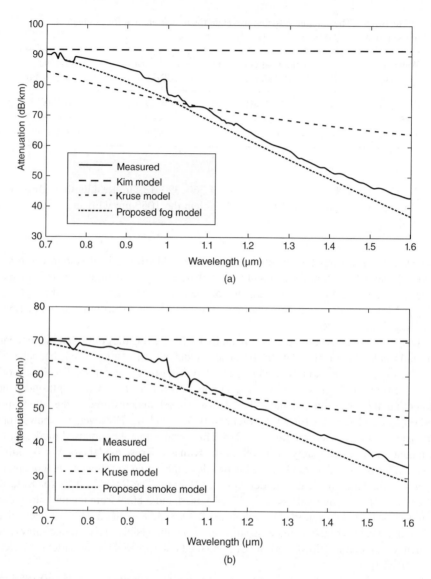

Figure 10.77 The measured smoke attenuation (dB/km) from the visible to NIR spectrum and the comparison with the selected empirical models: a) for $V = 0.185$ km, and b) for $V = 0.245$ km

to NIR SUT range. The model is better than the others in predicting smoke attenuation, but different in predicting fog attenuation.

10.4.10 Conclusions

This section discussed the effects of turbulence on the FSO link using Ethernet and Fast-Ethernet operation under controlled conditions in an indoor atmospheric chamber. This section has also presented, characterized and calibrated the chamber model, as theoretically and verified experimentally. Methods to generate turbulence and control its levels were discussed and practically implemented. Also discussed was the relation between the experimental environment and the outdoor FSO link, in order to ensure total reciprocity. The obtained results show that turbulence can severely affect the link availability of the Ethernet and Fast-Ethernet FSO systems, due to the sharp response in BER performance of the link and the nature of Ethernet (TCP/IP) connectivity. The analysis and demonstration presented in this chapter have shown a step forward to characterize the atmospheric channel in an indoor environment, which is by nature complex and unpredictable.

In addition, this section demonstrated the impact of fog and smoke on the FSO link performance using the continuous wavelength spectrum range of 600 nm $< \lambda <$ 1600 nm. A wavelength dependent model has been proposed for fog and smoke channels, which is valid in the visible to NIR range for the visibility range of 1 km. It was experimentally demonstrated that the most robust wavelengths windows (830, 940 and 1550 nm) could be adopted for fog conditions in order to minimize the FSO link failure. Furthermore, to validate the behavior of the proposed empirical model for selected wavelengths, this section has experimentally compared the continuous attenuation spectrum for the same fog and smoke conditions and found that the attenuation is almost linearly decreasing in both cases. The proposed model is calibrated for outdoor FSO links using the Kim model, thus validating the laboratory-based empirical model.

10.5 WLAN Systems and Fiber Networks

Dr. Riccardo Scopigno and Daniele Brevi
Istituto Superiore Mario Boella (ISMB), Torino, Italy

The purpose of this section is to present a promising family of network architectures which integrate optical fibers and Wi-Fi. Since this book mainly focuses on optical fibers, the first subsections (up to Chapter 10.5.2) are meant to provide the reader with the necessary background on wireless, before entering into a discussion on network solutions (Chapter 10.5.3).

To start from the very beginning, Wireless Local Area Networks (hence the acronym WLANs) are meant to provide local connectivity and are aimed at the rapid provision of nomadic (i.e. not under heavy mobility) and multiple-access (i.e. multi-user) connectivity to a network.

With some exceptions (such as their adaptation to the so-called "white spaces"), WLANs usually do not require a license, that is, they are usually set in the unlicensed frequency ranges of the radio spectrum: among them, the ISM[1] (industrial, scientific and medical) and U-NII[2] band are the WLAN quintessential ones.

---

[1] ISM bands are internationally reserved for industrial, scientific and medical purposes other than telecommunications. The ISM bands are defined by the ITU-R in the Radio Regulations (RR) 5.138, 5.150, and 5.280 [10.321]. The ISM band includes several channels spanning over a wide range (from 6 MHz to 244 GHz): individual countries are subject to slightly different rules, due to national radio regulations (in the United States of America, the usage of ISM bands is governed by FCC rules, while in Europe, the ETSI is responsible). Applications in these bands comprise, for instance: radio-frequency process heating, microwave ovens, and medical diathermy machines. Despite the initial intent, the ever growing use of these bands is that of short-range, low-power communications systems: cordless phones, Bluetooth devices, near field communication (NFC) devices, and WLANs.

[2] The Unlicensed National Information Infrastructure (U-NII) is an FCC [10.322] regulatory domain which defines four ranges in the 5 GHz band: 2 of them are adopted by IEEE 802.11: the 5.25-5.35 GHz (for indoor use) and the 5.47-5.725 GHz (both for indoor and outdoor). Also in the case of U-NII, national rules may restrict the number of available channels.

This implies that WLANs are far from being exempt from interference, either by other wireless networks (WLANs, WPANs, etc.) or by other devices emitting in the same range (including, for instance, radars, microwave ovens, garage door openers and toys). For this reason, WLANs are improving their cognitive and self-configuring features, so as to automatically detect possible interfering sources and to self-reconfigure the network, preventing disturbances as much as possible. This is the case for functions such as the Dynamic Frequency Selection (DFS) – for radar avoidance – and the Transmit Power Control (TPC) – for limiting mutual interference between WLAN networks.

Another differentiator of WLAN is broadband capacity: even if WLANs are used for manifold applications, their use par excellence remains the support of data connections, especially for browsing. This entails high data rates and, on the other hand, loosens the requirements on determining transmissions: in most cases, it is acceptable that some frames get lost and are re-transmitted or that they may undergo some transmission delays.

All in all, WLANs are distinctively characterized by being broadband- but not QoS-capable, Quality of Service, nomadic and local, prone to interference, but smart and reactive.

For the sake of precision, in the past some alternatives were proposed for WLANs (proprietary protocols or dismissed ones, as HyperLAN [10.315] or HyperLAN/2 [10.316–10.318]). However, currently, when we talk about WLANs, we inherently refer to the media access control (MAC) and physical layer (PHY) specified by the IEEE LAN/MAN Committee (IEEE 802) in the IEEE 802.11 standard [10.319]: they represent the worldwide utmost WLAN solutions and are commercially also known as *Wireless Fidelity* (Wi-Fi). This is from the name of the non-profit international association (Wi-Fi Alliance [10.320]) established in 1999 to certify the interoperability of wireless Local Area Network products based on IEEE. For this reason, hereafter, we will use Wi-Fi, WLAN and IEEE 802.11 in an interchangeable way.

Given their characteristics, WLANs are intrinsically complementary to Fiber Networks: the points of strength of the former well balance the weaknesses of the latter and vice versa. Table 10.16 summarizes these facts. Basically, WLANs are good candidates to complement

Table 10.16   Summary of the some complementary aspects of Fiber Networks and WLANs, which sustain their integration

	Fiber Networks	WLANs
Physical medium	Wired	Wireless (nomadic access)
Distance	Long Reach (depending on fiber type, transmitted power, modulation ...)	Short Reach (mainly indoor or hotspot – limited also by co-existence)
Type of connection	Point-to-point or managed tree (passive optical networks)	Multi-access
Robustness	Very high	Exposed to interference
Type of service	Intrinsic QoS	Lack of strict quality of service (QoS) but broadband (shared) access
Practical issues	Lack of flexibility in user's connection	Lack of scalability (number of users) Lack of mechanisms for QoS Radio issues (interferences/disturbs, fading, hidden terminals ...)

the wired networks and to provide them with a more flexible access; the other way around, if we should reach the Internet by only using Wi-Fi, with multiple hops over the WLAN, the final quality would dramatically deteriorate, so fiber networks may provide a powerful resource for the back-hauling of the end-user traffic to the Internet. This is the rationale for putting together two such heterogeneous media, and for this reason it is not uncommon to read about *Fiber-Wireless Access Networks* (aka FiWi, with a clear linking to Wi-Fi). We will focus on the Wi-Fi-based FiWi solutions.

Before focusing on the specific FiWi architectures which exploit Wi-Fi, it may be useful to widen the perspective so as to prevent too rigid interpretations. Even if a little outdated, a good survey of the possible diversified embodiments of FiWi is provided by [10.323]. The following points are aimed at the same purpose, including also, when possible, some recent technological opportunities:

- In a wider FiWi acceptation, the use of fiber for the connection of wireless access units gets a completely different embodiment, depending on the subtended wireless access technology. For example, when LTE is involved, less strict cost constraints (and higher expected profits) will be involved, because LTE is far from being as cheap as Wi-Fi and intrinsically supports QoS. In the case of LTE, the signal distribution by fibers is still expected to be beneficial but for a different reason, that is, to simplify the installation and management of base stations.
- Wi-Fi is not limited to non-licensed bandwidths. For instance, the IEEE 802.11 ([10.324] approved in February 2014) operates in the TV white space spectrum (in the VHF and UHF bands between 54 and 790 MHz) and, for this reason, is also referred to as White-Fi and Super Wi-Fi. With White-Fi, the scalability of Wi-Fi would become stressed (because a wider area would be reachable), while the occurrence of hidden terminals would become less likely. This is because in the UHF and VHF ranges, the attenuation by materials such as bricks and concrete is lower. This would impact on the different FiWi architectures described in Chapter 10.5.3.
- Potentially disruptive FiWi solutions could emerge with LiFi (or Light-Fi) [10.325–10.327], that is, the use of light short-range free-space for communications either in the visible (VLC – visible-light communications) or in the non-visible (nVLC) wavelength ranges. Literally, LiFi is a wireless medium, even if not radio. LiFi access would not be exposed to significant scalability problems, but rather to limitations in the transmission range which, however, would also largely prevent the occurrence of hidden terminals. Consequently, possible FiWi architectures involving LiFi should address completely different aims.

Once clarified that FiWi is much more than Fibers and (unlicensed) Wi-Fi, it can be asserted that the most interesting and flexible FiWi networks are the Wi-Fi ones. They could result in potentially great numbers and, from a scientific point of view, different alternative architectures are possible (Chapter 10.5.3). For this reason, the proposed FiWi analysis is here restricted to Wi-Fi–FiWi.

It is clear that an extensive dissertation on all the IEEE 802.11 functions would require itself a volume and is beyond the scope of this subchapter: a detailed in-depth review of the key PHY and MAC mechanisms can be found either in the standard [10.319] or in some recent books, such as [10.328]. Conversely, the purpose of this section is the analysis of those scenarios and architectures in which WLAN systems may represent a way for the complementation of wired network. Before entering into details about the FiWi network architectures involving Wi-Fi, a brief introduction will recap the main facts about Wi-Fi and its most recent evolution.

10.5.1   A Historical Perspective on IEEE 802.11 WLANs

IEEE 802.11 is not just a standard about physical and MAC layers, but rather a member of the large protocol family named IEEE 802 [10.329]. In other words, IEEE 802.11 is part of a series of specifications for networks of different ranges (Body/Personal-, Local-, Metropolitan-, Regional-, Wide-Area Networks) and different media (fiber, copper, radio …), covering both physical and data link components. For instance, IEEE 802.3 (aka Ethernet), IEEE 802.17 (aka resilient packet ring) and IEEE 802.11 are all protocols that have been developed within the same framework of IEEE 802. As such, all the protocols in the IEEE 802 family have some points in common and, as a result, their mutual interworking is easier, due to the same upper layers (i.e. LLC [10.330] data link sublayer), the same management functions (e.g. both 802.3 and 802.11 may use the same 802.1x [10.331] port-authentication mechanism) and the intrinsic interoperable framing and bridging from one medium to the other.

Consequently, also the interworking between Wi-Fi- and IEEE 802-based protocols for fibers will be straightforward and it makes definite sense to investigate how to put together Wi-Fi and fiber-networks based on Ethernet (IEEE 802.3). From another perspective, if also the integration of 802.11 and LTE is being investigated (for the offload of mobile networks), the integration with fiber networks would certainly be less troublesome for Wi-Fi.

Wi-Fi counts several versions, so discussing IEEE 802.11 is not unequivocal. The first version of the IEEE 802.11 standard dates back to 1997. Over the years, several new functions and capabilities have been added (and are still being developed or investigated), so as to allow WLANs technology to cope with several possible scenarios. As a result, the 802.11 technology has evolved a lot and, nowadays, it is a very powerful, versatile (… and huge) standard, suitable for very heterogeneous applications, such as data networking, telephony (VoIP), TV streaming, factory automation, digital divide, LTE network offload, low-consumption sensor networks, etc.

The topic of evolution of Wi-Fi is then multifaceted. In the following sections, the unlicensed operating channels of Wi-Fi, the evolution of its throughput and some other relevant evolutions are briefly recalled.

10.5.1.1   The unlicensed channels of Wi-Fi

Over time, the frequency ranges used by Wi-Fi have been widened to encompass, for example, the 5.9 GHz range for vehicular communications, the 60 GHz for IEEE 802.11ad, the white spaces for IEEE 802.11af, etc.

The two frequency ranges of Wi-Fi par excellence remain the two original ones at 2.4 GHz (ISM band) and 5 GHz (U-NII). In both ranges, the regional rules play a primary role: there are some channels which can be used in Europe and Japan, others in the USA, and others in China. For instance, in Europe we can use, in the ISM band, 13 channels, of whom only 3 do not overlap (Figure 10.78 – top); in the U-NII range, instead, there are 19 non-overlapping

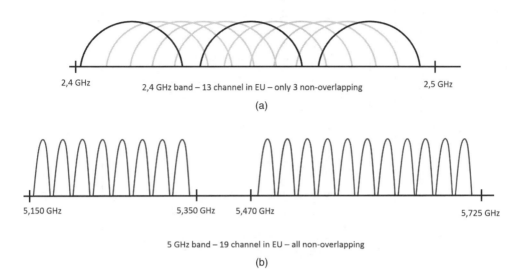

Figure 10.78   The distribution of channels of IEEE 802.11 in the 2.4 GHz (*top*) and 5 GHz (*bottom*) frequency ranges

channels (Figure 10.78 – *bottom*). Conversely, in the USA, we can have 11 channels (3 non-overlapping) in the ISM and, in the U-NII, 14 channels – with an extra range (5.7325–5.825) compared to Europe, but with a gap in the lower channels (Figure 10.78 shows the U-NII range in the different cases). In addition, regional laws rule out the maximum transmitted power which, in general, keeps between 2 and 30 dBm of EIRP.

It is clear that, from the viewpoint of FiWi, the ISM range would be preferable to cover slightly higher areas (in principle, lower frequencies, means wider spans), while the U-NII one would give the possibility to perform more effective radio planning, avoiding interference by nearby cells (due to the higher number of non-overlapping channels).

10.5.1.2   The supported transfer rates of Wi-Fi

Apart from the new flavors and *frequency ranges of Wi-Fi*, the main strategic development of IEEE 802.11, which promoted the evolution of WLAN systems, is the available transfer rate. In the first version, IEEE 802.11-1997 (using DSSS and FHSS techniques), the upper limit was 2 Mb/s, then with the exploitation of OFDM techniques, the data rates climbed to 54 Mb/s (IEEE 802.11a and g). Further progress has been made in the last few years; in the 802.11ac standard [10.332], which was ratified in December 2013, transfer rates as high as over 866 Mb/s are foreseen (per each spatial flow).

The most relevant steps in the growth of IEEE 802.11 speed are summarized in Table 10.17. Basically, the technological boosters of this increase were, over time, OFDM (which is intrinsically more powerful than the former FHSS and DSSS techniques), increasingly aggressive OFDM modulations (from BPSK up to 256 QAM, with different Viterbi coding rates), channel bonding (2 × 20 channels for IEEE 802.11n and up to 8 × 20 MHz with IEEE 802.11ac), the

Table 10.17 Summary of the evolution of IEEE 802.11 at the physical layer, including the nominal and actual transfer rates and the main technological innovations

	802.11	802.11b	802.11a	802.11g	802.11n	802.11ac
Year	1997	1999	1999	2003	2009	2014
Frequency Range	2.4 GHz	2.4 GHz	5 GHz	2.4 GHz	2.4–5 GHz	(2.4) – 5 GHz
Transfer Bit rates (Mb/s)	1, 2	1, 2, 5.5, 11	6, 9, 12, 18, 24, 36, 48, 54	1, 2, 5.5, 6, 9, 11, 12, 18, 22, 24, 36, 48, 54	From 6.5–150 Mb/s on a single spatial stream: depending on guard interval (800/400 ns) and channel width (20 vs. 40 MHz)	From 6.5–866 Mb/s on a single spatial stream: depending on guard interval (800/400 ns) and channel width (20 vs. 40 vs. 80 MHz)
Maximum Transfer Bit rate (Mb/s)	2	11	54	54	600 w/4 × MIMO	6930 w/8 × MU–MIMO
Max Net data-rate (Mb/s)	1	5–7	30–37	30–37	Impact of spatial multiplexing	Impact of spatial multiplexing
Physical modulation	FH-DSSS	HR-DSSS	OFDM (from BPSK $^1/_2$ to 64QAM$^{3/_4}$)	OFDM (from BPSK $^1/_2$ to 64QAM $^3/_4$)	OFDM (from BPSK $^1/_2$ to 64QAM $^3/_4$) 40 MHz ch. 4 × MIMO beamforming	OFDM (from BPSK $^1/_2$ to 256QAM 5/6) 160 MHz ch. 8 × muMIMO beamforming

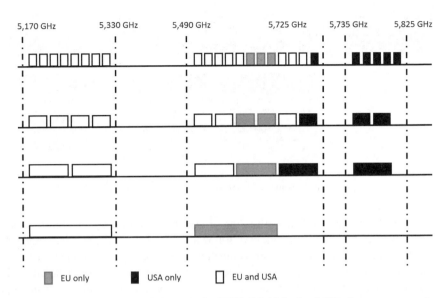

Figure 10.79   The current availability of channels for IEEE 802.11 in the 5 GHz frequency range in Europe and the USA

optimization of guard times among symbols and, more recently, spatial multiplexing and the smart use of multiple antennas (beamforming, MIMO and multi-user MIMO[3]).

Notably, the recent mechanisms based on massive channel bonding, have also impacted on the radio design, as when we adopt 40 MHz channels, we may have only 1 channel in the ISMB band and 9 in the U-NII; 80- and 160 MHz channels are possible only in the U-NII range, with differences between countries (Figure 10.79).

In 10 years, Wi-Fi multiplied its transfer capacity by an impressive factor as high as 400, as no other wireless technology before. Such an explosion of available transmission rates entailed the implementation of some tricks for co-existence among heterogeneous transfer rates (i.e. the discussion about TXOP in Chapter 10.5.2 and for the backward compatibility of following versions of the standard).

First, each IEEE 802.11 version defines multiple available transfer rates and encodings at the physical layer (Table 10.17). For example, with IEEE 802.11a, we could select the preferred rate, to better match the requested speed or the robustness, among 6, 9, 12, 18, 24, 36, 48 and 54 Mb/s. Even more importantly, every time that the standard was amended, the IEEE added

---

[3] MIMO means multiple-input and multiple-output and refers to the use of multiple antennas at both the transmitter and receiver to boost the communication performance. With the single-user MIMO (SU-MIMO), the additional antennas are exploited to widen the information available for the signal decoding: it mainly results in the improvement of physical layer, making the communications more reliable and supporting higher speeds.

With MU-MIMO (multi-user MIMO), the spatial diversity of signal is used for spatial multiplexing, thus the access point (AP) communicates with multiple users at the same time. On the downlink (MIMO broadcast channel), the base station sends different information streams to the users; on the uplink, the base station can simultaneously receive different information from the users.

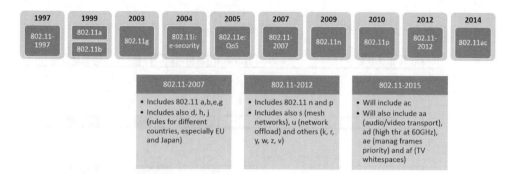

Figure 10.80   Some well-known milestones of IEEE 802.11 development

new rates but spent significant efforts on preserving backward compatibility, so as to let the legacy stations work in a network built following the rules of the amended standard.

The multi-rate and the backward compatibility have an impact on the transfer mode to be used in the initial part of the frames (the so-called (PLCP) Protocol Data Unit (PDU)). The reason is that PDUs need to be decoded by all stations, hence to be sent using a predefined and fixed transfer rate independent of the actual data rate. In fact, the PLCP specifies the rate (hence the decoding to be used) used in the inner part of the frame and its length, so PLCP needs to be understood by all stations for the correct reception and for the correct estimation of channel occupancy.

For example, in the 2.4 GHz range, IEEE 802.11g added some extended rate physicals (ERPs based on OFDM at 6, 9, 12, 18, 24, 36, 48 and 54 Mb/s). Nonetheless, this entails that the PLCP is sent either at 1 Mb/s in "compatibility mode" or, alternatively, at 6 Mb/s. This will obviously affect the medium efficiency. Something different, but conceptually similar, also happens to cope with backward compatibility in the case of channel bonding (IEEE 802.11n and IEEE 802.11ac).

10.5.1.3   Other evolutionary trends of Wi-Fi

IEEE 802.11 paid great attention to the connection speed. However, for the sake of completeness, the evolution of IEEE 802.11 had an impact on many more aspects. Since its first release, several other logical planes have been enriched: the MAC plane (due to the introduction of new medium access methods), and the control plane and the management and security planes (with optimized association and authentication processes). In this way, IEEE 802.11 was also made suitable for vehicular communications (communications among cars, with the former IEEE 802.11p, already subsumed by [10.319]), quality of service (QoS), robust authentication and enriched by several functions (spanning from the configuration of clients to some cognitive features and the adaptation of the target frequency ranges).

The evolution of 802.11 has also included milestones other than the speed. Some of them are briefly recalled in Figure 10.80. Following the typical approach of IEEE 802 standardization[4],

---

[4] According to the rules of the IEEE Standards Association, there is only one current standard which, for Wi-Fi, is denoted by IEEE 802.11 followed by the date of its publication. At the time of our writing, IEEE 802.11-2012

most of the amendments have been merged in the latest release of the IEEE 802.11 standard. A brief discussion will complete this section by highlighting those evolutionary aspects which are relevant to our analysis of FiWi architectures:

- First, even if the IEEE 802.11 MAC has evolved, it still preserves the initial purpose of being decentralized and simple. As a result, the base channel access was slightly modified to enable the management of statistical priorities, by the former IEEE 802.11e (currently in [10.319]). It still kept the random access principle of the so-called CSMA/CA (Carrier-Sense with Multiple Access), a listen-before-talk (LBT) method. More precisely, since the IEEE 802.11e was published, the channel access is managed by the so-called Hybrid Coordination Function (HCF) which subtends, in addition to the CSMA/CA with priorities (called EDCA), a centrally policed and QoS-capable method called HCCA (HCF Controlled Channel Access), with collision-free periods (CFP). Unfortunately, HCCA has never been implemented and Wi-Fi stations can still only use CSMA/CA. The current Wi-Fi network cannot guarantee the quality of transmissions, as we do not know how much and when a station can transmit and, in spite of the LBT technique, two stations may transmit at the same time and interfere, due to simultaneous decision to transmit. This also implies a sub-optimal channel exploitation, as further discussed in Chapter 10.5.2.
- Over time, from IEEE 802.11a to IEEE 802.11ac, increasingly higher throughputs have been reached. By reading the IEEE 802.11ac, we get the impression that the upper theoretical bound has almost been reached – considering the target transmission ranges of WLAN and the available frequency ranges. Coherent with this interpretation, the most recent IEEE 802.11 task groups seem to be moving their focus from maximum theoretical individual throughput to an increase of the aggregated throughput, benefitting from features such as spatial multiplexing (MIMO and MU-MIMO), power adaptation and cognitive features for achieving higher densities and using Wi-Fi in massively crowded areas (this is the objective of the new TG IEEE 802.11ax). These solutions are potentially relevant to *FiWi* and are also recalled in Chapter 10.5.3.
- Some other recent trends are targeting completely new features, which fall outside the scope of this chapter. For example:
  – IEEE 802.11ah will deliver low-rate transmissions for sub-1 GHz and license-exempt operation, for sensor networks and smart metering, even over wider distances (the issue is expected for 2016).
  – IEEE 802.11ad ([10.333] published in 2012) defined a Wi-Fi mode and physical layer for delivering very high throughput at 60 GHz over short distances, inheriting the mandate of wireless USB. It was promoted by the trade association named Wireless Gigabit Alliance (WiGig), which was subsumed by the Wi-Fi Alliance[5] in March 2013.

---

is the only version in publication; next one is expected to be the IEEE 802.11-2015. The standards are updated by means of amendments, which are created by task groups (TG). Both the task group and their finished document are denoted by 802.11 followed by a letter (or a couple of letters), such as IEEE 802.11a and IEEE 802.11ac. In order to create a new stable version, task group m (TGm) combines the previous version of the standard and all the published amendments not subsumed yet. New versions of the IEEE 802.11 were published in 1999, 2007 and 2012, as depicted in Figure 10.80.

[5] Wi-Fi Alliance is a trade association that promotes Wi-Fi technology and certifies Wi-Fi products if they conform to certain standards of interoperability. They own a trademark which manufacturers may use to brand certified products that have been tested for interoperability.

– Eventually, 802.11af (issued in 2014 [10.324]) extend wireless local area network (WLAN) operation in the TV white space spectrum (VHF and UHF) and complements Wi-Fi with the capability to cover longer ranges.

All these trends considered, we can see that there is an attempt to leverage Wi-Fi to fill some gaps in the possible distance/throughput targets of wireless communications (long-reach wireless sensor networks with 802.11ah, very-high-speed and short-range communications with 802.11ad, and long-range WLAN with 802.11af).

• Last but not least, a trend concerns the improvement of the interworking between Wi-Fi devices and between Wi-Fi and other networks, acting on authentication. In particular, in the IEEE 802.11-2012, the amendment IEEE 802.11u has been subsumed so as to provide Wi-Fi with additional tools for the network discovery and the management of QoS (service advertisement, mapping of IP priority over Wi-Fi classes, user traffic segmentation, etc.). These functions are particularly useful to simplify the Wi-Fi authentication procedures, merging them with those used by cellular networks. In more detail, Passpoint[6] [10.335] by the Wi-Fi Alliance (aka Hotspot 2.0) specifies procedures to seamlessly integrate Wi-Fi within 3G/4G networks, by putting together functions from the former IEEE 802.11u, authentication methods from the IEEE 802.11i (i.e. EAP-SIM and EAP-AKA, coming from the cellular world) and the encapsulation of IEEE 802.1x.

In a different use case, the Wi-Fi Direct [10.336] specification (still by Wi-Fi Alliance) permits an easy ad-hoc connection between two Wi-Fi devices, without the need of an access point. The pairing is accomplished by a proximity out-of-band connection (e.g. NFC, Bluetooth), with at least one of the devices fulfilling the Wi-Fi direct specifications and acting as a soft access point. The screen casting by Miracast [10.337] is enabled by Wi-Fi Direct. Both Passpoint and Wi-Fi direct may be relevant to FiWi architectures, but the former in particular, since a common authentication framework will facilitate the integration between wired and wireless accesses.

10.5.2   Relevant Operating Principles of WLAN Systems

The main *operating principle of Wi-Fi* medium access is its decentralized channel access. Each station which participates in the WLAN will compete for channel access with no need for a central coordinator arbitrating the accesses. Considering the goals of this section, we will skip most of the details and introduce only the prominent aspects.

10.5.2.1   Wi-Fi medium access control in a nutshell

The fundamental MAC mechanism of the IEEE 802.11 is the so-called Distributed Coordination Function (DCF) and employs a CSMA/CA (Carrier-Sense Multiple-Access with Collision

---

[6] In the last years the Wi-Fi Alliance started the creation of custom protocols subsets, with the aim of promoting some specific features: this for example happened with the Passpoint feature (certified by Wi-Fi Alliance and falling within the IEEE 802.11u specification), which uses EAP [10.334] authentication and the WPA2-Enterprise encryption (Wi-Fi Protected Access, formerly in the amendment IEEE 802.11i and now in [10.319]), to improve the interworking with external networks.

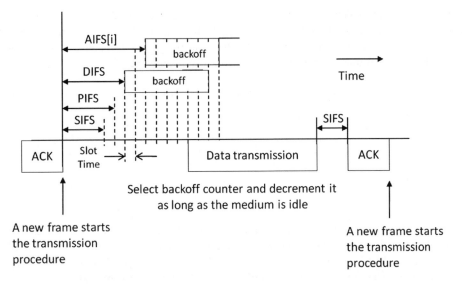

Figure 10.81　Time-diagram showing the back-off entities and operations, as defined by the CSMA/CA algorithms in the IEEE 802.11

Avoidance) with binary exponential back-off algorithm: the previously mentioned EDCA is just a variant of DCF, as briefly discussed below.

DCF is classified as the listen-before-talk (LBT), as a station wishing to transmit will have to listen for the channel status for a given DIFS interval[7].

If the channel is found to be busy, the station defers its transmission and, afterwards it will wait for an additional DIFS interval and set the back-off counter to a random value in the interval $[0, CW_i-1]$, which represents the current contention window size. The back-off timer is then decreased by 1, every time the medium is perceived as idle for as long as a so-called slot time (DIFS is less than half a slot time), frozen when a transmission is detected on the medium, and resumed when the channel is detected as idle again for a DIFS interval. When the back-off counter reaches 0, the station is allowed to transmit. This mechanism is shown in Figure 10.81.

Channel sensing is not sufficient to guarantee collision-less transmissions (two or more stations might count down simultaneously to 0). To cope with this, collisions are revealed (for unicast transmissions) thanks to the mandatory acknowledgement (ACK) control frames

---

[7] The main characteristic time intervals of the DCF are (1) Short Interframe Space (SIFS) - the small time interval between the data frame and its acknowledgment (also used for RTS-CTS frame exchange), (2) the slot time (identified in IEEE 802.11 by the parameter) $aSlotTime$, (3) DCF Interframe Space (DIFS), with DIFS = SIFS + 2∗ $aSlotTime$.

All these intervals depend on the physical layer, i.e. the frequency range. For instance, in IEEE 802.11a, IEEE 802.11n at 5GHz and in IEEE 802.11ac, SIFS = 16 μs and $aSlotTime$ = 9 μs (hence DIFS = 34 μs).

Instead, with IEEE 802.11g and IEEE 802.11n (at 2.4 GHz) in backward-compatibility mode, SIFS = 10 μs and $aSlotTime$ = 20 μs (hence DIFS = 50 μs). Eventually, with IEEE **802**.11g and IEEE 802.11n (at 2.4 GHz), without backward compatibility, SIFS = 10 μs, $aSlotTime$ = 9 μs, DIFS = 28 μs.

which are to be immediately sent back by the receiving station. In the event of a collision (i.e. missing ACK reply), the transmission is repeated after waiting for a random time picked up over a wider CW_i' interval. The length of the back-off time at the i-th consecutive attempt to transmit the same frame is determined by the following equation:

$$T_{BO} = random([0, CW_i - 1]) * aSlotTime, \tag{10.40}$$

$$CW_i = min(2^{i+m}, 2^{10}) \tag{10.41}$$

where:

T_{BO} Back-off time
CW_i Contention window

More explicitly, at each consecutive attempt to re-transmit a given frame, the interval is doubled, that is, it grows exponentially according to the binary exponential back-off (BEB) – from [0,15] to [0,1023], in the case of IEEE 802.11a. Obviously, CW_i does not vary for broadcast/multicast, since we cannot count on ACKs to check frame reception.

In order to prevent excessive transmission delays, the maximum number of consecutive retries is limited by two MAC variables, respectively for short and long frames (the respective default values are 7 and 4).

In addition, since each station can select its transfer rate, a slow station would severely degrade the performance of the high-rate stations [10.338,10.339] in a multi-rate WLAN. For this purpose, it is defined as an additional parameter (Transmission Opportunity – TXOP), which controls how long a station can keep the channel busy – with the additional benefit of permitting the transmission of TXOP-long bursts of multiple consecutive frames, so avoiding multiple contentions. EDCA only acts on DIFS (redefining it into different possible $EIFS_i$), CW_{min}, CW_{max} and TXOP to differentiate channel access among stations, by following the principle that the longer the waiting time, the lower the access likelihood.

10.5.2.2 The issue of hidden terminals

One typical problem of wireless medium access concerns hidden terminals (HTs), which occur when one station A can receive and sense another two stations (say B and C) which, by contrast, cannot sense each other (Figure 10.82). HTs intrinsically deteriorate Wi-Fi transmissions, because B and C, not being in the radio range of each other, cannot rely on CSMA to prevent their simultaneous transmissions (and collisions) on A. The occurrence of HTs may be related to distance, but also to obstructions. Even worse, when, due to some physical obstruction, the two stations (B and C) are close but cannot sense each other, the effect of their strong mutual interference will be disruptive (A will not be able to receive simultaneous transmissions either by B and C, in spite of their proximity).

IEEE 802.11 briefly introduced how to cope with HTs. It is the optional mechanism called RTS/CTS (Request to Send/Clear to Send). When RTS/CTS is enabled, a station will refrain from sending a data frame until the station completes an RTS/CTS exchange with the destination station. The source initiates the process by sending an RTS frame and the destination

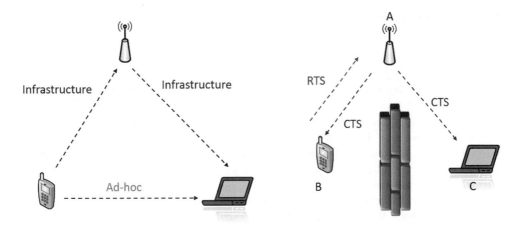

Figure 10.82 Some concepts of Wi-Fi networking: ad-hoc mode vs. infrastructure mode (left); occurrence of hidden terminals and RTS/CTS protocol exchange (right)

responds with a CTS (Figure 10.82 – right)). The source can send its data frame only after receiving the CTS, as the CTS also contains a time value that alerts other stations to hold off from accessing the medium while the station initiating the RTS transmits its data. In this way, collisions by HTs are prevented. Unfortunately, the RTS/CTS dramatically degrades the medium efficiency [10.340], especially for short frames.

10.5.2.3 Infrastructure and ad-hoc mode

Finally, a short recap on the so-called Wi-Fi access points, whose role may indeed be relevant to FiWi network architectures. If Wi-Fi is supposed to be used for a wireless access to a fiber network, each fiber will be connected to a gateway that will collect and forward wireless traffic to the wired network and vice versa. However this node might either work as a connection to the wired network or as a true access point (AP). Here comes the distinction.

A Wi-Fi network working in an infrastructure mode (or BSS, basic service set) is built on the role of an AP, which optionally announces the name of the network (the so-called SSID), authorizes or rejects the access, manages the network (including the encryption rules), but does not have any privilege (i.e. priority) in medium access with CSMA/CA. Conversely, all the devices on the network will have to communicate through the access point, generally a wireless router. For example, two devices with one close to the other, will not communicate directly, but will have to communicate indirectly through the wireless access point. This leads to a network that is more organized and controlled, especially for tailoring the management of QoS management. The main drawback consists in the overhead halving the network capacity for station-to-station transmissions (Figure 10.82, left).

APs are not mandatory for IEEE 802.11, as it is possible to have either an infrastructure network (with an AP) or an ad-hoc network (without AP). In the latter case, a "peer-to-peer" model is adopted and devices on the wireless network can connect directly to each other.

10.5.2.4 Wi-Fi: strengths and weaknesses

From these operating principles, the main characteristics of Wi-Fi can be summarized to highlight once more its points of strengths and its weaknesses.

The points of strengths stand in the unlicensed channels which Wi-Fi uses (in most cases) and in the simplicity of the approach which it proposes. This approach is decentralized (it may even work without any APs), does not required synchronization, is easy to configure (it also comprises several self-configuration features), mostly backward compatible and offers high-speed transfer rate. These characteristics upheld Wi-Fi success. In the last 10 years, the Wi-Fi market has continuously grown and forecasts state that in 2017 it will exceed $7 billion dollar. However, IP traffic over Wi-Fi will exceed wired traffic for the first time by 2018 (as reported by Cisco Systems).

In spite of its extraordinary market success, Wi-Fi also presents weaknesses which, more or less, can be seen as the drawbacks of its same points of strength.

First, given the use of a limited and unlicensed spectrum range, it is becoming increasing difficult to perform an effective network planning, as the available non-overlapping channels are so few. In the most recent standards (such as IEEE 802.11ac), channels can be used in groups (up to 8), so that the number of actually available channels becomes even more lowered. Finally, IEEE evolution is proposing the use of Wi-Fi in areas densely crowded by users and by APs. IEEE 802.11ax is fostering mechanisms for having many APs, the one close to the other, and with a lot of users for each. Altogether, this might become a mess; how to have so many nodes in the same radio range, while preventing their mutual interference?

At the moment, the only network tool available to address such scenarios is cognitive intelligence. Cognitive behavior can work separately in the nodes, when each node senses the medium and makes its optimization choices, or it can be performed via a decentralized sensing and a centralized control, as proposed in the LWAP solution (Section 10.5.3), which relies on a wired connection between the APs and the centralized controller. This introduces the potential role for FiWi architectures for solving Wi-Fi, issues by leveraging fiber connections to the APs.

The other major problems of Wi-Fi concern the lack of determinism in CSMA/CA and has effects at the same time on:

1. the QoS which can be delivered to the stations
2. the vulnerability by hidden stations
3. the scalability of the MAC (how the access worsens, as the number of stations grows)

It is well known that the Wi-Fi performance decreases as the traffic load approaches the maximum theoretical throughput (saturation) and/or the number of stations increases. In both cases, the number of simultaneous transmissions by nodes will increase, due to the more frequent events of simultaneous count-down of the back-off periods. This will cause frame-losses (especially for broadcast transmissions, which do not foresee either ACKs or retransmissions) and transfer delays (especially for unicast, due to multiple transmissions attempts). Several papers have described studies of the actual capacity of Wi-Fi, either theoretically or by simulations, both under saturation and far from it, and under different traffic scenarios (variable number of stations, different frame length, variable BER, with or without RTS/CTS, for unicast

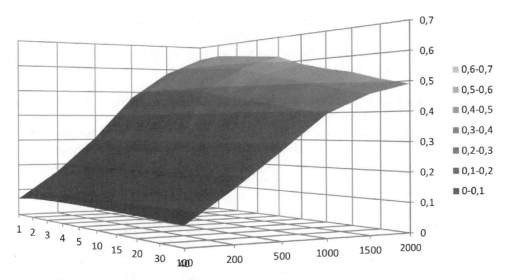

Figure 10.83 Normalized channel saturation for an IEEE 802.11 g network: ideal channel (BER = 0), R = 54 Mb/s, unicast traffic, number of stations between 1 and 140; frame length between 100 and 2000 bytes

or broadcast) [10.340–10.342]. The literature is self-consistent and all the papers show similar results, quite independent of the Wi-Fi version.

For example, Figure 10.83 shows a graph based on the data of [10.341], with the plot showing the normalized saturation rate (the maximum aggregated transfer rate, normalized by the nominal rate), for an IEEE 802.11g network, with an ideal channel (BER = 0), the rate fixed at 54 Mb/s, unicast traffic, a variable number of stations (from 1 to 140) and variable frame lengths (100–2000 bytes). The axes are not linear and there is a fixed distance between consecutive samples of the axes grids, independent of the actual distance between them.

It can be seen that the longer the frames, the better the percentage usage, due to lower impact of overheads (PLCP overhead and ACK frames transmitted at low rate, as discussed in Section 10.5.1). Moreover, the higher the number of stations, the lower the saturation rate, due to collisions. In general, normalized saturation, in spite of the ideal conditions, does not exceed 65% and, as the number of stations grows, drops below 50%.

Besides, the channel usage for lower rates (e.g. 6 Mb/s) is better than for higher (54 Mb/s), because the PLCP and ACK slow transmission has a smaller percentage impact at a lower rate (at 6 Mb/s, the saturation can exceed 80%), the results of which are not depicted here.

This preliminary analysis already shows that growing the nominal transfer rate does not proportionally increase the actual transfer rate, due to underlying MAC mechanisms. In addition, the higher the transfer rate, the heavier the impact of inefficiencies. This is the reason which a dedicated task group of IEEE 802.11 (IEEE 802.11ax) is pursuing high-efficiency WLANs (HEW). At the present time, the HEW scenarios are being defined and some ideas are already being collected. The authors have made their proposal with TD-uCSMA [10.343–10.345], that is a backward-compatible coordination functions using CSMA but with EDCA parameters variable over time, so that each station may exclusively have high priority for some time. TD-uCSMA leads to some kind of decentralized reservation and a QoS guarantee which may prevent collisions. Other proposals will pursue novel approaches for the high-density scenarios, with a large number of APs and of users at each AP, mainly enforcing cognitive techniques.

Especially while waiting for the development of IEEE 802.11ax, a valid alternative will be the adoption of FiWi architectures, where all the Wi-Fi nodes belong to the same wired network, so that their configuration can be centrally optimized by a controller, rather than negotiated, and the number of APs can grow, so as to limit the number of users using the same one.

The other main issues of Wi-Fi MAC are related to QoS and to hidden terminals. The former is quite intuitive, for if it does not know the number of stations, it cannot know what the saturation throughput and the percentage of throughput is available to it. Moreover, priority alone cannot work, for instance if many stations are using the same highest priority.

Concerning hidden terminals, few studies have addressed this issue (i.e. [10.346–10.348]), but it is emerging as a major threat. Also the lack of determinism and the impact of HTs could be at least partially solved by a clever FiWi network design, with a dense number of APs, preventing too heavy traffic on each of them and the occurrence of HTs. FiWi architectures will be discussed below.

10.5.3   Hybrid Fiber-Wireless Network Architectures: Wi-Fi-based FiWi Architectures

Fibers and wireless access are complementary and Wi-Fi can be a cheap, available worldwide and a free solution for capillary wireless access, without requiring any radio licenses.

Even more, in its evolution, Wi-Fi has always moved in the direction of a higher capacity and more flexible management of wireless access in crowded areas. As discussed in Section 10.5.2, Wi-Fi efficiency has always been limited – and, so far, it is still – by the CSMA/CA operating principles. CSMA/CA makes Wi-Fi non-QoS-capable, exposed to hidden terminals and acknowledged only in the case of unicast traffic. All of these characteristics need to be kept in mind when discussing the FiWi architectures based on Wi-Fi.

10.5.3.1   Wi-Fi in FiWi reference architectures

The first possible approach, which is now presented, literally is not a FiWi solution, but rather uses coaxial cables. We will show why this does not actually make sense in an RoF embodiment.

As shown in Figure 10.84, cables can be used only to extend the radio coverage of an AP. In this case, the perspective is reversed and the cables are used to extend the reach of an AP and not to distribute a fiber access by leveraging Wi-Fi. In practice, we plug in cables so that the antennas are as far away from the AP as possible, or even to split the signal of a single antenna into multiple remote antennas.

This approach was sometimes used at the dawn of Wi-Fi technology, when the cost of an AP could be significant, and especially in scenarios with low traffic (e.g. wireless coverage for warehouse automation). In such cases, the wireless network would unlikely to be overloaded

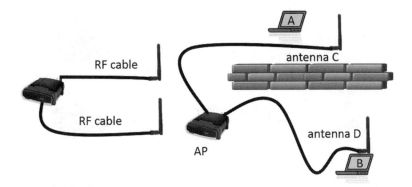

Figure 10.84 The simplest wired/wireless integration: cables used to extend wireless coverage by displacing the antennas: conceptual implementation (left) and potential worsening by hidden terminals (right)

and it would be acceptable to use the same network to collect the concurrent (low bit-rate) traffic from a wide area.

More importantly, this solution is dramatically affected by hidden terminals. Supposing we have multiple antennas (without MIMO intelligence) used to widen the span of radio coverage, there will be frequent cases (Figure 10.84 – right) when the received power from station A to the AP (through antenna C) and from station B to the AP (through antenna D) would be sufficient for frame reception, but A and B could not sense each other. The farther apart the antennas C and D are (or the more they are exposed to mutual obstructions), the more likely the occurrence of hidden terminals (in principle, the two antennas could themselves be in positions where they were not in the radio range of each other). It is clear that this configuration does not seem to be either effective or promising for FiWi and it would not be cost-effective to use fibers in the place of cables.

It would not make sense to use such "extended" APs for widening the access to a fiber network, at least for two reasons. First, the market and standardization trends of Wi-Fi go in the opposite direction, which means wider bands to the user, heavily crowded areas and dense coverage (as from the TG IEEE 802.11ax), while the solution proposes an inefficient bandwidth sharing. Even more, densely crowded areas will emphasize the effects of collisions by hidden terminals [10.347,10.348], which the architecture intrinsically suffers from and, on the other hand, the use of RTS/CTS protocol would not be sensible either, due to its detrimental effect on the bandwidth actually available in the wireless channel, compared to the bandwidth potentially available on the fiber side.

The most obvious FiWi architecture is, instead, the one depicted in Figure 10.85, where fibers are used to connect APs to a switch. A possible question pertains when it may make sense to use fibers instead of copper (UTP or coaxial cables). Under the hypothesis of heavy traffic, which may motivate the use of fibers, Gigabit Ethernet (1 GigE) and 10 Gigabit Ethernet (10 GigE) can be considered.

Concerning 1 GigE, with IEEE 802.3-2008, five physical layers standards have been defined using optical fiber (1000BASE-X), but also twisted pair cable (1000BASE-T), or shielded balanced copper cable (1000BASE-CX). With fibers we can cover up to 70 km (1000BASE-ZX @ 1550 nm), and at the cut-off distance of 100 m, we may opt either for a "good" UTP

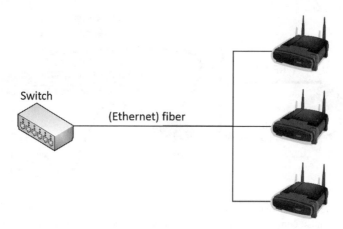

Switch

(Ethernet) fiber

Figure 10.85   FiWi architecture where fibers are used to interconnect APs and to overcome the weaknesses related to all-Wi-Fi architectures

(Cat.6 and 7 as from 1000BASE-TX) or for a cheap fiber which may go further (multi-mode fibers of 1000BASE-SX).

Also the same ranges apply to 10 GigE (IEEE 802.3-2008), up to 40 km over a single-mode fiber with 10GBASE-ER, up to 400 m with a multi-mode fiber 10GBASE-SR, and no more than 100 m with UTP cat.7 in 10GBASE-T. Instead, the most recent 100 GigE (IEEE 802.3-2012 [10.349], incorporating the former IEEE 802.3ba) foresees only fibers (for UTP only 40 Gb/s over 30 m and UTP cat.8).

It is clear that to fully exploit the most recent Wi-Fi releases, a broadband link is required to interconnect the APs. Especially with IEEE 802.11ac MU-MIMO, we may have more than 6 Gb/s traffic at each AP. Here copper would be sufficient only on short traits (with strict requirements on the quality of UTP cables), so fibers would be the natural choice.

Altogether, the circulating traffic and the possible distances are decisive factors for choosing an FiWi or a copper/Wi-Fi network. The more recent the Wi-Fi network, the higher the achievable throughput and, as a result, the more likely the adoption of FiWi architectures.

The architecture of Figure 10.85 can still be optimized for some aspects. In particular, under the hypothesis of having a very dense number of APs, multiple APs will use the same spectrum range and will have to dynamically adapt their transmission parameters, so as to prevent mutual interferences. For this purpose, there are at least three ways to reconfigure the network manually (i.e. by a network or element manager), to define some wireless protocol which makes the configuration dynamic or to have a dynamic (automatic) configuration which is centrally managed. The last option is the one which has met with market success and led to the definition of an architecture where the media access control (MAC) of an AP is split between the so-called wireless LAN controller and the access point (lightweight access point – LWAP) (Figure 10.86 – right). Timing-critical functions, such as the subatomic handshake and emitting beacons to the access point, are managed at the access point. Other network-critical functions, such as mobility management, authentication, VLAN segregation, RF management, wireless IDS and packet forwarding, are managed at the wireless LAN controller. Controller

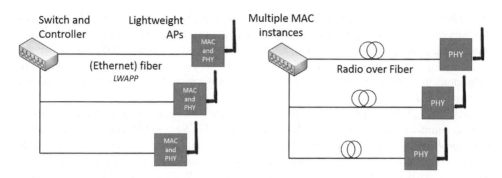

Figure 10.86   Other two FiWi architectures: fibers used to interconnect LWAPs (left) and RoF solution (right)

and LWAPs communicate via the LWAP Protocol (LWAPP), which is defined in the IETF RFC 5412 [10.350].

The benefits of the solution are manifold:

- *Dynamic RF planning*: Wireless LAN controllers have a built-in understanding of the signal strength that exists between lightweight access points within the same network. These controllers can use this information to create a dynamic optimal RF topology for the network, to be as stable as possible. The result is a dynamic wireless network that adapts to changing RF conditions in real time.
- *Load balancing of users*: The controller knows both the signal strengths of APs and the clients' probe requests to associate. Consequently, it can choose which access point should respond, considering the client's signal strength and the AP traffic load. For example, an adjacent access point may provide an equivalent service but at a lower signal strength, or otherwise an AP may be unloaded and selected, raising its transmitted power.
- *Simplified roaming*: The LWAPP permits to have an easy L3 roaming from one LWAP to the other, without the need for complex protocols, such as Mobile IP.
- *Added value services*: Location-based services are simplified, thanks to the intrinsic information about the Received Signal Strength Indication (RSSI) of LWAPs and clients. Voice services are enforced by the central coordination of EDCA parameters, of number of voice users (Call Admission Control – CAC) and of the RSSI.
- *Ease of deployment and of upgrade*: The controller programs all RF policies and wireless LAN policies onto the LWAP. All packets from the access points are placed into an LWAPP tunnel and subsequently sent to the wireless LAN controller, so there is no need to extend special VLANs to individual access points. Also the VLAN mapping (traffic filtering to one user to another) is centrally managed.

LWAP falls under the umbrella of Software Defined Networking (SDN), the paradigm which proposes to separate the data and control planes and to centralize the intelligence of the network so as to make it more flexible and proactive to network changes.

In the process of centralizing more and more the network intelligence, it has been proposed to exploit Radio over Fiber (RoF) in FiWi architectures. In this case, all the MAC functions would become centralized. It is as if the central coordinator hosts multiple APs, whose antennas are remotely installed (Figure 10.86 – right). The connection to the antennas is by fiber, so there is the possibility to place them quite far apart. For this reason we talk about RoF and, collated with the antenna, we will find a device transducing the signal from the optical to the radio domain and vice versa. The solution is also known as the Fiber Distributed Antenna System (DAS).

So far, fiber DAS has not met with great success for Wi-Fi. In general, RoF is mostly used for cellular communications (GSM to LTE) rather than Wi-Fi. We could find two manufacturers of DAS for Wi-Fi: Zinwave and Optical Zonu. The former sells a turn-key solution optimized for a simplified deployment in heterogeneous wireless scenarios. They describe case studies with a cellular network working with other technologies (i.e. Wi-Fi or TETRA). Their maximum supported frequency is 2.7 GHz, so that they operate only with 802.11 g. Optical Zonu, by contrast, claims a rich set of scenarios and IEEE 802.11a/b/g/n/ac, hence also the MIMO technology; their point of strength consists in the possibility to carry signals as wide as 6 GHz.

In principle, DAS and LWAP are similar and, even more, with DAS we might cut some more costs, since all the network intelligence would be in the same central device. However, by deeper analysis, we understand that the currently available DAS devices are limited to the physical layer, as they can transduce signals but they do not cover, for instance, the MAC management. As a result, the current DAS cannot replace the others (i.e. LWAP), but can only extend their signals. In most cases, we would opt for an LWAP and would consider DAS mainly for logistic reasons, for example when there is really sufficient room to accommodate the antenna and not the full access point or base station, which is more likely and sensible for cellular base stations rather than for the typically small Wi-Fi devices.

10.5.3.2   Overall FiWi architectures

What has been discussed so far refers to the Fiber-Wi-Fi integration. But the *overall FiWi architectures* also subtend service motivations. The two most promising scenarios for Wi-Fi evolution and growth seem to be the use of Wi-Fi for LTE off-load and the use in mesh networks; in both cases, there are also concrete opportunities for FiWi architectures.

The case of LTE off-load by Wi-Fi is already viable from a technical point of view and the solutions included in Passpoint already facilitate the task. From a market point of view, the off-load is emerging for extending the coverage to indoor areas where outdoor signals do not penetrate well, or to add network capacity in areas with very dense phone usage, such as train stations or stadiums. In this sense, Wi-Fi networks become complementary to the use of small GSM/UMTS/LTE cells (micro-, pico- and femto-cells[8]).

The reason for using Wi-Fi rather than small base stations is the possibility to use a Wi-Fi infrastructure which may be generally unloaded and already installed, for example for home browsing.

The off-load of LTE by Wi-Fi is particularly promising for those Service Providers which already deliver Wi-Fi at home and intend to become a Mobile Virtual Network Operator (MVNO), that is, an operator providing mobile services without building and operating its

---

[8] Even if there is not a unique standard definition, typically the range of a microcell is less than 2 km, a picocell is 200 m or less, and a femtocell is on the order of 10 m.

own network infrastructure (or possessing its own radio spectrum). MVNOs purchase network capacity from licensed mobile network operators, reselling this airtime to their own customers. However, if a MVNO already has a dense Wi-Fi coverage in a city, it may strongly rely on it to deliver its service (to nomadic users, not fast moving ones), thus paying less to the operator whose connections it is reselling. In addition, the more broadband the Wi-Fi accesses, the more the network can deliver, so the more significant the potential benefits. This also represents an opportunity for FiWi, in preventing the bottlenecks of Wi-Fi access over long distances.

Another interesting scenario is that of mesh networks, which are networks of routers whose connections are mainly wireless, with only a limited number of wired connections to the backbone. Mesh infrastructures carry data over large distances by splitting the distance into a series of short hops. Intermediate nodes not only boost the signal, but cooperatively pass data from point A to point B by making forwarding decisions based on their knowledge of the network. Wireless mesh networks have a relatively stable topology, except for the occasional failure of nodes or addition of new nodes. Wi-Fi has its own standard solutions (the former IEEE 802.11s, included in IEEE 802.11-2012 [10.319]).

Mesh networks might represent a way to further load a Wi-Fi network, hence to benefit from an FiWi architecture. Two facts may substantiate the statement.

First, it is possible not to degrade the QoS over multiple wireless hops; in fact, with Wi-Fi we also have multi-radio mesh networks, those mesh nets in which a unique frequency is used for each wireless hop. In multi-radio meshes, each link will have it dedicated CSMA collision domain, so that we can optimize the transmissions, without bandwidth degradation due to the mesh and without adding latency.

Second, the most recent Wi-Fi releases (IEEE 802.11ac) are characterized by very high aggregated throughput and by an efficient spatial multiplexing (multi-user MIMO), which is currently limited to no more than 8 flows. Altogether, mesh networks can help to fully exploit the available bandwidth (Figure 10.87). Each MIMO spatial flow may collect multiple users (sometimes also as device-to-device connections, as shown), thus significantly increasing the

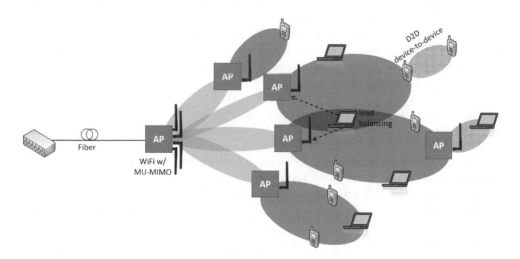

Figure 10.87   Example of a rich FiWi architecture, involving fibers and wireless mesh network

number of stations connected. As a result, each MIMO flow will be frequently loaded, due to the statistical multiplexing of the connected stations. Last but not least, there will also be the possibility that an end-user can load-balance its traffic over multiple paths, with benefits on the traffic management and quality of service experienced. Under the hypothesis of a heavily loaded Wi-Fi mesh network, a fiber connection is needed to prevent bottlenecks in the backhauling.

The scenarios of FiWi with mesh wireless network are significant, both for their flexibility and from a research point of view. Several interesting starting points can be found in the literature, about new possible research directions. Far from providing an exhaustive analysis, here some promising ideas are briefly recalled.

In general, independent of the specific fiber and underlying wireless technologies, [10.351] it proposes to manage the overall QoS of a FiWi network as a problem to be addressed by network virtualization, adhering to the most recent trends in networking [10.352].

A certain number of papers focus instead on the optimization of FiWi networks involving Wi-Fi and a *Passive Optical Network* (PON). For instance, [10.353] analyzes how to best configure the joint scheduling in the Wi-Fi network and in the PON, considering the impact of the number of optical and wireless stations on the overall traffic management. From the different perspective of network engineering in [10.354,10.355], this study shows how to position the PON nodes so to optimize the Wi-Fi mesh traffic, whereas in [10.356], the PON and Wi-Fi FiWi network becomes optimized for the energy consumption. All these studies could be easily generalized to other FiWi architectures, including the LWAP one.

Finally, there are a certain number of papers which focus on possible new service scenarios, such as PON FiWi for the sustainability of smart grid management [10.357], FiWi involving PON and WiMax [10.358], FiWi in next generation and high-throughput PON and Wi-Fi [10.359] optimization of PON FiWi for Video-on-Demand Services

The research on FiWi is a niche but scientifically very interesting and vital and, so far, mainly explored for PON. For this reason, it may be worth closing this section with some additional considerations of FiWi based on PON + Wi-Fi.

A passive optical network (PON) is a wired network that uses a point-to-multipoint topology built on fibers. Unpowered optical splitters are used to enable a single optical fiber to serve multiple premises. As depicted in Figure 10.88, a PON encompasses an optical line terminal (OLT), typically at the service provider's central. and a number of optical network units (ONUs). In most cases, downstream signals are broadcast to all premises, with encryption preventing eavesdropping, while upstream signals are combined using a multiple access protocol, usually time division multiple access (TDMA) or wavelength division multiple access (WDMA). Also IEEE 802.3, the Ethernet standard, has specific amendments for 1 Gb/s and 10 Gb/s PON (the so-called Ethernet PON, or EPON, respectively in IEEE 802.11ah and IEEE 802.11av).

PONs have both advantages and disadvantages over active networks, as they avoid the outdoor installation of electronic equipment and make it simple to also broadcast analog signals. On the other hand, since splitters cannot perform buffering, the ONU must carefully coordinate their transmissions, so as to prevent signals sent by customers from colliding with each other.

From the point of view of FiWi architectures, if we can rely on a cognitive and smart wireless management (as in the case of LWAP) and on a flexible EPON, we can propose a framework to optimize the throughput and minimize the transfer delay, as proposed in [10.359]. This would

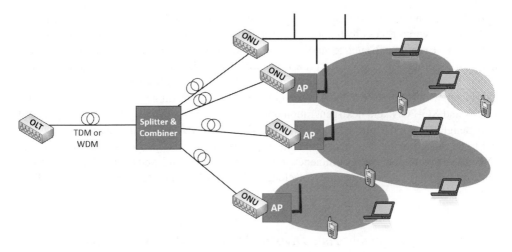

Figure 10.88   FiWi architectures involving a PON

indeed enforce the FiWi approach and adhere to the most recent approaches of Self-Optimized Networks (SON) [10.360] and Software-Defined Networks (SDN) [10.361].

This preliminary discussion, even if partial, should have already shed light on the scientific challenge, on the novelty and practical relevance of the FiWi solutions which, definitely, deserve future investigation.

10.6   Energy Efficiency Aspects in Optical Access and Core Networks

Dr. Paolo Monti, Dr. Lena Wosinska and Dr. Richard Schatz
KTH Royal Institute of Technology, Stockholm, Sweden
Dr. Luca Valcarenghi, Dr. Piero Castoldi
Scuola Superiore Sant'Anna, Pisa, Italy, Aleksejs Udalcovs, Institute of Telecommunications, Riga Technical University, Riga, Latvia

Improving the energy efficiency in telecommunication networks has been one of the most important research topics in the past few years. As a result, many energy-efficient algorithms have been proposed, some focusing only on energy savings, others also looking into the impact that energy saving has on other network performance metrics. The aim of this section is to provide an insight into some of the research work that has been done in the field of *green optical networks*. Two main topics are addressed: green optical access networks (Sections 10.6.1 and 10.6.2) and energy efficiency in optical core networks (Sections 10.6.3 and 10.6.4).

10.6.1   Energy Efficiency in Current and Next Generation Optical Access Networks

It has been widely recognized that reducing power consumption in data communication networks is becoming an important goal for reducing not only CO_2 emissions but also Operating Expenditure (OPEX). Optical access networks (OANs), such as Point to Point (P2P) optical access network and *Passive Optical Networks* (PONs), have been widely considered as the major wired access technologies of the future, with an expected massive deployment worldwide. Even though such technologies are more energy-efficient compared to other wired access technologies (e.g. xDSL), it is desirable to further reduce their energy consumption. The research community, network providers and standardization authorities have been targeting energy efficient solutions for both current OAN networks, including P2P, EPON, GPON, 10G-EPON and XG-PON, and next-generation optical access (NGOA) networks, such as Time and Wavelength Division Multiplexed PON (TWDM-PON or NG-PON2). This section deals with the issue of improving the energy efficiency of current and next-generation access networks (NGOA). In particular, energy efficiency in Time Division Multiplexed (TDM) Passive Optical Networks (PON) and in Time and Wavelength Division Multiplexed (TWDM) PON will be considered.

10.6.2   Energy Efficient Time Division Multiplexed Passive Optical Networks

Passive optical networks [10.362] are all-optical access networks in which the only active devices reside in the central office (i.e. the Optical Line Terminal – OLT) and at the customer premises (i.e. the Optical Networking Unit – ONU). PONs feature different topologies but the most popular is the tree topology in which an optical splitter (i.e. Remote Node – RN) distributes data from OLT to ONUs and vice versa (Figure 10.89). An ONU transceiver consists of a receiver and a transmitter. They transmit and receive at two different wavelengths: downstream wavelength around 1.5 μm for transmission from OLT to ONU and upstream wavelength around 1.3 μm from ONUs to OLT. The transmitter and receiver architectures are depicted in Figure 10.89. The OLT is equipped with an electronic Laser Driver (LD) and an optical laser diode transmitter (OTx). The OLT Burst Mode (BM) receiver consists

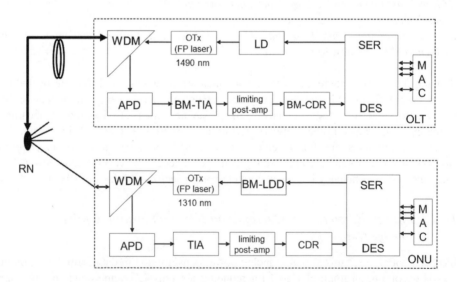

Figure 10.89   PON tree architecture and components

Table 10.18 Power consumption (mW) of discrete components in a front-end ONU receiver

Front-End Component	GEPON		GPON		10GEPON		10GPON	
	Avg	Range	Avg	Range	Avg	Range	Avg	Range
APD	2.6	2–3.75	2.6	2–3.75	2.6	2–3.75	2.05	0.5–3.75
TIA	83.4	56–112	83.4	56–112	123	105–160	123	105–160
LA	121	89–140	126	100–165	145	110–165	154	125–180
CDR	545	540–580	520	260–790	356		356	
SERDES	550	530–660	560	530–660	NA		NA	
Total Front-End		1302		1292				

of an Avalanche PhotoDiode (APD), an electronic BM TransImpedence Amplifier (TIA), an electronic limiting post-amplifier and an electronic BM Clock and Data Recovery (CDR) circuit. The electronic SERializer/DESerializer (SERDES) is utilized for serializing/deserializing electrical data for both downstream and upstream transmissions. The ONU architecture is similar to the OLT architecture, but features a BM laser driver (BM-LDD) in transmission, because of the tendency of its upstream data to burst.

Several standards have been proposed for passive optical networks, such as Gigabit capable Passive Optical Network (GPON ITU-T G.984) and Ethernet Passive Optical Networks (EPON IEEE 802.3-2008, Section 5). Currently, solutions for improving PON capacity through the utilization of WDM technologies and their reach (by placing active devices between the OLT and the ONUs) have been proposed [10.363].

The power consumption of PON equipments depends mainly on the electronic devices used to build them. Power consumption data are available from research papers [10.364] and components data sheets. Table 10.18 summarizes a survey that was conducted on power consumption of several ONU receiver components. The survey has been conducted by analyzing the data sheets available on-line from tens of vendors. In Table 10.18, data from discrete components are considered. In this case, CDR and SERDES are the ONU receiver components consuming the most in both 1 Gb/s EPON (GEPON) and GPON. For example, CDR and SERDES consume more than 80% of the entire front-end ONU receiver. Although data are not available for 10 Gb/s EPON and GPON, similar behavior can be expected.

However, the trend in electronics (and nowadays in optics as well) is to integrate several components. As shown in Table 10.19, integration allows a decrease in component power consumption. For example, the integration of CDR, LA and SERDES allows reduction of

Table 10.19 Power consumption (mW) of integrated components in a front-end ONU receiver

Integration	GEPON	GPON	10GEPON	10GPON
		Power (mW)		
CDR	545	520	356	356
CDR+LA	410	410	350	NA
CDR+SERDES	910	790	NA	
CDR+LA+ SERDES	610	610		

Table 10.20   Power consumption (mW) of ONU transceiver and ONU services of different PON systems

Variables	GEPON		GPON		10GEPON		10GPON	
	Avg	Range	Avg	Range	Avg	Range	Avg	Range
Transceiver	1350	1100–2500	1500	1040–2250	1600	1300–2300	1800	1800–1800
Back-End Circuit		2700		3150		5850		6750
Whole ONU (Services)		6000 (Ethernet Data Port+IPTV)		7000 (Triple Play+ Multicast Video)		13 000 (prediction)		15 000 (PoE on Gigabit Ethernet Port)

almost one half of the power consumption with respect to utilizing discrete components in 1Gb/s EPON.

Finally, the data collected in the survey confirm data already presented in the literature [10.365] for the overall ONU consumption (Table 10.20). Between 60–70% of the ONU power consumption is due to the PON transceiver and back-end circuit.

10.6.2.1   Approaches for saving energy in passive optical networks

Other approaches, other than maximizing device integration, have been proposed to *save energy in PONs* and in customer premises equipment (i.e. ONUs) in particular. This section provides a classification of the approaches proposed so far and underlines their characteristics. The classification is based on the EPON standard and on the two ISO/OSI layers EPON standards dealt with: the Physical and the Data Link layer. Thus, three classes of approaches are identified, as depicted in Figure 10.90: physical, data link and joint.

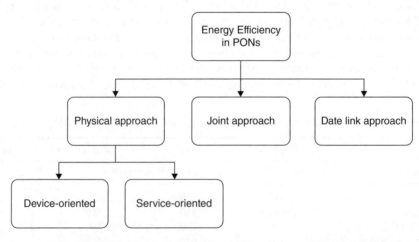

Figure 10.90   Taxonomy of approaches for energy-efficient PONs

Table 10.21 Current values for clock recovery in PONs

	T_{rec} (ms)							
	GEPON		GPON		10GEPON		10GPON	
	Avg	Range	Avg	Range	Avg	Range	Avg	Range
	Discrete components							
CDR	2.05	0.6–14	2.2	0.6–14	1.1	0.2–2	1.1	0.2–2
	Integrated components							
CDR+LA	1		1		2		NA	
CDR+DMUX	3		2		NA		NA	
CDR+LA+ DMUX	0.6		0.6		NA		NA	

Physical layer approaches include solutions aimed at reducing PON energy consumption by targeting the physical layer of the IEEE 802.3 protocol architecture (i.e. the Physical Medium Dependent (PMD) sublayer, the Physical Medium Attachment (PMA) sublayer and the physical coding sublayer (PCS)). They can be further divided into device- and service-oriented approaches. Device-oriented approaches aim at reducing the energy consumption of the devices, enabling the services provided by each sublayer. Device-oriented approaches include equipping transceivers with link rate adaptation (i.e. adaptive link rate (ALR) such as in copper Ethernet [10.366]), optimizing device energy consumptions [10.367], and utilizing new modulation formats for data transmission that are more energy efficient [10.368,10.369]. Service-oriented approaches aim at improving the performance of the service provided by a sublayer (e.g. the clock recovery within the PMA sublayer) to enable upper layer solutions (e.g. *sleep mode*) [10.370]. For example, current values for clock recovery T_{rec} are summarized in Table 10.21 for discrete and integrated components.

Data Link layer approaches target the Data Link layer of the IEEE 802.3 architecture (i.e. the MAC layer) and are based on the possibility of switching network elements to low power mode. Although the availability of a low power mode must be provided by the physical layer, such approaches can be classified as Data Link approaches, because they are based on extensions of the Multi Point Control Protocol (MPCP) and on modified Dynamic Bandwidth Allocation (DBA) algorithms. In principle, Data Link layer approaches require no physical layer modifications but low power mode support in the devices. Research in academia has proposed several methods for dynamically switching to sleep mode, both in EPON and GPON. Such approaches are based on sleep and periodic wake up of ONUs in EPON [10.371] and GPON [10.372]. Other approaches combine dynamic sleep mode with upstream and downstream scheduling, to minimize the impact of putting ONU to sleep on traffic delay [10.373].

In standard bodies, sleep mode has been standardized for Ethernet interfaces [10.366]. ITU-T G.sup 45 [10.374] proposes three types of power conservation standards, namely power shedding, dozing and sleeping (further divided into deep and fast sleep). The approaches mainly differ in the behavior of the ONU transmitter and receiver when traffic must be neither transmitted nor received at the ONU. In [10.365] it is showed that the combination of sleep mode (i.e. turning-off the Access Network Interface (ANI), that is ONU transmitter and receiver) and power shedding (i.e. shutting down unnecessary services at the User Network Interfaces (UNI), such as the voice (POTS) interface and the Ethernet interface as well as the related core services in the System on Chip (SoC)), has the potential of reaching 80% power savings. However, in the case of power shedding, all the packets arriving at the UNI are lost. Thus power shedding can be applied only during a long period of ONU inactivity.

Joint approaches are those that combine physical and data link layer approaches to reduce energy consumption. In general, approaches are classified as joint when they cannot be implemented separately. They include, for example, extensions to MPCP protocol for enabling sleep mode with physical layer approaches for remotely powering ONUs [10.375]. Other studies consider the combination of sleep mode and adaptive link rate to optimize power consumption while adapting to the real traffic demand [10.376]. Finally, some studies propose the combination of modified ONU architectures with dynamic sleep mode for improving the clock recovery after the ONU wake up [10.377]. In this way, the overhead time is minimized and the achievable energy efficiency increases.

10.6.2.2   Fundamental parameters affecting energy savings in EPONs

This section investigates the key parameters that affect PON energy savings, how they are related to one another and where work is needed to improve PON energy efficiency. EPONs with fixed downstream bandwidth allocation (FBA) and dynamic *sleep mode* in ONU are considered as a sample implementation. ONUs are turned ON during their assigned time slot, and then they are switched OFF. First of all, how much energy sleep mode can save and which are the key parameters affecting its performance are examined. P_a and P_s are the power consumption of ONU in active state (i.e. when ON) and sleep state (i.e. when OFF) respectively. T_{OH} denotes the overhead time for clock recovery after the ONU is turned ON at the ONU receiver. E and E' are the energy utilized by the ONU to receive the same D bits of data when no sleep mode and sleep mode are utilized, respectively:

$$E = T_c P_a \tag{10.42}$$

$$E' = (T_{sl} + T_{oh})P_a + (T_c - T_{sl} - T_{oh})P_s \tag{10.43}$$

where:

T_{sl} Slot time (i.e. the time allocated to each ONU for transmission/reception)
T_c Cycle time (i.e. the time between two consecutive transmissions/receptions by the same ONU).

Because there are N ONUs in the system and considering negligible the guard time between two consecutive time slots $T_{sl} = T_c/N$, the percentage of power savings can be expressed as:

$$\frac{E - E'}{E} = \frac{N-1}{N} + \frac{T_{oh}}{T_c}\frac{P_s}{P_a} - \frac{T_{oh}}{T_c} - \frac{N-1}{N}\frac{P_s}{P_a} \tag{10.44}$$

The percentage of power savings obtained as a function of the ratios T_{oh}/T_c and P_s/P_a if $N = 16$ and $T_c = 2$ ms, is depicted in Figure 10.91. It is shown that only by decreasing both

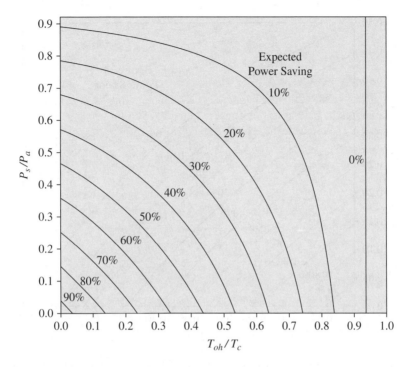

Figure 10.91 Power savings as a function of T_{oh}/T_c and P_s/P_a

the power consumed during sleep mode and the overhead time, high energy savings can be obtained. Even though components with very low energy consumption are developed, if they are not capable of recovering the clock quickly, high energy savings cannot be achieved and vice versa is also true.

Adaptive Link Rate (ALR) has been proposed in [10.376] to save energy in ONUs while supporting co-existence of 1 Gb/s and 10 Gb/s rates. In our study, the utilization of ALR and of sleep mode in 10 Gb/s EPON with 1G/10G ONUs are compared in terms of energy efficiency. As in the previous scenario, FBA is utilized for downstream transmission. The data to be transmitted downstream D are equal to the amount of data transmitted during 1 slot time at 1 Gb/s. The cycle time T_c is the same for both downstream transmission rates. $T_{slot1} = T_c/N$ is the time required to transmit/receive D data bits in 1 Gb/s EPONs. $T_{sl10} = T_c/(10\,N)$ is the time required to transmit/receive D data bits by 10G ONUs before going to sleep. P_{a1}, P_{a10} and P_{s10} are the power consumption of 1G ONU in active state, 10G ONU in active state and 10G ONU in sleep state, respectively. E_1 and E_{10} are the energies required to receive the same D data bits by 1G and 10G ONUs, respectively. If ALR is used, ONUs are assumed not to utilize sleep mode, that is, they are always ON during T_c. In addition, from the results of the survey in Table 10.20, we assume that $P_{a10} = 2P_{a1}$. E_1 and E_{10} can be written as:

$$E_1 = T_c P_{a1} \tag{10.45}$$

$$E_{10} = (T_{sl10} + T_{oh})P_{a10} + (T_c - T_{sl10} - T_{oh})P_{s10} \tag{10.46}$$

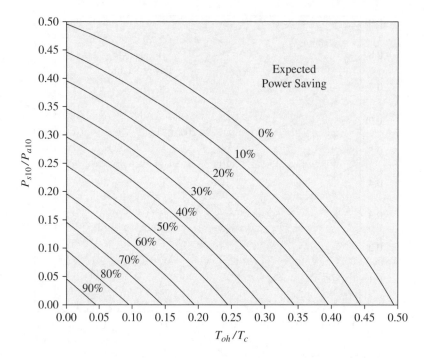

Figure 10.92 Power savings of sleep mode vs. adaptive link rate as a function of T_{oh}/T_c and P_{s10}/P_{a10}

Then power savings achievable by sleep mode with respect to adaptive link rate can be written as:

$$\frac{E_1 - E_{10}}{E_1} = \frac{10N - 2}{10N} + \frac{2T_{oh}}{T_c}\frac{P_{s10}}{P_{a10}} - \frac{2T_{oh}}{T_c} - \frac{20N - 2}{10N}\frac{P_{s10}}{P_{a10}} \tag{10.47}$$

Figure 10.92 shows the power savings in the form of a contour plot as a function of the ratios T_{oh}/T_c and P_{s10}/P_{a10}. If the power consumed during sleep mode P_{s10} by the ONU is greater than half of the power consumed during active state at 1 Gb/s (i.e., $P_{s10} > P_{a1}$), sleep mode is not beneficial with respect to adaptive link rate, even if overhead time is nil. Moreover, even in this case, both ratios must be minimized to obtain better performance with sleep mode than with ALR.

10.6.3 Energy Efficient Time and Wavelength Division Multiplexed Passive Optical Networks

Time- and Wavelength-Division Multiplexing (TWDM) PONs are currently under standardization by ITU-T as the Next Generation PON 2 (NG-PON2) [10.378]. Energy efficient schemes proposed for TDM PONs can still be applied to TWDM PON for the subset of Optical Network Units (ONUs) transmitting and receiving at the same wavelength. However, because of the presence of multiple transmitters and receivers at the Optical Line Terminal (OLT), novel opportunities arise for improving the OLT energy efficiency. However, some challenges must be faced for making energy-saving schemes effective.

The classification of energy-saving schemes at the OLT can be similar to the one presented in [10.379] and reported in the previous section for TDM PONs. Physical layer approaches aim at reducing the energy consumption of the OLT components to meet, for example, requirements similar to those specified by the Broadband Equipment Code of Conduct [10.380] for TDM PONs. Data link layer approaches are based on dynamically aggregating downstream (DS) and upstream (US) transmission to/from ONUs dynamically based on the traffic statistics, thus allowing to turn OFF unutilized OLT transceivers [10.381,10.382]. In general, dynamic aggregation is based on Dynamic Wavelength and Bandwidth Allocation (DWBA) executed at the OLT. Hybrid approaches are those that combine data link layer approaches with physical layer approaches that, instead of decreasing OLT energy consumption, enable data link layer solutions. An example is the development of fast tuning transceivers at the ONU, as the example presented in [10.383], to decrease the tuning time after a reconfiguration, thus decreasing the experienced delay and, potentially, increasing energy savings.

Many of the current schemes for saving energy at the OLT are based on data link layer approaches thus, in turn, on different DWBA schemes. The often considered TWDM PON architecture is similar to that proposed in [10.384], which follows the guidelines published in [10.378] (this architecture is better detailed in the next section). The main characteristic of such architecture is the possibility of the ONUs to transmit/receive data to/from any of the fixed-wavelength OLT transceiver (also called Optical Subscriber Unit, OSU as in [10.387]) by means of tunable transceivers.

As also stated in (10.385,10.389), the problem of minimizing the number of wavelengths (i.e. OSUs) necessary to serve all the demands by the ONUs is equivalent to a bin-packing problem. However, because bin-packing is an NP-hard problem, the general approach to implement the DWBA is to utilize two-phase scheduling schemes. In (10.385) and (10.387), dynamic wavelength minimization and assignment (DWA) is decoupled from dynamic bandwidth allocation (DBA) within a wavelength. In particular, DWA and DBA have two different periods, a longer one for DWA and a shorter one for DBA. The longer period for DWA is motivated by the fact the DWA might imply ONU transceiver tuning incurring in an additional delay. In (10.387), the DWA reassigns to different OSUs part of the ONUs unassigned during the latest DBA cycle contained in a DWA cycle (i.e. before DWA) is triggered).

In short, ONUs that on average have been assigned slots smaller than the requested ones are reassigned to (already active) OSUs whose slot is almost full but that can accommodate the average allocated slot size to the ONU. Otherwise, a new OSU must be activated. However, to synchronize the end of a DBA cycle with the end of a DWA cycle, the DWA cycle must be a multiple of the DBA cycle. In (10.387), the authors show that, in a specific ONU traffic variation scenario, the proposed DWBA algorithm effectively balances the OSU load and avoids too frequent re-assignment of ONUs. In addition, an experimental evaluation of the proposed scheme is presented in (10.388).

In [10.385], DWBA is combined with cyclic sleep at the ONU, thus improving energy efficiency at both the OLT and at the ONUs. The DWA (there called WMA) is based on a heuristic, run at fixed DWA cycles, in which the number of OSUs to be activated is decided based on the ratio between the sum of the data requested by all the ONUs and the amount of data a wavelength can transmit over the DWA cycle. The assignment of ONUs to active OSUs is done in a round-robin fashion by assigning an ONU to an OSU "… if the normalized load of that ONU is smaller than twice the remaining capacity of the OSU ..." DBA (there called TSA) is based on an upstream and downstream centric (UDC) algorithm, previously proposed by the authors in [10.386], where the transmission slot of an ONU is computed as a function of the buffer backlog of the downstream and the upstream traffic of the ONU.

The issue of synchronization between the DWA cycle and the DBA cycle is solved by introducing Adjusting Cycles (AdC). In AdC, when the remaining time to be assigned in a DWA cycle of a specific OSU becomes smaller than the maximum DBA cycle, the remaining time of the OSU is equally distributed among all the ONUs assigned to that OSU, proportionally to the ONU bandwidth requests. Results show that the proposed method reduces energy consumption to about 30% of the one without applying the proposed method. Moreover, it is shown that, for a specific tuning time and at a fixed network load, an optimal value of the WMA cycle time exists that minimizes the average frame delay. However, as also shown in [10.382], the average frame delay presents spikes at network loads corresponding to the OSU (de)activation, because the OSU (de)activation threshold is fixed at multiples of the load that can be served by a single OSU.

In [10.389], an analysis is performed of the impact of the allowed ONU level of aggregation into active OSUs on the average frame delay as a function of the ONU load. The considered number of available OSUs at the OLT is 64 and the total number of considered ONUs is 128. OSUs are assumed to support 10 Gb/s, while the maximum ONU data rate is assumed to be 1 Gb/s. If the ONU load is high (i.e. ~800Mb/s), between 50 and 75% of the OSUs must be active to guarantee an average frame delay of between 6 and 8 ms. If the ONU load is low (i.e. about 200 Mb/s), between 6 and 15% of OSUs must be active to guarantee an average frame delay below 10 ms. Moreover, in [10.389], a scheme is proposed for dynamic consolidation of ONUs to OSUs. The scheme, given an initial assignment of ONUs to OSUs, reassigns some ONUs to OSUs, minimizing the energy consumption (i.e. the number of active OSUs), load balancing among active OSUs, and minimizing ONU migration to avoid delay penalties due to ONU transceiver tuning time and signaling time. Results show that the proposed scheme achieves an energy consumption per bit very close to the optimal one computed through the solution of the integer liner programming problem of assigning the ONUs to the minimum number of OSUs, while satisfying Quality of Service (QoS) constraints.

Although all the aforementioned methods provide interesting solutions to the problem of minimizing the OLT energy consumption, they assume that the OLT energy consumption is directly proportional to the energy consumed by transceivers (i.e. OSUs) only. However, in real systems, there is bulk energy consumption at the OLT needed for powering shelves and racks. Such energy consumption is independent of the number of active transceivers. In the following sections, an evaluation is made of how much energy can be really saved by utilizing DWBA, also taking into account the bulk energy consumed at the OLT. Moreover, the impact of the tuning time on the average frame delay is also shown.

10.6.3.1 TWDM PON and OLT model

The considered TWDM PON architecture (Figure 10.93a) consists of OLTs featuring fixed transceivers (i.e. ports or OSUs) transmitting and receiving at different wavelengths installed in line cards (LCs), a wavelength multiplexer, a passive splitter, and several ONUs featuring tunable transceivers, as proposed in [10.384]. In addition, Optical Amplifiers (OAs) can be utilized as boosters and pre-amplifiers.

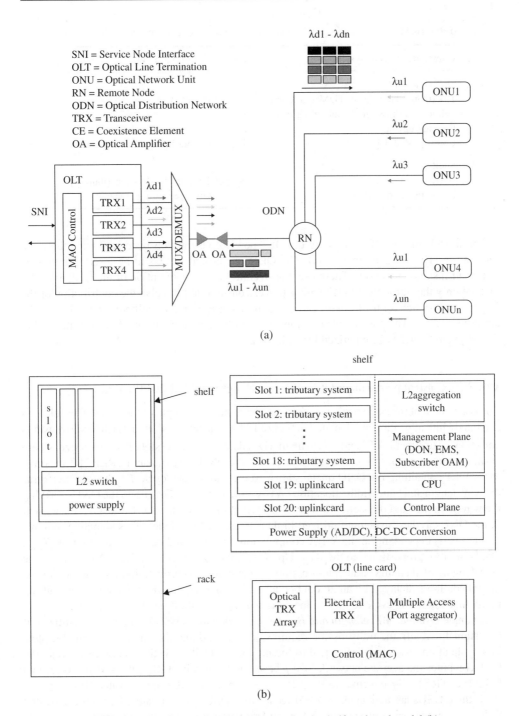

Figure 10.93 TWDM-PON model (a) and rack, shelf, and card model (b)

Table 10.22 OLT components and sub-subsystem power consumption values

Component or Sub-system	Power consumption [W]
OLT (line card)	
10×10Gb/s TRX Array TDMA coloured line card (burst mode, including FEC)	20 (per line card)
Port aggregator	0.5 (per port, per 1 Gb/s)
Control (MAC)	1 (per port, per line card)
Shelf	
Basic shelf power	90 (per 20 slots [18 slots for tributary])
L2 aggregation switch	1 (per 1Gb/s, per card, per shelf)

The model considered to compute the OLT power consumption is derived from the work reported in [10.390] (Figure 10.93b). The OLTs are hosted in LCs to be inserted into slots of a shelf installed in a rack. Therefore, the overall power consumption of the OLT must not include only the transceiver (TRX) array power consumption but also the contribution of the Layer 2 (L2) switch for traffic aggregation, the port aggregator, the shelf power supply, and the amplifiers, if any. The power consumption values of the components or sub-system utilized in the considered OLT is summarized in Table 10.22.

10.6.3.2 Control protocol and dynamic wavelength and bandwidth allocation

The considered network control protocol time chart and an example of the utilized wavelength and bandwidth allocation are depicted in Figure 10.94. The time is slotted. Each ONU transmits and receives upstream (US) and downstream (DS) data during time slots T_i^j of fixed size, where i is the slot index and j is the cycle index. In its slot, the ONUs also transmit a REPORT message to the OLT containing the ONU queue status. At the end of each cycle T^{cj}, the OLT performs the wavelength and bandwidth allocation (WBA) and then sends GATES to the ONUs carrying the wavelength(s) and the slot(s) in which they will transmit in the next cycle. GATEs are sent to the ONUs at the wavelength utilized in their latest scheduled slot. This signaling requires a time that is dependent on the WBA computation time T_{sch} and on the signaling time T_{gate}, that is, in turn, proportional to the RTT. Upon reception of the GATEs, the ONUs tune to the wavelengths of their first transmission slot. To maintain synchronization among the ONUs, the scheduling is delayed by an amount equal to the time required by the ONUs transmitting in the first slot of the cycle to tune to the assigned wavelength, if needed.

A simple *Dynamic Wavelength and Bandwidth Allocation* (DWBA) algorithm is utilized. It is assumed that a frame transmission requires one slot and that the number of available slots in a cycle at one wavelength is equal to the number of ONUs. At the end of each cycle, the OLT computes the average network load ρ based on the received REPORTs. If $\rho < 0.25$, one OSU (i.e. TRX) only is activated. If $0.25 \leq \rho < 0.5$, two OSUs are activated. If $0.5 \leq \rho < 0.75$, three OSUs are activated. If $\rho \geq 0.75$, all four OSUs are activated. OSUs are activated following an increasing wavelength ID order. Irrespectively of the queued frames, each ONU is assigned one slot. Then the remaining slots belonging to the active wavelengths are assigned to the ONUs proportional to the number of queued frames, by utilizing the largest remainder

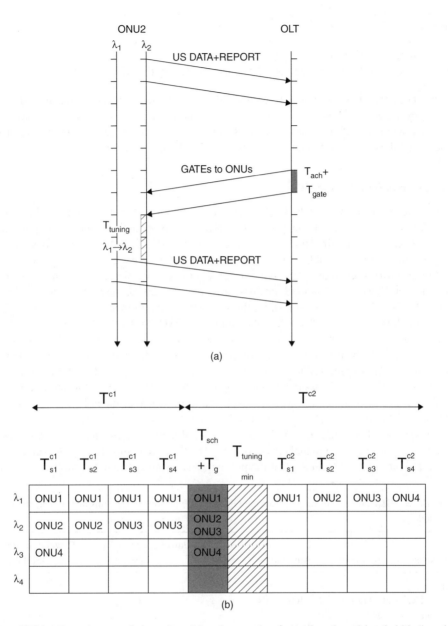

Figure 10.94 Control protocol time-chart (a) and example of wavelength and bandwidth (i.e. slot) assignment (b)

method (also known as Hare-Niemeyer method). Thus, the probability of assigning a slot i to an ONU n is computed as:

$$P_i(n) = Q(n) \Big/ \sum_{k=0}^{N-1} Q(k) \tag{10.48}$$

where:

   N Number of ONUs
   $Q(n)$ Number of frames queued at ONU n as reported by the REPORT message

It must be noted that if a frame arrives at an ONU during a cycle, it can be transmitted only in the next cycle, even if slots are available in the current one, because the WBA is performed at the beginning of each cycle. Slots are then scheduled and assigned a wavelength. A longest-first (in terms of slots assigned to one ONU) and first-fit scheme is utilized (Figure 10.94b).

10.6.3.3   Evaluation scenario

Performance evaluation is conducted by means of a custom built time-driven simulator coded in C. For simplicity, but without loss of generality, the propagation delay is assumed to be negligible (i.e. Round Trip Time $= 0$) and time is slotted. US and DS traffic is symmetric and generated based on a Bernoulli arrival process. That is, in each slot I, each ONU n has a probability $P_i(n)$ of generating/receiving a frame. Frame transmission is assumed to require one time slot. Because US and DS data transmissions are locked, the performance evaluation is valid for both US and DS data. In all the simulations, the number of ONUs is $N = 4$, the number of slots per cycle is $S = 4$, and the number of OSUs (TRXs) at the OLT is $L = 4$.

The power consumed by the OLT during ON and OFF periods is computed similar to that reported in [10.390] by assuming a direct proportion between the values reported in Table 10.23 and the considered configuration. The computed values are reported in Table 10.23. The 4×10 Gb/s TRX array is assumed to occupy two shelf slots (one for TX and one for RX) and the power it consumes is linearly proportional to the one consumed by the 10×10 Gb/s TRX array. Moreover, the power consumed by the TRX array is assumed to be proportional to the ports (i.e. TRXs) that are ON. Therefore, if all of them are OFF, the TRX array does not consume power. Even though the power consumed by the port aggregator is proportional to the aggregated bit rate, it is assumed that the port aggregator presents a bulk power consumption

Table 10.23   OLT power consumption during ON and OFF periods in the considered rack, shelf and card models

Component or Sub-system	Amount	Power consumption ON [W]	Power consumption OFF [W]
OLT			
4×10G TRX array TDMA coloured line card	2 line cards	16	0
Port aggregator	80 Gb/s	40	20
4xXG-PON Control MAC	8 ports	8	8
Shelf			
Basic shelf power	3 slots (2 TRX array, 1 MUX/DEMUX)	10	10
L2 Aggregation Switch	80 Gb/s	80	80
Total		154	118

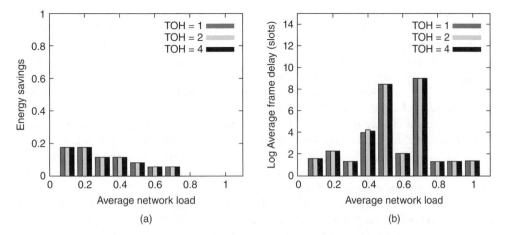

Figure 10.95   Energy savings (a) and average frame delay

of 20 W. The remaining power consumption is linearly proportional to the number of TRX that are ON. For all the other components, the power consumed represents a bulk power consumption that does not decrease if the TRX are OFF.

The evaluation parameters are the OLT average energy savings η and the average frame delay D. The OLT average energy savings are defined as the percentage of energy saved when the TRXs are dynamically turned ON and OFF with respect to the energy consumed when all the TRXs are always ON. The average frame delay is defined as the average delay experienced by a frame between its arrival and its transmission.

10.6.3.4   Results

Figure 10.95 shows the *energy savings* and the natural logarithm of the average frame delay (expressed in units of slots) as a function of the average network load and of three different values (expressed in units of slots) of the tunable TRX tuning time (called tuning overhead T_{OH}).

The results reported in Figure 10.95a show that, due to the bulk power consumption of the OLT, a maximum value of energy savings equal to about 20% can be achieved. Figure 10.95b confirms that frame delay increases if the average network load is close to the maximum load that can be supported by the active OSUs. This is mainly due to the impact of tuning overhead and to the sub-optimality of the WBA. The frame delay increase is higher for high average network loads. However, it appears that the T_{OH} does not impact on the average frame delay, at least not in the scenario considered here.

10.6.4   *Spectral and Energy Efficiency Considerations in Single Rate WDM Networks with Signal Quality Guarantee*

An important challenge for operators is to accommodate large traffic demands while maintaining, at the same time, the network power consumption at acceptable levels [10.391]. One way to achieve this objective is to make a more efficient use of the existing fiber resources, such as increasing the *spectral efficiency* (SE). On the other hand, a higher SE, achieved with higher transmission rates and/or smaller channel spacing, in turn may exacerbate the effect of optical transmission impairments [10.392], reducing the maximum distance that an optical signal can travel without regeneration, such as the transparent reach. This means that when a specific (QoT) level has to be ensured at the receiving node, the optical signal might need to be regenerated along the way [10.393]. 3R operations (i.e. re-amplification, re-timing, re-shaping) usually involve optical-to-electrical-to-optical (OEO) conversion, which translates into: (i) the need for extra equipment (i.e. 3Rs) at selected nodes in the network, and (ii) an increased overall power consumption. Therefore SE, transparent reach and power consumption are closely interrelated and an optimization process needs to carefully consider how these three parameters influence each other.

In the literature, this aspect is only partially addressed. The authors in [10.392] investigate (over a single fiber span) what is the maximum SE that still allows to keep the QoT above a certain threshold, but they do not look into the relationship between QoT and power consumption. This latter aspect is addressed in [10.393], where an efficient optical network design strategy with signal quality guarantee is proposed, but no investigation is made on how various spectral efficient solutions influence the system power consumption. The impact of coherent and non-coherent technologies on the power consumption in translucent networks is studied in [10.394], but no conclusions are drawn on the maximum achievable transparent reach. Finally, usually systems and components are modeled in terms of their power consumption, but no considerations are made on how efficiently they utilize the limited frequency resources of the transmission band they use [10.391,10.395].

This section explores how SE, transparent optical reach and power consumption influence each other in the same transmission system. More specifically, the objective is to evaluate the power cost per transmitted bit (W/bps) required to establish end-to-end connections while guaranteeing a specific QoT level. This is achieved by comparing the maximum SE and the corresponding power consumption of a number of transponders (TSP) and 3Rs for different modulation formats, used for transmission over one or more unamplified fiber spans when a given QoT level is required. The modulation formats under examination are 10 Gbps and 40 Gbps Non-Return-to-Zero On-Off Keying (NRZ OOK); 10 Gbps and 40 Gbps NRZ Differential Phase-Shift Keying (NRZ-DPSK); and 10, 40 and 100 Gbps coherent Dual Polarization Quadrature Phase-Shift Keying (DP-QPSK). SE (bps/Hertz) is dependent both on the number of bits per symbol used by a specific modulation format, and on the minimum allowable channel spacing, so as to avoid excessive signal degradation due to cross-talk from adjacent channels. The power efficiency (W/bps) for each modulation format is calculated using the total power consumption of the corresponding transponders and (when required) the energy cost for additional 3Rs used at intermediate nodes.

10.6.4.1 Simulation set-up

This section describes the assumptions used to simulate a fiber-optic link, and the configuration parameters of the link and the system components.

The RSoft's OptSim software is used for simulations. The software solves the non-linear Schrödinger equation using a time domain split-step algorithm. The Q-value of the received signal is used to measure the optical signal quality. The accuracy of the obtained Q-factor values strongly depends on the total number of simulated bits. In our numerical experiments, we used at least 1000 bits that yields a Q-factor uncertainty of less than 0.77 dB [10.396].

The scenario under examination is based on a five-channel wavelength division multiplexing (WDM) system that consists of a transmitter (Txi), a wavelength multiplexer (MUX), a booster amplifier, a standard single-mode fiber (SSMF) used for transmission, a chromatic dispersion compensation module (DCM) and a receiver (Rxi). The number of channels is chosen to take cross-talk from adjacent channels into account. The booster amplifier – an Erbium Doped Fiber Amplifier (EDFA) with a fixed 10 dBm output power – is placed after the multiplexer. In a realistic long-reach WDM system, there would also be fixed-gain inline optical amplifiers (to compensate for the attenuation in the SSMF and the DCM), and a pre-amplifier before the demultiplexer. However, since this study is focused on the maximum tolerable SE for signal transmission over a single span, these amplifiers are not considered. The optical dispersion compensation module is based on a dispersion compensating fiber (DCF). The SSMF and DCF lengths are 40 km and 8 km, respectively, and their characteristics are described in [10.397]. Optical dispersion compensation is omitted for the DP-QPSK modulation formats. Instead, electronic dispersion compensation is used at the receiver side.

Tx and Rx vary depending on the modulation format and the bit rate considered during each specific experiment. The OptSim block diagrams for the transmitter and receiver of NRZ-OOK and NRZ-DPSK transmitting and receiving units can be found in [10.392], while those for DP-QPSK are shown in [10.398]. The transmitters are driven by rectangular-shaped NRZ signals filtered through an electrical low-pass Bessel filter. For DP-QPSK, a Super-Gaussian optical filter is used after the transmitter. A similar optical filter is applied before detection at the receiver side for all modulation formats. The number of poles and the bandwidth of the optical and electrical filters are shown in Table 10.24. These parameters ensure the highest Q-values for the WDM channels arranged using the ITU-T Rec. G.694.1 frequency grids based on: 25 GHz channel spacing for the 10 Gbps NRZ-OOK, NRZ-DPSK, DP-QPSK, and 40 Gbps DP-QPSK; 50 GHz for the 100 Gbps DP-QPSK; and 200 GHz for the 40 Gbps NRZ-OOK and NRZ-DPSK. Such frequency intervals are common for the modulation formats and for the per-channel bit rates considered. The value of filter bandwidth is chosen as a compromise between the levels of linear cross-talk (using the channel spacing described above) and signal distortions (due to filtering). It is also assumed that noise is negligible. The OOK signals are detected using a single photodetector and the DPSK signals with two balanced photodetectors, respectively. The PIN photodiodes are assumed to have 80% quantum efficiency corresponding to 1 A/W response.

Table 10.24   Electrical and optical filter parameters

Modulation format and bit rate	1&3	2&4	5	6	7
Bessel electrical filter:					
• Number of poles	5	5	5	5	5
• −3 dB bandwidth, [GHz]	10	65	4	14	20
Super-Gaussian optical filter:					
• Order	1	2	2	2	2
• −3dB two-sided bandwidth, [GHz]	14	70	10	40	40

Key: 1 is 10 Gbps NRZ-OOK; 2: 40 Gbps NRZ-OOK; 3: 10 Gbps NRZ-DPSK; 4: 40 Gbps NRZ-DPSK; 5: 10Gbps DP-QPSK; 6: 40 Gbps DP-QPSK; 7: 100 Gbps DP-QPSK.

A coherent receiver is used to decode DP-QPSK. The electrical receiver filter is assumed to be similar to the electrical transmitter filter. Finally, it is assumed that the major system impairment that limits the spectral efficiency is the linear interchannel cross-talk, even when the numbers of channels increases beyond five. The fixed 10 dBm output power for the 5-channel system (2 mW per channel) and the fixed link distance of 40 km were therefore chosen conservatively so that the influence of non-linear cross-talk and of the receiver noise would be small. This assumption simplifies the simulations considerably and makes the results more transparent compared to a fully realistic case where the channel spacing, the output power and the link distance would have to be optimized and varied depending on the number of WDM channels. The current assumptions may overestimate the power consumption at a low number of channels, where the channel link distance could be chosen to be longer than 40 km, and may underestimate the power consumption at a high number of channels where non-linear impairments may start to become the limiting factor.

10.6.4.2   Results and discussion

In this section, the tradeoff between SE, power consumption and transparent reach is evaluated. The focus of the study is on systems without *Forward Error Correction* (FEC) and with a required *Bit Error Rate* (BER) level of less than 10^{-9}.

The value of the power consumption of TSPs and 3Rs computed for the different modulation format is summarized in Table 10.25. Their functionalities and schematics are the same as those presented in [10.402], with the exception of the FEC capabilities, that are omitted in this study.

The purpose of the table is to assess the power efficiency of each transponder and 3R options. The power values are calculated based on the data from [10.402] and then benchmarked using other research papers and data sheets. The value of the power consumption of the 100G transponder (with FEC) comes from the work in [10.401]. In this study, an additional 20% is added to the consumption values of both TSP and 3Rs to also account for the so-called management power [10.391,10.402].

The table shows that if the transmission distance is limited to one SSMF span (i.e. if no 3Rs are needed) for low capacities (i.e. below 30 Gbps), the 10 Gbps NRZ-OOK and NRZ-DPSK transponders are the most power-efficient (i.e. in terms of W/bps). On the other hand, for capacities larger than 80 Gbps, the coherent 100G DP-QPSK becomes more convenient.

Table 10.25   Power consumption of transponders and 3Rs

No. and type	Equipment	Power, [W] (no FEC)	TSP found in [Reference]
1: 10G NRZ-OOK	TSP/3R	22.4/20.8	34.0 [10.391], 35.0 [10.399]
2: 40G NRZ-OOK	TSP/3R	69.4/42.8	–
3: 10G NRZ-DPSK	TSP/3R	19.7/16.6	–
4: 40G NRZ-DPSK	TSP/3R	69.8/43.6	85.0 [10.400]
5: 10G DP-QPSK	TSP/3R	40.6/58.4	–
6: 40G DP-QPSK	TSP/3R	120.4/144.8	113.0 [10.391]
7: 100G DP-QPSK	TSP/3R	132.1/158.5	139.0 [10.401], 188.0 [10.391]

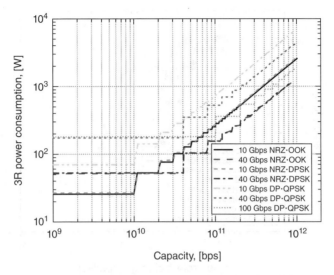

Figure 10.96   Power consumption of different 3Rs types as a function of the transmitted capacity

It can also be concluded that coherent 10G and 40G DP-QPSK transponders have the least power-efficiency. The situation is different for 3Rs (Figure 10.96), where those utilizing direct detection are more power-efficient than the coherent 3Rs. 10G NRZ-OOK or 10G NRZ-DPSK 3Rs have the lowest power consumption at capacities lower than 20 Gbps. At larger capacities, 40G NRZ-OOK or 40G NRZ-DPSK 3Rs are the most power-efficient. Note that in Figure 10.96, the line in red (40G NRZ-OOK) and the one in pink (40G NRZ-DPSK) overlap.

If the capacity of a system is limited by the available optical bandwidth, the spectral efficiency of a modulation format becomes an important parameter. It depends both on the number of bits each symbol can carry and on the minimum possible channel spacing that it is possible to use without significantly degrading the signal. Figure 10.97 presents the degradation of the Q-value as a function of the spectral efficiency over a distance of 40 km, the medium span length for in-line optical amplifiers in WDM optical networks [10.395].

To determine the maximum spectral efficiency of each modulation format, the minimum tolerable channel spacing is defined as giving BER $< 10^{-9}$ for the most affected channel. This corresponds to a minimum Q-factor value of 6 (=15.56 dB on the electrical side). The channel spacing is restricted to a 3.125 GHz granularity of the central channel frequency in accordance with an ITU-T G.694.1 fixed grid. The maximum spectral efficiency that is obtained for the different modulation formats is presented in the first column of Table 10.26.

Once the value of the maximum spectral efficiency is fixed, based on the maximum acceptable BER value, it is interesting to evaluate what is the maximum transmission capacity (and the respective power consumption) that a given modulation format can offer over the entire C-band (4.4 THz). This gives an idea of how efficiently the total available bandwidth of the C-band is used. Figure 10.98 shows that for each modulation format, the power consumption increases linearly with the capacity provided as more and more wavelength channels are utilized until the maximum spectral efficiency given in Table 10.26 is reached.

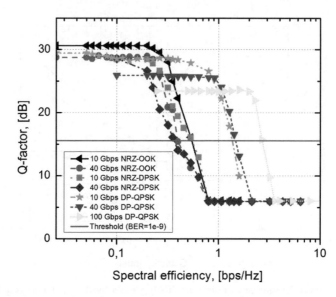

Figure 10.97   Q-factor values detected for the worst channel as a function of spectral efficiency

A 100G DP-QPSK transponder gives the lowest overall power consumption in combination with high spectral efficiency. The number of needed transponders and the total (transponder) power for maximum spectral efficiency of each modulation format are also given in Table 10.26.

The maximum spectral efficiency for each modulation format has so far been calculated assuming a link span of 40 km. If the same maximum SE value is used over a distance longer than 40 km, signal regeneration will be required to keep the *BER* value below 10^{-9}. When more than one fiber span is used, the total power consumption will then mainly depend on the number and the power consumption of the 3Rs, and no longer on the transponder power consumption.

Figure 10.99 presents power efficiency values (i.e. W/bps) over a given distance for all the considered modulation formats. 3Rs are assumed to be deployed at the end of each fiber span (i.e. every 40 km to ensure the required BER value over the entire end-to-end connection).

Table 10.26   SE, capacity and power consumption values when BER $\leq 10^{-9}$

No. and type	SE [bps/Hz]	Capacity [Tbps]	No. of TSP	TSP power [kW]
1: 10G NRZ-OOK	0.53	2.34	234	6.20
2: 40G NRZ-OOK	0.53	2.36	59	4.94
3: 10G NRZ-DPSK	0.36	1.59	159	3.77
4: 40G NRZ-DPSK	0.36	1.60	40	3.36
5: 10G DP-QPSK	1.28	5.64	564	27.57
6: 40G DP-QPSK	1.40	6.16	154	22.35
7: 100G DP-QPSK	2.67	11.80	118	18.70

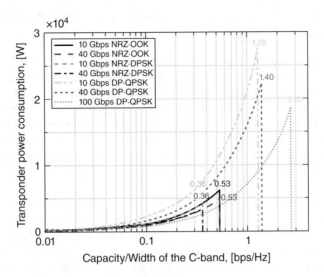

Figure 10.98 Power consumption values for the different types of transponders in function of the transmitted capacity over C-band (4.4 THz)

If the transmission distance is longer than 80 km, the lowest energy per bit is with 40 Gbps NRZ-OOK and with 40 Gbps NRZ-DPSK transmission technologies. For a 1000 km point-to-point connection, using 100 Gbps DP-QPSK instead of 40 Gbps NRZ-OOK transponders increases the power consumption by more than 35% for each transmitted bps.

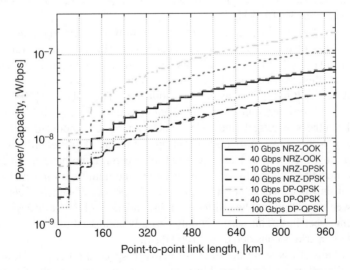

Figure 10.99 Power per bit/s as a function of the point-to-point distance

10.6.5   Spectral versus Energy Efficiency in Mixed-Line Rate WDM Systems with Signal Quality Guarantee

Network operators are currently facing a challenge that is two-fold [10.391,10.403]. On the one hand, the power consumption of the network infrastructure amounts to approximately 12% of the total power necessary to accommodate the Internet traffic. This value is expected to reach 20% by 2020 [10.402]. On the other hand, the Internet traffic is constantly growing (i.e. 38% annually, at least for the IP backbone [10.405]) and becoming increasingly heterogeneous with respect to the application supported, for example, a mix of low-bit-rate and high-bit-rate services [10.403]. This scenario translates into the need for a capacity upgrade of the network infrastructure, while limiting as much as possible the increase in the total power consumption [10.391].

The *Mixed-Line Rate* (MLR) concept can be helpful in this regard [10.406–10.408]. MLR provides operators with a cost-efficient solution to increase the network capacity. This in fact makes it possible to allocate additional high speed wavelength channels (e.g. 40 Gbps and/or 100 Gbps) next to the legacy ones (10 Gbps), while at the same time allowing for a contained power consumption increase due to the deployment of transponders using highly efficient modulation formats (i.e. with a high number of bits per symbol) [10.409]. In a MLR scenario, wavelength channels are organized into several groups (here referred to as *sub-bands*), each of which consists of wavelength channels operating on the same bit rate, for example, there can be groups of channels transmitting at 10 Gbps, 40 Gbps and 100 Gbps [10.391].

The design of MLR optical networks can be optimized with respect to a number of performance parameters. One aspect that is important to consider to further increase the network capacity is *spectral efficiency* (SE). There are a number of ways in which spectral efficiency can be improved. One is to increase the percentage of wavelength channels supporting high rates (e.g. operating at 100 Gbps) on a single fiber link. Another possibility (not mutually exclusive with the first one) is to reduce the frequency spacing (i.e. the *subband spacing*) between the various sub-bands, where usually 200 GHz intervals are used between 10G/40G and 40G/100G sub-bands [10.391].

As was the case for the single-line rate case, an improved spectral efficiency may in turn reduce the maximum distance that an optical signal can travel without regeneration, that is, the *transparent reach* [10.408]. In other words, also in an MLR scenario, there is a clear trade-off between spectral efficiency, transparent reach and power consumption in the network.

In the literature, this trade-off is only partially addressed. The authors in [10.410] investigate the relationship between energy and cost minimization while designing MLR optical networks. In [10.408], the authors investigate the trade-off between bit rate, modulation format and cost of a transponder with respect to its transparent optical reach. However, in this literature, no investigation is made on how the relative width of a certain sub-band in an MLR system influences the network power efficiency. In addition, the authors do not address the more general relationship between achievable spectral efficiency, transmission capacity, optical reach and power consumption when different transponders working at different bit rates co-exist in the same fiber link. This aspect is only partially addressed in [10.394], where only a single-line rate scenario is considered (i.e. 100 Gbps).

This section aims at assessing the trade-off between spectral efficiency, power consumption and optical reach in an MLR-based wavelength division multiplexing (WDM) transport network. More specifically, the objective of this work is to evaluate the power consumption per transmitted bit (W/bps) required to establish end-to-end connections, while guaranteeing a specific *quality of transmission* (QoT) level. This is accomplished by evaluating a number of spectral efficient MLR solutions, where the number of wavelength channels allocated to each sub-band used for transmitting over one or more unamplified fiber spans is varied, while keeping the required BER level at the receiving node fixed. Spectral efficiency (bps/Hertz) is dependent on the number of bits per symbol used by a specific modulation format; on the number of wavelength channels within the same sub-band and their minimum allow-able channel spacing; and on the minimum frequency spacing between sub-bands working at different rates. The values for the channel spacing (i.e. within the same sub-band) and the sub-band spacing (i.e. between sub-bands) are chosen to avoid excessive signal degradation due to cross-talk from adjacent channels and sub-bands, respectively. The power efficiency (W/bps) for each modulation format is calculated using the total power consumption of the corresponding transponders and (when required) the energy cost for additional 3Rs used at intermediate nodes.

10.6.5.1 System architecture and simulation set-up

This section describes the assumptions for the MLR WDM link used to investigate the rela-tionship between the minimum value of the *sub-band spacing* and the signal quality at the receiver. This section then concludes by presenting the set-up used for the RSoft OptSim simulation software.

The MLR WDM link under examination is shown in Figure 10.100. The transmission distance is limited to one fiber span. The WDM system consists of five transmitters that may be different, depending on the rates considered in a specific experiment (i.e. 10G/10G/100G/10G/10G, 40G/40G/100G/40G/40G or 10G/10G/100G/40G/40G, respectively); a wavelength multiplexer (MUX); a booster amplifier; a single-mode transmission fiber (SMF); a chromatic dispersion compensation module (DCM); a wavelength demultiplexer (DEMUX); and five receivers (Rx_i) that exactly match the sequence and the order of the transmitters. In a realistic long-reach WDM system, there would also be a number of fixed-gain inline optical amplifiers (to compensate for the attenuation in the SSMF and the DCM), and a pre-amplifier before the demultiplexer. However, since this study is focused on the maximum tolerable *spectral efficiency* (SE) for a signal transmitted over a single span, these amplifiers are not considered.

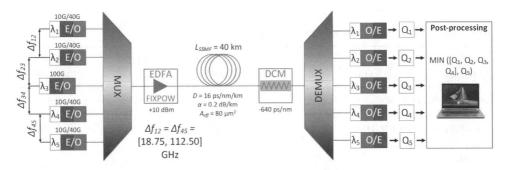

Figure 10.100 Layout of a Multi Line Rate WDM fiber-optical link

When evaluating the signal quality at the receiver, it is assumed that the following effects are predominant: (i) linear cross-talk of the second channel (λ_2) as a result of the traffic on the first (λ_1) and the third (λ_3) channel (in a completely symmetric way the third (λ_3) and fifth (λ_5) channels affect the cross-talk levels of the fourth channel (λ_4)); and (ii) linear cross-talk of the third channel (λ_3) as a consequence of the traffic on the second (λ_2) and the fourth (λ_4) channel.

The Synopsys' RSoft OptSim 5.3 software was used to perform the required simulations. The software solves the non-linear Schrödinger equation using a time domain split-step algorithm. The Q-value of the received signal is used to measure the optical signal quality. The accuracy of the obtained Q-factor values strongly depends on the total number of simulated bits. In our simulations, we used at least 2500 bits that yields a Q-factor uncertainty range below 0.5 dB.

The transmitter and the receiver are selected, depending on the modulation format and the bit rate considered during each specific experiment. The OptSim block diagrams for the transmitter and receiver of NRZ-OOK and NRZ-DPSK transmitting and receiving units can be found in [10.392], while the transmitter and receiver description for the DP-QPSK is provided in [10.411]. The transmitters are driven by rectangular-shaped NRZ signals filtered through an electrical low pass Bessel filter. For DP-QPSK, a Super-Gaussian optical filter is used after the transmitter. A similar optical filter is applied before detection at the receiver side for all modulation formats. The number of poles and the bandwidth of the optical and electrical filters are shown in Table 10.27. These filter parameters were chosen to provide the maximum Q-value for the WDM system under examination when the channels are arranged using 25 GHz channel spacing for the 10 Gbps NRZ-OOK; 50 GHz for the 100 Gbps DP-QPSK; and 200 GHz for the 40 Gbps NRZ-DPSK. In this scenario, the noise level is low and the obtained values of the Q-factor are assumed to be mainly limited by interchannel cross-talk. Such frequency intervals are common for the modulation formats and for the per-channel bit rates considered.

As a booster, we used an Erbium Doped Fiber Amplifier (EDFA) with a fixed output power level (i.e. 10 dBm) to compensate for the signal attenuation in the transmission fiber and in the DCM. As a transmission fiber, we used a standard single-mode fiber (SSMF), while for chromatic dispersion (CD) compensation, we considered a dispersion compensating fiber (DCF). The SSMF and DCF lengths are 40 km and 8 km, respectively. The fiber parameters used in this work are the same as those in [10.412]. On the receiver side, OOK signals were detected using a single PIN photodetector and DPSK signals with two balanced photodetectors,

Table 10.27 Optical and electrical filter parameters

Bit rate and modulation format	10 Gbps NRZ-OOK	40 Gbps NRZ-DPSK	100 Gbps DP-QPSK
Bessel electrical filters			
• Number of poles	5	5	5
• −3 dB bandwidth, [GHz]	10	65	20
Super-Gaussian optical filters			
• Order	1	2	2
• −3 dB two-sided bandwidth, [GHz]	14	70	40

respectively. The PIN photodiodes were assumed to have 80% of quantum efficiency. A coherent receiver was used to decode DP-QPSK signals.

10.6.6 Results and Discussion

In this section, the trade-off between achievable spectral efficiency, power consumption and transparent reach is evaluated. The focus of the study is on a system without *Forward Error Correction* (FEC) and with a required *Bit Error Rate* (BER) level of less than 10^{-9}. First, the power consumption of transponders and 3Rs is presented. Then, the minimum sub-band spacing value is determined by studying the degradation of the Q-factor of the received signal due to interchannel cross-talk over a single fiber span of 40 km. Finally, the power efficiency (W/bps) as a function of the distance is studied for a point-to-point MLR-based WDM transmission system (possibly covering multiple fiber spans) configured for maximum spectral efficiency.

The power consumption values of transponders (TSPs) and regenerators (3Rs) for each of the line rates considered (i.e. 10, 40 and 100 Gbps) are summarized in Table 10.28. They are calculated based on the data presented in [10.402], with the exception of the FEC capabilities that are omitted in this study. Table 10.28 also presents the power efficiency values (W/bps) computed for transmitting/regenerating 1 Tbps of aggregated capacity. These data show that for transmission capacities larger than 80 Gbps, it is more efficient to use 100 Gbps transponders. However, the situation is slightly different for 3Rs where, for transmission capacities larger than 200 Gbps, 40 Gbps 3Rs have lower power consumption than the 100 Gbps ones.

In order to mitigate the effects that the 10 Gbps channels have on the 100 Gbps channels, in MLR networks 1 or more 40 Gbps channels are usually placed between them [10.406]. All these sub-bands are then separated by the sub-band spacing. The smaller the spacing, the higher is the spectral efficiency of the system. Figure 10.101 shows the degradation of the Q-factor value of the received signal as a function of the value of sub-band spacing. The values are obtained by simulating the system presented in Figure 10.100. The different sub-band spacing values are obtained by gradually reducing the frequency intervals separating λ_3 from λ_2, and λ_3 from λ_4. For the 10–100 Gbps curve, λ_1, λ_2, λ_4 and λ_5 are all assumed to operate at 10 Gbps, while $\Delta f_{1,2}$ and $\Delta f_{3,4}$ are both set to 18.75 GHz [10.409]. For the 40–100 Gbps curve, λ_1, λ_2, λ_4 and λ_5 operate at 40 Gbps, while $\Delta f_{1,2}$ and $\Delta f_{3,4}$ are both set to 112.5 GHz

Table 10.28 TSPs and 3Rs power consumption and power efficiency

Bit rate and modulation format		Power consumption, [W]	Power efficiency, [nW/bps]@ 1 Tb transmitted
10 Gbps	TSP	22.4	2.7
NRZ-OOK	3R	20.8	2.6
40 Gbps	TSP	69.8	2.1
NRZ-DPSK	3R	43.6	1.3
100 Gbps	TSP	132.1	1.6
DP-QPSK	3R	158.5	1.8

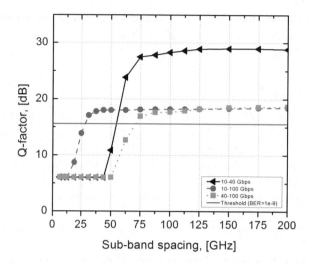

Figure 10.101 Degradation of the Q-factor value of the received signal as a function of the value of sub-band spacing

[10.409]. The values for the 10–40 Gbps curve were computed in an earlier work [10.392] and are reported here.

From the results presented in Figure 10.101, it can be deduced that if the BER level is required to be below 10^{-9}, the minimum sub-band spacing values are 31.25 GHz between the 10 Gbps and the 100 Gbps sub-band, 75 GHz between the 40 Gbps and the 100 Gbps sub-band, and 62.5 GHz the 10 Gbps and the 40 Gbps sub-band.

The minimum sub-band spacing for a number of line rate combinations was so far calculated assuming a link span of 40 km. If the same sub-band spacing values are used over distances longer than 40 km, signal regeneration will be required to keep the BER at the receiving node below 10^{-9}. It is then important to assess the relationship between power efficiency (W/bps) and the end-to-end connection distance of a number of MLR solutions, where the number of wavelength channels in each sub-band is varied.

In this particular set of experiments, the entire C-band ($\Delta F = 4.4$ THz) is divided into three sub-bands, one for each line rate (i.e. ΔF_{10G} for 10 Gbps, ΔF_{40G} for 40 Gbps, and ΔF_{100G} for 100 Gbps) (Figure 10.102). The number of channels allocated varies according to the width of each sub-band and depends on the value of the channel spacing used within each sub-band (i.e. 18.75 GHz for the 10 Gbps sub-band, 112.5 GHz for the 40 Gbps sub-band, and 37.5 GHz for the 100 Gbps sub-band). The width of the 10 Gbps sub-band can have any value (i.e. point B in Figure 10.102 can move freely along A–D). Once the width of ΔF_{10G} is fixed, the amount of frequency resources allocated to ΔF_{40G} and ΔF_{100G} may vary, depending on the values of the X and Y parameters (Figure 10.102), for example, if X = 1 and Y = 7, then ΔF_{100G} is allocated 7 times more frequency resources than ΔF_{40G}.

Figure 10.103 presents spectral and power efficiency results as a function of the width of each sub-band and also for different end-to-end connection length values (i.e. L). Figure 10.103a shows the utilization of the C-band (i.e. spectral efficiency) as a function of ΔF_{10G}, ΔF_{40G} and ΔF_{100G}. In terms of spectral efficiency, using a lot of channels working at high rate modulation

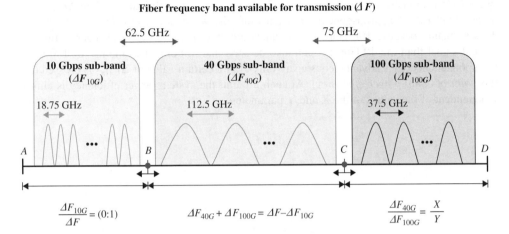

Figure 10.102   Sub-band and channel allocations in 10-40–100 Gbps MLR WDM system

formats (i.e. $X = 1$, $Y = 7$) has a clear advantage over solutions where the number of 100 Gbps channels is reduced (i.e. $X = 7$ and $Y = 1$). On the other hand, the use of 100 Gbps channels might present some drawback in terms of power efficiency when the end-to-end connection distance increases, as explained next.

If the end-to-end distance is limited to one fiber span, the most power efficient solution is that of taking advantage as much as possible of the 100 Gbps transponders (i.e. $X = 1$, $Y = 7$). This can be expected from the results presented in Figure 10.103a. However, the situation changes with longer end-to-end distances. Figures 10.103c and d present power efficiency results in the case of $L = 190$ km and $L = 1000$ km (i.e. the shortest and the longest link length

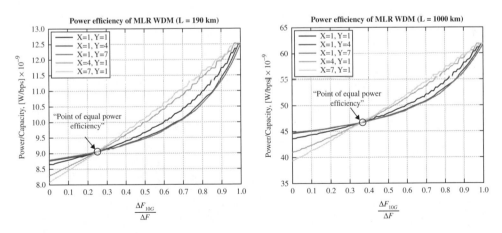

Figure 10.103   Spectral efficiency over the C-band (a), and power efficiency as a function of the width of each sub-band and of the end-to-end connection lengths: $L = 40$ km (b), $L = 190$ km (c) and $L = 1000$ km (d)

in the COST 239 network topology [10.413]), respectively. In both figures, when $\Delta F_{10G} = 0$, using many 100 Gbps transponders translates into the worst power efficiency performance. This is mainly because 100 Gbps transponders are less energy-efficient in regenerating the optical signal than the 40 Gbps. It can also be seen that in both cases the use of 100 Gbps transponders becomes advantageous again only after a certain value of ΔF_{10G} is crossed (i.e. the *point of equal power efficiency*). At such a point, the system power efficiency is almost independent of the values of the X and Y parameters.

11

Optical Data-Bus and Microwave Systems for Automotive Application in Vehicles, Airplanes and Ships

11.1 Communication in Transportation Systems

Dr. Kira Kastell
Frankfurt University of Applied Sciences, Germany

Communication in transportation systems can be manifold. There are many different means of transportation: cars, busses, trains, airplanes and ships will be regarded here. As different as these are, they are used for different periods of time, from minutes to days. The communication behavior of the users also differs from voice over data transfer to streaming, and they have different preferences in terms of features of their communication system. The combinations of these aspects introduce a broad variety of use cases and scenarios. Here it becomes apparent that no single communication system exists which is suitable for all purposes.

After introducing some common communication systems, their use in transportation systems is discussed. This will lead to different planning constraints for the various scenarios. It will become clear that no single network is able to fulfil all requirements and that it is far too expensive to construct a new network which can fulfil them all. Therefore, the hybrid use of existing networks in combination with slight adaptations towards the special needs of different transportation systems, will be the solution to communication in transportation systems.

Communication in transportation systems consists of a multitude of different physical properties, protocols, concepts and networks. It includes the communication to the transportation system which is wireless because of the mobility required for the means of transportation. But it also consists of the communication network needed to connect the passengers to that outside *wireless system*, when it is a means of mass transportation. These inside systems may

Optical and Microwave Technologies for Telecommunication Networks, First Edition. Otto Strobel.
© 2016 John Wiley & Sons, Ltd. Published 2016 by John Wiley & Sons, Ltd.

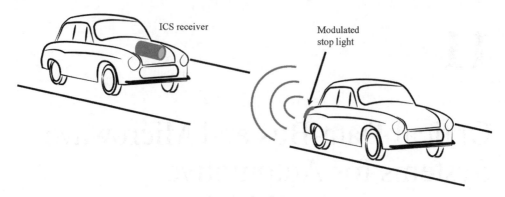

Figure 11.1 Car-to-car communication with optical sensors

also be wireless, wired, or a mixture of both, depending on the extension of the means of transportation and its structure and construction material.

Besides personal communication, there may also be communication for *data exchange* between the components of the system to control the form of transportation, for example sensors for engine, brake, distance measurement, etc. The different systems run different applications from very different sources, so a common standard would be desirable, but the different systems in use have rarely been designed to work together. Nevertheless, interworking is crucial to enhance overall communication quality. This interworking allows the data exchange of sensors and personal communication systems, enabling traffic flow control, accident avoidance, fleet management, driver's assistance, and many more. The sensors themselves may also use wireless transmission of electromagnetic waves as RADAR or optical sensors as in Visible Light Image Sensor Communication (VL-ISC), for example for pre-crash safety [11.1].

The use of personal communication systems allows the exchange of data between different users and control entities. For example, the status of a traffic light or an obstacle can be communicated from car to car by short-range communication or to a central server by mobile communication. The central entity can then inform cars entering the vicinity of the traffic light or obstacle. Nowadays, this is already popular for road traffic and railways. But these two already have very different restrictions, for example accident avoidance. They differ in structure of the means of transport, structure of the available communication systems, available sensors and measurement devices for data collection.

11.1.1　Communication Needs in Transportation Systems

Communication in transportation systems consists of three domains: communication of the means of transportation with the outside world, communication within the means of transportation and machine-to-machine communication for control purposes. Even though the three domains will interact, they have very different requirements.

11.1.1.1　Communication to the outside world

Communication to the outside world has to be wireless when using *radio access networks*. For aerial and maritime communication, satellite networks may be used. In principle, satellite

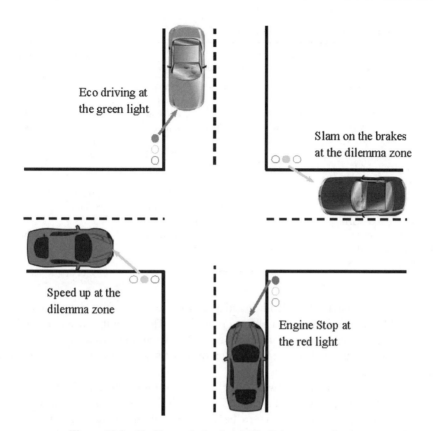

Figure 11.2   Traffic regulation by visible light communication

networks can also be used for terrestrial transportation systems, but need a line of sight to the means of transport. In inner city, forest and tunnels, this cannot be guaranteed. Therefore, for cars, busses and trains, most often terrestrial networks are used. These networks consist of fixed transceiver stations, for example a base station or an access point, which serve a certain area with their antenna(s). To cover a larger area, more transceiver stations have to be rolled out. The network then needs protocols to handle the transfer of ongoing communication from one transceiver station to another while the means of transportation is in motion. These protocols are known as *handover procedures*.

The use of existing communication networks for communication in transportation systems is most common, although there are amendments to standards if a certain scenario becomes popular. Table 11.1 lists existing networks with their main features, with the choice of an appropriate network for a specific scenario. It includes IEEE 802.11p [11.2] as one of the amendments especially suited to vehicular use. IEEE 802.11p is part of the family of standards P1609 for *Wireless Access in Vehicular Environments* (WAVE) [11.3], defining system architecture, management structure, security mechanisms, communications models, interfaces, physical access and services. WAVE is an attempt to foster a common base for a broad range of applications in the transportation environment or so-called intelligent transportation systems

Table 11.1   Overview over radio access networks

Network	Frequency	Data rate	Bandwidth	Delay	Security key	Speed limit
GSM	900 MHz–1900 MHz	384 kbps	200 kHz	<500 ms	64 bit	500 km/h
UMTS	2 GHz	2 Mbps	5 MHz	<200 ms	2 × 128 bit	500 km/h
LTE	700 MHz–2.6 GHz	300 Mbps	1.4 MHz–20 MHz	5 ms	2 × max. 256 bit	500 km/h
WLAN	2.4 GHz, 5 GHz	1 Mbps–600 Mbps	20 MHz–22 MHz	0.5 ms–10 s	Max. 128 bit	Not specified
IEEE 802.11p	5.85 GHz–5.925 GHz	54 Mbps	10 MHz	50 ms–100 ms	Better than WLAN	200 km/h
Mobile WiMAX	2G Hz–11 GHz	63 Mbps	1.25 MHz–20 MHz	20 ms–1.2 s	256 bit	125 km/h
Satellite (maritime/ aviation)	Around 1500 MHz	Up to 1.6 Mbps	Min. 384 kHz	Some hundred ms	network dependent	1000 km/h

(ITSs). The GSM standard has also been adapted for use in the railway environment. This is called GSM-R and is included in phase 2+ of GSM. Therefore it is listed under GSM in Table 11.1. Maritime communication has special networks at lower frequencies, for example the Marine Radiotelephone Service, but today most often satellite radio access is used [11.4], for example via Inmarsat.

These are the most common networks to choose from for connecting the means of transportation with the outside world. For maritime and airplane communication, the choice is very limited, as large distances have to be covered, thus the options are better for terrestrial transportation systems. Here the choice of the network depends on many criteria. Among the criteria there can be different weights representing their different importance for that particular case or user. Common criteria include availability, capacity, reliability of coverage and equipment, data rate, delay, bandwidth, security and the speed of the mobile device.

From the users' perspective, cost of use will play a role, but from the operators' point of view, infrastructure and maintenance expenses are important when calculating the revenue. But among the available networks, seldom will a single network fulfill all requirements of the different scenarios, even within only one transportation system. This leads to the use of a combination of multiple networks, known as a *hybrid network*. One of the benefits of a hybrid network is that the user will not notice that there are multiple networks involved. He or she just makes a handover from one transceiver station in one network to another transceiver station in another network. Thus, building fast, secure and reliable handover protocols is essential for the set-up of a hybrid network.

11.1.1.2   Communication inside the means of transportation

The choice of network for communication inside the means of transportation depends on the means of the transportation itself, the number of users and the kind of service or type of communication. Sometimes there is no need for an additional network, such as for mobile

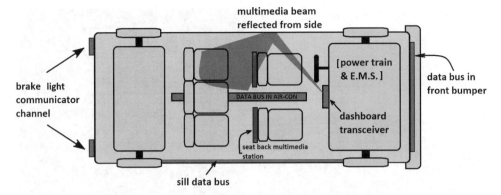

Figure 11.3 Possible optical wireless channels in a car

phone use in cars and trains. If the speed or the overall dimension of the means of transportation is high, if there is a high number of users, or if the means of transportation is well insulated against radio wave, for example constructed from metal or shielded with metallic film, it is better to place an antenna for reception and transmission outside and connect the antenna to a distribution system inside. This distribution system can consist of a single transceiver with an antenna, a relay or a complete network. The latter can be the same type of network used for outside communication or a completely different one. We could imagine a WLAN inside a train which connects to the outside via LTE. But the distribution system can also be wired, either to cover longer distances with small losses or to connect integrated displays and other built-in devices.

Modern means of transportation already have a communication network on board, as they are all controlled and partly maintained and serviced electronically. This communication network is an *in-car network* only, without connection to the outside world. Data may be read by diagnostic equipment but there is no permanent connection. These built-in communication networks may be used for the transport of additional data. Any communication equipment fixed and firmly installed in the means of transportation could be connected to the onboard network. Problems may occur if the onboard network is not dimensional for the extra amount of data and of course it still lacks connection to the outside network.

Therefore, for entertainment, voice, data and streaming, WLAN becomes popular. It is easy to install and easy to connect to an outside network. For larger means of mass transportation, such as ships, airplanes or trains, a single router may not be sufficient and a complete network consisting of several hot spots must be configured. The routers themselves can be connected as either wireless or wired. The specific infrastructure of the access routers depends on the topology of the transportation device [11.5,11.6]. Especially for trains, the cars of which may be reassembled quite frequently, a self-configuration of the onboard network will be essential [11.7]. Interference will not play a major role in passenger compartments. They are well protected against interference from the outside radio communication because of the metallic structure of such means of transport or metallization for protection against high pressure. Instead, this metallization limits the propagation range of the WLAN and makes careful configuration a must. In addition to WLAN, for short-distance communication, optical wireless solutions are emerging [11.8].

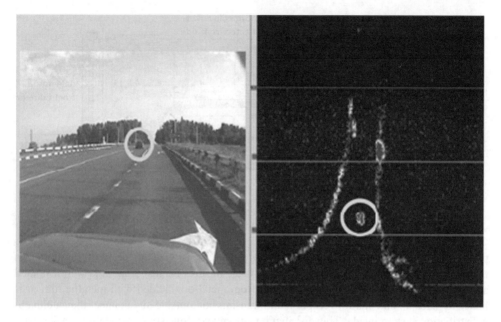

Figure 11.4 Radar measurement

11.1.1.3 Communication for control and maintenance

Sensors and actuators for control are used in every part of the means of mass transportation; prominent places are engine or breaks. The sensors monitor temperature, pressure, abrasion, etc. The data transmission needs to be wireless where moving parts are involved, for example tires and axles. As sensors are also often exposed to a harsh environment in terms of temperature or electromagnetic interference, for example caused by the engine, optical fiber is preferred if moving parts do not play a role. Fiber avoids severe interference caused by ignition sparks which affect the common frequency bands of wireless communication. It also resists high temperatures and is therefore able to provide high data rates under difficult environmental conditions [11.5].

The onboard communication network in modern means of mass transportation most often has a total length of more than 1 km for cars and will reach nearly 20 km for busses. Airplanes and ships have even more onboard wiring because of the extensive distances involved. Here also the weight of the wiring plays a role and again optical fiber has an advantage over copper and other metals. Their immense length results from the replacement of mechanical sensors by electronic ones and the integration of advanced electronic systems in the means of mass transportation. Regarding cars, among others there are automotive radar, driver assistance systems (e.g. for parking), route guidance systems, automatic tire pressure control, and the electronic maintenance system.

But there is an upper limit to the amount of wires that can be placed onboard, as the wires need to be placed somewhere, preferably out of reach of the users and not consuming space designated for passengers. In cars this will be between the automobile's body and the car interior lining or below the floor. As the wires will be packed closely together, insulation will play an ever increasing role. Here optical fibers also have an advantage. We can enhance the bandwidth or data rate of the wires by means of better materials, better modulation and multiplexing, pushing the limit. But this cannot guarantee unlimited transmission options. As the space is still limited, wireless communication will gain more importance for transmission of data in places that do not suffer from high-frequency interference on a regular basis. But for wireless communication, the metal constructed parts inside the means of transportation will contribute substantial attenuation. For communication along the complete length of a ship, optical fibers would be the choice. These may connect routers for wireless coverage of single sections and build a core network of the ship. A combination of both will be most beneficial in terms of coverage and the amount of wiring.

11.1.2   Communication with Transportation Systems

The structure of a means of transportation is more or less static. Thus, planning an *onboard network* has only to take into consideration this structure, the amount of data, and the places where to deliver it. The communication with the outside world is far more challenging because of mobility. The environment may be very different for different means of mass transportation and may change frequently.

Transportation takes place on the ground by car, bus, train, in the air by airplane, or at sea by ship. In any environment, the metal construction or metallization of the body highly attenuates electromagnetic waves. Therefore the recommendation is to place an *antenna* on the outside of the body to avoid attenuation. The signal from this antenna can be redistributed inside the body, or the signals from inside may be multiplexed to form one signal to be transmitted to the outside world by the antenna. But electromagnetic waves may reach the means of transportation from very different networks. The choice of the network depends on the environment in which the means of transportation is. This could be on land or at sea, as well as in the air.

11.1.2.1   Terrestrial environment

The *terrestrial environment*, especially ground-based motion on land, is the most diverse among the three. The topology differs from flat to hilly or mountainous. There are also very different kinds of clutter from forests, either with needles or leaves or both over field and lawn to rocks, cities of different sizes with different buildings, lakes and more. To cluster them, we distinguish between rural, suburban and urban environments. A *rural environment* has very few buildings and mostly is referred to as lawn with a few obstacles. A forest may build a class of its own. It behaves somehow rural but with higher attenuation than the open field. What both have in common is that there are very few obstacles and the environment seldom changes abruptly. Shadowing may still occur because of hilly topology, but rarely from obstacles, making propagation easy to calculate.

A *suburban environment* consists of smaller buildings, usually with a fair amount of space between them, and streets that tend to be comparatively wide. Usually the attenuation is higher than in rural environments and changes somewhat more frequently. Shadowing occurs more often and may become a severe problem for cars turning around the corner of a building.

An *urban environment* is even worse in terms of attenuation, as buildings are higher and compared to that streets seem more narrow. This makes them occur as cuttings into a block of buildings. Reflection increases because of the exterior of skyscrapers consisting of glass and metallized material.

The way cars or busses, and also trains move through this environment, is different from each other. Cars have a higher mobility than trains in terms of degrees of freedom on where to move. They are bound to streets but these have frequent crossings, diverse slopes and typically there is a lot of acceleration and deceleration during their movement. The signals will often be shadowed, as antennas cannot always be placed in line-of-sight, especially in an urban environment. Therefore, the received field strength changes rapidly. Trains operate in the same environment, but their tracks pose more restrictions. They do not turn abruptly and the slope of the track cannot be more than a few degrees. This is why there are lots of tunnels and bridges and also the tracks are often placed in cuttings and forest aisles, which behave like tunnels from a propagation point of view. Therefore, a train passing through a forest may be compared to a car passing through an inner city. This means rural terrain in a planning tool may not behave as rural but more as urban. Therefore, other propagation models will be required.

Tunnels have two effects: On the one hand, they act as a waveguide for frequencies used in public mobile communications, but on the other hand, as the tunnel is merely wider than the track, the train nearly fills its cross-section, shrinking the area of propagation to a minimum. Besides this, guidance of communication signals in a tunnel is difficult. Antennas inside suffer from high pressure and limited space, and antennas outside may only bring a small part of their signal into the tunnel. So, in general, tunnels are attenuators. Bridges are the opposite, having perfect propagation conditions as there are rarely any obstacles nearby. But antennas must be mounted well above ground and therefore may waste energy or reach so far that they cause severe interferences over a large distance.

In addition to the nature of the tracks, these are also equipped with catenary, the impact of which is two-fold. Sparking caused by a passing train will interfere with the signal and the electromagnetic field of the catenary is a constant source of interference. This is why propagation along tracks is worse than in areas without catenary.

In all terrestrial scenarios, speed is an important parameter as it causes the *Doppler shift*. The Doppler shift plays an increasing role with the increase of speed and has a non-negligible impact on communication with high-speed trains. In addition, communication systems have an upper speed limit up to which they are standardized (Table 11.1). A summary on mobility models for transportation can be found in [11.9].

11.1.2.2　Environment at sea

The clutter for *terrestrial maritime communication* is the sea, or more generally water. This will mainly cause *free-space attenuation* but also reflection on the water, which is a very good reflector. However, as a ship may be several thousand kilometers from land, the distance between transmitter and receiver can become very high. Therefore, low frequencies are dersirable, as free-space attenuation increases with the frequency. Because of the long distances to be covered, satellite communication is attractive for at-sea communication. This is feasible as power consumption does not rely on a small battery, as power can be generated by the ship's engine. This helps to cover the distance to a satellite, which may be as high as 36 000 km (geostationary orbit). The free-space attenuation will be very high and require more powerful transmitters than terrestrial communication, but the set-up of the network would be much easier, with fewer handovers and fewer stations.

11.1.2.3   Aerial environment

An *aerial environment* is mostly free of obstacles. Therefore, free-space propagation is the dominant effect. Reflection or even absorption does not occur. Nevertheless, there is a strong impact of attenuation because of the larger distance between transmitter and receiver. Intercontinental flights typically cruise at an altitude of 10 km. As attenuation is the square of the distance, this may cause attenuation of 100 dB and more. Combined with the large transmission distance is the high speed of the airplane of up to 1000 km/h, which causes a severe *Doppler shift*. The high speed also requires large areas to be covered by a single transceiver station. Therefore satellite communication will be very attractive for airplanes, as three geostationary satellites can cover the complete surface of the Earth. Besides, as shown in Table 11.1, the maximum speed in terrestrial network standards is 500 km/h.

11.1.3   Hybrid Networks for use in Transportation Systems

Transportation systems exist in very different environments, in which different communication networks are already available. For the choice of a suitable network or a combination of suitable networks to build a *hybrid network*, the different communication needs for different means of transportation have to be discussed.

11.1.3.1   Different means of transportation

We can distinguish between means of mass transportation and means of individual transportation. In the first case, passengers have spare time to use for working, reading, playing, streaming or other entertainment. And usually there are more passengers than in a means of individual transportation. This will create a higher demand for data exchange in the means of mass transportation. Therefore, they usually have an onboard distribution network for entertainment, which may be wired partly but which also allows access with personalized mobile devices over a radio interface.

Cars on streets require increasingly more different sensors for driver assistance systems than other means of transportation, because they have more degrees of freedom in their movement. They can change direction more abruptly and encounter more mobile obstacles. Trains need to monitor the track in front but cannot react to obstacles appearing from the side as they cannot leave the track. Ships are much slower and will also encounter obstacles, but larger distances allow more time for observation of obstacles. Airplanes are a bit like cars, but there are far fewer airplanes in the air than cars in a street, so the monitoring has a different quality. Here the critical factor is the speed of the airplane so that, as with a ship, obstacles should be identified at first sight while they are still several kilometers away. For cars, the range of a few tens of meters around the car is more interesting. Obstacles farther apart may only become important if they cause a traffic jam that may be avoided using a completely different route. But then the position and the kind of the obstacle need not be known very precisely.

The last big difference is the dimension of the means of transportation and its inner structure. This depends whether a wireless network may be suitable for stand-alone onboard coverage or if a wired network will do so, or if the latter is needed as support.

11.1.3.2 Networks inside a means of transportation

Inside a means of transportation there may be *hybrid networks*. These most likely will not consist of different wireless networks but will be a mixture of one wireless and one or two wired networks. Recent standardization [11.10] proposes moving relay nodes for *terrestrial mobile communication*, which collect the data streams from inside the means of mass transportation and pass them to the antenna mounted on the outside. With this antenna, the means of mass transportation appears as one single device to the outside network [11.11]. Mobility management is simplified as this moving relay node can perform a group handover instead of one individual handover per device.

Then there is a terrestrial mobile communication network with a moving relay node, so that the individual devices have less overhead with handovers. Alternatively, the wireless signal may be distributed by *WLAN* inside the means of mass transportation. The WLAN access point or router may directly communicate with the moving relay node. A moving WLAN relay will not be feasible, as WLAN cells are very small so that lots of handovers would be needed for higher speed. Besides, there will not be enough cells in the outside world, as WLAN usually is based on a hotspot structure and not for providing seamless coverage of a larger area.

Both WLAN and mobile communication networks need to be configured inside the means of mass transportation. The configuration has to take into account the layout of the interior and its attenuation characteristics. In most means of transportation, the configuration will be fixed, but this is different in railway environments. Trains are reassembled of different coaches after each journey or at the end of the day. Here, self-configuration of the onboard network plays an important role. The different coaches may be connected in a wired [11.7] or wireless [11.12] fashion.

For smaller interiors, as in cars, optical wireless solutions are proposed [11.8]. Because of the high attenuation, the links can only cover smaller distances. Nevertheless, optical communication plays an important role inside this means of transportation, as among the wired networks, optical fiber networks are dominant [11.13].

11.1.3.3 Hybrid networks for connection with the outside world

The connection to the outside world has to be wireless. The choice of the network depends on network availability and service requirements. Typically, the following networks are candidates [11.14]: *GSM* (phase 2+), *UMTS* (especially HSPA (high-speed packet access)) and *WiMAX*, IEEE 802.11. Alternatively, *radio-satellite communication* may be a candidate for environments with clear visibility of the sky (e.g. no tunnels). Fast Low-latency Access with Seamless Handoff – Orthogonal Frequency Division Multiplexing (*FLASH-OFDM*) in the

Polymer Optical Fiber
POF

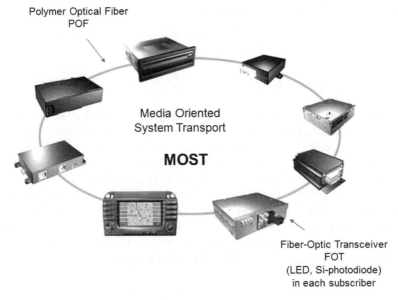

Media Oriented
System Transport

MOST

Fiber-Optic Transceiver
FOT
(LED, Si-photodiode)
in each subscriber

Figure 11.5 MOST network with ring topology

450 MHz band with a data rate of up to 5.1 Mbit/s is used in railway environments for entertainment purposes, as its reliability may be limited as well as its availability. *Radio over fiber* can also be used for distribution, whereas the last mile still stays wireless.

The choice of the corresponding network is critical as transportation systems include a huge variety of applications and environmental conditions, which have to be taken into consideration for communication purposes. The main decision factors are type of transportation, determining the number of passengers and the traveling speed, the services and applications to be transmitted, especially their data rate and security issues, and the environment. Most often the data rate is the bottleneck in transmission planning of connections to the outside world. High data rates often require stable channel conditions, which can only be assumed with lower speeds. The higher the speed, the more important localization and location prediction is. As the speed of the transportation may vary, different networks may be suited along a line of travel. The communication should not be affected by the change of the network, so a hybrid network with seamless handovers needs to be built. Depending on the constellation of the hybrid network, new protocols have to be developed.

Besides seamless handover from one radio access network to another in terms of latency, signaling conversion and mapping, integrated security is an important factor. As handovers have to take place rapidly, in most handover protocols authentication is not integrated, neither for networks nor for users. The authentication of the user is important, not only for access but for billing and accounting. Without this, providers will not allow access to networks which charge users a usage fee. For the users, it is important to only access networks in which they trust. So, mutual authentication is mandatory if the handover takes place between different radio access networks, even more so if more than one provider is involved in the handover.

11.2   Radar for Transportation Systems

Prof. Dr. Vladimir Rastorguev, Dr. Andrey Ananenkov, Engineer Anton Konovaltsev,
Vladimir Nuzhdin, Engineer Pavel Sokolov
Moscow Aviation Institute
National Research University, Russia

With the development of the motor industry worldwide, the intensity and density of traffic
on the roads at the simultaneous growth of the average speed of movement have significantly
increased. For this reason the number of *road traffic accidents* (RTAs) has sharply increased.
As a result, the RTA has become one of the most serious problems worldwide, and solutions
in which way the safety of road traffic can be increased are searched for.

Considering causes of RTAs, the following main types can be marked: transport vehicle fault,
driver inattention, insufficient information awareness of the driver regarding the road traffic
situation. For example, as per data of the US agency NTHSA [11.16, 11.23], the mentioned
RTAs are most frequently associated with driver inattention (68%), with the failure to keep
sufficient distance (19%), as well as with insufficient information obtained by the driver during
movement.

The degree of information awareness of the driver regarding the situation on the road directly
depends on the conditions of optical visibility, such as:

- Road and roadside illumination
- Availability of qualitative traffic-lane markings and guide signs
- Degree of visibility through windscreen
- Weather conditions (heavy snow and rainfall, fog), smog, dust and other impeding factors

As a rule, these negative factors accumulate, which increases RTA risk and at certain
conditions makes them the most probable cause of an accident. This includes, for example, at
night together with poor illumination, fog may be present, and in winter constantly splashed
and ice-covered glass, including headlight glass, as well as blinding light of oncoming vehicles.
Also included is heavy snowfall and short daylight hours, etc.

Of the entire vehicle park, those vehicles may be marked, which undergo the influence of
these negative factors most frequently. For example, an ordinary car, owned by a private driver,
is operated primarily in daylight hours; then, on the other side, a semitrailer or inter-urban bus is
operating during all seasons, and very often at night, when roads are less crowded. Besides, for
a whole range of special transport vehicles (ambulances, police cars, fire trucks, special cargo
transportation trucks) an all-weather and round-the-clock movement needs to be performed,
and with high speeds (>60 km/h) under the conditions of all the above-mentioned factors.

This task is most actual during vehicle operation along winter highways, especially in the
far north, where the probability of snowfall in winter is very high, as well as in sand and desert
areas, where sand and dust storms often occur without warning.

The task of improving information awareness of the driver regarding the traffic situation
was put forward long ago, and to solve it, many countries offered programs on the creation
and implementation of various systems which could help the driver under the conditions of
limited visibility, and which could replace him fully in the future [11.17,11.19–11.21,11.23].
In Russia, Germany, USA, Japan and France, starting from the 1990s, a lot of research and

development was being carried out in order to create various systems and sensors which allow the reduction of the probability of RTAs.

Basically, specialists in the area of automobile electronics concentrate their efforts in two directions: the development of information support of the road infrastructure and the creation of a "reasonable" vehicle, which are both closely connected. Perhaps, today, the information technologies of a "reasonable" vehicle are developed most productively. Let us consider some of them.

For example, specialists try to solve this task by installing "clever" optical means on the vehicle, as well as by implementing *infrared night-vision systems*, both active and passive. For example, in the year 2000, General Motors (GM) together with Raytheon Systems created the passive infrared system "NightVision" [11.16,11.23,11.27]. This system, registering the natural thermal radiation of the object, gives the driver the opportunity to control the road situation at a distance twice the length that is illuminated by most modern headlights in the far-light mode. However, in countries with high average air temperature (over 15–20°C), the system "Night Vision" may turn out to be inoperable.

In the year 2009, Mercedes-Benz started to install the active infrared system "Night View Assist Plus" on its vehicles, which ensures a qualitative presentation of the road and present pedestrians at a distance of up to 250 m [11.25]. However, the use of non-cooled thermal cameras in such systems makes their work dependent on weather conditions. For example, at few foggy ambient conditions, the visibility range amounts to 500–1000 m, whereas at strong fog or a sand storm, it may decrease down to dozens and even to a few meters.

A shortage of these solutions is evident – out of all the negative factors, reducing the optical visibility, they only help at insufficient illumination. Therefore, the known methods of solving the task of traffic safety increase by means of installing optical and IR sensors on the vehicle, operating in different light ranges, are ineffective under severe weather conditions, including smoke and dust, etc.; ultrasonic sensors, on the other hand, are limited as far as long ranges are concerned.

Therefore, the main way to solve the actual task of increasing the vehicle movement safety under the conditions of unavailable or limited optical visibility, is the use of a radar sensor in the millimeter range of wavelengths. Its operation depends neither on the time of day nor on the weather conditions (snow, rain, fog), nor on smoke or dust.

Over the last decade, the main research and development in the area of creation of a vehicle radar were focused on investigating the following frequency ranges [11.18–11.22]: 24 GHz – for a *close range radar* (SRR, up to 50 m), and 76–77 GHz for a *wide range radar* (LRR, 150–200 m). The active research in the area of using the range of 79 GHz has only recently started [11.25]. At the same time, the mastering of a new frequency range of 77–81 GHz is performed with the aim to reduce the mass and dimensions of the equipment, while retaining high information capability of the image. However, the cost of microwave modules in this frequency range is far too high, which reduces the commercial effect of its implementation.

Up to now, the functions of vehicle radar was directed primarily towards increasing the driving convenience and safety. In particular, this is true for radar sensors of the adaptive system of maintaining the preset speed (*adaptive cruise-monitoring*). Moreover, there are systems which warn of collision danger in the traffic lane and of "dead regions", facilitation systems which help when changing traffic lanes, systems that warn of objects moving in the transverse direction behind the vehicle, and parking aids. For example, Audi offers a technology, possessing the possibility of "predicting" an accident – Audi pre sense [11.24].

So, the adaptive cruise-monitoring includes not only the function Stop&Go and safety system "Audi pre sense front", but also the function of automatically maintaining the distance (at a speed from 0 up to 250 km/h). Delphi Automotive offers an electronically scanning radar ESR (Electronically Scanning Radar), operating in two modes, in the near and far distance ranges [11.20,11.24]. During operation in the near range (SRR mode), at a distance of up to 60 m, ESR allows detection of vehicles and pedestrians, and in the far range (LRR mode), at a distance of up to 174 m, it defines the distance and speed of up to 64 objects in the way of a moving vehicle.

Undoubtedly, a wide adoption of vehicle radars was facilitated by the success of researchers and engineers from various countries in the development of modern technologies for the creation of radar microwave modules in the form of Monolithic Microwave Integrated Circuit (MMIC). It is based on the use of SiGe and GaAs technologies, and in the development of high-speed analog to digital converters (ADC), and digital to analog converters (DAC) for microcontrollers.

Most types of vehicle radar currently on the market, either solves the task of preventing a collision with an obstacle within the traffic lane (e.g. Audi pre-sense system radars and "DIS-TRONIC PLUS" system radar made by Mercedes-Benz [11.23,11.24]), or have a limited (up to 3–6 degrees) resolving capacity in the azimuth plane (e.g. Delco firm radar [11.23,11.24]). Therefore, it may be stated that no vehicle radar presently on the market allows the formation of a high-quality and high-information *panoramic radar image* (RI) with an azimuth resolution of 0.7–1 degrees and a resolution of the distance of approximately 1 m; both are necessary for a detailed generation of RI and safe vehicle driving under the conditions of limited or unavailable optical visibility.

If the azimuth resolution of a radar sensor does not allow an unequivocal discrimination as to whether an unknown obstacle is inside the traffic lane in front of the vehicle or, for example, is parked on the roadside, the system will continuously demand a stop, or a reduction of the speed in order to solve the task of safe driving. In the 21st century, most roads have objects of infrastructure – signs, posts, guards, etc. According to research [11.17], these objects provide a sufficient level of reflected signal. Therefore, the azimuthal resolution of the radar sensor must be sufficient to distinguish between objects located in the traffic lane and outside of the traffic lane, at a distance exceeding the breaking distance.

This section presents the results of many years of research by scientists of the Radio-receiver Devices Department of the Moscow Aviation Institute (National Research University) – MAI (NRU), directed at the creation of an experimental sample of a panoramic radar sensor of the vehicle, looking in a forward direction. Owing to the high information awareness of the RI, this sensor is called the *"Automobile Radio Vision System"* (ARVS) [11.17,11.33]. The ARVS is a small, integrated, all-weather, information and measuring system of the vehicle, which has nothing comparable on the world market.

The experimental sample of the ARVS, operating in the millimeter (MM) range of the wavelength, is intended to form a panoramic radar image (RI) of the road situation in front of the vehicle. Under conditions of insufficient optical visibility, the driver can observe on the passenger compartment indicator screen such things as impediments, boundaries, other roads, passing cars, oncoming vehicles, as well as vehicles parked at the roadside, within a working distance that accounts for the dynamics of the movement of his own vehicle.

It may be stated that the use of ARVS will allow increased road safety in a significant way, partly due to automatic driving the vehicle in critical situations and due to the reduction of the influence of the human factor on road safety.

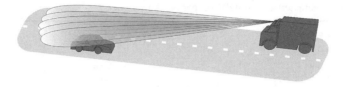

Figure 11.6   Radiovision task geometry

11.2.1   ARVS Main Features

The automobile RVS shall form a panoramic radar image of the road situation and adjacent locality on the monitor screen in the front coverage area in real time (Figure 11.6).

The investigations directed at the creation of ARVS were started by specialists of the Radio-receiver Devices Department of the MAI (NRU) in 1995. In 1998, the first experimental mock-up of the ARVS with a pulse modulation of the probing signal, operating over the range of 39 GHz, was built and tested [11.33,11.34]. However, this mock-up allowed seeing only vehicles and large impediments in a stable way at distances of up to 50–150 m, with a resolution of 1.5 × 6 m and in the azimuth a coverage area of ±45°. The main disadvantage of this ARVS was a narrow dynamic range, not able to see the roadside and, hence define the road boundaries. That is why since 2001 experimental mock-ups, with a continuous probing signal with a linear frequency modulation (LFM), are allowed for such radar, operating in the range of 39 and 77 GHz. [11.34]. The use of LFM meant raising the ARVS power budget by 40 dB and improving its resolution to 1.5 m × 2 m. The use of a new antenna for the leaking mode, of digital signaling processors and of a liquid crystal indicator, allowed a reduction of the mass and the dimension of the system, that according to these parameters came close to IR vehicle systems.

Numerous laboratory tests and experimental research have shown that ARVS allows forms a panoramic RI on the monitor screen in the preset angle sector at distances of 5 m to 250 m in the coordinate's *azimuth-distance*, with a required resolution and in real time. This provides the driver with clear observation of the highway boundary on the RI, moving and standing vehicles (both oncoming and passing cars), other objects and impediments. The joint radar and optical images of different sections of a standard road (Figure 11.7) clearly illustrate ARVS operation.

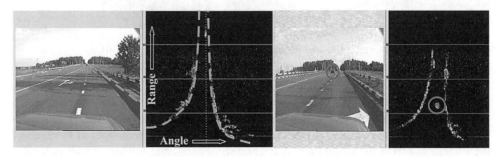

Figure 11.7   Examples of radar and optical images in ARVS

Table 11.2   Main characteristics of ARVS experimental samples

Operating frequency, GHz	39 or 77
Action distance, m	5–250
Image details on distance (resolution), m	≤1,5
Resolution on azimuth, degree	≤1
Radiated power, mW	50–70
Antenna:	slotted-guide (39 GHz) or leaky-wave (77 GHz)
Antenna scanning sector in the azimuth plane, degree	60 (39 GHz) 30 (77 GHz)
Effective width of antenna directivity diagram, degree	
– in azimuth plane:	≤1
– in elevation plane	20
Image coordinates	azimuth-distance
Number of image information points	166 × 512 for one frame
Power consumption, W	<100 (embedded network 12 V–24 V)
Information update frequency, Hz	10

Digital signal processing in ARVS allows not only formation of a radar image, but also measuring the roadway, the distance to left and right of the roadside, performing a virtual marking of road lanes and forming distance assessments for impediments located in the road "lane" of the driver's vehicle.

All the ARVS experimental samples are ecologically safe, since the radiated power of the system is a few times lower than the radiated power of a GSM standard mobile telephone. Table 11.2 shows the main characteristics of ARVS experimental samples.

The experimental sample ARVS (frequency 77 GHz) consists of two compact, space-saving and high-technology modules (Figure 11.8):

Figure 11.8   External (a) and internal (b) ARVS modules (77 GHz)

The external module is located in the area of radiotransparency of the vehicle and includes antenna and combined (receiver and transmitter) modules. The *antenna module* is implemented in the form of a leaky wave antenna with electromechanical scanning. This module was developed by scientists of the Institute of Radio Engineering and Electronics of UAS, under the leadership of Doctor Evdokimov A.P. [11.38]. The external module, which is located outside of the cab, includes a thermal regulation subsystem, using Peltier elements, which operates over a temperature range of ±50°C. The operating temperature range for the external ARVS module is −10°C up to +40°C for the internal one, and +5°C up to +40°C. The antenna view sector axis is rigidly connected with the direction of the vehicle movement. The ARVS internal module consists of a specialized computer and an indicator panel (Figure 11.8b). The weight of the ARVS experimental sample (for the frequency of 77 GHz) is about 4 kg, which allows implementing the sensor in a portable version.

The ARVS pertains to a class of radars of a minute distance, which defines a range of features in processing and presenting (indicating) the radar information. For example, a background RI of the subjacent Earth's surface and local objects (LO) is informative for solving the navigation task and is presented to the user.

The secondary processing products (marked oncoming and passing cars, with a prognosis of their movement, road lane boundaries, interconnecting with GPS receivers and *digital terrain maps* – DTM) may be present in the form of vector masks, imposed on the background RI. The following features distinguishing the ARVS from radars of other classes are:

1. Purpose features
2. Features associated with the specific dimensions of typical targets and the scene to be observed
3. Observation geometry features
4. Features associated with the wide range of the reflecting capacity of the scenes to be observed within the range of distances
5. Features of mass use
6. Features of electromagnetic compatibility
7. Ecological features

11.2.1.1 Purpose features

The necessity of following dynamic objects at minute (with regard to the traditional location) distances requires a high rate of obtaining and updating information on the ARVS indicator, compatible with the rate of image upgrading in television systems. At the same time, it is necessary to ensure the availability of a convenient and friendly interface, allowing the instant and adequate reaction of the driver to the changing situation. Delays in information processing must be less than the time of the RI frame formation (operation in the "soft" real time mode – with delays of less than a time frame, 0.1 s). In the experimental sample of the ARVS, the achieved rate of obtaining a frame of RI amounts to 10 Hz.

11.2.1.2 Features associated with the characteristic dimensions of typical targets and the scene to be observed

In so far as the task number for ARVS is to detect impediments in its own road lane (a vehicle, a man, etc.), the resolving capacity of the radar sensor should allow localizing these objects within their proper road lane up to a distance of not less than the length of the braking distance.

Therefore, for a safety corridor of the length of for example 2 m, the resolution capacity on the azimuth for the ARVS should be about 1 m at a distance of 60 m (or not worse than 1 degree). The resolution capacity for distance should allow detecting adjacent roads (or crossroads), which requires a resolution at the level of 1 to 2 m.

In the built sample of the ARVS, the resolution on the azimuth amounts to 0.7 degree and a distance of 1 to 1.5 m. Therefore, at characteristic vehicle sizes from 3 up to 18 m, practically all the targets for the ARVS are spaciously distributed (multi-pointed) and require involvement of cluster procedures for the analysis.

The radiometric resolution of the ARVS should allow steady detection of low contrast boundaries of the surfaces, different in the value of a *specific absolute cross-section* (SACS) [3.25] of less than 10 dB. Such a resolution is necessary to detect natural boundaries of the road with no available artificial road-guards.

All the above-mentioned characteristics must be ensured in a wide (30–90 degrees) azimuth scan sector in front of a vehicle. Such high characteristics of *resolution* ensure a unique detailed (information awareness) RI in the ARVS and are not characteristic of most radar stations of other types.

11.2.1.3 Observation geometry features

In so far as the ARVS is located in the vehicle (Figure 11.6), the hoisting height of the electric center of the antenna system above the surface (road lane) is very small and is practically situated (depending on the vehicle type) in the interval of 0.4–2 m. This feature causes small angles of an electromagnetic wave drop on the surface to be observed (the so-called *grazing* (laid) *viewing angles*). At the same time, all the objects and obstacles within the specified distance and coverage area are observed (and must be automatically marked out) on the background of reflections from the underlying surface.

A small distance of objects located from the ARVS, at a high direction of the antenna system, leads to the fact that the targets are situated not only in the far area of the antenna $R > 2D^2/\lambda$ [11.15,11.26], but also in the intermediate area of Fresnel, as well as in the near area. This feature of observation needs to be taken into account when calculating the direct diagram width (DDW) and boost (amplification) factor of the antenna system. In the permitted frequency range for the ARVS, an asphalt road coating, above which the ARVS antenna system is always situated, tends to represent a reflecting mirror surface [11.17], which causes an electromagnetic wave multipathing in the ARVS. It is known that an electromagnetic wave multipathing of signals leads to significant fluctuations of the intensity of the received signal [11.17]. For the ARVS, this problem is eliminated in a natural way – due to the vehicle movement; however, algorithms of impediments detection must take into account the significant fluctuations of the signal intensity from the impediments towering above the road lane.

11.2.1.4 Features associated with the wide range of the reflecting capacity of the scenes to be observed within the range of distances

As far as the ARVS must inform the driver of all impediments in the movement direction, multiply exceeding the length of the breaking distance, the dead zone of the radar sensor must end not further than 3 m from the front bumper of the vehicle. Even in the case of locating the sensor on the vehicle roof, the size of the dead zone should not exceed 1.5–2 m. Taking into account the ARVS working distance of 150–250 m and the fourth degree of the signal power dependence on the distance, the range of the received signal power variation due to a change of distance will be not less than 80 dB. The range of value change of the *Radar Cross-Section* (RCS) of targets is about 30 dB; however, besides obstacles, the ARVS should detect thin road boundaries up to a value of the RCS: –20 dB (at the selection element square of 1–3 m).

Therefore, the dynamic range of the ARVS received signals amounts to a value of 130 dB. Any signal above this range should not disable the equipment, and also not increase errors of defining impediments coordinates. The operation with significant dynamic ranges of input signals is a characteristic for some types of radars (i.e. meteorology radar [11.29,11.32]). However, a typical technical solution for this problem, such as the use of multi-channel (or logarithmic) receivers, is a rather expensive method. Therefore, in order to compress the dynamic range of the received signal, other methods should be used in the ARVS receiver (Chapter 11.2.2).

11.2.1.5   Features of mass use

The key features of the ARVS are low cost, small weight and dimensions of radar equipment, and its low power consumption. The ARVS equipment should not disturb the aerodynamics and design of the vehicle. Most known radars with comparable characteristics of detailed representation of the situation and accuracy of the angles and distances definition are inadmissibly costly and bulky for their mass use on vehicles. When constructing the ARVS, neither the accuracy, the detail nor the rate of data updating were neglected – otherwise, the entire venture of creating the ARVS becomes senseless. However, we can boldly limit the range of operating distances for the system down to 250 m, or even down to 150 m. The limitation of an operating distance is the only resource for reducing ARVS dimensions, mass and cost down to acceptable values for the mass user. In addition, this requirement (the reduction of the sensor price) makes specialists refrain from using the many solutions in traditional radars.

11.2.1.6   Features of electromagnetic compatibility (EMC)

The need of mass use makes it necessary to look at the problem of the *EMC* in a different manner – the number of simultaneously operating ARVSs will be significant in the near future. Good support for solving this task is the frequency range for such devices, selected in Europe, – 77 GHz (the second allowed range, – 24 GHz), which ensures a relatively fast attenuation for radiation of the radar in the atmosphere and in precipitation. However, within the operating radius, a few such radar sensors will inevitably operate. They should not create fatal mutual interferences and also not raise measuring errors for each other. Additional possibilities for solving this problem lie in the area of processing algorithms and accounting (synchronization) of operating modes of neighboring ARVS as per the technologies of interaction.

11.2.1.7   Ecological features

When there is continuous ARVS operation in the immediate neighborhood of people, it is necessary to ensure a minimal low level of *microwave radiation*. On the one hand, the earlier introduced limitation for the distances of operating range already limits the level of the required radiation. In addition, protection mechanisms must represent a narrowly-directed type of radiation and the location of the antenna module outside of the body of the vehicle. The issues of the controllable reduction of the radar radiation level in the habitable environment have already been raised in [11.31]. Besides, non-specific but very strict demands on stability of the images to be formed and accuracy of defining the coordinates of observation objects are placed.

In conclusion of this brief review of specific ARVS features, it should be noted that up until recently, time radar sensors, meeting the whole set of the above-mentioned requirements, were not available. A full-fledged ARVS has not been presented on the market so far.

11.2.2   Features of ARVS Equipment Construction

As was shown in the previous section, compliance with the contradictory requirements for the ARVS is not a simple technical task. Considering the features of constructing the ARVS, we use as an example the experimental sample, which was created at the department of radio-receiver devices of MAI (NRU). This ARVS specimen represents a panoramic forward-looking radar sensor of ultra-small distance with a *linear frequency modulation* (LFM) probing signal and homodyne reception method [11.37]. Using this sensor, a large scope of experimental research on EM waves scattering under characteristic road conditions was performed [11.33–11.35].

The ARVS structural chart is presented in Figure 11.9 and functionally consists of the following main units:

- combined antenna
- UHF block
- receiving box block (RCB)
- digital signal processing unit (DSP)
- information representation and control unit
- power supply unit

A combined ARVS antenna is made in the form a narrow-band travelling-wave antenna with a width of the directional (radiation) pattern (DPA) in the *azimuth plane* of less than 1 degree. The DPA width in the elevation plane amounts to 20 degrees. The antenna gain amounts to $G = 400$. To provide scanning, a rotation of the antenna is performed by a step motor, with a vision frequency of 10 Hz. To monitor the start of the scanning sector in the azimuth plane, an "Antenna position sensor" is installed.

The UHF block consists of three individual modules, a transmitter module, an attenuator module and a balance mixer module. The transmitter module contains a high-frequency generator (HFG) with amplifiers and ensures the power of the probing signal of 100 mW. The transmitter block ensures the frequency modulation of the probing signal by means of an "*LFM modulator*". A periodic linearly growing signal is used as a modulating signal (Figure 11.11). The balance mixer module is intended to multiply an input signal, reflected from the target, by a reference signal of the transmitter with the power of 1 mW, which was preliminarily obtained from the directional coupler.

The digital signal processing unit (DSP) consists of the following three modules: an analog-digital converter (ADC), a clock pulse generator (CPG) and an interface device to an on-board

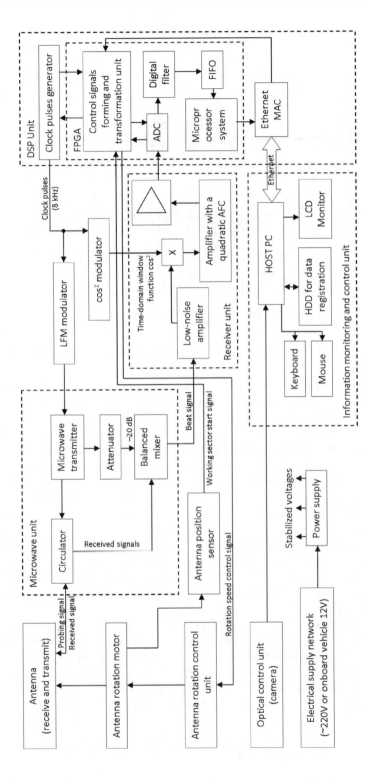

Figure 11.9 Generalized ARVS structural diagram

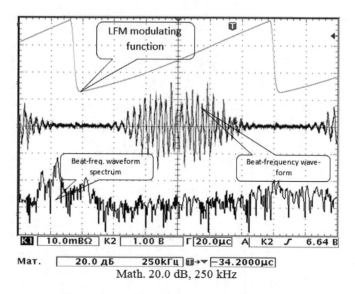

Мат. [20.0 дБ   250кГц ⓘ→▾ ‾34.2000μc]
Math. 20.0 dB, 250 kHz

Figure 11.10   Oscillogram of signals in the ARVS

digital computer. The CPG ensures formation of all service and command (control) signals and frequencies.

A COS modulator is used to form a time-domain window, by which a beat-frequency waveform is multiplied. The use of a time window suppresses an incidental amplitude modulation of the VCO and reduces the level of side lobes of the beat-frequency waveform spectrum. A signal from the oscillator (modulator) is presented in the oscillogram in Figure 11.10.

An amplifier with a quadratic amplitude-frequency characteristic is used to compensate for the dependence of the amplitude of the beat-frequency waveform on the distance to the target, from which a reflected signal arrives. The use of this amplifier allows reduction of the dynamic range of the signal, fed to the analog to the digital converter (ADC). The ARVS power supply unit provides the possibility of a long ARVS operation, both from the grid of the AC 220 V~ 50 Hz and from the on-board electric grid of the vehicle.

The on-board digital computer calculates a spectrum of the beat-frequency waveform of each reflected signal (see the oscillogram in Figure 11.10) and thereby one column as per the distance of the RI to be created is formed. By accumulating individual columns of the RI during antenna scanning in the *azimuth plane*, the computer forms a primary RI in *azimuth distance* coordinates.

In the considered ARVS with an LFM-probing signal, the information of the distance to the target is contained in the beat-frequency waveform, and the intensity of the beat-frequency waveform is proportional to the value of the target RCS. To define the beat-frequency waveform, it is necessary to perform a spectral analysis of the received signal. This is accomplished by means of fast Fourier transform (FFT).

The resolving capacity (resolution) on distance is defined by the following ratio:

$$\delta R = \frac{c}{2 \cdot W \cdot F_M} \cdot \delta f_R \geq \frac{c}{2 \cdot W \cdot F_M} \cdot F_M = \frac{c}{2 \cdot W} \tag{11.1}$$

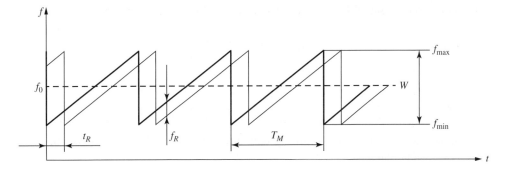

Figure 11.11   Law of radiated and received signals frequency change in the ARVS

Thus, the maximum resolution of the ARVS with the FM on distance is defined only by a deviation of frequency and depends neither on the signal carrier frequency f_0, nor on the modulation frequency F_M. According to the above formulated requirements of the ARVS, the resolving capacity on distance must be not less than 1 m. Hence, from Eq. (11.1), the necessary value of the frequency deviation amounts to $W = 150$ MHz.

Considering the main energy relations in the ARVS, the law of the change of the frequency of radiated and received signals for a periodic LFM is presented in Figure 11.11, for a particular case of availability within the ray of the antenna diagram of only the single-point scatterer.

To reduce the cost of the entire solution, the ARVS receiver module starts from the mixer (Figure 11.9). As a result, the receiver module has a high noise factor $N_D = 20$.

When using the FFT as a spectral analysis tool, it may be considered that the signal detection will be performed at the outlet of the linear filter with the band pass, defined by a value, which is reciprocal to the time of processing the digital sampling:

$$\frac{N_{FFT}}{2f_D} = T_{PROCESS} \tag{11.2}$$

where:

f_D Sample rate
N_{FFT} Number of digital samples of FFT

At the correct matching of parameters of the digital analysis system and the period of repetition of modulating function $T_{PROCESS} = T_M$, noise bandwidth Δf_N of the equivalent analyzing filter will be equal F_M: $\Delta f_N = F_M$.

During localization of point targets, the power of the reflected signal at the receiver inlet is defined by the main *radiolocation formulae* [11.15, 11.26]:

$$P_{RCV} = \frac{P_{TR} \cdot G_A^2 \cdot S_{EF} \cdot \lambda^2 \cdot \eta^2}{(4 \cdot \pi)^3 \cdot R^4} \tag{11.3}$$

The noise component of the system is defined by the following expression:

$$P_N = k \cdot T_\Sigma \cdot \Delta f_N \cdot N_D \qquad (11.4)$$

where:

k Boltzmann constant

Then, the signal-to-noise ratio of the system during location of point targets is defined by the following expression:

$$q = \frac{P_{RCV}}{P_N} = \frac{P_{TR} \cdot G_A^2 \cdot S_{EF} \cdot \lambda^2 \cdot \eta^2}{(4 \cdot \pi)^3 \cdot R^4 \cdot (k \cdot T_\Sigma \cdot \Delta f_N \cdot N_D)} \qquad (11.5)$$

The absolute cross-section for prolonged surfaces is defined by the following relation [11.35]:

$$S_{EF} = \sigma_0 \cdot S_{GEOM}, \qquad (11.6)$$

where:

S_{GEOM} Geometrical area of the surface to be irradiated, which depends on the antenna diagram and distance up to that area

σ_0 *Specific absolute cross-section* (SACS)

The geometrical size b of the irradiated area width is defined by the following ratio $b = R \cdot \Delta\alpha$, where R is the distance up to the area to be irradiated, and $\Delta\alpha$ is the antenna diagram width in the azimuth plane. The longitudinal size m (since at the grazing (laid) viewing angles in the ARVS, the distance strobe projection on the surface practically coincides with the element of resolution on distance). As a result, the *signal-to-noise ratio* during the location of the prolonged surface is defined by the following relation:

$$q = \frac{P_{RCV}}{P_N} = \frac{P_{TR} \cdot G_A^2 \cdot \lambda^2 \cdot \eta^2 \cdot \sigma_0 \cdot 1.5 \cdot \Delta\alpha}{(4 \cdot \pi)^3 \cdot R^3 \cdot k \cdot T_\Sigma \cdot N_D \cdot \Delta f_N} \qquad (11.7)$$

Charts for the dependence of the signal-to-noise ratio during location of point targets with RCS 1 and 100 m^2 and prolonged surface with the scattering factor 0.01 are presented in Figure 11.12.

Based on the charts in Figure 11.12, it can be seen that the signal-to-noise ratio when using the selected parameters of the ARVS is as follows: $q > 10$ at operating distances of ARVS that allows observing targets visually. Such a signal-to-noise ratio is not sufficient for the operation of automatic detectors, therefore detection over the entire range of distances is performed for impediments ($S_{EF} \geq 1$ m^2), and for the road boundaries – only for distances of 100–150 m.

ARVS receiver device construction features are associated with a wide dynamic range of reflected signals and the necessity to minimize the entire system cost. It is evident, that the

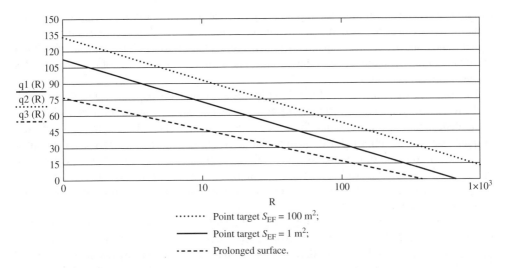

$$\cdots\cdots\cdots \quad \text{Point target } S_{EF} = 100 \text{ m}^2;$$
$$\text{————} \quad \text{Point target } S_{EF} = 1 \text{ m}^2;$$
$$\text{- - - - -} \quad \text{Prolonged surface.}$$

Figure 11.12   Dependence of the signal-to-noise ratio q on the distance R (in m)

dynamic range of signals at the receiver device inlet depends on the values P_{MAX}^{RCV} and P_{MIN}^{RCV} – maximal and minimal power of the received signals. At the same time, the maximal power of the received signals corresponds to the signals, scattered by point objects, located at minimal distances ($R = 1.5$–2 m) and having the maximal effective reflection area (RCS 10^2 m^2– 10^3 m^2). The maximal effective reflection area, S_{EF}^{MAX}, amounts to about 100 m^2 and corresponds to the reflections from heavy haulers; however, it can multiply increase due to interference (caused by the multi-ray distribution of signals) [11.17].

The minimal intensity level of the received signals will correspond to the signals, reflected from extended surfaces during grazing incidence of electromagnetic waves, for example reflections from the roadside and surfaces adjacent to the road. In this case, the effective area of reflection can be assessed according to the formula:

$$S_{EF}^{MIN} = \frac{c}{2W} \cdot R \cdot \Delta\theta_{AZ} \cdot \sigma_0 \tag{11.8}$$

where:

- $2W$ Probing signal frequency band
- c Velocity of light
- R Distance to the reflecting element
- $\Delta\theta_{AZ}$ Antenna diagram effective width in the azimuth plane

Using the radiolocation formula in Eq. (11.3), for the power of the received signal [11.15, 11.26], it is not difficult to assess the dynamic range of the received signals. Regarding the ARVS considered, the dynamic range exceeds the value of 110 dB.

It is evident, that the implementation of a broadband receiver (the band on the carrier frequency exceeds 220 MHz) with such a dynamic range represents a difficult technical assignment. However, in this case the problem of creating an amplifying device with a wide dynamic range can be avoided by use of a homodyne method. A multiple reduce of the received signal bands and frequency expanding is achieved. This allows us to obtain a substantial narrowing of the dynamic range of the signals to be amplified at the outlet of the last cascades of the amplifying device.

To reduce the dynamic range of the signals to be amplified, the dependence of the intensity of the received signals on the distance or frequency can be used, by means of introducing non-linear dependence of the *amplitude frequency characteristic* (AFC) of the amplifier-corrector. A quadratic (on the amplitude) frequency characteristic is implemented in the most simple way. The introduction of the quadratic dependence of the linear amplifier AFC equalizes the power of signals, received from targets located at different distances and having the same value of S_{EF}. At the same time, the dynamic range of all the signals narrows down to the dynamic range of the scene to be observed.

However, besides the marked positive effect of the dynamic range narrowing, a problem of the increase of the *Side-Lobe Level* (SLL) of the LFM signal uncertainty function emerges. This SLL increase creates substantial interferences – a spatial screen flare by distance, following the signal, which corresponds to the reflection from the object. To eliminate this effect, as well as the parasitic amplitude modulation of the VCO, the receiver module of the ARVS uses the function of a "strobe pulse":

$$\cos^2\left(\pi \cdot \left(\frac{t - T_M/2}{T_M}\right)\right) \tag{11.9}$$

This results in an amplitude modulation of the received beat-frequency waveform. It is not difficult to show, that in this case the spectrum envelope of the beat-frequency waveform will have a form of the following function:

$$S(\omega) = \frac{\sin(\omega - \omega_B)}{2(\omega - \omega_B)} \cdot \frac{1}{1 - \left(\frac{T_M(\omega - \omega_B)}{2\pi}\right)^2} \tag{11.10}$$

where:

ω_B Beat-frequency

Unlike the function:

$$\frac{\sin\left(\frac{(\omega - \omega_B)T_M}{2}\right)}{\frac{(\omega - \omega_B)T_M}{2}} \tag{11.11}$$

the spectrum envelope of the beat-frequency waveform at the receiver device outlet $S(\omega)$ – has a wider main lobe and side lobes, which fall quicker – inversely to the third degree of frequency. So, the introduction of a quadratic AFC leads to the compensation of the factor:

$$\frac{1}{1 - \left(\frac{(\omega - \omega_B)T_M}{2\pi}\right)^2} \tag{11.12}$$

and therefore, after the multiplier and subsequent amplifier-corrector (AC) with a quadratic AFC, the spectrum envelope of the beat-frequency waveform will have the function:

$$S(\omega) \cdot K_{AmpC}(\omega) = T_M \frac{\sin\left((\omega - \omega_B)^T M/_2\right)}{(\omega - \omega_B)^T M/_2} \tag{11.13}$$

As a result, an undesired side lobe level growth will not be observed.

It is natural that the introduction of the window function leads to some reduction of the ARVS potential. We have to pay for this reduction of the potential by the decrease of the side lobes level and by the narrowing of the dynamic range of signals, coming to the inlet of the Radar ARVS digital part.

11.2.2.1 Parameters of digital processing in the ARVS

The output signal of the receiver is an analog beat-frequency waveform, which contains the information about the distance to the target and the assessment of its RCS. To extract this information, the output signal must be processed in an appropriate manner and presented in a convenient form for the user. The processing of the output signal and transformation of the received information include the following: digitization of the analog signal, finding of the signal spectrum, and output of the radar image to the user interface.

The analysis of the parameters of ARVS has shown that the spectrum of the analog beat-frequency waveform at the receiver output must occupy the frequency range of 80 kHz– 2.5 MHz at $F_M = 8$ kHz (frequency of repetition of the law of modulation of the transmitter by saw-shaped voltage and $W - 150$ MHz – selected deviation for the transmitter frequency). The frequency range was calculated for distances R, lying in the range of 2–200 m.

According to earlier specified requirements, the ADC converter must ensure covering the full dynamic range of the input signals (40 dB) at the sample rate of not lower than 4 MHz. The lower limit of the sample rate is selected as per the sampling theorem (Kotelnikov/Nyquist), according to which a sampling frequency must be at least twice the value of the upper boundary of the spectrum frequency of the output beat-frequency waveform. However, in practice, in association with the drop of prices for ADC, it is technically easier to perform digitization at a remarkably higher frequency, at the same time having simplified requirements for parameters of antialiasing filter.

The excessive value of the sample rate of the ADC converter leads to a great excess of the sample sequence at its output. Therefore, to increase the computer speed and to reduce the volume of data processed in the FFT unit, a procedure of signal *digital filtering* (DF) and subsequent samples decimation has been used. The form of AFC of such DF must be close to ideal (rectangular), and the cut-off frequency must meet the requirements of the sampling theorem.

In the experimental sample of the ARVS, a non-recursive filter is used, that is the DF with a finite impulse response (FIR-filter). According to the type of selectivity, this filter is a low-pass filter (LPF). This fact is defined by the range of the following ARVS operating frequencies: 80 kHz–2.5 MHz.

The DF operation algorithm assumes a large number of operations of addition and multiplication tasks. This circumstance leads to a significant increase of the digit capacity of the DF output values compared to the input word length.

The filter is synthesized by means of the specialized software (SSW) "Quartus FIR Compiler" of the firm Altera. The metrics for the synthesis code are as follows:

- window type: Hanning window
- input signal digit capacity: 10
- output signal digit capacity: 16
- digit capacity of internal representation of IC coefficients: 12
- number of coefficients to be taken into account: 80
- sample rate: 20 MHz
- decimation factor: 4
- cutoff frequency: 2.5 MHz

The AFC of the synthesized DF is presented in Figure 11.13. The values of frequencies on the abscissa axis are normalized with regard to the sampling rate (20 MHz). The attenuation value in decibels is along on the ordinate axis.

As follows from the AFC analysis, that the requirements to the SLL value are met: the level of the first side lobe amounts to –43 dB, and the transition bandwidth amounts to about 380 kHz.

The ARVS sample, considered as an example, meets contradictory requirements, formulated in Chapter 11.2.2, and allows forming radar data frames with a high rate corresponding to the

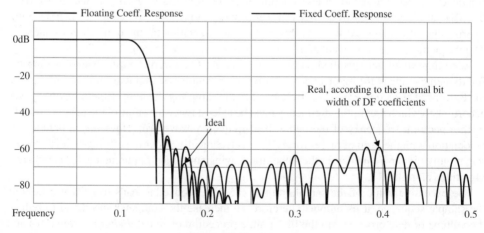

Figure 11.13   DF AFC

road situation in the radar responsibility area. The specifics of the obtained frames of radar data are recorded in the lines corresponding to azimuth channels. The beat-frequency waveforms, the frequency of the spectrum can be observed, which is proportional to the object distance. In order to present these data on the driver (operator) indicator, it is necessary to perform a preliminary processing of them.

11.2.3   Main Tasks and Processing Methods of Radar Data in the ARVS

The main features of the algorithms of forming an RI in the ARVS are considered in [11.40]. That is why this section will consider algorithms of RI processing. All the algorithms of information processing in the ARVS can be divided into two groups: the first one is the algorithm of the processing of a radar signal from the A to D converter output and of RI forming, the second one is the algorithm of the processing of an already formed RI. Algorithms of radar signal processing in the ARVS are performed by the following operations:

- frame-by-frame (512×256 points) Fast Fourier Transform (FFT)
- real time on-screen display of RI
- ARVS operation control

The tasks of the secondary processing of a formed RI in the ARVS are reduced to the following:

- Conversion of distance-azimuth coordinates of the initial RI into the polar coordinate system
- Marking out of immobile objects on the traffic lane and assessment of their current coordinates
- Marking out, assessment of current coordinates and speeds, as well as a classification of the moving objects (motorcar, average-dimensional and large-dimensional trucks, motorcycles, bicycles, pedestrians). Prognosis of their movement
- Checking of criteria of a dangerous situation and forming of the information warning of the danger
- Forming of the symbolic output information, easily perceived by the person, ensuring a clear and definite explanation of the sense of the warning signal
- Marking of a road edge and a dividing boundary (if there is one) between the oncoming traffic bands
- Classification of the type of edge – soft verge, metallic or concrete barrier, noise shields
- Definition of the current road width, passing and oncoming traffic bands
- Road curvature prognosis
- Road profile prognosis
- Correlation of these data with the digital map of the world (during interconnecting with GPS)
- Marking of isolated roadside objects (lighting columns, distance columns, etc.) and assessment of their current coordinates
- Reduction of data to a single moment of time (with an account to the diversity of obtaining primary data from different azimuths)
- Functional connection with other information sensors of the vehicle (speedometer, turn sensor)

It should be noted that in order to facilitate an RI operator perception, it is desirable to increase the rate of the information update on the indicator. The task of increasing the representation rate is solved by the interpolation of a primary RI and the restructuring of an image on the screen. For that, the RI processing algorithm must automatically and continuously form the assessment of the vehicle's proper speed vector with regard to immobile local objects and waysides. This assessment can be implemented using speedometer data or by means of the ARVS similar to a known *correlation velocity meter* (CVM) [11.16,11.28], using data analysis from several straddling azimuth channels.

Based on the obtained velocity vector, the interpolated RI (obtained with a smaller rate) can be displayed on the indicator a few times (until obtaining the next frame) with an appropriate displacement, ensuring continuity of the movement of the displayed picture. It is expedient to eliminate the area in front of the vehicle from the area of the analysis of a correlation velocity meter in which the appearance of objects moving with regard to the Earth's surface is probable. The account of the movement speed for image forming can define the scale of the image to be displayed: as the speed increases, an RI to a greater distance is displayed on the indicator.

The use of a brightness RI in the transformed coordinates, significantly simplifies the perception of the information by an operator, since the distortions of the geometrical form of the location objects are eliminated, which are characteristic for the RI in non-transformed coordinates (distance-azimuth) (Figure 11.14).

The frame of a real RI of a free section of an automobile road (Figure 11.14), illustrates the convenience of observing the road boundaries in the form of parallel direct lines. However, when performing the transformation, not all the information obtained from the ARVS is displayed (since the close area in Figure 11.14, drawn in detail, during the transformation on the left is collapsed into a single point), and the screen area is used less effectively. However, ensuring the correspondence of the resulting RI to the coordinate system, natural for an operator, is of prevailing importance.

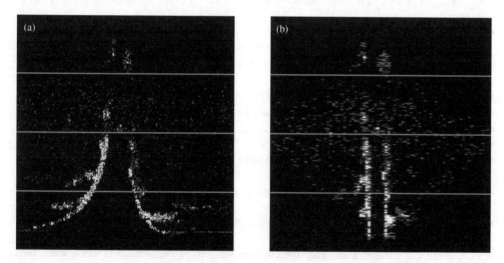

Figure 11.14   RI in distance-azimuth coordinates (a) and transformed image (b)

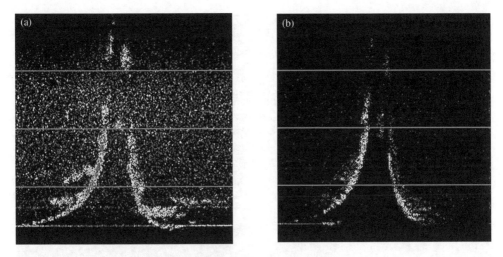

Figure 11.15 RI before (a) and after (b) the noise filtering

Limitations for an accessible radar potential lead to the availability in the radar data array of thermal noises of the receiver. To reduce their intensity, it is expedient to use filtering procedures (Figure 11.15).

In order to transmit a wide dynamic RI range, it is suitable to use the algorithm of quantization on intensity, which allows marking (Figure 11.16) individual objects with a significant RCS, for example, oncoming vehicles, and thereby forcing the driver to pay immediate attention to them [11.39].

To ensure a continuous monitoring of the vehicle position on the roadway, it is necessary to impose vector lines and digital data – image processing markers on the primary RI (Figure 11.17).

As mentioned above, the driver has to know the position of his own vehicle on the roadway, the roadway width and its curvature, the position of any objects on that road, and respectively, the distance to them and whether they are in the way of traffic. As far as the case of movement

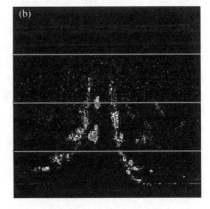

Figure 11.16 Use of a pseudo color palette for marking the objects with a significant RCS

Figure 11.17   RI with superimposed processing markers

on highways is concerned, it can initially be suggested that along the entire distance of the ARVS action (up to 250 m) the roadway is straightforward or has insignificant curvatures. At the same time, as a first approximation, turn-offs, crossroads, etc. are not considered. Besides, not all the objects on the road should be considered dangerous, that is why it is expedient to introduce the notion "*safe traffic corridor*" (STC).

The STC is the road area in which objects will be considered potentially dangerous, and it is necessary to define the distance to them exactly. The STC sizes are chosen proceeding from the size and type of the actual vehicle (determining its maneuverability), its velocity of travel, meteorological conditions and type of road surface coating. As a result, the algorithm of RI processing with account of the adopted approximations consists of the following operations:

- detection, marking and construction of boundaries (roadsides) of the roadway on the RI, i.e. measuring and providing the information about distances up to the left and right roadsides to the driver
- calculation, construction and visualization of the STC boundaries
- detection of dangerous objects within the STC (the closest ones on distance), marking them on the RI with distance measuring and providing the driver with this information

Figure 11.17 presents an example of using this algorithm:

1. The road boundaries are shown by dashed-dotted lines
2. The safety corridor is shown by dotted lines
3. The distance up to the roadside – upper digits in the bright insert
4. The distance to the vehicle which is moving in front (lower digits on the bright insert)

In conclusion, it is necessary to note that in this case the quality of information is closely associated with its perceptibility by the operator. Moreover, this parameter is largely subjective.

Hence, the most acceptable way of evaluating the quality or perceptibility of the information in the concept, formed by the developed software (SW), is an expert assessment.

As a result of the theoretical and experimental ARVS research performed, the following conclusions can be drawn:

1. Using the created ARVS sample, the experimental research of the scattering signal charac-teristics in typical road situations has been made. Based on the experiments is the analysis of typical situations obtained by the RI, which emerge during the movement of vehicles along weakly loaded highways.
2. As a result of the roadway and adjacent locality RI research performed, it is shown that these RI are characterized by highly informative properties, sufficient for a reliable definition of the main parameters of the roadway, such as its width (the distance to the left and right roadsides), available turns, available guards, available movement of obstacles and their parameters.
3. The assessment of average RCS values of the following objects of natural and artificial origin, typical for radio-wave imaging systems has been made: roadsides, woodlands and scrublands, metallic enclosures.
4. It is shown that the RI, obtained by means of the ARVS, can be effectively used for con-trolling vehicle movement under the conditions of limited or unavailable optical visibility.

The ARVS experimental samples provide the following:

- Measurement and reflection of highway and obstacle parameters for safe vehicle driving. On the RI it also forms a safe movement corridor and vehicle path of motion to be predicted by taking into account of driver's own speed and angular position of the vehicle wheels
- Information about distances to obstacles, located in the path of motion. Warning the driver of dangerously approaching obstacles by means of suitable signals
- Ensures the possibility of a special-purpose vehicle movement in the motorcade at a specified distance and at a specified velocity of travel
- Besides the information and warning properties, it possesses the unique possibility of implementing an interactive vehicle control mode through the analysis of the road situation by the driver, according to the radar image, displayed on the digital display screen

The ARVS may be used for heavy trucks navigation (e.g. mine trucks in the opencast mines). The main ARVS consumer advantages are as follows:

- high radar information update speed, comparable with optical cameras and television sys-tems
- high ARVS spatial resolution (on azimuth and distance) substantially enhances the possi-bilities of detecting and identifying small-size objects, for example, a man on the road at the available reflections from an underlying surface
- high information capability of radar images in the ARVS and their stability allows imple-menting the main parameters measurement of vehicle movement (trajectory, speed and coordinates of passing and oncoming vehicles)
- high ARVS interference immunity suppresses incidental and industrial interferences effectively

- small dimensional sizes and radio-ecological cleanness (the low level of the average radiated power, comparable with the power of mobile phone transmitters) are positively conceived by the consumers of this system: the vehicle and road servicing drivers

There is a possibility of implementing the ARVS with the operating action distance (up to 500 m and even 1000 m) without any radical changes, which enhances the area of application of these systems. In such a design, the ARVS can be used for ship navigation, in part to solve the following tasks:

- passing under bridge spans or other constrictions
- passing through locks
- high-precision navigation within a harbor
- detection of partly-submerged objects
- overcome the problems of the "dead" zone of commercial navigation ship radar

11.2.4   Main Problems and Tasks of ARVS Development

In spite of the successes achieved, the radio-wave imaging systems creators do face a whole range of vital problems of both an economic character (associated with the reduction in price of the technology), and principal ones, natural to the selected method of a space view.

The following can be marked among the latter:

- The problem of *interference marks* from the objects, located above the roadway
- The problem of *shadowing objects* by other objects

The problem of interference marks from objects located above the roadway consists in the observation of the indicator with marks from the spreads such as posters and road signs, bridges and overpasses above the road, which are not a threat to movement. This problem can be solved, for example, by reducing the width of the antenna diagram on the elevation angle. However, such a decision will require control of the orientation of the antenna ray on the elevation angle. The most probable solution to this problem consists of using the information from the cartographic database and global positioning systems. A combination of such data will allow a decision to be made as to whether or not to ignore such noise signals.

The problem of shadowing of objects by other objects is known to everybody by analogy with optical systems. However, the feature of the millimeter-range radar sensor is an emergence of the channel of electromagnetic waves propagation under the vehicle bottom (which was observed experimentally). In this case, a motor car shadowed by other vehicles and therefore not visible by optical means, can be observed on a radar image.

The following can be marked among the main directions of development of modern vehicle radars:

1. Creation of compact antenna systems for vehicle radars, adequate to the tasks to be solved, in particular regarding the following:
   - high scanning rate (mechanical or electronic) of the narrow antenna diagram in the azimuth plane or creation of a multi-ray antenna diagram in the azimuth plane

- low level of the antenna diagram side lobes with the aim of preventing the capture of false objects (e.g. road guards) in the course of vehicle movement
2. Development of fast-acting algorithms for real-time processing of the received signals, solving the task of stationary and mobile objects recognition on the road and ensuring adequate (for the driver) reproduction of the radar image
3. Optimal interconnecting of various sensors – both autonomous ones (e.g. optical, infrared, radar ones) and external ones (e.g. GPS receivers and digital fields). The aim of interconnecting is to optimize the vehicle safety system construction architecture, to increase the accuracy and reliability of measurements (e.g. in case of loss of signals of one of the sensors), to reduce the system cost owing to elimination of sensor doubling properties. With regard to our task, the following sensors can be interconnected on board the vehicle:
   - homogeneous (one-type) sensors, e.g. radar sensors of great and small distance
   - heterogeneous (different ones on the physical operation principle) sensors, e.g. a radar sensor of long-range action and an infrared or optical sensor

The ARVS, infrared and video camera interconnecting is thought to be most effective.

11.2.5 Conclusions

Not claiming to play the leading role in ensuring safety, the modern radar sensor is capable of detecting the oncoming, passing and immobile vehicles at a distance of up to 200 m–250 m in a guaranteed manner under any weather conditions. People and large animals in the traffic area up to a distance of 100–150 m can be identified, defining their coordinates and direction of movement.

Interconnecting the radar sensor with the cartographic support and global positioning systems, optical and other sensors, will allow compensating weaknesses of individual technologies, forming a single actual data stream to aid in making decisions about the traffic situation.

The creation of a radar sensor with a continuous panoramic field of view allows the evaluation of maneuvers of other traffic participants and to predict approaching danger. The adoption of ARVS (panoramic radar sensors) will allow solving principal difficulties in the assessment of road situations, which the designers of the automatic vehicle movement control systems have at the present time.

References

[1.1]   Einstein, A., Zur Quantentheorie der Strahlung. *Phys. Zeitschr.* 18: (1917)
[1.2]   Maiman, T.H., Optical and microwave-optical experiments in ruby. *Phys. Rev. Lett.* 4: (1960)
[1.3]   Quist, T M. et al., Semiconductor Maser of GaAs. *Appl. Phys. Lett.* 1: (1962)
[1.4]   Hayashi, I. et al., Junction lasers which operate continuously at room temperature. *Appl. Phys. Lett.* 17(3): 109 (1970)
[1.5]   Kao, C.K. and Hockham, G.A., Dielectric-fiber surface waveguides for optical frequencies. *Proc. IEEE.* 113(7): 1151 (1966)
[1.6]   Börner, M., Deutsches Patent No. 1254513: Mehrstufiges Übertragungs system für in Pulscodemodulation dargestellte Nachrichten.
[1.7]   Kapron, F.P. et al., Radiation losses in glass optical waveguides. *Appl. Phys. Lett.* 17(10): 423 (1970)
[1.8]   Adams, W.G. and Day, R.E., The action of light on selenium. *Proc. R. Soc.* 25: 113 (1876)
[1.9]   Melchior, H. et al., Photodetectors for optical communication systems. *Proc. IEEE.* 58(10): 1466 (1970)
[1.10] Miller, S., Integrated optics: an introduction. *Bell Syst. Techn. J.* 48(7): 2059 (1969)
[1.11] Gruß, R., Übertragung von Laserstrahlen durch die Atmosphäre. *Ntz Nachrichtentechn. Z.* 22(3): 184 (1969)
[1.12] Marstaller, A., private communication
[1.13] Strobel, O., *Lichtwellenleiter-Übertragungs-und Sensortechnik* (VDE Verlag, 3rd Edition) (2014)
[1.14] MOST Cooperation. *MOST Brand Book, vol. 1.1,* August 2006. Retrieved December 7, 2009, from www.mostcooperation.com (2007)
[1.15] Majkner, R., *Overview – Lightning Protection of Aircraft and Avionics.* Sikorsky Corp. Retrieved (2003)
[1.16] FlexRay Consortium, www.FlexRay.com, last accessed date July 8 (2009)
[1.17] Kenward, M., Plastic fiber homes in/on low-cost networks. *Fiber Systems,* 5(1): (2001)
[1.18] Pei, Y. et al., LED modulation characteristics in a visible-light communication system, *Optics and Photonics Journal,* 3, 139–142 (2013)
[1.19] Langer, K.S., *Hochratige drahtlose Datenübertragung mit LED.* Heinrich-Hertz-Institut, Fraunhofer Institute for Telecommunications, Berlin, private communication (2010)
[1.20] Iizuka, N., Some comments on merged draft from the viewpoint of the VL-ISC, IEEE 802.15-08-0759-02-0vlc, Project: IEEE P802.15 Working Group for Wireless Personal Area Networks (WPANs) (2008)
[1.21] Son, M. et al., doc.: IEEE 802.15-08-0759-02-0vlc, Project: IEEE P802.15 Working Group for Wireless Personal Area Networks (WPANs) (2008)
[1.22] Denny Both, Piranha.dl 3d Animation, Berlin: https://www.facebook.com/pages/piranhadl-3D-animation/131721586891915?ref=hl
[1.23] Bletzer, H., Esslingen University of Applied Sciences, private communication.
[1.24] Herter, E. and Graf, M., *Optische Nachrichtentechnik.* München/Wien: Carl Hanser, (1994)
[1.25] Martin, R., *History of Optical Communication,* Esslingen University of Applied Sciences, private communication.

[2.1]   Hecht, E., *Optics, Band IH*. Addison Wesley, Berlin/San Francisco (2002)

[2.2]   Tosco, F., *Fiber Optic Communication Handbook*, TAB Books McGraw-Hill, New York, p. 148 (1981)

[2.3]   see [2.1] p. 385.

[2.4]   Pohl, R.W., *Optik und Atomphysik*. Berlin/Heidelberg/New York: Springer-Verlag, (1976)

[2.5]   see [2.1] p. 390.

[2.6]   Reisinger, A.R. et al., Coherence of a ROOM-TEMPERATURE CW GaAs/GaAlAs injection laser. *IEEE J. Quant. Electr.* QE-15(12): 1382 (1979)

[2.7]   Auch, W., Mohr, R. and Strobel, O., *Entwicklung und Erprobung eines miniaturisierten Faserkreisels*. BMFT Schlussbericht. Förderungskennz. 514-8891-LFL 8185, p. 118 (1985)

[2.8]   Azzam, R.M.A. and Bashra, N.M., *Ellipsometry and Polarized Light*. Amsterdam: North Holland Publ. Comp. (1977)

[2.9]   see [2.1] p. 325.

[2.10] Hering, E., Martin, R. and Stohrer, M., *Physik für Ingenieure*. Düsseldorf: VDI-Verlag, Springer Heidelberg, Dodrecht London/New York, p. 579.

[3.1]   Kibler, T. et al., Optical data buses for automotive applications, *J. Ltw. Techn.* 22: 2184–2199 (2004)

[3.2]   Cohen, L.G. et al., Low loss quadruple-clad single mode lightguides with dispersion below 2 ps/km.nm over the 1.28 pm–1.65 p.m wave length range. *Electron. Lett.* 18(24): 1023 (1982)

[3.3]   Francois, P.L., Dispersion-free single mode doubly-clad fibers with small pure bend losses. *Electron. Lett.* 18(19): 818 (1982)

[3.4]   Miya, T. et al., fabrication of low dispersion single-mode fibers over a wide spectral range. *IEEE J. Quant. Electron.* 19: 885 (1983)

[3.5]   Gowar, J., *Optical Communication Systems*, Prentice Hall, New York/London, p. 643

[3.6]   Unger, H.G., *Optische Nachrichtentechnik*, Teil I. Heidelberg: Hüthig Verlag, p. 92 (1983)

[3.7]   Geckeler, S., *Lichtwellenleiter für die optische Nachrichtenübertragung*. Berlin/Heidelberg/NewYork/London: Springer-Verlag, p. 151 (1987)

[3.8]   Wilson, J. and Hawkes, J.F.B., *Optoelectronics – An Introduction*. Prentice Hall International, pp. 8–10.

[3.9]   see [2.2] p. 170.

[3.10] see [3.6] p. 164.

[3.11] see [3.7] p. 158.

[3.12] Lutzke, D., *Lichtwellenleitertechnik*. München: Pflaum Verlag (1986)

[3.13] Miller, S.E. et al., research toward optical-fiber transmission systems. *Proc. IEEE*, 6112: 1703 (1973)

[3.14] Gloge, D. et al., Multimode theory of graded core fibers. *Bell Syst. Techn. J.* 52: 1563 (1973)

[3.15] Grau, G., *Quantenelektronik*. Braunschweig: F. Vieweg Verlag (1978)

[3.16] Abramowitz, M. and Stegun, I.A., *Handbook of Mathematical Functions*. Dover Publications (1970)

[3.17] Roßberg, R., SEL-Alcatel (now Bell-Labs Germany) Stuttgart, private communication.

[3.18] Mohr, F., SEL-Alcatel (now Bell-Labs Germany) Stuttgart, private communication.

[3.19] Jeunhomme, L. and Moniere, M., Polarisation-maintaining single-mode fiber cable design. *Electr. Lett.* 16(24): 921 (1980)

[3.20] Mohr, R. and Scholz, U., Active polarization stabilization systems for use with coherent transmission systems or fiber optic sensors. *Proc. 9th ECOC Geneva*, p. 313 (1983)

[3.21] Not, R., Endless polarisation control in coherent optical communications. *Electr. Lett.* 22(15): 772 (1986)

[3.22] Hadjifotiou, A., *The Prospects for High Power Optical Transmission Systems*. Standard Telecom. Lab. Report, R 600/TES/84/30.

[3.23] Cotter, D., Optical non-linearity in fibers: a new factor in systems. *BRTL Techn. J.* 1(2): 17 (1983)

[3.24] Stolen, R.H., Nonlinearity in fiber transmission, *Proc. IEEE*. 68(10): 1232 (1980)

[3.25] Cotter, D., Optical non-linearity in fibers may affect advanced transmission systems. *Laser Focus*, June: 77 (1984)

[3.26] see [3.46], p. 171.

[3.27] Bondiek, R. and Freyhardt, W.E., Rückstreu-Dispersionsmessgerät für die Glasfaserproduktion. *ntz Nachrichtentech. Z.* 36(7): 438 (1983)

[3.28] see [2.2] p. 286.

[3.29] Fischbach, J.-U., Kaiser, M. et al., *Optoelektronik-Bauelemente der Halbleiteroptoelektronik*. Grafenau: expert verlag, p. 218 (1982)

[3.30] Bludau, W., Gündner, H.M. and Kaiser, M., *Systemgrundlagen und Messtechnik in der optischen Übertragungstechnik*. Stuttgart: Teubner, p. 125

[3.31] see [3.6] p. 171 (1985).

[3.32] Marcuse, D., Calculation of bandwidth from index profiles of optical fibers. 1. *Theory. Appl. Opt.* 18(12): 2073 (1979)

[3.33] see [3.6] p. 194/195.

[3.34] see [3.20] p. 30.

[3.35] Geckeler, S., Length-dependence of bandwidth of graded index optical waveguides. *Siemens Forsch.-u. Entw.-Bet.* 12(5): 313 (1983)

[3.36] Rosenberger, D. and Geckeler, S., *Optische Informationsübertragung mit Lichtwellenleitern*. Grafenau: expert-verlag; Berlin u. Offenbach: VDE Verlag GMBH, p. 29 (1982)

[3.37] Li, T., Structures, parameters, and transmission properties of optical fibers. *Proc. IEEE.* 68(10): 1175 (1980)

[3.38] see [2.2] p. 321.

[3.39] Marcuse, D., *Principles of Optical Fiber Measurements*. New York: Academic Press (1981)

[3.40] Cohen, L.G. et al., Dispersion and bandwidth spectra in single-mode fibers. *IEEE J. Quant. Electr.* QE-18(1): 49 (1982)

[3.41] see [2.2] p. 333.

[3.42] see [3.26] p. 106.

[3.43] Cancellierei, G. et al., Scanning technique for investigating optical fiber dispersion, *Opt. Quant. Electr.* 11: 305 (1979)

[3.44] Gliemroth, G. et al., Gläserne Telefondrähte. *Schott Information*, 2L: 1 (1976)

[3.45] see [2.2] p. 321.

[3.46] Bludau, W., *Lichtwellenleiter in Sensorik und optischer Nachrichtentechnik*. Berlin/Heidelberg/New York: Springer-Verlag, p. 133, (1998)

[3.47] Nagasawa, Y. et al., High performance dispersion compensating fiber module. *Fujikura Technical Review* (2001)

[3.48] Mahlke, G. and Gössing, P., *Finer-Optic Cables*, Publicis MCD Corporate Publishing, Erlangen/Munich.

[3.49] Chbat, M.W., Managing polarization mode dispersion. *Photonics Spectra*, June: 100 (2000)

[3.50] Mollenhauer, L.F. and Stolen, R.H., Solitons in optical fibers. *Fiberoptic Technol.* April: 193 (1982)

[3.51] Malyon, D.J. et al., Demonstration of optical pulse propagation over 10 000 km of fiber using recirculating loop. *Electr. Lett.* 27(2): 120 (1991)

[3.52] Mollenhauer, L.F. and Smith, K., Soliton transmission over more than 6000 km in fiber with loss periodically compensated by raman gain. Invited Paper 18A1-3, *IOOC '89*, Kobe, Japan (1989)

[3.53] Mollenhauer, L.F. et al., Demonstration of soliton transmission at 2.4 Gbit/s over 12 000 km. *Electr. Lett.* 27(2): 178 (1991)

[3.54] Andrekson, P.A. et al., Soliton data transmission at 32 Gbit/s over 90 km. *Proc. Opt. Ampl. and their Appl.*, Snowmass, CO, PdP2 (1991)

[3.55] Marshal, I.W. et al., 20 Gbit/s 100 km non-linear transmission with semiconductor source. *Proc. OFC '90*, San Francisco, CL. PD6 (1990)

[3.56] Bergano, N.S. et al., Bit-error-rate measurements of a multi thousand kilometer fiber-amplifier transmission system using a circulating loop. *Proc. Opt. Ampl. and their Appl.*, Snowmass, CO, ThA5 (1991)

[3.57] Nakazawa, M. et al., 10 Gbit/s soliton data transmission over one million kilometers. *El. Lett.* 27(14): 1270 (1991)

[3.58] Yamada, E. et al., 10 Gbit/s single-path soliton transmission over 1000 km. *El. Lett.* 27(14): 1289 (1991)

[3.59] Nakazawa, M., Optical soliton communication using erbium-doped fiber amplifiers. *Proc. ECOC/IOOC'91*, Paris, 2: 150 (1991)

[3.60] Flannery, D., Raman amplifiers: powering up for ultra-long-haul. *Fiber Systems*, 5(7): 48 (2001)

[3.61] Georges, T.J., Soliton transport gives backbones more speed. *Fiber Systems*, 4(5): 51 (2000)

[3.62] Udem, T. et al., Uhrenvergleich auf der Femtosekundenskala. *Physik J.* 1(2): 39–45 (2002)

[3.63] Wedding, B., Reduction of bit error rate in high speed optical systems due to optimized electrical pulse shaping. *Proc. ECOC '88*, Brighton, 1: 187 (1988)

[4.1] Guinau, M. et al., Optical fiber advances and next gen. backbone, *WP6024*, June: Corning Inc. (2012)

[4.2] *J-Fiber News at Photonics West* (2013)

[4.3] Mazzarese, D., *Manufacturing Multimode Fiber*, OFS/Fitel (2003)

[4.4]   Bergen, v. A. and Breuls, T., PCVD The ultimate technology for production of high bandwidth multimode fibers, *International Wire & Cable Symposium* (1998)

[4.5]   Golowich, S.E. et al., MM Optical Fiber having improved Refractive Index Profile, US Patent 6,292,612 B1, 09, Lucent Technologies Inc. (2001)

[4.6]   Kouzmina, I. et al., CPC protective coating, *WP3703*, July: Corning Inc (2010)

[4.7]   Nextrom and DSM have successfully demonstrated High Speed Optical Fiber Draw Process, Vantaa, Finland, May (2012)

[4.8]   Scott Glaesemann, G. et al., *The Mechanical Reliability of Corning Optical Fiber in Small Bend Scenarios*, WP, December: Corning Inc. (2007)

[4.9]   Vydra, J., Sattmann, R., Tenovom, H., Chang, H.K. et al., Next generation fiber manufacturing for the highest performing conventional single mode fiber, *Optical Society of America*, OFS (2005)

[4.10] Kuyt, G., Bertaina, A. and Davies, M., Trends in FTTH/FITH optical fibre and cable and associated standardization. *FTTH Conference*, London, Feb. (2013)

[4.11] IEC 60793-2-50 Ed. 4.0: Optical fibres – Part 2-50: Product specifications – Sectional specification for class B single-mode fibres (2012)

[4.12] IEC 60793-2-10: Optical fibres – Part 2-10: Product specifications – Sectional specification for category A1 multimode fibres (2011)

[4.13] ITU-T Rec. G.652: Characteristics of a single-mode optical fibre and cable (2009) http://www.itu.int/ITU-T/recommendations/index.aspx?ser=G.

[4.14] ITU-T Rec. G.653: Characteristics of a dispersion-shifted singlemode optical fibre and cable (2010)

[4.15] ITU-T Rec. G.655: Characteristics of a non-zero dispersion-shifted single-mode optical fibre and cable (2009)

[4.16] ITU-T Rec. G.656: Characteristics of a fibre and cable with non-zero dispersion for wideband optical transport (2010)

[4.17] Roberts, K., O'Sullivan, M., Wu, K.T., Sun, H., Awadalla, A. et al., Performance of dual-polarization QPSK for optical transport systems, *J. Lightw. Tech.*, 27(16): (2009)

[4.18] Matthijsse, P. and Griffioen, W., Matching optical fiber lifetime and bend-loss limits for optimized local loop fiber storage. *Opt, Fiber Tech.* 11: (2005)

[4.19] ITU-T Rec. G.657: Characteristics of a bending loss insensitive single mode optical fibre and cable for the access network (2009)

[4.20] Mazzarese, D., New cable designs avert conduit congestion. *Lightwave*, 30(5): (2013)

[4.21] White, W., Blyler, L., Park, M. and Ratnagiri, R., Manufacture of perfluorinated plastic optical fibers. *Proc. OFC'04*, paper ThI1: (2004)

[4.22] Ishigure, T., Koike, Y. and Fleming: J., Optimum index profile of the perfluorinated polymer-based GI polymer optical fiber and its dispersion properties. *J. Lightw. Tech.* 18(2): (2000)

[4.23] Walter, A. and Schaefer, G., Chromatic dispersion variations in ultra-long-haul transmission systems arising from seasonal soil temperature variations. *Proc. OFC'02*, WU4 (2002)

[4.24] *FTTH Handbook, Edition 5.* Fibre to the Home Council Europe (2012)

[4.26] Crawford, D., *Fiber Optic Cable Dig-Ups – Causes and Cures*. Network Reliability and Interoperability Council, Chicago (1993)

[4.27] *Reliability of Fiber Optic Cable Systems: Buried Fiber Optic Cable, Optical Groundwire Cable, All Dielectric, Self Supporting Cable*. Alcoa Fujikura Ltd. (2001)

[4.28] Lundergan, M. and Dallas, K., Field-aging study shows strength of optical ground wire cable. *Lightwave*, 14(10): (1997)

[4.29] *DrakaElite High Temperature Acrylate Single-Mode Fiber.* Data sheet (2009)

[4.30] Sun, Z., Xiong, Z., Xu, D., Xu, Z. and Wan, B., Development of high temperature fiber and ETFE tight buffered fiber. *Proc. 58th IWCS/IICIT* (2009)

[4.31] Mitra, A., Kouzmina, I. and Lopez, M., Thermal stability of the CPC fiber coating system. *WP4250*, Corning Inc. (2010)

[4.32] Rowland S., Prevention of dry-band arc damage on ADSS cables. *IEEE Trans. Dielectrics and Electrical Insulation*, 13(4): (2006)

[4.33] Murata, H. *Handbook of Optical Fibers and Cables, Second Edition.* Marcel Dekker, Inc. (1996)

[4.34] Mrotek, J., Matthewson, J. and Kurkjian, C., Diffusion of moisture through optical fiber coatings. *J. Lightw. Tech.* 19(7): (2001)

[4.35] Gebizlioglu, O.S., Time- and temperature-dependent material behavior and its impact on low-temperature performance of fiber optic cables. *Materials Research Society Symposium Proceedings*, 531: Symposium DD: Reliability of Photonics Materials and Structures, San Francisco, CA (1998)

[4.36] Tanaka, S., Kameo, Y. and Tsuneishi, K., Long-term reliability of layer-type tight structure cable, *J. Lightw. Tech.* 4(8): (1986)

[4.37] Borzycki, K., Temperature dependence of polarization mode dispersion in tight-buffered optical fibers. *J. Telecommunications and Information Tech.* 1: (2008)

[4.38] Product Catalog. Optical Cable Corporation. Available from www.occfiber.com

[4.39] Optical Fibre Cables. Catalog, Tele-Fonika Kable (2012) http://www.tfkable.com/en/documents-and-files/product-catalogs.html

[4.40] IEC 60304: Standard Colours for Insulation Ffr Low-Frequency Cables and Wires (1982)

[4.41] ANSI/TIA-598-C-2005: Optical Fiber Cable Color Coding (2005)

[4.42] NFPA 90A: Standard for the Installation of Air-Conditioning and Ventilating Systems. National Fire Protection Association (2012) http://www.nfpa.org/codesand-standards/document-information-pages

[4.43] NFPA 262: Standard Method of Test for Flame Travel and Smoke of Wires and Cables for Use in Air-Handling Spaces. National Fire Protection Association (2011) http://www.nfpa.org/codes-and-standards/document-information-pages

[4.44] UL 1666: Standard Test for Flame Propagation Height of Electrical and Optical-Fiber Cables Installed Vertically in Shafts (2007)

[4.45] EN 50399:2011: Common test methods for cables under fire conditions. Heat release and smoke production measurement on cables during flame spread test. Test apparatus, procedures, results (2011)

[4.46] Kawataka, J. et al., Novel optical fiber cable for feeder and distribution sections in access networks, *J. Lightwn Tech.* 21(3): (2003)

[4.47] Okada, N., Latest optical fiber cable technologies for further expansion of FTTH network. *Fujikura Technical Review* (2013)

[4.48] Optical Fiber Ribbon Slotted Core Cable (Polyethylene Sheath Structure for Duct Application). Data sheet, Furukawa Electric Co. (2002)

[4.49] Air-Blown Micro Cable: Stranded Loose Tube All-Dielectric, Central Tube All Dielectric. Data sheet, CommScope (2011)

[4.50] Bludau, W. and Roßberg, R., Characterization of laser-to-fiber coupling techniques by their optical feedback. *Appl. Opt.* 21(11): 1933 (1982)

[4.51] Bludau, W; Roßberg, R.: Low-loss laser-to-fiber coupling with neglible optical feedback. *J. Lightw. Tech.* LT-3(2): 294 (1985)

[4.52] Wenke, G. and Zhu, Y., Comparison of efficiency and feedback characteristics of techniques for coupling semiconductor lasers to single-mode fiber. *Appl. Opt.* 22, 23: 3837 (1983)

[4.53] Young, M., *Optics and Lasers*. Springer, Berlin/Heidelber/ New York, p. 5.

[4.54] Hentschel, C., *Fiber Optics Handbook*. Hewlett-Packard GmbH, Böblingen, p. 51 (1988)

[4.55] Saruwatari, M. and Nawata, K., Semiconductor laser to single-mode fiber coupler. *Appl. Opt.* 18(11): 1847 (1979)

[4.56] Kuwahara, H. et al., Reflected light in the coupling of semiconductor lasers with tapered hemispherical end fibers. *Appl. Opt.* 22(17): 2732 (1983)

[4.57] Khoe, G.D. et al., Efficient coupling of laser diode to tapered monomode fibers with high-index end. *Electr. Lett.* 19(6): 205 (1983)

[4.58] Roßberg, R., Kopplungselemente für LWL. *Systeme Seminar Nachrichtenübertragung mit Lichtwellenleitern an der Technischen Akademie Esslingen* (1981)

[4.59] Köster, W., Efficient laser-to-fiber coupling. *Laser Focus*, 18: 103 (1982)

[4.60] Roßberg, R., *Kompendium 8 der telekom praxis: Kabeltechnik (I), Lichtwellenleiter-Verbindungstechniken*: Stecker und Spleiße. Berlin: Schiele und Schön (1991)

[4.61] Di Vita, P. and Rossi, U., Theory of power coupling between multimode optical fibers. *Opt. Quant. Electr.* 10: 107 (1978)

[4.62] Geckeler, S., *Das Phasenraumdiagramm, ein vielseitiges Hilfsmittel zur Beschreibung der Lichtausbreitung in Lichtwellenleitern*. Siemens Forsch.- u. Entwickl.-Ber. 10(3): 162 (1981)

[4.63] Marcuse, D., Loss analysis of single mode-fiber splicep. *Bell Syst. Techn. J.* 56(5): 703 (1977)

[4.64] see [4.53] p. 139

[4.65] Young, W.C. et. al., A transfer moulded connector with insertion losses below 0.3 db without index match. *Proc. 6th ECOC*, York, p. 310 (1980)

[4.66] Tomlinson, W.J., Wavelength multiplexing in multimode optical fibers. *Appl. Opt.* 16(8): 2180 (1977)

[4.67] Kawasaki, B. et al., Biconical-taper single-mode fiber coupler. *Opt. Lett.* 6(7): 327 (1981)

[4.68] Sharma, A. et al., Tracing rays through graded-index media: a new method. *Appl. Opt.* 21(6): 984 (1982)

[4.69] Parriaux, O. et al., Distributed coupling on polished single-mode optical fibers. *Appl. Opt.* 20(14): 2420 (1981)

[4.70] Nayar, B.M. and Smith, D.R., Monomode-polarization-maintaining fiber directional couplers. *Opt. Lett.* 8(10): 543 (1983)

[4.71] Bricheno, T. and Fielding, A., Stable low-loss single-mode couplers. *Electr. Lett.* 20(6): 230 (1984)

[4.72] Roßberg, R., LWL-Verschmelzkoppler. Verbindungstechnik '91 für elektronische und elektrooptische Geräte und Systeme, *VDI-Ber.* 876: 231 (1991)

[4.73] Köster, W., Wellenlängenmultiplex mit Monomodefasern durch Energietransfer zwischen den Mantelfeldern. *Frequenz.* 38(11): 273 (1984)

[4.74] Yataki, M.P. et al., All fiber polarising beamsplitter. *Electr. Lett.* 21(6): 249 (1985)

[4.75] Bülow, H. and Roßberg, R. *Performances of Twisted Fused Fiber*, EDFA (1985)

[4.76] Mortimore, D.B., Low-loss 8 × 8 single-mode star coupler. *Electr. Lett.* 21(11): 502 (1985)

[4.77] Köster, W., Optische Schalter. *Elektronik*, 15(15): 53 (1980)

[4.78] Roßberg, R., Optische Anordnung für Systeme der optischen Nachrichtentechnik. *Deutsche Patentanmeldung*, P 3910711.6. 3: 4 (1989)

[4.99] Mitsunaga, Y., Katsuyama, Y. and Ishida, Y., Thermal characteristics of jacketed optical fibers with initial imperfection. *J. Lightw. Tech.* 2(1): (1984)

[5.1] Hundsperger, R.G., *Integrated Optics: Theory and Technology*. Berlin/Heidelberg/New York: Springer (1982)

[5.2] Alferness, R.C., Recent advances in integrated optics. *Proc. ECOC/IOOC'91*, Paris, 2: 101 (1991)

[5.3] Leonberger, F.J. et al., LiNbO₃ and LiTaO₃ integrated optic components for fiber optic sensors. *Proc. Optical Fiber Sensors*. Berlin/Heidelberg: Springer (*Proc. in Physics*, 44: Ed. Arditty, H.J. et al.) (1989)

[5.4] Fuest, R. et al., Integriert-optisches Michelson-Interferometer mit Quadraturdemodulation in Glas zur Längenmessung. *Proc. Sensor'91*, Nürnberg, I: 107, (1991)

[5.5] Kersten, R. Th., Integrated optical circuit engineering III (Innsbruck). *Proc. SPIE 651*, Bellingham, WA (1986)

[5.6] Nolting, H.P. and Ulrich, R., *Integrated Optics*. Berlin: Springer (1985)

[5.7] Sohler, W., Bauelemente der integrierten Optik: eine Einführung. *Laser u. Optoelektr.* 4: 323 (1986)

[5.8] Voges, E. and Neyer, A., Integrated-optic devices an LiNbO3 for optical xommunication. *J. Lightw. Techn.* LT-5(9): 1229 (1987)

[5.9] Wulf-Mathies, C., Integrierte Optik für faseroptische Sensoren. *Laser und Optoelektr.* 21(1): 57 (1989)

[5.10] Schmuck, H. and Strobel, O., Stabilization of a fiber-optic mach-ehnderinterferometer used as an intensity modulator. *J. Opt. Commun.* 7(3): 86 (1986)

[5.11] Izutsu, M. et al., 0 to 18 GHz travelling-wave optical waveguide intensity modulator. *Tranp. IECE Jap.* E63(11): 817 (1980)

[5.12] Fukuda, M. and Noda, J., Optical properties of titanium-diffused linb03 strip waveguides and their coupling-to-a-fiber characteristicp. *Appl. Opt.* 19(4): 591 (1980)

[5.13] Auracher, F. and Keil, R., Design considerations and performance of machzehnder waveguide modulators. *Wave Electronics*, 4: 129 (1980)

[5.14] see [2.1] p. 792.

[5.15] von Helmholt, C.H. et al., Bragg-witch/SSB-modulator with integrated-optic single-mode waveguides. *Proc. 2nd Europ. Conf. an Integrated Optics, IEE Con. Publ.*, Florence, 227: 132 (1983)

[5.16] Suematsu, Y. et al., Fundamental transverse electric field (tep)-mode selection for thin film asymetric light guides. *Appl. Phys. Lett.* 21(6): 291 (1972)

[5.17] Eberhard, D. and Bülow, H., *Single-Mode Channel Waveguide Polarizer on LiNb03*. See [5.6] p. 202.

[5.18] Strobel, O., Simultaneous wavelength and power stabilization of an AAI-AS semiconductor laser. *Proc. ECOC '84, Stuttgart*. Berlin u. Offenbach: VDE Verlag GMBH, p. 146 (1984)

[5.19] Regener, R.A. and Sohler, W., efficient second-harmonic generation in ti:linb03 channel waveguide resonators. *J. Opt. Soc. Am. B.* 5(2): 267 (1988)

[6.1]   Winstel, G. and Weyrich, C., *Optoelektronik I.* Berlin/Heidelberg/New York: Springer (1981)

[6.2]   Bleicher, M., *Halbleiter-Optoelektronik.* Heidelberg: Hüthig (1986)

[6.3]   Conradt, R., *Auger-Rekombinationen in Halbleitern. Festkörperprobleme XII.* Braunschweig: Vieweg (1972)

[6.4]   see [3.5] p. 265.

[6.5]   Harth, W. and Grothe, H., *Sende- und Empfangsdioden für die Optische Nachrichtentechnik.* Stuttgart: Teubner, p. 26 (1984)

[6.6]   Takeshita, T., High-power operation in 0.98 p,m strained layer in ingaas-gaas single quantum-well ridge waveguide laser. *IEEE Photonics Techn. Lett.* 2(12): 849 (1980)

[6.7]   see [2.101] p. 669.

[6.8]   Davies, I.G.A. et al., Optical fiber technology semiconductor light sources. *Electr. Commun.* 56(44): 338 (1981)

[6.9]   Sugiyama, K. and Saito, H., GaAsSb-AlGaAsSb double heterojunction lasers. *Japan J. Appl. Phys.* 1 l(S): 1057 (1972)

[6.10] Burrus, C.A. and Miller, B.I., Small area double heterostructure AlGaAs electroluminescent diode sources for optical-fiber transmission lines. *Opt. Commun.* 4(4): 307 (1971)

[6.11] Kressel, H. and Ettenberg, M., A new edge emitting (AlGa)As Heterojunction LED for fiber-optic communications. *Proc. IEEE 63,* 9: 1360 (1975)

[6.12] Ettenberg, M. et al., Very high radiance edge-emitting LED. *IEEE J. Quant. Electr.* QE-12(6): 360 (1976)

[6.13] Lee, T.P. and Dentai, A.G., Power and modulation bandwidth of GaAs-AlGaAs high-radiance leads for optical communication systems. *IEEE J. Quant. Electr.* QE-14(3): 150 (1978)

[6.14] Panish, M.B., Heterostructure injection lasers. *Proc. IEEE 64,* 10: 1512 (1976)

[6.15] Agrawal, G.P. and Dutta, N.K., *Long-Wavelength Semiconductor Lasers.* New York: Van Nostrand Reinhold, p. 128 (1986)

[6.16] Asada, M. et al., The temperature dependence of the threshold current of GaInAsP/InP DH Lasers. *IEEE J. Quant. Electr.* QE-17(5): 611 (1981)

[6.17] Kobayashi, T. et al., Thermal diagnosis of dark lines in degraded GaAs-GaAlAs double heterostructure lasers. *Japan. J. Appl. Phys.* 14(H.4): 508 (1975)

[6.18] Chinone, N. et al., Long-term degradation of GaAs-Gal-,AlXAs DH Lasers due to facet erosion. *J. Appl. Phys.* 48(3): 1160 (1977)

[6.19] Tietze, U. and Schenk, C., *Halbleiter-Schaltungstechnik.* Berlin/Heidelberg/New York: Springer (1980)

[6.20] *Guide to Thermoelectric Heatpumps.* Marlow Industries, Inc.

[6.21] Petermann, K. and Arnold, G., Noise and distortion characteristics of semiconductor lasers in optical fiber communication systems. *IEEE J. Quant. Electr.* QE-18(4): 543 (1982)

[6.22] Großkopf, G. and Küller, L., Measurement of nonlinear distortions in index- and gain-guiding GaAlAs lasers. *J. Opt. Commun.* 1(1): 15 (1980)

[6.23] Boers, P.M. et al., Dynamic behaviour of semiconductor lasers. *Electr. Lett.* 11(10): 206 (1975)

[6.24] see [6.5] p. 98.

[6.25] see [3.5] p. 402.

[6.26] Unger, H.G., *Optische Nachrichtentechnik*, Teil II. Heidelberg: Hüthig Verlag, p. 339 (1985)

[6.27] Kressel, H. (Ed.), Semiconductor devices for optical communications. *Top. in Appl. Phys.* 39: 224 (1980)

[6.28] Botez, D., Single mode AlGaAs diode lasers. *J. Opt. Commun.* 1(2): 42 (1980)

[6.29] Wölk, C. et al.: Criteria for designing V-groove lasers. *IEEE J. Quant. Electr.* QE-17(5): 756 (1981)

[6.30] Chinone, N. et al., Highly efficient (GaAl)As buried-hetero structure lasers with buried optical guide. *Appl. Phys. Lett.* 35(7): 513 (1979)

[6.31] Peled, S., Near- and far-field characterization of diode lasers. *Appl. Opt.* 19(2): 324 (1980)

[6.32] Ikegami, T., Reflectivity of mode at facet and oscillation mode in double-heterostructure injection lasers. *IEEE J. Quant. Electr.* QE-8(6): 470 (1972)

[6.33] Streifer, W. et al., Longitudinal mode spectra of diode lasers. *Appl. Phys. Lett.* 40(4): 305 (1982)

[6.34] Strobel, O., *Wellenlängen-und Leistungsstabilisierung.* Dissertation TU Berlin, D83 (1986)

[6.35] Epworth, R.E., Modal noise – causes and cures. *Laser Focus*, Sept: 109 (1981)

[6.36] Köster, W. and Strobel, O., The influence of modal noise on a digital 1.3 MM optical transmission system. *J. Opt. Commun.* 5(H.1) (S): 27 (1984)

[6.37] Miles, R.O. et al., Low-frequency noise characteristics of channel substrate planar GaAlAs laser diodes. *Appl. Phys. Lett.* 38(11): 848 (1981)

[6.38] Domann, G., Fiber-gyro with integrated-optic phase modulator and spectrally broad light sources. *Proc. Opt. Fiber Sensors*, Stuttgart. Berlin u. Offenbach: VDE Verlag GMBH, p. 259 (1984)

[6.39] Strobel, O. and Schmuck, H., Optischer Überlagerungsempfang-eine Übersicht. *Frequenz* 41(8): 201 (1987)

[6.40] Auch, W. et al., Fiber-optic gyro with polarization preserving fiber. *Symp. Gyro* Techn. 1983, *Deutsche Gesellschaft f. Ortung u. Navigation (DGON) Conf. Proc.*, Stuttgart (1983)

[6.41] Okoshi, T. and Kikuchi, K., Heterodyne-type optical communication. *J. Opt. Commun.* 2(3): 82 (1981)

[6.42] Ito, T., Intensity fluctuation in each longitudinal mode of a multimode AlGaAs laser. *IEEE J. Quant. Electron.* QE-13(8): 574 (1977)

[6.43] Schawlow, A.L. and Townes, C.H., Infrared and optical masers. *Phys. Rev.* 112(6): 1940 (1958)

[6.44] Henry, C.H., Theory of the linewidth of semiconductor lasers. *IEEE J. Quant. Electr.* QE-18(2): 259 (1982)

[6.45] Tucker, R.S., High-speed modulation of semiconductor lasers. *J. Lightw. Tech.* LT-3(6): 1180 (1985)

[6.46] Strobel, O., Automatic side mode suppression of an InGaAsP BH Laser. *Frequenz*, 43(5): 133 (1989)

[6.47] Lee, T.P. et al., Measured spectral linewidth of variable gap-cleaved coupled-cavity lasers. *Electr. Lett.* 21(2): 53 (1985)

[6.48] Lee, T.P., Wavelength-tunable and single-frequency semiconductor lasers for photonic communications networks. *IEEE.* Oct: 42 (1989)

[6.49] Lee, T.P., Recent advances in long-wavelength semiconductor lasers for optical fiber communication. *IEEE Proc.* 79(3): 253 (1991)

[6.50] Best, H., private communication.

[6.51] Schimpe, R. et al., FM noise of index-guided GaAlAs diode lasers. *Electr. Lett.* 20(5): 206 (1984)

[6.52] see [2.2] p. 501.

[6.53] Arnold, G. and Petermann, K., Intrinsic noise of semiconductor lasers in optical communication systems. *Opt. and Quant. Electr.* 12(3): 207 (1980)

[6.54] Miles, R.O. et al., Low frequency noise characteristics of channel substrate planar gaalas laser diodes. *Appl. Phys. Lett.* 38(11): 848 (1981)

[6.55] Petermann, K., *Laser Diode Modulation and Noise*. Dodrecht/Boston/London: Kluver Academic Publishers (1991)

[6.56] Simon, J.C., Semiconductor laser amplifier for single mode optical fiber communications. *J. Opt. Commun.* 4(2): 51 (1983)

[6.57] Kobayashi, T and Kimura, T., Semiconductor optical amplifiers. *IEEE Spectrum*, May(S.26): (1984)

[6.58] Melchior, H. et al., Photodetectors for optical communication systems. *Proc. IEEE 58*, 10: 1466 (1970)

[6.59] Pearsall, T.P., Photodetectors for Optical Communication. *J. Opt. Commun.* 2(2): 42 (1981)

[6.60] see [3.5] p. 344.

[6.61] see [6.26] p. 456.

[6.62] see [6.5] p. 160.

[6.63] Petermann, K., *Vorlesung: Einführung in die Optische Nachrichtentechnik*, TU Berlin, SS (1984)

[6.64] Lee, T.P. et al., InGaAs/InP p-i-n photodiodes for lightwave communications at the 0.95–1.65 p,m wavelength. *IEEE J. Quant. Electr.* QE-17(2): 232 (1981)

[6.65] Trommer, R., Photodetectors for optical communication in the 1 to 1.6 pm wavelength range. Nuclear instruments and methods. *Physics Research A*, 253: 400 (1987)

[6.66] Trommer, R. et al., Planar small area ingaas/inp photodiodes for wavelengths of 1.3 pm and 1.55 j.m. *Siemens Research and Development Report*, 14(6): 280, (1985)

[6.67] Ando, H. et al., Characteristics of germanium avalanche photodiodes in wavelength region of 1–1.6 p.m. *IEEE J. Quant. Electr.* QE-14(11): 804 (1978)

[6.68] Trommer, R., InGaAs/InP avalanche photodiodes with very low dark current and high multiplication. *Proc. ECOC '83*, H. Melchior and A. Sollberger (Eds), Elsevier Science Publishers (North Holland), p. 159 (1983)

[6.69] Ebbinghaus, G. et al., Small area ion implanted p+n germanium avalanche photodiodes for a wavelength of 1.3 gm. *Siemens Research and Development Report* 14(6): 284 (1985)

[6.70] Bludau, W., *Halbleiter-Optoelektronik*. München/Wien: Hanser, p. 180 (1995)

[6.71] JohannesH.-H., *Vorlesung: Polytronik*, TU Braunschweig, SS2014.

[6.72] Dinu, D. et al., Temperature and bias voltage dependence of the mppc detectors. *IEEE Nuclear Science Symposium*, Knoxville, TN (2010)

[6.73] Cehovski, M. et al., Silicon photomultiplier for laser detection of objects in the near vehicle environment. *IEEE Conference on Transparent Optical Networks (ICTON)*, Cartagena (2013)

[6.74] Seifert, S. et al., Simulation of silicon photomultiplier signals. *IEEE Transactions Nuclear Science*, 56(6): December (2009)

[6.75] Ward, M.A. and Vacheret A., *Impact of After-pulse, Pixel Crosstalk and Recovery Time in Multi Pixel Photon Counter Response*, University of Sheffield, July (2008)

[6.76] Oide, H. et al., Study of afterpulsing of MPPC with waveform analysis. *International Workshop on New Photon-Detectors PD07*, Kobe University Japan, June (2007)

[6.77] Ubukata, T. Isoshima, T. and Hara, M., Wavelength-programmable organic distributed-feedback laser based on a photoassisted polymer-migration system, *Adv. Mat.* 17: 1630–1633 (2005)

[6.78] Karnutsch, C. and Gýrtner, C. et al., Low threshold blue conjugated polymer lasers with first- and second-order distributed feedback. *Appl. Phys. Lett.* 89: 201108 (2006)

[6.79] Riedl, T. and Rabe, T. et al., Tunable organic thin-film laser pumped by an inorganic violet diode laser. *Appl. Phys. Lett.* 88: 241116 (2006)

[6.80] Ge, C., Lu, M., Jian, X., Tan, Y. and Cunningham, B.T., Large-area organic distributed feedback laser fabricated by nanoreplica molding and horizontal dipping *Opt. Expr.* 18: 12980–12991 (2010)

[6.81] Sorokin, P.P. and Lankard, J.R., Stimulated emission observed from an organic dye, chloro-aluminum phthalocyanine. *IBM J. Res. Dev.* 10: 162 (1966)

[6.82] Soffer, B.H. and McFarland, B.B., Continuously tunable, narrow-band organic dye lasers. *Appl. Phys. Lett.* 10: 266 (1967)

[6.83] Lawandy, N.M., Balachandran, R.M., Gomes, A.S.L. and Sauvain, E. Laser action in strongly scattering media. *Nature*, 368: 436–438 (1994)

[6.84] Berggren, M., Dodabalapur, A., Slusher, R.E. and Bao, S., Organic lasers based on Förster transfer. *Synthetic Metals*, 91: 65–68 (1997)

[6.85] Kogelnik, H. and Shank, V., Stimulated emission in a periodic structure. *Appl. Phys. Lett.* 18: 152 (1971)

[6.86] Weinberger, M.R and Langer, G. et al., Continuously color-tunable rubber laser. *Adv. Mat.* 16(2); 130–133 (2004)

[6.87] Riedl, T. and, Rabe, T. et al., Tunable organic thin-film laser purıped by an inorganic violet diode laser. *Appl. Phys. Lett.* 88: 241116 (2006)

[6.88] Yang, Y., Turnbull, G.A. and Samuel, I.D.W., Hybrid optoelectronics: A polymer laser pumped by a nitride light-emitting diode. *Appl. Phys. Lett.* 92: 163306 (2008)

[7.1] Wedding, B. and Heidemann, R., Dispersion penalties in a 5 Gbit/s optical transmission system using directly modulated DFB laser. *Proc. EFOC/LAN '88*, p. 65, Amsterdam (1988)

[7.2] Wedding, B. et al., 2.24 Gbit/s 151-km optical transmission system using high-speed integration silicon circuits. *IEEE J. Lightw. Tech.* 8(2): 227 (1990)

[7.3] Wedding, B., Reduction of bit error rate in high speed optical systems due to optimized electrical pulse shaping. *Proc. ECOC '88*, 1: 187 (1988)

[7.4] Personick, S.D., Receiver design for digital fiber optic communication systems I + II. *Bell Syst. Techn. J.* 52(6): 843 (1973)

[7.5] Personick, S.D., Receiver design for optical fiber systems. *Proc. IEEE*, 65(12): 1670 (1977)

[7.6] Muoi, T.V., Receiver design for high-speed optical fiber-systems. *IEEE J. Lightw. Tech.* LT-2(3): 243 (1984)

[7.7] see [1.13] p. 171.

[7.8] Esman, R.D. and Williams, K.J., Measurement of harmonic distortion in microwave photodetectors. *IEEE Photon. Techn. Lett.* 2(7): 502 (1990)

[7.9] see [1.13] p. 173.

[7.10] Fußgänger, K., SEL-Alcatel (now Bell-Labs Germany) Stuttgart: *A Review of Key Technologies for Fiber-Optic Broadband Distribution Systems*, private communication.

[7.11] Goell, J.E., An optical repeater with high-impedance input amplifier. *Bell Syst. Tech. J.* 53(4): 629 (1974)

[7.12] Shikada, M. et al., Long distance gigabit range optical fiber transmission experiments employing DFB-LDs and InGaAs-APDs. *IEEE J. Lightw. Tech.* LT-5(10): 1488 (1987)

[7.13] Hullet, J.L. and Moustakas, S., Optimum transimpedance broadband optical preamplifier design. *Opt. Quant. Electr.* 13(1): 65 (1981)

[7.14] Grau, G., *Optische Nachrichtentechnik*. Berlin/Heidelberg/New York: Springer (1981)

[7.15] see [3.5] p. 471.

[7.16] Krumholz, O., Signal/Rausch-Verhältnis bei Avalanche-Photodioden. *Wiss. Ber. AEG-Telefunken*, 44(2): 80 (1971)

[8.1]   Payne, D.N. et al., Fiber optical amplifiers, *Proc. OFC '90*. Tutorial paper ThFl, p. 335, San Francisco (1990)

[8.2]   Kimura, Y. et al., 46.5 dB Gain in Er3+-doped fiber amplifier pumped by 1.48 jm GaInAs laser diodes. *Electr. Lett.* 25(24): 1656 (1989)

[8.3]   Gnauck, A.H. et al., 1 Tbit/s km transmission experiment at 16 Gbit/s using conventional fiber. *Electr. Lett.* 25: 1695 (1989)

[8.4]   Suzaki, T. et al., 10 Gbit/s optical transmitter module with MQW DFB LD and DMT driver IC. *Electr. Lett.* 26(2): 151 (1990)

[8.5]   Fujita, S. et al., 10 Gbit/s 100 km optical fiber transmission experiment using high-speed MQW DFB-LD and back-illuminated GaInAs APD. *Electr. Lett.* 25(11): 702 (1989)

[8.6]   Vodhanel, R.S. et al., 10 Gbit/s 48.5 km DPSK transmission experiment with direct modulation of a 1530 nm DFB laser and 1310 nm optimized fiber. *Proc. ECOC '90*, Amsterdam, p. 41 (1990)

[8.7]   Heidemann, R. et al., 5 Gbit/s transmission system experiment over 111 km of optical fiber. *Electr. Lett.* 23(19): 1030 (1987)

[8.8]   Wedding, B. and Pfeiffer, Th., 20 Gbit/s optical pattern generation, amplification and 115 km fiber propagation using optical time division multiplexing. *Proc. ECOC'90*, Amsterdam, p. 453 (1990)

[8.9]   Izadpanah, H. et al., Optical amplification of high speed signals up to 100 gbit/s with erbium-doped fiber amplifiers. *Proc. ECOC '90*, Amsterdam, p. 1033 (1990)

[8.10] Edagawa, N. et al., 10 Gbit/s, 1500 km transmission experiment using 22 ER-doped fiber amplifier repeaters. *Proc. ECOC/IOOC '91*, Paris, 3: 76 (1991)

[8.11] Kahn, J.M. et al., 4 Gbit/s PSK homodyne transmission system using phase-locked semiconductor lasers. *Proc. OFC '90*, San Francisco CA, PD 10 (1990)

[8.12] Saito, S. et al., An over 2200 km coherent transmission experiment at 2.5 gb/s using erbium-doped-fiber in-line amplifiers. *J. Lightw. Tech.* 9(H.2,S): 161 (1991)

[8.13] Saito, S. et al., 2.5 Gbit/s 400 km coherent transmission experiment using two in-line erbium-doped-fiber amplifiers. *Proc. OFC '90*, San Francisco CA, WO. (1990)

[8.14] Park, Y.K. et al., 1.7 Gb/s-419 km transmission experiment using shelf-mounted FSK coherent system and packaged fiber amplifier modules. *IEEE Photon. Techn. Lett.* 2(12): 917 (1990)

[8.15] Takachio, N. et al., Chromatic dispersion equalization in an 8 Gbit/s 202 km optical CPFSK transmission experiment. *Proc. IOOC '89*, Kobe, Japan, 20PDA-13 (1989)

[8.16] Wedding, B. et al., 10 Gbit/s to 260 000 subscribers using optical amplifier distribution network. *Proc. ICC Supercom'92*, Session 340.6, Chicago IL, June (1992)

[9.1]   see [3.5] p. 497.

[9.2]   see [3.5] p. 501.

[9.3]   Krimmel, H., SEL-Alcatel (now Bell-Labs Germany) Stuttgart, private communication.

[9.4]   Bünning, H., Güte einer analogen optischen Breitbandübertragungsstrecke. *Frequenz*, 37(9): 241 (1983)

[9.5]   Childs, R.B. et al., Predistortion linearization of directly modulated DFB lasers and external modulators for AM video transmission. *Proc. OFC '90*, paper WH6, San Francisco CA, p. 79 (1990)

[9.6]   Darcie, T.E. and Bodeep, G.E., Lightwave subcarrier CATV transmission systems. *IEEE Transact. and Microw. Theory and Tech.* 38(5): 524 (1990)

[9.7]   Hölzler, E. and Holzwarth, H., *Pulstechnik*. Berlin: Springer (1982)

[9.8]   Kersten, R.Th., *Einführung in die Optische Nachrichtentechnik*. Berlin/Heidelberg/New York: Springer-Verlag, p. 354, 387 (1983)

[9.9]   Ogawa, K., Considerations for single-mode fiber systems. *Bell Syst. Techn. J.* 61(8): 1919 (1982)

[9.10] see [9.8] p. 351.

[9.11] see [7.26] p. 588.

[9.12] Veith, G., Physical limitations and advanced technologies of high-speed fiber optic transmission systems. *Laser u. Optoelektr.* 22(6): 63 (1990)

[9.13] SEL-Datenblatt 152-81-0387 für Leitungsausrüstung LA140GF2.

[9.14] Kaiser, N., SEL-Alcatel (now Bell-Labs Germany) Stuttgart, private communication.

[9.15] Drullmann, R. and Kammerer, W., Leitungscodierung und betriebliche Überwachung bei regenerativen Lichtleitkabelübertragungssystemen. *Frequenz*, 34(2): 45 (1980)

[9.16] Köster, W. and Mohr, F., Optische Zweiwegübertragung. *Elektr. Nachrichtenwesen.* 55(4): 342 (1980)

[9.17] Köster, W., Einfluss des Rückstreulichts auf die Nebensprechdämpfung in bidirektionalen Übertragungssystemen. *Frequenz*, 37(4): 87 (1983)

[9.18] Fußgänger, K. and Roßberg, R., Uni and bidirectional 4 X × 560 Mbit/s transmission systems using WDM devices an wavelength-selective fused single-mode fiber couplers. *IEEE J. Select. Areas in Commun.* 8(6): 1032 (1990)

[9.19] Sano, K. et al., A 4-wavelength optical multi/demultiplexer for WDM subscriber loop systems using analog baseband video transmission. *J. Lightw. Tech.* LT-4: 631 (1986)

[9.20] Stern, M. et al., Three-channel high-speed transmission over 8 km installed 1300 nm optimized single-mode fiber using 800 nm CD laser and 1300/1500 nm LED transmitters. *Electr. Lett.* 24(3): 176 (1988)

[9.21] Cisco, *Visual Networking Index: Forecast and Methodology* 2013–2018, White Paper, June (2014)

[9.22] *Nu-Wave Optima^TM, A Flexible Optical Transport Platform that Offers Industry's Most Advanced 100G Technology to Maximize the Combination of Capacity and Reach for Optical Networking*, White Paper, October (2012)

[9.23] Essiambre, R.-J. and Tkach, R.W., Capacity trends and limits of optical communication networks, *Proc. of the IEEE*, 100: 1035–1055 (2012)

[9.24] Nölle, M., Schubert, C. and Freund, R. Techniques to realize flexible optical terabit per second transmission systems. *Proc. SPIE 8646, Optical Metro Networks and Short-Haul Systems V*, 864602, December 26 (2012)

[9.25] R.W. Chang, R.W., *Orthogonal Frequency Multiplex Data Transmission System*, US Patent No. 3488445 (1966)

[9.26] Jooley, J.W. and Tukey, J.W., An algorithm for the machine calculation of complex Fourier series. *Mathematics of Computation*, 19: 297–301 (1965)

[9.27] Lowery, A.J., Du, L. and Armstrong, J., Orthogonal frequency division multiplexing for adaptive dispersion compensation in long haul WDM systems. *Proc. Optical Fiber Communication Conference (OFC)*, paper PDP39 (2006)

[9.28] Shieh, W. and Athaudage, C., Coherent optical orthogonal frequency division multiplexing, *Electronics Letters*, 42: 587–589 (2006)

[9.29] Schuster, M., Spinnler, B., Bunge, C.A. and Petermann, K., Spectrally efficient OFDM transmission with compatible single-sideband modulation for direct detection. *Proc. European Conference on Optical Communication (ECOC)*, paper PO75 (2007)

[9.30] Peng, W.-R., Wu, X., Arbab, V.R., Shamee, B., Yang, J.-Y. et al., Experimental demonstration of 340 km SSMF transmission using a virtual single sideband OFDM signal that employs carrier suppressed and iterative detection techniques. *Proc. Poptical Fiber Communication Conference (OFC)*, paper OMU1 (2008)

[9.31] Ma, Y., Shieh, W. and Yi, X., Characterisation of nonlinearity performance for coherent optical OFDM signals under influence of PMD. *Electronics Letters*, 43(17): 934–35 (2007)

[9.32] Shieh, W., Yi, X. and Tang, Y., Experimental demonstration of transmission of coherent optical OFDM systems. *Proc. Optical Fiber Communication Conference (OFC)*, paper OMP2 (2007)

[9.33] Jansen, S., Morita, I., Takeda, N. and Tanaka, H., 20-Gb/s OFDM transmission over 4160-km SSMF enabled by RF-pilot tone phase noise compensation. *Proc. Optical Fiber Communication Conference (OFC)*, paper PDP15 (2007)

[9.34] Armstrong, J., OFDM for optical communications. *J.Lightw. Tech.* 27(3): 189–204 (2009)

[9.35] Garcia Gunning, F.C., Ibrahim, S.K., Frascella, P., Gunning, P. and Ellis, A.D., High symbol rate OFDM transmission technologies, *Proc. Optical Fiber Communication Conference (OFC)*, paper OTHD1 (2010)

[9.36] Liu, X., Chandrasekhar, S., Zhu, B. and Peckham, D.W., Efficient digital coherent detection of a 1.2-Tb/s 24-carrier no-guard-interval CO-OFDM signal by simultaneously detecting multiple carriers per sampling. *Proc. Optical Fiber Communication Conference (OFC)*, paper OWO2 (2010)

[9.37] Hillerkuss, D. et al., Simple all-optical FFT scheme enabling Tbit/s real-time signal processing. *Optics Express*, 18(9): 9324–9340 (2010)

[9.38] Takiguchi, K. et al., Integrated-optic eight-channel OFDM demultiplexer and its demonstration with 160 Gbit/s signal reception. *Electronics Letters*, 46(8): 575–576 (2010)

[9.39] Chen, H. et al., All-optical sampling orthogonal frequency-division multiplexing scheme for high-speed transmission system. *J. Lightw. Tech.* 27(21): 4848–4854 (2009)

[9.40] Schwarz, A., Bruns, S., Voigt, J., Winzer, K., Zimmermann, G. et al., Silicon photonic implementation of a scalable O-OFDM demultiplexer. *IEEE Photonics Technology Letters*, 25(20): 1977–1980 (2013)

[9.41] Da Ros, F., Nölle, M., Meuer, C., Rahim, A., Voigt, K., et al., Experimental demonstration of an OFDM receiver based on a silicon-nanophotonic discrete Fourier transform filter. *IEEE Photonics Conference (IPC)*, (2014)

[9.42] Elschner, R., T. Richter, T., Kato, T., Watanabe, S. and Schubert, C., Distributed coherent optical OFDM multiplexing using fiber frequency conversion and free-running lasers *Proc. Optical Fiber Communication Conference (OFC)*, paper PDP5C.8 (2012)

[9.43] Richter, T., Schmidt-Langhorst, C., Elschner, R., Kato, T., Tanimura, T. et al., Coherent in-line substitution of OFDM subcarriers using fiber-frequency conversion and free-running lasers. *Proc. Optical Fiber Communication Conference (OFC)*, paper Th5B.6 (2014)

[9.44] Dischler, R. and Buchali, F., Transmission of 1.2 Tb/s continuous waveband PDM-FDM-FDM signal with spectral efficiency of 3.3 bit/s/Hz over 400 km of SSMF. *Proc. Optical Fiber Communication Conference (OFC)*, paper PDPC2 (2009)

[9.45] Ma, Y. et al., 1-Tb/s per channel coherent optical OFDM transmission with subwavelength bandwidth access. *Proc. Optical Fiber Communication Conference (OFC)*, paper PDPC1 (2009)

[9.46] Hillerkuss, D., Schmogrow, R., Schellinger, T., Jordan, M., Winter, M. et al., 26 Tbit/s line-rate super-channel transmission utilizing all-optical fast Fourier transform processing. *Nature Photonics*, 5: 364–371 (2011)

[9.47] Nölle, M., Molle, L., Gross, D. and Freund, R., Transmission of 5×62 Gbit/s DWDM Coherent OFDM with a Spectral Efficiency of 7.2 Bit/s/Hz using Joint 64-QAM and 16-QAM Modulation. *Proc. Optical Fiber Communication Conference (OFC)*, paper OMR4 (2010)

[9.48] Cai, J.-X., Cai, Y., Davidson, C.R., Foursa, D.G., Lucero, A. et al., Transmission of 96×100G prefiltered PDM-RZ-QPSK channels with 300% spectral efficiency over 10,608 km and 400% spectral efficiency over 4,368km. *Proc. Optical Fiber Communication Conference (OFC)*, paper PDPB10 (2010)

[9.49] Gavioli, G., Torrengo, E., Bosco, G., Carena, A., Curri, V. et al., Investigation of the impact of ultra-narrow carrier spacing on the transmission of a 10-Carrier 1Tb/s superchannel. *Proc. Optical Fiber Communication Conference (OFC)*, paper OThD3 (2010)

[9.50] Cigliutti, R., Nespola, A., Zeolla, D., Bosco, G., Carena, A. et al., Ultra-long-haul transmission of 16×112 Gb/s spectrally-engineered DAC-generated Nyquist-WDM PM-16QAM channels with 1.05×(symbol-rate) frequency spacing. *Proc. Optical Fiber Communication Conference (OFC)*, paper OTh3A.3 (2012)

[9.51] Schmogrow, R., Meyer, M., Wolf, S., Nebendahl, B., Hillerkuss, D. et al., 150 Gbit/s real-time Nyquist pulse transmission over 150 km SSMF enhanced by DSP with dynamic precision. *Proc. Optical Fiber Communication Conference (OFC)*, paper OM2A.6 (2012)

[9.52] Yan, M., Tao, Z., Yan, W., Li, L., Hoshida, T. and Rasmussen, J.C., Experimental comparison of no-guard-interval-OFDM and Nyquist-WDM superchannels. *Proc. Optical Fiber Communication Conference (OFC)*, paper OTh1B.2 (2012)

[9.53] Jansen, S., Morita, I., Forozesh, K., Randel, S., van den Borne, D. and Tanaka, H., Optical OFDM, a hype or is it for real? *Proc. European Conference on Optical Communication (ECOC)*, paper Mo.3.E.3 (2008)

[9.54] Freund, R., Nölle, M., Schmidt-Langhorst, C., Ludwig, R., Schubert, C. et al., Single- and multi-carrier techniques to build up Tb/s per channel transmission systems. *International Conference on Transparent Optical Networks (ICTON)*, paper Tu.D1.4 (2010)

[9.55] Barbieri, A., Colavolpe, G., Foggi, T., Forestieri, E., and Prati, G., OFDM versus single-carrier transmission for 100 Gbps optical communication. *J. Lightw. Tech.* 28(17): 2537–2551 (2010)

[9.56] Witting, M. et al., Status of the European data relay satellite system. *Proc. of ICSOS*, October 9–12 (2012)

[9.57] Weinert, C. and Freund, R., Heavenly Communication by Light. *European Research and Innovation Review*, 16: April (2011)

[9.58] Perlot, N., Dreischer, T., Weinert, C.M. and Perdigues, J., Optical GEO feeder link design. *Future Network & Mobile Summit*, Berlin (2012)

[9.59] Kästner, M. and Kriebel, K.-T., Alpine cloud climatology using long-term NOAA-AVHRR satellite data. *Deutsches Zentrum für Luft- und Raumfahrt (DLR), Institut für Physik der Atmosphäre*, Report No. 140 (2000)

[9.60] Rose, T.S., et al., Gamma and proton radiation effects in erbium-doped fiber amplifiers: active and passive measurements. *J, Lightw. Tech.* 19(12): 1918–1923 (2001)

[9.61] Andrews, L.C. and Phillips, R.L., *Laser Beam Propagation through Random Media. Second Edition*, Spie Press (2005)

[9.62] Mata Calvo, R., Becker, P., Giggenbach, D., Moll, F., Schwarzer, M. et al., Transmitter diversity verification on ARTEMIS geostationary satellite. *Proc. SPIE 8971, Free-Space Laser Communication and Atmospheric Propagation XXVI*, 897104, March 6 (2014)

[9.63] Gonté, F., Courteville, A. and Dändliker, R., Optimization of single-mode fiber coupling efficiency with an adaptive membrane mirror. *Opt. Eng.* 41(5): 1073–1076, May 1 (2002)

[9.64] Churnside, J.H., Aperture averaging of optical scintillations in the turbulent atmosphere. *Applied Optics*, 30: 1982–1994, 1991.

[9.65] DIN; VDE; DKE (Hrsg.), Optische Strahlungssicherheit und Laser 2. *DIN-VDE-Taschenbücher*, Band 526(2) Auflage (2010)

[9.66] Lange, R. and Smutny, B., Optical inter-satellite links based on homodyne BPSK modulation: Heritage, status, and outlook, *Free-space Laser Communication Technologies XVII. Proc. of SPIE*, 5712 (2005)

[9.67] Hemmati, H. (Ed.), *Deep-space Optical Communications*. Wiley-Interscience, 2006.

[9.68] Fletcher, A.S., Royster, T.C., List, N. and Shoup, R., Optical M-PPM signaling with binary forward error correction. *Military Communication Conference*, MILCOM 2010: 779–784 (2010)

[9.69] Alonso, A., Reyes, M. and Sodnik, Z., Performance of satellite-to-ground communications link between ARTEMIS and the optical ground station. *Proc. SPIE 5572, Optics in Atmospheric Propagation and Adaptive Systems VII*, 372: November 11 (2004)

[9.70] Boroson, D.M., Robinson, B.S., Murphy, D.V., Burianek, D.A., Khatri, F. et al., Overview and results of the lunar laser communication demonstration. *Proc. SPIE 8971, Free-Space Laser Communication and Atmospheric Propagation XXVI*, 89710S, March 6 (2014)

[9.71] Baack, C. et al., Zukünftige Lichtfrequenztechnik in Glasfasernetzen. *Nachrichtentech. Z.* 35(11): 686 (1982)

[9.72] Yamamoto, Y. and Kimura, T., Coherent optical fiber transmission systems. *IEEE J. Quant. Electr.* QE-17(6): 919 (1981)

[9.73] Okoshi, T. and Kikuchi, K., Heterodyne-type optical communication. *J. Opt. Commun.* 2(3): 82 (1981)

[9.74] Okoshi, T. et al., Computation of bit-error rate of various heterodyne and coherent-type optical communication schemes. *J. Opt. Commun.* 2(3): 89 (1981)

[9.75] Stein, S. and Jones, I., *Modern Communication Principles with Application to Digital Signalling*. New York: McGraw-Hill (1967)

[9.76] Fritzsche, C., private communication.

[9.77] Linke, R., Coherent lightwave transmission experiment using amplitude and phase modulation at 400 Mbit/s and 1 Gbit/s data rates. *Proc. OFC/OFS'85, p. 87*, San Diego, CA (1985)

[9.78] Hodgkinson, T., Preliminary results for 1.5 gm coherent optical fiber transmission experiments. *IEE Conf. Coherence in Optical Systems '82*, London (1982)

[9.79] Schmuck, H. et al., Widely tunable narrow linewidth erbium doped fiber ring laser. *Electr. Lett.* 27(23): 2117 (1991)

[9.80] Okoshi, T., Heterodyne and coherent optical fiber communications: recent progress. *IEEE Trans. Microw. Theory and Techn.* MTT-30, 8: 1138 (1982)

[9.81] Saito, S. et al., S/N and error rate evaluation for an optical fsk heterodyne detection system using semiconductor lasers. *IEEE J. Quant. Electr.* QE-19, 2: 180 (1983)

[9.82] Wölfelscheider, H. and Kist, R., Intensity-independent frequency stabilization of semiconductor laser using a fiber optic fabry-perot-resonator. *J. Opt. Commun.* 5(2): 53 (1984)

[9.83] Matthews, M.R. et al., Packaged frequency-stable tunable 20 kHz linewidth 1.5 gm InGaAsP external cavity laser. *Electr. Lett.* 21(3): 113 (1985)

[9.84] Kasper, B. L. and Burrus; C.A., Balanced dual detector receiver for optical heterodyne communication at Gbit/s rates. *Electr. Lett.* 22(8): 413 (1986)

[9.85] Goldberg, L. et al., Spectral characteristics of semiconductor lasers with optical feedback. *IEEE J. Quant. Electr.* 18(4): 555 (1982)

[9.86] Jopson, R.M. et al., Bulk optical isolator tunable from 1.2 gm to 1.7 pm. *Electr. Lett.* 21(18): 783 (1985)

[9.87] Okoshi, T. et al., Polarisation-diversity receiver for heterodyne/coherent optical fiber communications. *Proc. IOOC'83*, Tokyo (1983)

[9.88]   Mohr, F.A. and Scholz, U., Polarisation control for an optical fiber gyroscope. *Proc. FORS ' 82*, Cambridge MA, p. 163 (1982)

[9.89]   Kimura, T. and Yamamoto, Y., Progress of coherent optical fiber communication systems. *Opt. Quant. Electr.* 15: 1 (1983)

[9.90]   Alferness, R., Guided-wave devices for optical communication. *IEEE J. Quant. Electr.*, QE-17(6): 946 (1981)

[9.91]   Shikada, M. et al., 100 Mbit/s ASK heterodyne detection experiment using 1.3 gm DFB laser diodes. *Electr. Lett.* 20(H.4,S): 164 (1984)

[9.92]   Sueta, T. and Izutsu, M., High speed guided-wave optical modulators. *J. Opt. Commun.* 3(2): 52 (1982)

[9.93]   Kikuchi, K. et al., Bit-error rate of PSK heterodyne optical communication system and its degradation due to spectral spread of transmitter and local oscillator. *Electr. Lett.* 19(H.11,S): 417 (1983)

[9.94]   Wyatt, R. et al., 1.52 ltm PSK heterodyne experiment featuring an external cavity diode laser local oscillator. *Electr. Lett.* 19(14): 550 (1983)

[9.95]   Hodgkinson, T. et al., Demodulation of optical DPSK using in-phase and quadrature detection. *Electr. Lett.* 21(19): 867 (1985)

[9.96]   Veith, G., Physical limitations and advanced technologies of high speed fiber optic transmission systems. *Laser u. Optoelektr.* 22(6): 63 (1990)

[9.97]   Shibutani, M. et al., Ten channel coherent optical FDM broadcasting system. *OFC Housten, TX*, paper THC2 (1989)

[9.98]   Wagner, R.E. et al., 16-channel coherent broadcast network at 155 Mbit/s. *OFC Housten, TX*, paper PD12 (1989)

[9.99]   Alcatel-Lucent, Bell Labs breaks optical transmission record, 100 Petabit per second kilometer barrier (2009) http://phys.org/news173455192.html/

[9.100]   Sano, A. et al., 69.1-Tb/s (432 × 171-Gb/s) C- and Extended L-Band Transmission over 240 km Using PDM-16-QAM Modulation and Digital Coherent Detection, OSA/OFC/NFOEC (2010)

[9.101]   Hillerkuss, D. et al., 26 Tbit s^{-1} line-rate super-channel transmission utilizing all-optical fast Fourier transform processing. *Nature Photonics*, 5: (2011)

[9.102]   Sakaguchi, J. et al., Space divisionmultiplexed transmission of 109-tb/s data signals using homogeneous seven-core fiber. *J. Lightw. Tech.* 30(4): (2012)

[9.103]   Nippon Telegraph and Telephone Corporation, Fujikura Ltd., Hokkaido University, Technical University of Denmark, World Record One Petabit per Second Fiber Transmission over 50-km, NTT Press Pelease (2012) http://www.ntt.co.jp/news2012/1209e/120920a.html#a1

[9.104]   Poletti, F. et al., Towards high-capacity fibre-optic communications at the speed of light in vacuum, *Nature Photonics*, DOI: 10.1038/NPHOTON.2013.45 (2013)

[10.1]   Kenward, M., Plastic fiber homes in/on low-cost networks. *Fiber Systems*, 5(H.1,S): 35 (2001)

[10.2]   Walter, G., Wu, C., Then, H., Feng, M. and Holonyak Jr., N., Tilted-charge high speed (7 GHz) light emitting diode. *Appl. Phys. Lett.*, 94: 231125 (2009); http://dx.doi.org/10.1063/1.3154565

[10.3]   Hofstetter, R., Schmuck, H. and Heidemann, R., Dispersion effcts in optical mm-wave systems using self-heterodyne method for transport and generation. *IEEE Transactions on Microwave Theory and Techniques*, 2(43,H.9): (1995)

[10.4]   Schmuck, H., Comparison of optical mm-wave system concepts with regard to the chromatic dispersion. *Electr. Lett.* 31(H.21,S): 1848 (1995)

[10.5]   Rigby, P., Its optics, its wireless and its built for speed. *Fiber Systems*, 4(7,S9) 53 (2007)

[10.6]   Pohl, W., Bluetooth: Technik und Einsatzgebiete. *KES* 1(S): 43 (2001)

[10.7]   siehe [9.76] S. 212 ff.

[10.8]   Sykes, E., Modelling sheds light an next-generation networks. *Fiber Systems*, 5(3,S): 58 (2001)

[10.9]   Weiershausen, W. et al., Realization of next generation dynamic WDM networks by advanced OADM Design. *Proc. European Conference and Networks and Optical Communications (NOC 2000)*, S.199 (2000)

[10.10]   *Definition of Terms Version 3.0*. FTTH Council (2011)

[10.11]   Brown, L., *G.fast for FTTdp*. Joint ITU/IEEE Workshop on Ethernet – Emerging Applications and Technologies, Genewa, Switzerland, September 22 (2012)

[10.12]   ITU-T Rec. G.652, *Characteristics of a Single-Mode Optical Fibre and Cable* (2009) Available from http://www.itu.int/ITU-T/recommendations/index.aspx?ser=G.

[10.13] ITU-T Rec. G.657, *Characteristics of a Bending Loss Insensitive Single Mode Optical Fibre and Cable for the Access Network* (2009)

[10.14] Baabali, N., *Futureproofing Broadband Networks in Europe.* PLNOG12, Warsaw, Poland, March 4 (2014)

[10.15] Lam, C.F., *Passive Optical Networks: Principles and Practice.* Elsevier, Inc. (2007)

[10.16] *FTTH Handbook, Edition 6.* FTTH Council Europe (2014) Available from http://www.ftthcouncil .eu/EN/home/forms/form-handbook.

[10.17] ITU-T Rec. ITU-T G.983.1, *Broadband Optical Access Systems Based on Passive Optical Networks (PON)* (2005)

[10.18] ITU-T Rec. ITU-T G.984.1, *Gigabit-Capable Passive Optical Networks (GPON): General Characteristics* (2008)

[10.19] IEEE 802.3-2012, *IEEE Standard for Ethernet* (2012) Available from http://standards.ieee.org/ findstds/standard/802.3-2012.html

[10.20] *ONU-2024-21 GEPON Optical Network Unit with 24-port Fast Ethernet Switch.* Data sheet, ZyXEL Communications Corp (2009)

[10.21] *V2824 L2 Fast Ethernet Switch with G-PON.* Data sheet, Dasan Networks, Inc. (2011)

[10.22] ITU-T Rec. G.986, *1 Gbit/s Point-to-Point Ethernet-Based Optical Access System* (2010)

[10.23] *Fibrain Passive Optical Network Products 2012/13.* Catalog version 1.4, Elmat. (2012) Available from http://en.fibrain.com/catalogos/fibrainpon.pdf

[10.24] *Corning SMF-28e+ Optical Fiber.* Data sheet PI1463, Corning Inc. (2013)

[10.25] *BendBright Single-Mode Optical Fiber.* Data sheet, Draka Communications (2010)

[10.26] ITU-T Rec. G.983.3. *A Broadband Optical Access System with Increased Service Capability by Wavelength Allocation* (2001)

[10.27] *FTTH Solutions: High Performance Fiber-to-the-Home Networks.* Catalog, BKtel Communications GmbH (2012)

[10.28] ITU-T Rec. G.987.1, *10-Gigabit-Capable Passive Optical Networks (XG-PON): General Requirements* (2010)

[10.29] ITU-T Rec. G.987.2, *10-Gigabit-Capable Passive Optical Networks (XG-PON): Physical Media Dependent (PMD) Layer Specification* (2010)

[10.30] ITU-T Rec. G.989.1, *40-Gigabit-Capable Passive Optical Networks (NG-PON2): General Requirements* (2013)

[10.31] ITU-T Rec. G.987.4, *10 Gigabit-Capable Passive Optical Networks (XG-PON): Reach Extension* (2012)

[10.32] ITU-T Rec. G.984.3, *Gigabit-Capable Passive Optical Networks (G-PON): Transmission Convergence Layer Specification* (2014)

[10.33] ITU-T Rec. G.987.3, *10-Gigabit-Capable Passive Optical Networks (XG-PON): Transmission Convergence (TC) Specifications – Amendment* 1 (2012).

[10.34] *Catalog Ed. 6.* FOCI Fiber Optic Communications, Inc. (2012) Available from http://www.foci.com.tw/ pdf/FOCI_Full_catalogue_V6.0%2010222012.pdf

[10.35] Potelle, P.-Y., *French Regulatory Approach for NGA.* SEE Regulatory Forum, Sarajevo, November 5 (2010)

[10.36] ITU-T Rec. G.694.2, *Spectral Grids for WDM Applications: CWDM Wavelength Grid* (2003)

[10.37] *GATESURF GS Series ONU.* Data sheet, Ofisgate (2010)

[10.38] Lee, C-Ch., et al., WDM-PON Experiences in Korea. *Journal of Optical Networking,* 6(5): 451–464 (2007)

[10.39] Kim, B. and Kim, B.-W., WDM-PON development and deployment as a present optical access solution. *Proc. OFC'2009,* San Diego, CA, pp. 22–26 March 22, paper OThP5 (2009)

[10.40] ITU-T Rec. G.984.5: *Gigabit-Capable Passive Optical Networks (G-PON): Enhancement Band – Appendix I and II* (2007)

[10.41] Kuyt, G., Bertaina, A. and Davies, M., Trends in FTTH/FITH optical fibre and cable and associated standardization. *FTTH Conference,* London, February 19 (2013)

[10.42] ITU-T Rec. L.66, *Optical Fibre Cable Maintenance Criteria For In-Service Fibre Testing In Access Networks* (2007)

[10.43] Johnston, D., 12 Ways to Stop the Next Sandy. *Newsweek,* November 18 (2012)

[10.44] Hardy, S., Will FTTN advances delay FTTH? *Lightwave,* 30(4): (2013)

[10.45] Kawataka, J. et al., Novel optical fiber cable for feeder and distribution sections in access networks, *J. Lightw. Tech.* 21(3): (2003)

[10.46] Okada, N., Latest optical fiber cable technologies for further expansion of FTTH Network. *Fujikura Technical Review* (2013)

[10.47]   *Future Guide Optical Fibers and Cables.* Catalog, Fujikura (2013)

[10.48]   *Furukawa FTTx Solution.* Catalog, Furukawa Electric (2011)

[10.49]   Pesovic, A., *Pluggable OTDR Streamlines Troubleshooting.* AlcatelLucent TechZine, October 14 (2012) Available from http://www2.alcatel-lucent.com/blogs/techzine/2012/pluggable-otdr-streamlines-troubleshooting/

[10.50]   IEC 61753-041-2 (Ed.) 1.0, Fibre optic interconnecting devices and passive components – Performance standard – Part-041-2: Non-connectorised single-mode FTTx reflector device for Category C – controlled environment.

[10.51]   ITU-T Rec. G.987, *10-Gigabit-Capable Passive Optical Network (XG-PON) Systems: Definitions, Abbreviations and Acronyms* (2012)

[10.52]   IEEE 802.3bk: *Standard for Ethernet. Amendment: Physical Layer Specifications and Management Parameters for Extended Ethernet Passive Optical Networks* (2013)

[10.53]   IEEE P1904.1/D3.3, *Draft Standard for Service Interoperability in Ethernet Passive Optical Networks (SIEPON)* (2013)

[10.54]   Shen, Ch., FTTx in China – Current status and future prospects. *Joint ITU/IEEE Workshop on Ethernet – Emerging Applications and Technologies*, Geneva, September 22 (2012)

[10.55]   ITU-T Rec. G.988, *ONU Management And Control Interface (OMCI) Specification* (2012)

[10.56]   Brown, A., IEEE P1904.1 Standard for Service Interoperability in Ethernet Passive Optical Networks (SIEPON) – Project Overview. *IEEE 802.3 Plenary Meeting*, Waikoloa, Hawaii, March 11–16 (2012)

[10.57]   IEEE 803.2av-2009, *IEEE Standard for Information technology – Telecommunications and information exchange between systems – Local and metropolitan area networks – Specific requirements. Part 3: Carrier Sense Multiple Access with Collision Detection (CSMA/CD) Access Method and Physical Layer Specifications. Amendment 1: Physical Layer Specifications and Management Parameters for 10 Gb/s Passive Optical Networks* (2009)

[10.58]   *V8240: G-PON OLT with Multi-service Chassis.* Data sheet, Dasan Networks, Inc.

[10.59]   *V5812G: G-PON OLT System.* Data sheet, Dasan Networks, Inc. (2010)

[10.60]   *H640GV: GPON Gigabit ONT with VoIP Service.* Data sheet, Dasan Networks, Inc. (2013)

[10.61]   *H640GW: GPON Gigabit ONT with VoIP & Wi-Fi Service.* Data sheet, Dasan Networks, Inc. (2013)

[10.62]   *FIBRAIN FSR-R2: Router with multiple VLAN support.* Data sheet, Fibrain Co., Ltd (2012)

[10.63]   Ziemann, O., Zamzow, P.E. and Daum, W., *POF Handbook: Optical Short Range Transmission Systems.* Springer Verlag (2008)

[10.64]   Zubia, J. and Arrue, J., Plastic optical fibers: An introduction to their technological processes and applications. *Optical Fiber Technology*, 7(2): 101–140 (2001)

[10.65]   AGC Asahi Glass Co. Ltd., Fontex specifications from http://www.lucina.jp/eg_fontex/

[10.66]   Chromis Fiberoptics, Inc., GigaPOF specifications from http://chromisfiber.com/120ld.php

[10.67]   Murofushi, M., Low loss pefluorinated POF. *Proc. 6th International Conference on Plastic Optical Fibers and Applications*, Paris (1996)

[10.68]   Koike, Y and Koike, K., Progress in low-loss and high-bandwidth plastic optical fibers. *J. Polymer Science Part B: Polymer Physics*, 49(1): 2–17 (2011)

[10.69]   Optimedia, Inc., OM-giga specifications from http://www.optimedia.co.kr/eng_optimedia_main_b_01.htm

[10.70]   Lee, B.G., Kuchta, D.M., Doany, F.E., Schow, C.L., Baks, C. et al., 120-Gb/s 100-m Transmission in a Single Multicore Multimode Fiber Containing Six Cores Interfaced with a Matching VCSEL Array. *Photonics Society Summer Topical Meeting Series*, Piscataway NJ (2010)

[10.71]   Garito, A.F., Wang, J. and Gao, R., Effects of random perturbations in plastic optical fibers. *Science*, 281(14): 962–967 (1998)

[10.72]   Jiménez, D., Arrue, J., Aldabaldetreku, G. and Zubia, J., Numerical Simulation of Light Propagation in Plastic Optical Fibres of Arbitrary 3D Geometry. *Optics Express*, 13(11): 4012–4036 (2005)

[10.73]   Djordjevich, A. and Savović, S., Investigation of mode coupling in step index plastic optical fibers using the power flow equation. *IEEE Photonics Technology Letters*, 12(11): 1489–1491 (2000)

[10.74]   Mateo, J., Losada, M.A. and Zubia, J., Frequency response in step index plastic optical fibers obtained from the generalized power flow equation. *Optics Express*, 17(4): 2850–2860 (2009)

[10.75]   Ishigure, T., Kano, H. and Koike, Y., Which is more serious factor to the bandwidth of GI POF: Differential mode attenuation or mode coupling? *J. Lighw. Tech.*, 18(7): 959–965 (2000)

[10.76] van den Boom, H.P.A., Li, W., van Bennekom, P.K. Tafur Monroy, I. and Khoe, G.-D., High-capacity transmission over polymer optical fiber. *IEEE J.Selected Topics in Quantum Electronics*, 7: 461–470 (2001)

[10.77] Shin-Etsu Quarz Products Co., Ltd., Fluosil Preform fibers specifications from www.sqp.co.jp/e/seihin/catalog/pdf/e_4-2.pdf

[10.78] Koeppen, C., Shi, R.F., Chen, W.D. and Garito, A.F., Properties of plastic optical fibers. *J. Optical Society of America B*, 15(2): 727–739 (1998)

[10.79] White, W.R., Products and technologies for commercial deployment of gigabit pof networks. *Proc. 21st International Conference on Plastic Optical Fibers and Applications*, Atlanta GA (2012)

[10.80] Wirsing, M., Weberpals, F., Schunk, N., Stich, M., Janse Van Vuuren, D. and Wittl, J., Optical transceiver for today's and next generation POF systems. *Proc. 21st International Conference on Plastic Optical Fibers and Applications*, Atlanta GA (2012)

[10.81] Loquai, S., Winkler, F., Wabra, S., Hartl, E., Ziemann, O. and Schmauss, B., High-speed large-area optical receivers for next generation 10 Gbit/s data transmission over large-core 1 mm polymer optical fiber. *Proc.21st International Conference on Plastic Optical Fibers and Applications*, Atlanta GA (2012)

[10.82] Ab-Rahman, M.S., Harun, M.H. and Guna, H., Low-cost cascaded 1×4 polymer optical fiber coupler for multiplexing wavelengths. *Proc. 9th Malaysia International Conference on Communications*, Kuala Lumpur, Malaysia (2009)

[10.83] Lin, B., Zhai, Y. and Zhuang, Q., Research of a wavelength division multiplexer applied in LAN plastic optical fiber. *Proc. 6th SPIE International Symposium on Advanced Manufacturing and Testing Technologies*, Xiamen, China (2012)

[10.84] Haupt, M. and Fischer, U.H.P., Multi-colored WDM over POF system for triple-play. *Proc.14th Microoptics Conference*, Brussels, Belgium (2008)

[10.85] Park, J.W., Lee, H.S., Lee, S.S. and Son, Y.S., Passively aligned transmit optical subassembly module based on a WDM incorporating VCSELs. *IEEE Photonics Technology Letters*, 22(24): 1790–1792 (2010)

[10.86] Atef, M., Swoboda, R. and Zimmermann, H., Gigabit transmission over pmma step-index plastic optical fiber using an optical receiver for multilevel communication. *Proc.36th European Conference on Optical Communications*, Torino, Italy (2010)

[10.87] Aznar, F., Sánchez-Azqueta, C., Celma, S. and Calvo, K.B., Gigabit receiver over 1 mm SI-POF for home area networks. *J. Lightw. Tech.* 30(16): 2668–2674 (2012)

[10.88] Ziemann, O., Bunge, C.A., Kruglov, R., Vinogradov, J., Loquai, S. and Werzinger, S., Gigabit solution on PMMA-POF and the related European standardization. *Proc. 38th European Conference on Optical Communications*, Amsterdam (2012)

[10.89] Kato, S., Fujishima, O., Kozawa, T. and Kachi, T., High speed GaN-based green LED for POF. *Technical Journal R&D Review of Toyota CRDL*, 40(3): 7–10 (2005)

[10.90] Shi, J.W., Lin, C.W., Chen, W., Bowers, Sheu, J.K. et al., Very high-speed gan-based cyan light emitting diode on patterned sapphire substrate for 1 gbps plastic optical fiber communication. *Proc. Optical Fiber Communication Conference*, Los Angeles CA (2012)

[10.91] Junger, S., Tschekalinskij, W. and Weber, N., POF WDM transmission system for multimedia data. *Proc.11th International Conference on Plastic Optical Fibers and Applications*, Tokyo, Japan (2002)

[10.92] van den Boom, H.P.A., Li, W., van Bennekom, P.K., Tafur Monroy, I. and Khoe, G.D., High-capacity transmission over polymer optical fiber. *IEEE Journal on Selected Topics in Quantum Electronics*, 7(3): 461–470 (2001)

[10.93] Bartkiv, L., Poisel, H. and Bobitski, Y., Wavelength demultiplexer with concave grating for GI-POF systems. *Proc. 13th International Conference on Plastic Optical Fibers and Applications*, Nuremberg (2004)

[10.94] Kamiya, M., Ikeda, H. and Shinohara, S., Wavelength-division-multiplexed analog transmission through plastic optical fiber for use in factory communications. *IEEE Transactions on Industrial Electronics*, 49(2): 507–510 (2002)

[10.95] Yonemura, M., Kawasaki, A., Kagami, M., Ito, H., Terada, K., et al., 250 Mbit/s bi-directional single plastic optical fiber communication system. *Technical Journal R&D Review of Toyota CRDL*, 40(3): 18–23 (2005)

[10.96] Yang, J., van den Boom, H.P.A. and Koonen, A.M.J., Low cost high capacity data transmission over plastic optical fibre using wavelength multiplexed quadrature amplitude modulation. *Proc. IEEE/LEOS Benelux Symposium*, Brussels (2007)

[10.97] Ab-Rahman, M.S., Guna, H., Harun, M.H. and Jumari, K., A novel star topology POF-WDM system. *Proc. IEEE Symposium on Business, Engineering and Industrial Applications*, Langkawi, Malaysia (2011)

[10.98] Kruglov, R., Vinogradov, J., Ziemann, O., Loquai, S. and Bunge, C.A., 10.7-Gb/s Discrete Multitone Transmission Over 50-m SI-POF Based on WDM Technology. *IEEE Photonics Technology Letters*, 24(18): 1632–1634 (2012)

[10.99] Koonen, A.M.J., van den Boom, H.P.A., Tafur Monroy, I. and Khoe, G.D., High capacity multi-service in-house networks using mode group diversity multiplexing. *Proc. Optical Fiber Communication Conference*, Los Angeles, CA (2004)

[10.100] Nespola, A., Camatel, S., Abrate, S., Cárdenas, D. and Gaudino, R. Equalization techniques for 100 Mb/s data rates on SI-POF for optical short reach applications. *Proc. Optical Fiber Communication Conference*, Anaheim, CA (2007)

[10.101] Pinchas, M., Advantages and disadvantages of blind adaptive equalizers compared with the non adaptive and non blind approach. *The Whole Story Behind Blind Adaptive Equalizers/Blind Deconvolution. Bentham Books*, pp. 182–185 (2012)

[10.102] Breyer, F., Hanik, N., Randel, S. and Spinnler, B., Investigation on electronic equalization for step-index polymer optical fiber systems. *Proc. IEEE/LEOS Benelux Symposium*, Eindhoven, The Netherlands (2006)

[10.103] Zeolla, D., Antonino, A., Bosco, G. and Gaudino, R., DFE versus MLSE electronic equalization for Gigabit/s SI-POF transmission systems. *IEEE Photonics Technology Letters*, 23(8): 510–512 (2011)

[10.104] Nespola, A., Straullu, S., Savio, P., Zeolla, D., Ramírez Molina, J.C., et al., A new physical layer capable of record gigabit transmission over 1 mm step index polymer optical fiber. *J. Lightw. Tech.* 28(20): 2944–2950 (2010)

[10.105] Loquai, S., Kruglov, R., Schmauss, B., Bunge, C.A., Winkler, F. et al., Comparison of modulation schemes for 10.7 gb/s transmission over large-core 1 mm PMMA polymer optical fiber. *J. Lightw. Tech.* 31(13): 2170–2176 (2013)

[10.106] Randel, S., Breyer, F., Lee, S.C.J. and Walewski, W., Advanced modulation schemes for short-range optical communications. *IEEE Journal on Selected Topics in Quantum Electronics*, 16(5): 1280–1289 (2010)

[10.107] Okonkwo, C.M., Tangdiongga, E., Yang, H., Visani, D., Loquai, S. et al. Recent results from the EU POF-PLUS project: multi-gigabit transmission over 1 mm core diameter plastic optical fibers. *J. Lightw. Tech.* 29(2): 186–193 (2011)

[10.108] Breyer, F., Lee, S.C.J., Randel, S. and Hanik, N., Comparison of OOK- and PAM-4 modulation for 10 Gbit/s transmission over up to 300 m polymer optical fiber. *Proceedings of the Optical Fiber Communication Conference*, San Diego CA (2008)

[10.109] Decker, P.J., Pavan, S.K. and Ralph, S.E., Multilevel modulation of VCSEL-based POF links. *Proc. 21st International Conference on Plastic Optical Fibers and Applications*, Atlanta GA (2012)

[10.110] Breyer, F., Lee, S.C.J., Randel, S. and Hanik, N., PAM-4 signalling for gigabit transmission over standard step-index plastic optical fiber using light emitting diodes. *Proc. 34th European Conference on Optical Communications*, Brussels (2008)

[10.111] Lee, S.C.J., Breyer, F., Randel, S., Gaudino, R. et al., Discrete multitone modulation for maximizing transmission rate in step-index plastic optical fibers. *J. Lightw. Tech.* 27(11): 1503–1513 (2009)

[10.112] Neokosmidis, I., Kamalakis, T., Walewski, J.W., Inan, B. and Sphicopoulos, T., Impact of nonlinear LED transfer function on discrete multitone modulation: analytical approach. *J. Lightw. Tech.* 27(22): 4970–4978 (2009)

[10.113] Loquai, S., Kruglov, R., Bunge, C.A., Ziemann, O., Schmauss, B. and Vinogradov, J., 10.7-Gb/s Discrete multitone transmission over 25-m bend-insensitive multicore polymer optical fiber. *IEEE Photonics Technology Letters*, 22(21): 1604–1606 (2010)

[10.114] Yang, H. Lee, S.C.J., Tangdiongga, E., Okonkwo, C.M. and van den Boom, H.P.A. et al., 47.4 Gb/s transmission over 100 m graded-index plastic optical fiber based on rate-adaptive discrete multitone modulation' *J. Lightw. Tech.* 28(4): 352–359 (2010)

[10.115] Visani, D., Okonkwo, C.M., Loquai, S., Yang, H., Shi, Y. et al., Beyond 1 Gbit/s transmission over 1 mm diameter plastic optical fiber employing dmt for in-home communication systems. *J. Lightw. Tech.* 29(4): 622–628 (2011)

[10.116] Im, G.H., Harman, D.D., Huang, G., Mandzik, A.V., Nguyen, M.H. and Werner, J.J. 51.84 Mb/s 16-CAP ATM LAN standard. *IEEE Journal on Selected Areas in Communications*, 13(4): 620–632 (1995)

[10.117] Wieckowski, M., Jensen, J.B., Tafur Monroy, I., Siuzdak, J. and Turkiewicz, J.P., 300 Mbps transmission with 4.6 bit/s/Hz spectral efficiency over 50 m PMMA POF link using RC-LED and multi-level carrierless amplitude phase modulation. *Proc. Optical Fiber Communication Conference*, Los Angeles, CA (2011)

[10.118] Wei, J.L., Geng, L., Cunningham, D.G., Penty, R.V. and White, I.H., Gigabit NRZ, CAP and optical OFDM systems over POF links using LEDs. *Optics Express*, 20(20): 22284–22289 (2012)

[10.119] Stepniak, G. and Siuzdak, J., Transmission beyond 2 Gbit/s in a 100 m SI POF with multilevel CAP modulation and digital equalization. *Proc. Optical Fiber Communication Conference*, Anaheim, CA (2013)

[10.120] Kruglov, R., Loquai, S., Bunge, C.A., Schueppert, M., Vinogradov, J. and Ziemann, O., Comparison of PAM and CAP Modulation schemes for data transmission over SI-POF. *IEEE Photonics Technology Letters*, 25(23): 2293–2296 (2013)

[10.121] Huiszoon, B., Pals, L.Q.M. and Ortego Martínez, E.J., On the need and deployment of in-building POF networks: an industry point of view. *Proc. 20th International Conference on Plastic Optical Fibers and Applications*, Bilbao, Spain (2011)

[10.122] Guignard, P., Guillory, J., Chanclou, P., Pizzinat, A., Bouffant, O. et al., Multiformar home networks using silica fibres. *Proc. 38th European Conference on Optical Communications*, Amsterdam (2012)

[10.123] Mateo, J., Oca, A., Losada, M.A. and Zubia, J., Domestic multimedia network based on POF. *Proc. 17th International Conference on Plastic Optical Fibers and Applications*, Santa Clara,= CA (2008)

[10.124] Koonen, A.M.J. and Tangdiongga, E., Photonic home area networks. *J. Lightw. Tech.*, 32(4): 591–604 (2014)

[10.125] Koonen, A.M.J., van den Boom, H.P.A., Ortego Martínez, E., Pizzinat, A. et al., Cost optimization of optical in-building networks. *Optics Express*, 19(26): B399–B405 (2011)

[10.126] Mateo, J., Losada, M.A. and López, A. Application of the plastic optical fibre in domestic multimedia networks. *Proc. 3rd International Conference on Transparent Optical Networks (Mediterranean Winter)*, Angers, France (2009)

[10.127] Koike, Y., Status of high-speed plastic optical fibre towards giga house town. *Proc. 18th International Conference on Plastic Optical Fibers and Applications*, Sidney, Australia (2009)

[10.128] Lee, Y.K. and Lee, D., 'Broadband access in Korea: experience and future perspective. *IEEE Communications Magazine*, 41(12): 30–36 (2003)

[10.129] D. Cárdenas, A. Nespola, S. Camatel, S. Abrate, R. Gaudino, The rebirth of large-core plastic optical fibers: some recent results from the EU-project POF-ALL. *Proc. Optical Fiber Communication Conference*, San Diego, CA (2008)

[10.130] MOST Cooperation, Appendix D: Frame structure and boundary (Informative). *MOST Specification Rev. 3.0 E2*: 211–215 (2010)

[10.131] Anderson, D.E. and Beranek, M.W., 777 optical LAN technology overview. *Proc. 48th IEEE Electronic Components and Technology Conference*, Seattle WA (1998)

[10.132] Boeing 787 Dreamliner, http://www.newairplane.com/787/

[10.133] Lumexis Corporation, http://lumexis.com

[10.134] Aeroanautical Radio Inc. ARINC, http://www.arinc.com

[10.135] Truong, T.K., Commercial airplane fibre optics: needs, opportunities, challenges. *Proc. 19th International Conference on Plastic Optical Fibers and Applications*, Tokyo, Japan (2010)

[10.136] MOTIFES project, http://cordis.europa.eu/search/index.cfm?fuseaction=proj.document&PJ_RCN=5305159

[10.137] Cirillo, J., High Speed Plastic Networks (HSPN): a new technology for today applications. *IEEE Aerospace and Electronic Systems Magazine*, 11(10): 10–13 (1996)

[10.138] MIL-STD 1678, Fiber optic cabling systems requirements and measurements. *USA Department of Defense* (2009)

[10.139] RTCA/DO 160E, Environmental conditions and test procedure for airborne equipments. *Radio Technical Comission for Aeronautics* (2004)

[10.140] Cherian, S., Spangenberg, H. and Caspary, R., Integrating polymer optical fibers in civil aircraft: environmental requirements and challenges. *Proc. IEEE Avionics Fiber-Optics and Photonics Technology Conference*, Denver, CO (2010)

[10.141] Cherian, S., Spangenberg, H. and Caspary, R., Vistas and challenges for polymer optical fiber in comercial aircraft. *Proc. 19th International Conference on Plastic Optical Fibers and Applications*, Tokyo, Japan (2010)

[10.142] Cherian, S., Spangenberg, H. and Caspary, R. Power budget and system performance analysis of the POF link for future avionic applications. *Proc. IEEE Avionics Fiber-Optics and Photonics Technology Conference*, San Diego, CA (2011)

[10.143] Richards, D.H., Losada, M.A., Antoniades, N., López, A., Mateo, J. et al., Modeling methodology for engineering SI-POF and connectors in an avionics system. *J. Lightw. Tech.* 31(3): 468–475 (2013)

[10.144] Japanese Industrial Standard, Test methods for structural parameters of all plastic multimode optical fibers. *JIS. C:* 6862 (1992)

[10.145] Poisel, H., Ziemann, O., Berkovych, I. and Luber, M., Some remarks on POF flammability issues. *Proc. 21st International Conference on Plastic Optical Fibers and Applications*, Atlanta GA (2012)

[10.146] HomeGrid Forum, World's first service trial of ITU-T G. over plastic optical fibre (POF). *Technical Report* (2012)

[10.147] 802.15.3c, IEEE. *Wireless Medium Access Control (MAC) and Physical Layer (PHY) Specifications for High Rate Wireless Personal Area Networks (WPANs)*. http://www.ieee802.org/15/, 1–203 (2009)

[10.148] 802.16, IEEE. *Standard for Wireless Metropolitan Area Networks*. http://ieee802.org/16/ (2012)

[10.149] Reis, A.B., Sargento, S., Neves, F. and Tonguz, O.K., Deploying roadside units in sparse vehicular networks: what really works and what does not. *IEEE Transactions on Vehicular Technology*, 63(6): 2794–2806 (2014)

[10.150] Caragliu, A., Del Bo, C. and Nijkamp. P., *Smart Cities in Europe. Series Research Memoranda 0048*. VU University Amsterdam. Faculty of Economics, Business Administration and Econometrics (2009)

[10.151] Hirata, A., Takahashi, H., Yamaguchi, R., Kosugi, T., Murata, K. et al., Transmission characteristics of 120-GHz-band wireless link using radio-on-fiber technologies. *J. of Light*. 26(15): August, 2338–2344 (2008)

[10.152] Ng'oma. A., *Radio-over-Fibre Technology for Broadband Wireless Communication Systems*. Doctoral Thesis (2005)

[10.153] Ng'oma, A., Lin, C.-T., He, L.-Y.W., Jiang, W.-J., Annunziata, F. et al., 31 Gbps RoF system employing adaptive bit-loading OFDM modulation at 60 GHz. *Optical Fiber Communication Conference (OFC)*, 1–3 (2011).

[10.154] Nirmalathas, A., Novak, D., Lim, C. and Waterhouse. R.B., Wavelength reuse in the WDM optical interface of a millimeter-wave fiber-wireless antenna base station. *Transactions on Microwave Theory and Techniques*. 49(October): 2006–2012 (2001).

[10.155] Pereverzev, A. and Ageyev, D. Design method access network radio over fiber. *12th International Conference on the Experience of Designing and Application of CAD Systems in Microelectronics (CADSM)*, 288–292 (2013).

[10.156] Gowda, A.S., Dhaini, A.R., Kazovsky, L.G., Yang, H., Abraha, S.T. and Ng'oma, A., towards green optical/wireless in-building networks: radio-over-fiber. *J. Lightw. Tech*. 32: 3545–3556 (2014)

[10.157] Stohr, A., Kitayama, K. and Jager, D., Full-duplex fiberoptic RF subcarrier transmission using a dual-function modulator/photodetector. *IEEE Transactions on Microwave Theory and Techniques*, 47(July): 1338–1341 (1999).

[10.158] Stohr, A., Cojucari, O., van Dijk, F., Carpintero, G., Tekin, T. et al., Robust 71–76 GHz radio-over-fiber wireless link with high-dynamic range photonic assisted transmitter and laser phase-noise insensitive SBD receiver. *Conference on Optical Fiber Communication OFC*, 1–3 (2014).

[10.159] Stöhr, A., Babiel, S., Cannard, P.J., Charbonnier, B., van Dijk, F. et al., Millimeter-wave photonic components for broadband wireless systems. *IEEE Transactions on Microwave Theory and Techniques*, 58(11): November, 3071–3082 (2010)

[10.160] Tran, A.V., Tucker, R.S. and Boland. N.L., Amplifier Placement Methods for Metropolitan WDM Ring Networks. *J. Lightw, Tech*. 22(11); November, 2509–2522 (2004)

[10.161] Bragg. A.W., Which network design tool is right for you? *IT Professional*, 23–32 (2000)

[10.162] Zanella, A., Bui, N., Castellani, A., angelista, L. and Zorzi, M., Internet of things for smart cities. *IEEE Internet of Things J*. 1(1): February 22–32 (2014)

[10.163] Seeds, A.J. and Williams, J., Microwave photonics. *J. Lightw. Tech*. 24(12): (December, 4628–4641 (2006)

[10.164] Islam, A.R. and Bakaul, M., Simplified millimeter-wave radio-over-fiber system using optical heterodyning of low cost independent light sources and RF homodyning at the receiver. *International Topical Meeting on Microwave Photonics*, December: 1–4 (2009).

[10.165] Zhu, B., Taunay, T.F., Yan, M.F., Fini, J.M., Fishteyn, M. et al., Seven-core multicore fiber transmissions for passive optical network. *Optics* Express, 18(11): May, 11117–11122 (2010)

[10.166] Balakrishna, C., Enabling technologies for smart city services and applications. *Sixth International Conference on Next Generation Mobile Applications, Services and Technologies*, 223–227 (2012)

[10.167] Chang, C.H., Liang, T.C. and Huang, C.Y., DWDM self-healing access ring network with cost-saving, crosstalk-free and bidirectional OADM in single fiber. *Optics Communications*, 4518–4523 (2009)

[10.168] Lim, C., Nirmalathas, A., Novak, D., Tucker, R.S. and Waterhouse, R.B., Wavelength-interleaving technique to improve optical spectral efficiency in millimeter-wave WDM fiber-radio. *The 14th Annual Meeting of the IEEE LEOS*, 54–55 (2001)

[10.169] Lim, C., Nirmalathas, A., Novak, D., Waterhouse, R. and Yoffe, G., Millimeter-wave broad-band fiber-wireless system incorporating baseband data transmission over fiber and remote LO delivery. *J. Lightw. Tech.* 18(10): October, 1355–1363 (2000).

[10.170] Lim, C., Nirmalathas, A., Bakaul, M., Gamage, P., Lee, K. et al., Fiber-Wireless Networks and Subsystem Technologies. *J. Lightw. Tech.* 28(4): FebruarY, 390–405 (2010)

[10.171] Lim, C., Attygalle, M., Nirmalathas, A., Novak, D. and Waterhouse. R. Analysis of optical carrier-to-sideband ratio for improving transmission performance in fiber-radio links. *IEEE Transactions on Microwave Theory and Techniques*, 54(5): May, 2181–2187 (2006).

[10.172] Sun, C., Huang, J., Xiong, B. and Luo. Y., Low phase noise millimeter-wave generation by integrated dualwavelength laser diode. *Conference on Optical Fiber communication OFC*: 1–3 (2010)

[10.173] Zhang, C., Duan, J., Hong, C., Guo, P., Hu, W. et al., Bidirectional 60-GHz RoF system with multi-Gb/s *M*-QAM OFDM single-sideband modulation based m-qam ofdm single-sideband modulation based on injection-locked lasers. *IEEE Photonics Technology Letters*, 3(4): February, 245–247 (2011)

[10.174] Lin, C.-T., Ho, C.-H., Huang, H.-T. and Cheng, Y.-H., 84-Gbps 64-QAM 2×2 MIMO RoF system at 60 GHz employing single-sideband single-carrier modulation. *Conference on Optical Fiber Communication (OFC)*, 1–3 (2014)

[10.175] Mello, D.A.A., Schupket, D.A., Scheffelt, M. and Waldman, H., Availability maps for connections in WDM OPTICAL NETWORKS. *5th International Workshop on Design of Reliable Communication Networks*, 77–84 (2005).

[10.176] Evans, D. The Internet of things – how the next evolution of the Internet is changing everything. *Cisco Internet Business Solutions Group (IBSG)* (2011)

[10.177] Parekh, D., Yang, W., Ng'Oma, A., Fortusini, D., Sauer, M. et al., Multi-Gbps ask and QPSK-modulated 60 GHz RoF link using an optically injection locked VCSEL. *Optical Fiber Communication (OFC), National Fiber Optic Engineers Conference* (2010)

[10.178] Wake, D., Nkansah, A., Gomes, N., de Valicourt, G., Brenot, R. et al., A Comparison of radio over fiber link types for the support of wideband radio channels. *J. Lightw. Tech.* 28(16): August, 2416–2422 (2010)

[10.179] Wake, D., Lima, C.R. and Davies, P.A., Optical generation of millimeter- wave signals for fiber-radio systems using a dual-mode DFB semiconductor laser. *IEEE Transactions on Microwave Theory and Techniques.* 43: September, 2270–2276 (1995)

[10.180] Khorov, E., Lyakhov, A., Krotov, A. and Guschin, A. A survey on IEEE 802.11ah: An enabling networking technology for smart cities. *Computer Communications*, xx–xx. (2014)

[10.181] Wong, E., Prasanna, A.G., Lim, C., Lee, K.L. and Nirmalathas, A., Simple VCSEL base-station configuration for hybrid fiber-wireless access networks. *IEEE Photonics Technology Letters*, 21: April, 534–536 (2009)

[10.182] Giannetti, F., Luise, M., Reggiannini. R., Mobile and personal communications in the 60 GHz band: a survey. *Wireless Personal Communications*, 207–243 (1999)

[10.183] Gordejuela-Sánchez, F., Juttner, A. and Zhang, J., A Multiobjective Optimization Framework for IEEE 802.16e Network Design and Performance Analysis. *IEEE Journal of Selected Areas in Communications* 27(20): February, 202–216 (2009)

[10.184] Velez, F.J., Correira, L.M. and Brazio, J.M., Frequency Reuse and System Capacity in Mobile Broadband Systems: Comparison between the 40 and 60 GHz Bands. *Wireless Personal Communications*, 1–24 (2001)

[10.185] Lecoche, F., Charbonnier, B., Frank, F., Dijk, F.V., Enard, A. et al., 60 GHz bidirectional optical signal distribution system at 3 Gbps for wireless home network. *International Topical Meeting on Microwave Photonics*, October 1–3 (2009)

[10.186] Timofeev, F., Bennett, S., Griffin, R., Bayvel, P., Seeds, A. et al., High spectral purity millimetre-wave modulated optical signal generation using fiber grating lasers. *Electronics Letters*, 34: April, 668–669 (1998)

[10.187] Campuzano, G., Aldaya, I. and Castañón. G., performance of digital modulation formats in radio over fiber systems based on the sideband injection locking technique. *ICTON Mediterranean Winter Conference ICTON-MW*, December: 1–5 (2009)

[10.188] Castañón, G., Sarmiento, A.M., Ramírez, R. and Aragón-Zavala, A., Software tool for network reliability and availability analysis. *Wire Journal International*, 74–81 (2009)

[10.189] Castañón, G., Campuzano, G. amd Tonguz, O., High reliability and availability in radio over fiber networks." *OSA Journal of Optical Networking*, 7(6): 603–616 (2008)

[10.190] Grosskopf, G., Rohde, D., Eggemann, R., Bauer, S., Bornholdt, C. et al., Optical millimeter-wave generation and wireless data transmission using a dual-mode laser. *IEEE Photonics Technology Letters*, 12: December, 1692–1694 (2000)

[10.191] Agrawal, G.P., *Fiber-Optic Communication Systems*. John Wiley and Sons, Inc. (2002)

[10.192] G.694.1, *Recommendation ITU-T. Spectral Grids for WDM Applications: DWDM Frequency Grid.* http://www.itu.int/rec/T-REC-G.694.1/ (2012)

[10.193] G.694.2, Recommendation ITU-T. *Spectral Grids for WDM Applications: CWDM Wavelength Grid.* http://www.itu.int/rec/T-REC-G.694.2 (2003)

[10.194] Schaffers, H., Komninos, N., Pallot, M., Trousse, B., Nilsson, M. and Oliveira, A., Smart cities and the future internet: Towards cooperation frameworks for open innovation. *The Future Internet, Lect. Notes Comput. Sci.* 6656: 431–446 (2001)

[10.195] Toda, H., Nakasyotani, T., Kurit, T. and Kitayama, K. WDM mm-wave-band radio-on-fiber system using single supercontinuum light source in cooperation with photonic up-conversion. *IEEE International Topical Meeting on Microwave Photonics*, 161–164, (2004).

[10.196] Toda, H., Yamashita, T., Kuri, T. and Kitayama, K., Demultiplexing using an arrayed-waveguide grating for frequency-interleaved DWDM millimeter-wave radio-on-fiber systems. *J. Lightw. Tech.*, 21(8): August, 1735–1741 (2003)

[10.197] Song, H.-J., Ajito, K., Hirata, A., Wakatsuki, A., Muramoto, Y., et al., 8 Gbit/s wireless data transmission at 250 GHz. *Electronics Letters*, 45(22): October, 1121–1122 (2009)

[10.198] Hitachi, Ltd. *Hitachi's Vision for Smart Cities.* White Paper, 1–36 (2013)

[10.199] Aldaya, I., Gosset, C., Campuzano, G., Giacoumidis, E. and Castañón, G., Cost-efficient OFDM generation at 60-GHz by heterodyne technique with direct modulation and envelope detector. *15th International Conference on Transparent Optical Networks (ICTON)*, 1–5 (2013)

[10.200] Aldaya, I., Campuzano, G., Gosset, C. and Castañón, G., Simultaneous generation of WDM PON and RoF signals using a hybrid mode-locked laser. *International Conference on Transparent Optical Networks (ICTON)*, 1–4 (2014)

[10.201] Insua, I.G., Plettemeier, D. and Schffer, C.G., Simple remote heterodyne radio over fiber system for Gbps wireless access. *J. Lightw. Tech.* 28: February, 1–1 (2010)

[10.202] Baliga, J., Ayre, K., Sorin, R., Hinton, W. and Tucker. R.S., Energy consumption in access networks. *Conference on Optical Fiber communication/National Fiber Optic Engineers Conference*, 1–3 (2008)

[10.203] Baliga, J., Ayre, K., Hinton, W., Sorin, R. and Tucker, R.S., Energy consumption in optical IP Networks. *J. Lightw. Tech.* 27(13): July, 2391–2403 (2009)

[10.204] Beas, J., Castañón, G., Aldaya, I., Aragón-Zavala, A. and Campuzano, G., Millimeter-wave frequency radio over fiber systems: a survey. *IEEE Surveys and Tutorials*, 15(4): March, 1593–1619 (2013)

[10.205] Carroll, J., Whiteaway, J. and Plumb, D. Distributed feedback semiconductor lasers. *IET* (1998)

[10.206] Iness, J. and Mukherjee, B., New optical amplifier placement schemes for broadcast networks. *European Transactions on Telecommunications and Related Technologies*, 117–124 (2000)

[10.207] Liu, J., Chien, H.-C., Fan, S.-H., Chen, B., Jianjun Yu, S. et al., Bidirectional 1.25-Gbps wired/wireless optical transmission based on single sideband carriers in fabryperot laser diode by multimode injection locking. *J. of Lightw. Tech.* 23: September, 1325–1327 (2011)

[10.208] Wells, J., Faster than fiber: the future of multi-G/s wireless. *IEEE Microwave Magazine*, 104–112 (2009)

[10.209] Yao, J., Microwave photonics. *J. of Lightw. Tech.* February: 314–335 (2009)

[10.210] Zhensheng, J., Jianjun, Y., Georgios, E. and Chang, C.G., Key enabling technologies for optical–wireless networks: optical millimeter-wave generation, wavelength reuse, and architecture. *J. of Lightw. Tech.* 25(11): November, 3452–3471 (2007).

[10.211] Kanonakis, K., Tomkos, I., Krimmel, H., Schaich, F., Lange, C. et al., An OFDMA-based optical access network architecture exhibiting ultra-high capacity and wireline-wireless convergence. *IEEE Communications Magazine*, 71–78 (2012)

[10.212] Kim, C.H., Impact of various noises on maximum reach in broadband light source based high-capacity WDM passive optical networks. *Optics Express*, 1–6 (2010)

[10.213] Kim, H.S., Pham, T.T., Won, Y.Y. and Han, S.K., Bidirectional WDM-RoF transmission for wired and wireless signals. *Communications and Photonics Conference and Exhibition*, November: 1–2 (2009)

[10.214] Kitayama, K.I., Architectural considerations of radio-on-fiber millimeter-wave wireless access systems. *URSI International Symposium on Signals, Systems, and Electronics*, 378–383 (1998)

[10.215] Kitayama, K.I., Stohr, A., Kuri, T., Heinzelmann, R., Jager, D. and Takahashi, Y. An approach to single optical component antenna base stations for broad-band millimeter-wave fiberradio access systems. *IEEE Transactions on Microwave Theory and Techniques*, 48: December, 2588–2595 (2000)

[10.216] Kuri, T., Kitayama, K. and Takahashi, Y., 60-GHz-band full-duplex radio-on-fiber system using two-RF-port electroabsorption transceiver. *IEEE Photonics Technology Letters*, 12: April, 419–421 (2000)

[10.217] Kuri, T., Kitayama, K. and Takahashi, Y., A single light source configuration for full-duplex 60-GHz-band radio-on fiber system. *IEEE Transactions on Microwave Theory and Technique*. 51: February 431–439 (2003)

[10.218] Mohamed, M., Hraimel, B., Zhang, X., Sakib, M.N. and Wu, K., Frequency quadrupler for millimeter-wave multiband OFDM ultrawideband wireless signals and distribution over fiber systems. *IEEE/OSA Journal of Optical Communications and Networking*, 428–438 (2009)

[10.219] Radziunas, M., Glitzky, A., Bandelow, U., Wolfrum, M., Troppenz, U. et al., Improving the modulation bandwidth in semiconductor lasers by passive feedback. *IEEE Journal of Selected Topics in Quantum Electronics*, 13: January/February, 136–142 (2007)

[10.220] M.1645, ITU-R. *Framework and Overall Objectives of the Future Development of IMT-2000 and Systems beyond IMT-2000*. http://www.itu.int/ITU-R (2000)

[10.221] Monterrey, Escuela de Gobierno y Política Pública del Tecnológico de. *Smart Cities Summit Guadalajara*. February (2014) http://www.gda.itesm.mx/agenda/evento.php?cual=1440

[10.222] Ghazisaidi, N., Maier, M. and Assi, C.M., Fiber-Wireless (FiWi) Access networks: a survey. *IEEE Communications Magazine*, 160–167 (2009)

[10.223] Mohamed, N., Idrus, S. and Mohammad, A., Review on system architectures for the millimeter-wave generation techniques for RoF communication link. *IEEE International RF and Microwave Conference* (2008)

[10.224] Peng, P.C., Feng, K.M., Chiou, H.Y., Peng, W.R., Chen, J.J. et al., Reliable architecture for high capacity fiber-radio systems. *Optical Fiber Technology*, 236–239 (2007)

[10.225] Hartmann, P., Qian, X., Wonfor, A., Penty, R. and White, I., 1–20 GHz directly modulated radio over MMF link. *International Topical Meeting on Microwave Photonics*, October (2005)

[10.226] Shih, P.T., Lin, C.T., Huang, H.S., Jiang, W., Chen, J. et al., 13.75-Gb/s OFDM signal generation for 60-GHz RoF system within 7-GHz license-free band via frequency sextupling. *35th European Conference on Optical Communication (ECOC)*, 1–2 (2009)

[10.227] Chang, Q., Fu, H. and Su, Y., Simultaneous generation and transmission of downstream multiband signals and upstream data in a bidirectional radio-over-fiber system. *IEEE Photonics Technology Letters*, 20(3): February, 181–183 (2008)

[10.228] Shafik, R.A., Rahman, S. and Islam, R.A., On the extended relationships among EVM, BER and SNR as performance metrics. *4th International Conference on Electrical and Computer Engineering*, 408–411 (2006)

[10.229] Braun, R., Grosskopf, G., Rohde, D. and Schmidt, F., Optical Millimetre-wave generation and transmission experiments for mobile 60 GHz band communications. *Electronics Letter*, 32. March, 626–628 (1996)

[10.230] Daniels, R.C., Robert, J. and Heath, W., 60 GHz wireless communications: emerging requirements and design recomendations. *IEEE Vehicular Technology Magazine*, 41–50 (2007)

[10.231] Sambaraju, R., Herrera, J., Marti, J., Westergren, U., Zibar, D. et al., Up to 40 Gb/s wireless signal generation and demodulation in 75–110 GHz band using photonic techniques. *IEEE Topical Meeting on Microwave Photonics (MWP)*, 1–4 (2010)

[10.232] Shaddad, R., Mohammada, A.B., Al-Gailani, S.A., Al-hetar, A.M., Elmagzoub, M.A., A survey on access technologies for broadband optical andwireless networks. *J. Network and Computer Applications*, 459–472 (2014)

[10.233] Braun, R.P., Tutorial: Fibre radio systems, applications and devices. *24th European Conference on Optical Communication*, September: 87–119 (1998)

[10.234] Ghafoor, S. and Hanzo, L., Sub-carrier-multiplexed duplex 64-QAM radio-over-fiber transmission for distributed antennas. *IEEE Communications Letter*, 1–4 (2011)

[10.235] Korotky, S.K., Price-points for components of multi-core fiber communication systems in backbone optical networks. *J. Optical Communications and Networking*, 4(5): May, 426–435 (2012)

[10.236] Yong, S.K. and Chong, C.-C., An overview of multigigabit wireless through millimeter wave technology: potentials and technical challenges. *EURASIP Journal on Wireless Communications and Networking*, 1–10 (2007)

[10.237] Koenig, S., Antes, J., Lopez-Diaz, D., Kallfass, I., Zwick, T. et al., High-speed wireless bridge at 220 GHz connecting two fiber-optic links each spanning up to 20 km. Optical Fiber Communication Conference, 1–3 (2014)

[10.238] Lekamge, S. and Marasinghe, A., Developing a smart city model that ensures the optimum utilization of existing resources in cities of all sizes. *IEEE Computer Society*, 202–207 (2013)

[10.239] Pato, S., Pedro, J. and Monteiro, P., Comparative evaluation of fibre-optic architectures for next-generation distributed antenna systems. *International Conference on Transparent Optical Networks*, 1–4 (2009)

[10.240] Rebhi, S., Barrak, R. and Menif, M., Optic/RF Co-design For Oudoor RoF system at 60 GHz. *Mediterranean Microwave Symposium (MMS)*, 1–13 (2013)

[10.241] Sarkar, S., Dixit, S. and Mukherjee, B., Hybrid Wireless-Optical Broadband-Access Network (WOBAN): a review of relevant challenges. *J. Lightw. Tech.* 25(11): 3329–3340 (2007)

[10.242] Wong, S.-W. Campelo, D.R., Cheng, N., Yen, S.-H., Kazovsky, L. et al., Grid reconfigurable optical-wireless architecture for large-scale municipal mesh access network. *IEEE Global Telecommunications Conference.* GLOBECOM, 1–6 (2009)

[10.243] Baykas, T., Sum, C.-S., Lan, Z., Wang, J., Rahman, M.A. and Kato, H.H.S., IEEE 802.15.3c: The first IEEE wireless standard for data rates over 1 Gb/s. *IEEE Communications Magazine*, 49: 114–121 (2011)

[10.244] Chattopadhyay, T., A millimeter-wave radio-over-fiber system for overcoming fiber dispersion-induced signal cancellation effect. *Optoelectronics*, 293–296 (2011)

[10.245] Kawanishi, T., Ultra high-speed fiber wireless transport. *Conference on Optical Fiber Communication OFC*, 1–3 (2013)

[10.246] Kuri, T., Toda, H. and Kitayama, K., Novel demultiplexer for dense wavelength-division-multiplexed millimeter-wave-band radio-over-fiber systems with optical frequency interleaving technique. *IEEE Photonics Technology Letters*, 19(24): December, 2018–2020 (2007)

[10.247] Kuri, T., Toda, H., Vegas Olmos, J.J., Kitayama, K., reconfigurable dense wavelength-division-multiplexing millimeter-waveband radio-over-fiberaccess system technologies. *J. Lightw. Tech.* 28(160): August, 2247–2257 (2010)

[10.248] Ismail, T.M. and Seeds, A., Linearity enhancement of a directly modulated uncooled dfb laser in a multi-channel wireless-over-fibre system. *IEEE MTT-S International Microwave Symposium Digest*, 7–10 (2005)

[10.249] Nakasyotani, T., Toda, H., Kuri, T, and Kitayama, K., wavelength-division-multiplexed millimeter-waveband radio-on-fiber system using a supercontinuum light source. *J. Lightw. Tech.* 24: January, 404–410 (2006)

[10.250] Ohno, T., Nakajima, F., Furuta, T. and Ito, H., A 240-GHz active mode-locked laser diode for ultra-broadband fiber-radio transmission systems. *Conference on Optical Fiber Communication (OFC)*, 1–3 (2005)

[10.251] Rappaport, T.S., Murdock, J.N. and Gutierrez, F., State of the art in 60-GHz integrated circuits and systems forwireless communications. *Proc. IEEE*, 1390–1436 (2011)

[10.252] Pham, T.T., Lebedev, A., Beltrán, M., Yu, X., Llorente, R. and Monroy, I.T., Combined single-mode/multimode fiber link supporting simplified in-building 60-GHz gigabit wireless access. *Optical Fiber Technology*, 226–229 (2012)

[10.253] Pham, T.T., Kim, H.S., Won, Y.Y. and Han, S.K., Bidirectional 1.25-Gbps wired/wireless optical transmission based on single sideband carriers in fabry–pérot laser diode by multimode injection locking. *J. Lightw. Tech.* 27(13): July, 2457–2464 (2009)

[10.254] The Fiber Optic Association, Inc. Guide to fiber optic network design. *FOA Technical Bulletin*, 1–30 (2011)

[10.255] Yodprasit, U., Carta, C. and Ellinger, F., 11.5-Gbps 2.4-pJ/bit 60-GHz OOK demodulator integrated in a SiGe BiCMOS technology. *Proc. 8th European Microwave Integrated Circuits Conference*, 1–4 (2013)

[10.256] Shaw, W., Wong, S., Cheng, N. and Kazovsky, L.G., MARIN hybrid optical-wireless access network. *Optical Fiber Communication*, 1–3 (2006)

[10.257] Sun, X., Chan, C.K., Wang, Z., Lin, C. and Chen, L.K., A singlefiber bi-directional WDM self-healing ring network with bidirectional OADM for metro-access applications. *IEEE Journal on Selected Areas in Communications.* 25: April, 18–24 (2007)

[10.258] Yang, Y., Lim, C. and Nirmalathas, A., investigation on transport schemes for efficient high-frequency broadband OFDM transmission in fibre-wireless links. *J. Lightw. Tech.* 32(2): 267–274 (2014)

[10.259] Liu, Z., Sadeghi, M., de Valicourt, G., Brenot, R. and Violas, M., Experimental validation of a reflective semiconductor optical amplifier model used as a modulator in radio over fiber systems. *IEEE Photonics Technology Letters*, 23:(February, 576–578 (2011)

[10.260] Pi, Z. and Khan, F., An introduction to millimeter-wave mobile broadband systems. *IEEE Communications Magazine*, 101–107 (2011)

[10.261] Paraskevopoulos, A., Vučĉić, J., Voss, S.H., Swoboda, R. and Langer, K.D., Optical wireless communication systems in the mb/s to gb/s range, suitable for industrial applications. *Mechatronics, IEEE/ASME Transactions.* 15: pp. 541–547 (2010)

[10.262] Popoola, W.O. and Ghassemlooy, Z., BPSK subcarrier intensity modulated free-space optical communications in atmospheric turbulence. *J. Lightw. Tech.* 27: 967–973 (2009)

[10.263] Ntogari, G., Kamalakis, T. and Sphicopoulos, T., Analysis of indoor multiple-input multiple-output coherent optical wireless systems. *J. Lightw. Tech.* 30: 317–324 (2012)

[10.264] Aladeloba, A.O., Phillips, A.J. and Woolfson, M.S., Improved bit error rate evaluation for optically pre-amplified free-space optical communication systems in turbulent atmosphere. *IET-Optoelectronics*, 6: 26–33 (2012)

[10.265] Dordova, L. and Wilfert, O., Calculation and comparison of turbulence attenuation by different method, *Radioengineering*, 19: 162–163 (2010)

[10.266] Braua, B. and Barua, D., Channel capacity of MIMO FSO under strong turbulance conditions. *Int. J. Electrical & Computer Sciences*, 11: 1–5 (2011)

[10.267] Tsiftsis, T.A., Sandalidis, H.G., Karagiannidis, G.K. and Uysal, M., Optical wireless links with spatial diversity over strong atmospheric turbulence channels. *IEEE Transactions on Wireless Communications*, 8: 951–957 (2009)

[10.268] Gappmair, W., Further results on the capacity of free-space optical channels in turbulent atmosphere. *Communications, IET*, 5: 1262–1267 (2011)

[10.269] Perlot, N., Duca, E., Horwath, J., Giggenbach, D. and Leitgeb, E., System requirements for optical HAP-satellite links. *6th International Symposium on Communication Systems, Networks and Digital Signal Processing*, pp. 72–76 (2008)

[10.270] Rajbhandari, S., Perez, J., Le-Minh, H. and Ghassemlooy, Z., A fast ethernet FSO link performance under the fog controlled environment. *ECOC*, Geneva, pp. 1–3 (2011)

[10.271] Perez, J., Ghassemlooy, Z., Rajbhandari, S., Ijaz, M. and Minh, H., Ethernet FSO Communications link performance study under a controlled fog environment. *IEEE Communications Letters*, pp. 1–3 (2012)

[10.272] Strickland, B.R., Lavan, M.J., Woodbridge, E. and Chan, V., Effects of fog on the bit-error rate of a free space laser communication system. *J. Applied Optics*, 38: 424–431 (1999)

[10.273] Su, K., Moeller, L., Barat, R.B. and Federici, J.F., Experimental comparison of performance degradation from terahertz and infrared wireless links in fog. *Journal of Opt. Soc. Am. A*, 29: 179–184 (2012)

[10.274] Colvero, C.P., Cordeiro, M.C.R. and von der Weid, J.P., Real-time measurements of visibility and transmission in far mid and near-IR free space optical links. *IEEE Electronics Letters*, 41: 610–611 (2005)

[10.275] Muhammad, S.S., Flecker, B., Leitgeb, E. and Gebhart, M., Characterization of fog attenuation in terrestrial free space links. *J, Optical Engineering*, 46l: 066001–066006 (2007)

[10.276] Awan, M.S., Horwath, L.C., Muhammad, S.S., Leitgeb, E., Nadeem, F. and Khan, M.S., Characterization of fog and snow attenuations for free-space optical propagation. *Journal of Communications*, 4: 533–545 (2009)

[10.277] Andrews, L.C. and Phillips, R.L., *Laser Beam Propagation through Random Media, Second Edition.* Washington: SPIE Press (2005)

[10.278] Osche, G.R., *Optical Detection Theory for Laser Applications, First Edition.* Wiley-Interscience (2002)

[10.279] Gagliardi, R.M. and Karp, S., *Optical Communications, Second Edition.* New York: John Wiley (1995)

[10.280] Zhu, X. and Kahn, J.M., Free-space optical communication through atmospheric turbulence channels. *IEEE Transactions on Communications*, 50: August, 1293–1300 (2002)

[10.281] Karp, S., Gagliardi, R.M., Moran, S.E. and Stotts, L.B., *Optical Channels: Fibers, Clouds, Water and the Atmosphere.* New York: Plenum Press (1988)

[10.282] Goodman, J.W., *Statistical Optics.* New York: John Wiley (1985)

[10.283] Kaushal, H., Jain, V.K. and Ka, S., Effect of atmospheric turbulence on acquisition time of ground to deep space optical communication system. *Int. J. Electrical and Computer Engineering*, 4: 730–734 (2009)

[10.284] Majumdar, A.K. and Ricklin, J.C., *Free-Space Laser Communications: Principles and Advances.* New York: Springer (2008)

[10.285] Ghassemlooy, Z., Popoola, W.O., Ahmadi, V.A. and Leitgeb, E.A., MIMO free-space optical communication employing subcarrier intensity modulation in atmospheric turbulence channels. *Communications Infrastructure: Systems and Applications in Europe.* 16: 61–73 (2009)

[10.286] Navidpour, S.M., Uysal, M. and Kavehrad, M., BER performance of free-space optical transmission with spatial diversity. *Wireless Communications, IEEE Transactions*, 6: 2813–2819 (2007)

[10.287] Popoola, W.O., Ghassemlooy, Z., Lee, C. and Boucouvalas, A., Scintillation effect on intensity modulated laser communication systems – a laboratory demonstration. *Optics & Laser Technology*, 42: 682–692 (2010)

[10.288] Ghassemlooy, Z., Tang, X., Rajbhandari, S., Popoola, W.O. and Lee, C.G., Coherent optical binary polarization shift keying heterodyne system in the free space optical turbulence channel. *IET Microwaves, Antennas and Propagation*, 5: 1031–1038 (2011)

[10.289] Wang, Z., Zhong, W.-D., Fu, S. and Lin, C., Performance comparison of different modulation formats over free-space optical (FSO) turbulence links with space diversity reception technique. *Photonics J., IEEE*, 1: 277–285 (2009)

[10.290] Ijaz, M., Adebanjo, O., Ansari, S., Ghassemlooy, Z., Rajbhandari, S. et al., Experimental investigation of the performance of OOK-NRZ and RZ modulation techniques under controlled turbulence channel in FSO systems. 11th Annual Postgraduate Symp. on the Convergence of Telecommunications, Networking and Broadcasting, 21–22 June, Liverpool, UK, pp. 296–300 (2010)

[10.291] Kaushal, H., Kumar, V., Dutta, A., Aennam, H., Jain, V.K. et al., Experimental study on beam wander under varying atmospheric turbulence conditions. *IEEE Photonics Technology Letters*, 23: 1691–1693 (2011)

[10.292] Abou-Rjeily, C. and Slim, A., Cooperative diversity for free-space optical communications: transceiver design and performance analysis. *Communications, IEEE Transactions*, 59: 658–663 (2011)

[10.293] Vetelino, F.S., Young, C., Andrews, L. and Recolons, J., Aperture averaging effects on the probability density of irradiance fluctuations in moderate-to-strong turbulence. *Appl. Opt.*, 46: 2099–2108 (2007)

[10.294] Kim, I.I., McArthur, B. and Korevaar, E., Comparison of laser beam propagation at 785 nm and 1550 nm in fog and haze for optical wireless communications. *Proc. SPIE 4214*, Boston MA (2001)

[10.295] Naboulsi, M.A., Sizun, H. and Frédérique, D.F., Wavelength selection for the free space optical telecommunication technology. *SPIE*, 5465: 168–179 (2004)

[10.296] Kruse, P.W., McGlauchlin, L.D. and McQuistan, E.B., *Elements of Infrared Technology: Generation, Transmission and Detection.* J. Wiley and Sons, New York (1962)

[10.297] Corrigan, P., Martini, R., Whittaker, E.A. and Bethea, C., Mid-infrared lasers and the Kruse-Mie theorem in fog for free-space optical communication applications. *Conference on Lasers, Electro-Optics, Quantum Electronics and Laser Science. CLEO/QELS*, pp. 1–2 (2008)

[10.298] Nadeem, F., Javornik, T., Leitgeb, E., Kvicera, V.and Kandus, G., Continental fog attenuation empirical relationship from measured visibility data. *Radio Engineering*, 19(4): (2010)

[10.299] Grabner, M. and Kvicera, V., On the relation between atmospheric visibility and optical wave attenuation. *Mobile and Wireless Communications Summit, 2007. 16th IST*, pp. 1–5 (2007)

[10.300] Yang, Z.N.J., Shi, C., Liu, D. and Li, Z.A., Microphysics of atmospheric aerosols during winter haze/fog events in Nanjing. *J. Environmental Science*, 31(7): 1425–1431 (2010)

[10.301] WMO, *Gauide to Meteorological instruments and methods of observation world meteorological organisation, ITU, Geneva, Swetzerland*, 2006.

[10.302] Bohren, C.F. and Huffman, D.R., *Absorption and Scattering of Light by Small Particles.* John Wiley and Sons, New York (1983)

[10.303] Weichel, H., Laser beam propagation in the atmosphere. *Bellingham: SPIE Optical Engineering*, TT3: (1990)

[10.304] Kruse, L.D.M.P.W. and McQuistan, E.B., Elements of infrared technology: Generation, transmission and detection. J. Wiley and Sons, New York (1962)

[10.305] Grabner, M. and Kvicera, V., The wavelength dependent model of extinction in fog and haze for free space optical communication., *J. Optics Express*, 19: 3379–3386 (2012)

[10.306] Kartalopoulos, S.V., *Free Space Optical Networks for Ultra-Broad Band Services.* John Wiley & Sons, New Jersey. pp. 35–36 (2011)

[10.307] Blaunstein, N., Arnon, S., Zilberman, A. and Kopeika, N., *Applied Aspects of Optical Communication and LIDAR*. Boca Raton: CRC Press, London, pp. 53–55 (2010)

[10.308] Grabner, M. and Kvicera, V., Fog attenuation dependence on atmospheric visibility at two wavelengths for FSO link planning. *Loughborough Antennas and Propagation Conference (LAPC), IEEE*, 193–196 (2010)

[10.309] Hale, G.M. and Querry, M.R., Optical Constants of Water in the 200-nm to 200-µm Wavelength Region. *Journal of Applied Optics*, 12: 555–563 (1973)

[10.310] Rheims, J.K.J. and Wriedt, T., Refractive-index measurements in the near-IR using an Abbe refractometer. *J. Meas. Sci. Technol.* 8(6): 601–605 (1997)

[10.311] Pierce, R.M., Ramaprasad, J. and Eisenberg, E.C., Optical attenuation in fog and clouds. *Proc. SPIE 4530*, 58: (2001)

[10.312] Prokes, A., Atmospheric effects on availability of free space optics systems. *J. Optical Engineering*, 48: 066001–10 (2009)

[10.313] Ricklin, J.C., Hammel, S.M., Eaton, F.D. and Lachinova, S.L., Atmospheric channel effects on free-space laser communication. *J. Optical and Fiber Communications Research*, 3: 111–158 (2006)

[10.314] Naboulsi, M.A., d. Forne, F., Sizun, H., Gebhart, M., Leitgeb, E. et al., Measured and predicted light attenuation in dense coastal upslope Fog at 650, 850 and 950 nm for free space optics applications. *J. Optical Engineering*, 47: 036001-1–036001-14 (2008)

[10.315] ETSI EN 300 652 v1.2.1 (Broadband Radio Access Networks (BRAN); HIgh PErformance Radio Local Area Network (HIPERLAN) Type 1; Functional specification, European Telecommunications Standard Institute, Sophia Antipolis, France (1997/8)

[10.316] ETSI TR 101 031 v2.2.1 Broadband Radio Access Networks (BRAN); HIgh PErformance Radio Local Area Network (HIPERLAN) Type 2; Requirements and architectures for wireless broadband access, European Telecommunications Standard Institute, Sophia Antipolis, France (1999–01)

[10.317] ETSI TS 101 475 v1.3.1 Broadband Radio Access Networks (BRAN); HIgh PErformance Radio Local Area Network (HIPERLAN) Type 2; Physical (PHY) layer, European Telecommunications Standard Institute, Sophia Antipolis, France (2001–12)

[10.318] ETSI TR 101 683 v1.1.1 Broadband Radio Access Networks (BRAN); HIgh PErformance Radio Local Area Network (HIPERLAN) Type 2; System overview, European Telecommunications Standard Institute, Sophia Antipolis, France (2000–02)

[10.319] IEEE 802.11-2012 IEEE Standard for Information technology – Telecommunications and information exchange between systems Local and metropolitan area networks – Specific requirements Part 11: Wireless LAN Medium Access Control (MAC) and Physical Layer (PHY) Specifications, the Institute of Electric and Electronics Engineers, New York (2012)

[10.320] Wi-Fi Alliance. Available from http://www.wi-fi.org

[10.321] The Radio Regulations, ITU Radiocommunication Sector, Geneva (2012)

[10.322] FCC 13-22 (February), *In the Matter of Revision of Part 15 of the Commission's Rules to Permit Unlicensed National Information ET Docket No. 13-49 Infrastructure (U-NII) Devices in the 5 GHz Band*, Federal Communications Commission, Washington DC (2013)

[10.323] Ghazisaidi, N., Maier, M. and Assi, C.M. Fiber-wireless (FiWi) access networks: a survey, Communications Magazine, *IEEE*. 47(2): 160–167 (2009)

[10.324] IEEE 802.11af-2013 IEEE Standard for Information technology – Telecommunications and information exchange between systems – Local and metropolitan area networks – Specific requirements – Part 11: *Wireless LAN Medium Access Control (MAC) and Physical Layer (PHY) Specifications* Amendment 5: Television White Spaces (TVWS) Operation, IEEE, New York (2013)

[10.325] Tsonev, D., Videv, S. and Haas, H., Light fidelity (Li-Fi): towards all-optical networking, Institute for Digital Communications, Li-Fi R&D Centre, The University of Edinburgh (2014) Available from: http://www.see.ed.ac.uk/drupal/hxh

[10.326] Li-Fi Consortium. Available from http://www.lificonsortium.org

[10.327] IEEE 802.15.7-2011 IEEE Standard for Local and Metropolitan Area Networks – Part 15.7: Short-Range Wireless Optical Communication Using Visible Light, IEEE, New York (2011)

[10.328] Bing, B., *Broadband Wireless Multimedia Networks*. John Wiley & Sons, Hoboken NJ (2013)

[10.329] IEEE 802-2014 IEEE Standard for local and metropolitan area networks – Overview and architecture, IEEE, New York (2014)

[10.330] IEEE 802.2-1998 (ISO/IEC 8802-2:1998) IEEE Standard for Information technology – Telecommunications and information exchange between systems – Local and metropolitan area networks – Specific requirements – Part 2: Logical Link Control, IEEE, New York (1998)

[10.331] IEEE 802.1X-2010 IEEE Standard for Local and metropolitan area networks – Port-based network access control, IEEE, New York (2010)

[10.332] IEEE 802.11ac-2013 IEEE Standard for Information technology – Telecommunications and information exchange between systems—Local and metropolitan area networks – Specific requirements – Part 11: Wireless LAN Medium Access Control (MAC) and Physical Layer (PHY) Specifications – Amendment 4: Enhancements for very high throughput for operation in bands below 6 GHz, IEEE, New York (2014)

[10.333] IEEE 802.11ad-2014 IEEE Standard for Information technology – Telecommunications and information exchange between systems – Local and metropolitan area networks – Specific requirements – Part 11: Wireless LAN Medium Access Control (MAC) and Physical Layer (PHY) Specifications – Amendment 3: Enhancements for very high throughput in the 60 GHz band, IEEE, New York (2014)

[10.334] IETF RFC 3748 (2013). Aboba, B. et al. *Extensible Authentication Protocol, Internet Engineering Task Force.* Available from http://www.ietf.org

[10.335] Wi-Fi Certified Passpoint. Available from http://www.wi-fi.org/discover-wi-fi/wi-fi-certified-passpoint

[10.336] Wi-Fi Certified Wi-Fi Direct. Available from http://www.wi-fi.org/discover-wi-fi/wi-fi-direct

[10.337] Wi-Fi Certified Miracast. Available from http://www.wi-fi.org/discover-wi-fi/wi-fi-certified-miracast

[10.338] Heusse, M., Rousseau, F., Berger-Sabbatel, G. and Duda, A., Performance anomaly of 802.11b. *Proceedings of the Twenty-Second Annual Joint Conference of the IEEE Computer and Communications. IEEE Societies (IEEE INFOCOM)*, March 30–April 3, 2: 836–843 (2003)

[10.339] Abu-Sharkh, O. and Twefik, A.H., Throughput evaluation and enhancement in 802.11 WLANs with access point. *Proc. IEEE Vehicular Technology Conference, 61st IEEE VTC-Spring*, 30 May–1 June, 2: 1338–1341 (2005)

[10.340] Bianchi, G., Performance analysis of the IEEE 802.11 distributed coordination function. *IEEE Journal on Selected Areas in Communications*, 18(3): 535–547 (2000)

[10.341] Szczypiorski, K. and Lubacz, J., Saturation throughput analysis of IEEE 802.11g (ERP-OFDM) networks. *Proc. 12th IFIP International Conference on Personal Wireless Communications (PWC 2007).* (2007)

[10.342] Robinson, J.W. and Randhawa, T.S., Saturation throughput analysis of IEEE 802.11e enhanced distributed coordination function, *IEEE Journal on Selected Areas in Communications*, 22(5): 917–928 (2004)

[10.343] Vesco, A. and Scopigno, R., Time-division access priority in CSMA/CA. *Proc. 20th IEEE International Symposium on Personal, Indoor and Mobile Radio Communications (IEEE PIMRC 2009)*, pp. 2162–2166 (2009)

[10.344] Scopigno, R. and Vesco, A., A distributed bandwidth management scheme for multi-hop wireless access network. *Proc. 7th Wireless Communications and Mobile Computing Conference (IEEE IWCMC 2011)*, pp. 534–539 (2011)

[10.345] Vesco, A., Scopigno, R. and Masala, E., Supporting triple-play communications with TDuCSMA and first experiments. *Proc. IEEE Wireless Communications and Networking Conferences (IEEE WCNC 2014)* (2014).

[10.346] Cozzetti, H.A., Campolo, C., Scopigno, R. and Molinaro, A., Urban VANETs and hidden terminals: evaluation through a realistic urban grid propagation model. *Proc. IEEE International Conference on Vehicular Electronics and Safety (IEEE ICVES 2012)* (2012)

[10.347] Cozzetti, H.A., Brevi, D., Scopigno, R., Ferrari, P., Sisinni, E. and Flammini, A., MS-Aloha: preliminary analysis of its suitability for wireless automation. *Proc. 17th IEEE International Conference on Emerging Technologies and Factory (IEEE ETFA)* (2012)

[10.348] Jang, B. and Sichitiu, M.L., IEEE 802.11 saturation throughput analysis in the presence of hidden terminals, *IEEE/ACM Transactions on Networking*, 20(2): April, 557–570, (2012)

[10.349] IEEE 802.3-2012 IEEE Standard for Information technology – Telecommunications and information exchange between systems – Local and metropolitan area networks – Specific requirements – Part 3: Carrier sense multiple access with collision detection (CSMA/CD) access method and physical layer specifications, IEEE, New York (2012)

[10.350] IETF RFC 5412 Calhoun, P. et al. Lightweight access point protocol. *Internet Engineering Task Force* (2010) Available from http://www.ietf.org

[10.351] Dai, Q., Shou, G., HuY. and Guo, Z., A general model for hybrid fiber-wireless (FiWi) access network virtualization. Proc. IEEE International Conference on Communications (Workshops), June: 858–862, (2013)

[10.352] Khan, A., Zugenmaier, A, Jurca, D. and Kellerer, W., Network virtualization: a hypervisor for the Internet? *IEEE Communications Magazine*, 50(1): January, 136–143 (2012)

[10.353] Md Fadlullah, Z., Nishiyama, H., Kawamoto, Y. Ujikawa, H., Suzuki, K.-I. and Yoshimoto, N., Cooperative QoS control scheme based on scheduling information in FiWi access network. *IEEE Transactions on Emerging Topics in Computing*, 1(2): December, 375–383, (2013)

[10.354] Zheng, Z., Wang, J. and Wang, X., ONU Placement in fiber-wireless (FiWi) networks considering peer-to-peer communications. *Proc. IEEE Global Telecommunications Conference (GLOBECOM 2009)*, November 30–December 4, pp. 1–7 (2009)

[10.355] Bhatt, U.R. and Chouhan, N., ONU placement in fiber-wireless (FiWi) Networks. *Nirma University International Conference on Engineering (NUiCONE*, November 28–30, pp. 1–6 (2013)

[10.356] Liu, Y., Guo, L. and Yang, J., Energy-efficient topology reconfiguration in green fiber-wireless (FiWi) access network. *22nd Wireless and Optical Communication Conference (WOCC)*, May 16–18, pp. 555–559 (2013)

[10.357] Maier, M. and Leévesque, M., Dependable fiber-wireless (FiWi) access networks and their role in a sustainable third industrial revolution economy, *IEEE Transactions on Reliability*, 63(2): June, 386–400, (2014)

[10.358] Dhaini, A.R., Ho, P.H. and Jiang, X., QoS control for guaranteed service bundles over Fiber-Wireless (FiWi) broadband access networks. *J. Lightw. Tech.*, 29(10): May, 1500–1513 (2011)

[10.359] Aurzada, F., Lévesque, M., Maier, M. and Reisslein, M., FiWi access networks based on next-generation PON and gigabit-class WLAN technologies: a capacity and delay analysis. *IEEE/ACM Transaction on Networking*, 22(4): 1176–1189 (2014)

[10.360] Sheng, M., Yang, C., Zhang, Y. and Li, J., Zone-based load balancing in LTE self-optimizing networks: a game-theoretic approach. *IEEE Transactions on Vehicular Technology*, 63(6): July, 2916–2925 (2014)

[10.361] Ahmed, R. and Boutaba, R., Design considerations for managing wide area software defined networks. *IEEE Communications Magazine*, 52(7): July, 116,123 (2014)

[10.362] Lam, C., *Passive Optical Networks: Principles and Practice*, Academic Press (2007)

[10.363] http://fsanweb.com/archives/category/next-generation-pon-task-group-ng-pon.

[10.364] Aleksić, S., Energy efficiency of electronic and optical network elements. *IEEE JSTQE*, 17(2): March–April, 296–308 (2011)

[10.365] Trojer, E. and Eriksson, P., Power saving modes for GPON and VDSL. *Proc. 13th Euro. Conf. Netw. & Optical Commun. (NOC), Austria*, June 30–July 3 (2008)

[10.366] IEEE Std 802.3az-2010 (Amendment to IEEE Std 802.3-2008), October 27, pp. 1–302 (2010)

[10.367] Yen, S-H. et al., Photonic components for future fiber access networks. *IEEE Journal on Selected Areas in Communications*, 28(6): August (2010)

[10.368] Tucker, R.S., Green optical communications – Part II: Energy limitations in networks. *IEEE JSTQE*, 17(2): March–April, 261–274 (2011)

[10.369] Kilper, D.C., et al., Fundamental limits on energy use in optical networks. *ECOC* (2010)

[10.370] Valcarenghi, L. et al., Impact of modulation formats on onu energy saving. *ECOC* (2010)

[10.371] Kubo, R. et al., Adaptive power saving mechanism for 10 Gigabit class PON systems. *IEICE Transactions on Communications*, E93-B(2): February, 280–288 (2010)

[10.372] Smith, T., Tucker, R.S. and Hinton, K., A.V. Tran: Implications of sleep mode on activation and ranging protocols in PONs. *Proc. LEOS*, pp. 604–605 (2008)

[10.373] Yan, Y., Wong, S-W., Valcarenghi, l., She-Hwa Yen, S-H. et al., Energy management mechanism for Ethernet Passive Optical Networks (EPONs), *ICC* (2010)

[10.374] GPON Power Conservation, ITU-T G-series Recommendations – Supplement 45(G.sup45): 05 (2009)

[10.375] Freude, W., et al., Optically powered fiber networks, *Optics Express*, 16(26): (2008)

[10.376] Kubo, R., et al., Study and demonstration of sleep and adaptive link rate control mechanisms for energy efficient 10G-EPON, *IEEE/OSA JOCN*, 2(9): September, 716–729 (2010)

[10.377] Wong, S-W., Valcarenghi, L., Yen, S-H., Campelo, D.R., Yamashita, S., Kazovsky, L., Sleep mode for energy saving pons: advantages and drawbacks, GreenComm2. *IEEE GLOBECOM Workshops* (2009)

[10.378] ITU-T Rec. G.989.1, March (2013)

[10.379] L. Carenghi, L. et al., Energy efficiency in passive optical networks: where, when, and how? *IEEE Network*, 26(6): 61–68 (2013)

[10.380] European Commission Joint Research Centre, *Broadband Equipment Code of Conduct – Version 5*, December 20 (2013)

[10.381] Valcarenghi, L. et al., Energy saving in TWDM(A) PONs: challenges and opportunities, *ICTON* (2013)

[10.382] Valcarenghi, L. et al., Impact of energy efficient schemes on virtualized TWDM PONs, AF3G.1, ACP (2013)

[10.383] Taguchi, K. et al., 100-ns λ-selective burst-mode transceiver for 40-km reach symmetric 40-Gbit/s WDM/TDM-PON, Mo.4.F.5, *ECOC* (2013)

[10.384] Luo, Y. et al., IEEE/OSA. *J. Lightw. Tech.* 31(4): 587–593 (2013)

[10.385] Dixit, A. et al., Novel DBA algorithm for energy efficiency in TWDM-PONs. p. 6.10, *ECOC* (2013)

[10.386] Dixit, A. et al., ONU power saving modes in next generation optical access networks: progress, efficiency and challenges. *Optics Express*. 20(26): B52–B63 (2012)

[10.387] Yoshida, T. et al., Automatic load-balancing DWBA algorithm considering long-time tuning devices for λ-tunable WDM/TDM-PON, We.2.F.5, *ECOC* (2013)

[10.388] Kaneko, S. et al., First system demonstration of hitless λ-tuning sequence for dynamic wavelength allocation in WDM/TDM-PON, W3G.6, *OFC* (2014)

[10.389] Wang, R. et al., Energy Management in NG-PON2, Tu3C.4, *OFC* (2014)

[10.390] Technical Assessment and Comparison of Next-Generation Optical Access System Concepts, OASE_WP4_D4.2.2.doc, p. 66, June 23, 2013, www.ict-oase.eu/public/files/OASE_WP4_D4_2_2.pdf (latest access May 6, 2014).

[10.391] Vizcaino, J.L. et al., Energy efficiency analysis for flexible-grid OFDM-based optical networks. *The Int. J. of Comp and Telecomm Netw*, 56(10): July, 2400–2419 (2012)

[10.392] Udalcovs, A. and Bobrovs, V., Investigation of Spectrally Efficient Transmission for Unequally Channel Spaced WDM Systems with Mixed Data Rates and Signal Formats. *8th IEEE, IET Int. Sym. on CSNDSP*, pp. 1–4, (2012)

[10.393] Cavdar, C. et al., Design of green networks with signal quality guarantee. *IEEE ICC*, pp. 3025–3030 (2012)

[10.394] Rizzelli, G. et al., Energy efficient traffic-aware design of on-off multi-layer translucent optical network. *J. of Comp Netw*, 56(10): July, 2443–2455 (2012)

[10.395] Heddeghem, W.V. et al., Power consumption modeling in optical multilayer networks. *J. of Photon Netw Comm*, 24: 86–102 (2012)

[10.396] RSoft Design Group Inc, *Optsim User Guide*. USA, pp. 404 (2008)

[10.397] Udalcovs, A. et al., Evaluation of SMP-induced optical signals distortions in ultra-dense mixed-WDM system. *Int. Conf. on FGCT*, pp. 180–184 (2012)

[10.398] Fujitsu, The path to 100G. Available: http://www.fujitsu.com/downloads/TEL/fnc/whitepapers/Pathto100G .pdf

[10.399] Ciena, F10-T 10G Transponder. Data sheet, pp. 1–2 (2011) Available: http://www.ciena.com/products/f10-t/tab/features/

[10.400] MRV, LambdaDriver – DWDM 40Gbps transponder (TM-40GT8) (2011) Available: http://www.mrv.com/datasheets/LD/PDF300/MRV-LD-TM-40GT8_HI.pdf

[10.401] ADVA, FSP 3000 coherent transponders. Fact sheet (2012) Available: http://www.advaoptical .com/~/media/Innovation/Efficient%20100G%20Transport/100G%20Coherent%20Transponder.ashx

[10.402] Morea, A. et al., Power management of optoelectronic interfaces for dynamic optical networks. *ECOC'11*, We.8.K.3.pdf, pp. 1–3 (2011)

[10.403] Nag, A. and Tornatore, M., Transparent optical network design with mixed line rates, *2nd Int. Symp. on ANTS*, pp. 1–3 (2008)

[10.404] SMART 2020: Enabling the low carbon economy in the information age. GESI'08 (2008) [Online]. Available: http://www.theclimategroup.org/assets/resources/publications/Smart2020Report_lo_res.pdf

[10.405] Korotky, S.K., Traffic trends: Drivers and measures of cost-effective and energy-efficient technologies and architectures for backbone optical networks. *Proc. OFC*, Los Angeles CA, Paper OM2G1 (2012)

[10.406] Rival, O. and A. Morea, A., Resource requirements in mixed-line rate and elastic dynamic optical networks, *OFC/NFOEC*, pp. 1–3 (2012)

[10.407] Christodoulopoulos, K. et al., Planning mixed-line-rate WDM transport networks, *ICTON*, pp. 1–4 (2011)

[10.408] Nag, A. et al., Optical network design with mixed line rates and multiple modulation formats. *J. Lightw. Tech.* 28(4): February, 466–475 (2010)

[10.409] Udalcovs, A. et al., Power efficiency of WDM networks using various modulation formats with spectral efficiency limited by linear crosstalk, *Optics Communications* (2014)

[10.410] Chowdhury, P., Tornatore, M., Nag, A., Ip, E., Wang, T. and Mukherjee, B., On the Design of Energy-Efficient Mixed-Line-Rate (MLR) Optical Networks. *J. Lightw. Tech.*, 30(1): Januaryy, 130–139 (2012)

[10.411] RSoft Design Group Inc: Coherent PM-QPSK versus RZ-DQPSK and DPSK for high bitrate systems. Available: http://www.rsoftdesign.com/products.php?sub=System+and+Network&itm=OptSim&det=Application+Gallery&id=50

[10.412] Udalcovs, A. et al., Evaluation of SMP-induced optical signals distortions in ultra-dense mixed-WDM system. *Int. Conf. on FGCT*, pp. 180–184 (2012)

[10.413] Batchelor, P. et al., Study on the implementation of optical transparent transport networks in the European environment – results of the research project COST 239, *Photonic Network Communications*, 2(1): 15–32 (2000)

[11.1] Son, M., doc.: IEEE 802.15-08-0759-02-0vlc, Project: *IEEE P802.15 Working Group for Wireless Personal Area Networks* (*WPANs*) (2008)

[11.2] IEEE, Part 11: Wireless LAN Medium Access Control (MAC) and Physical Layer (PHY) Specifications Amendment 6: Wireless Access in vehicular Environments. *IEEE 802.11p Published Standard* (2010)

[11.3] IEEE, IEEE Trial-use standard for wireless access in vehicular environments (WAVE) – multi-channel operation, *Trial-Use Std. 1609.4* (2006)

[11.4] Afonin, A. and Shevtsov, V., Features of development of satellite networks of broadband access (BBA) in the Ka-range. *Proc. ICTON 2011, 13th International Conference on Transparent Optical Networks* (*ICTON*) (2011)

[11.5] Strobel, O., Rejeb, R. and Lubkoll, J., Communication in automotive systems: Principles, limits and new trends for vehicles, airplanes and vessels. *Proc. ICTON, 12th International Conference on Transparent Optical Networks* (*ICTON*) (2010)

[11.6] Kastell, K., Challenges and improvements in communication with vehicles and devices moving with high-speed. *Proc. ICTON, 13th International Conference on Transparent Optical Networks* (*ICTON*) (2011)

[11.7] Allemandou, P., Conceptual design, implementation and performance tests for WLAN coverage in trains. Diploma Thesis, TU Darmstadt (2005)

[11.8] Green. R.J., Optical wireless with application in automotives. *Proc. ICTON, 12th International Conference on Transparent Optical Networks* (*ICTON*) (2010)

[11.9] Kastell, K., Novel mobility models and localization techniques to enhance location-based services in transportation systems. *Proc. ICTO, 14th International Conference on Transparent Optical Networks* (*ICTON*). (2012)

[11.10] 3GPP, TR 36.836, Technical Specification Group Radio Access Network; Mobile Relay for Evolved Universal Terrestrial Radio Access (E-UTRA). tech. rep., http://www.3gpp.org/, accessed December 20 (2013)

[11.11] Sui, Y., Vihriala, J., Papadogiannis, A., Sternad, M., Yang, W. and Svensson, T., Moving cells: a promising solution to boost performance for vehicular users. *IEEE Communications Magazine*, 51(6): 62–68 (2013)

[11.12] Bianchi, G., Blefari-Melazzi, N., Grazioni, E., Salsano, S. and Sangregorio, V., Internet Access on Fast Trains: 802.11-based onboard wireless distribution network alternatives. *12th IST Mobile & Wireless Communications Summit*, pp. 15–18 (2003)

[11.13] MOST Cooperation MOST brand book vol. 1.1, August (2007) Retrieved December 7, 2009, from www.mostcooperation.com

[11.14] Fokum, D.T. and Frost, V.S., A survey on methods for broadband internet access on trains. *IEEE Communications Surveys & Tutorials*, 12(2): 171–185 (2010)

[11.15] *Radar Handbook, Third Edition*. Merrill I. Skolnik (Ed.), Copyright © 2008 by the McGraw-Hill Companies, ISBN 978-0-07-148547-0 (2008)

[11.16] *Communication in Transportation Systems*. Editor Otto Strobel, IGI Global, USA, ISBN 978-1-4666-2976-9 (2013)

[11.17] Nujdin, V.M., Sulimov, Yu.O., Sidorov, N.V., Klyucharev, Yu. M. and Ananenkov, A.E., Influence of re-reflections from the Earth surface on the luminance image, formed by a vehicle radar. *Radiotekhnika*, 3: (2001)

[11.18] Schneider, M., Groß, V. et al., Automotive 24 GHz Short Range Radar Sensors with Smart Antennas. *German Radar Symposium 2002*, Bonn, September (2002)

[11.19] Kühnle, G., Mayer, H., Olbrich, H. et al., Low-cost long-range radar for future driver assistance systems. *Auto Technology*, 4: 2–5 (2003)

[11.20] Kawakubo, A. and Tokoro. S. et al., Electronically-scanning millimeter-wave RADAR for forward objects detection. *SAE Congress*, Detroit MI, 127–134 (2004)

[11.21] Asano, Y., Ohshima, S. et al., Proposal of millimeter-wave holographic radar with antenna switching. *Int. Microwave Symposium 2001*, Phoenix AZ, May (2001)

[11.22] Schmidt, R.O., Multiple emitter location and signal parameter estimation, *Proc. RADC Spectrum Estimation Workshop*, RADC-TR-79-63, Rome Air Development Center, Rome NY, October 1979, p. 243 (reprinted in IEEE Trans. Antennas Propag., AP-34: 276–280 (1986)

[11.23] Schneider, M., Automotive radar – status and trends. robert bosch gmbh, corporate research, *Proceedingd of German Microwave Conference – GeMiC 2005*, University of Ulm, Germany, April 5–7 (2005)

[11.24] Sisoeva, S., Intelligent assistants and road sensors. Functions – more "iron"– less. *Components and Technologies*, Moscow, N 1 (2012)

[11.25] Bridzolara, D., Directions of development of automotive radar: the band 79 GHz. *Organization and Road Safety*, Moscow, No. 4 (2013)

[11.26] Shelukhin, O.I., *Radiosystems of close action*, Moscow: Publishing House. *Radio i svyaz* (in Russian), ISBN 5-256-00337-2 (1989)

[11.27] Sorokin, S., *Selection of thermal equipment, Safety Algorithm*, No. 5 (2011)

[11.28] Rastorguev, V. and Shnajder, V., Radiometric sensor of movement speed of vehicles. *Proc. 12th International Conference on Transparent Optical Networks – ICTON'2010*, Munich, Germany, June 27–July 1. IEEE Catalog Number: CFP10485-USB, Print ISBN: 978-1-4244-7797-5 Digital Object Identifier: 10.1109/ICTON.2010.5549089 (2010)

[11.29] Shtager, E.A., Radio scattering on bodies of complicated form. Moscow: Publishing House, *Radio i svyaz* (in Russian) (1986)

[11.30] Ananenkov, A.E., Nujdin, V.M., Rastorguev, V.V. and Skosirev, V.N. Features of assessment of detection characteristics in small-distance radars. *Radiotekhnika*, 11: (2013)

[11.31] Ananenkov, A.E., Konovaltsev, A.V., Nujdin, V.M., Rastorguev, V.V. and Skosirev, V.N., Management of radiated power: A necessary direction of development of radars of land transport systems. *Proceedings of the 15th International Conference on Transparent Optical Networks – ICTON'2013*, Cartagena, Spain, June 23–27, Digital Object Identifier: 10.1109/ICTON.2013.6602737, ISSN: 2161-2056 (2013)

[11.32] Doviak, R., and Zrnich, D., Doppler radars and meteorological observations, L. Publishing House, *Gidrometeoizdat* (in Russian) (1998)

[11.33] Ananenkov, A.E., Konovaltsev, A.V., Nujdin, V.M., Rastorguev, V.V., Sidorov, N.V. and Sulimov, Yu. O., Research of a vehicle forward looking radar with a frequency modulation. *Proc. 11th international conference Microwave engineering and telecommunication technologies, KryMiKo-2001*, Sevastopol, Ukraine, September 7–12, ISBN 966-7968-00-6 (2001)

[11.34] Ananenkov, A.E., Konovaltsev, A.V., Nujdin, V.M., Rastorguev, V.V., Sidorov, N.V. and Sulimov, Yu. O., MM range ARVS as a means of investigating into the objects of natural and artificial origin. *Elektronika i informatika*, 4: (2002)

[11.35] Ananenkov, A.E., Konovaltsev, A.V., Nujdin, V.M., Rastorguev, V.V. and Skosirev, V.N., Features of dispersion of wideband probing signals in Radio Vision Systems of the MM – wavelength, *Symposium on Sensors for Driver Assistance Systems, September 21+22*, Technische Akademie Heilbronn, Germany (2006)

[11.36] Kukhorev, A.A., Nujdin, V.M. and Sokolov, P.V., Development of a digital module of the autoradar, ensuring warning of vehicles and eliminating factors, restraining further commercialization. *Proc. All-Russia Youth Scientific and Innovation Competition-Conference, Elektronika-2006*, Zelenograd, Russia, November 30 (2006)

[11.37] Ananenkov, A.E., Konovaltsev, A.V., Kukharev, A.V., Nujdin, V.M., Rastorguev, V.V. and Skosirev, V.N., Features of constructing a highly informative system of radiovision of MM-range of wave length with frequency modulation. *Proc. 2nd Russian Scientific and Technical Conference Radioheightmetrics – 2007*, Kamensk-Uralskiy, Russia, October 5–9 (2007)

[11.38] Yevdokymov, A.P. and Kryzhanovskiy, V.V., Diffraction radiation antennas for SHF and EFH radiosystems. A.Ya. Usikov Institute of Radiophysics and Electronics of the NAS of Ukraine, *International Conference on Antenna Theory and Techniques*, September 17–21, Sevastopol, Ukraine, pp. 59–64 (2007)

[11.39] Ananenkov, A., Konovaltsev, A., Nujdin, V., Rastorguev, V., and Sokolov, P. Characteristics of radar images in radio vision systems of the automobile. *Proc. International Conference on Transparent Optical*

Networks – ICTON-MW'08, Marrakech, Morocco, December 11–13 – IEEE Catalog Number: CFP0833D-CDR, ISBN: 978-1-4244-3485-5, Library of Congress: 2008910892 (2008)

[11.40] Ananenkov, A., Konovaltsev, A., Kukhorev, A., Nujdin, V. and Rastorguev. V., Features of formation of radar-tracking and optical images in a mobile test complex of radiovision systems of the car. *J. Telecommunications and Information Technology*, Warsaw, 1: 29–33 (2009)

Index

Optical and Microwave Technologies for Telecommunication Networks, First Edition. Otto Strobel.
© 2016 John Wiley & Sons, Ltd. Published 2016 by John Wiley & Sons, Ltd.